Applied Life
Data Analysis

Applied Life
Data Analysis

WAYNE B. NELSON

Consultant, Schenectady, New York

formerly
Corporate Research and Development
General Electric Company

A JOHN WILEY & SONS, INC., PUBLICATION

Copyright © 2004 by John Wiley & Sons, Inc. All rights reserved.

Published by John Wiley & Sons, Inc., Hoboken, New Jersey.
Published simultaneously in Canada.

For general information on our other products and services please contact our Customer Care
Department within the U.S. at 877-762-2974, outside the U.S. at 317-572-3993 or fax 317-572-4002.

Wiley also publishes its books in a variety of electronic formats. Some content that appears in print,
however, may not be available in electronic format.

Library of Congress Cataloging-in-Publication is available.

ISBN 0-471-64462-5

Printed in the United States of America.

10 9 8 7 6 5 4 3 2 1

*In Grateful Memory
of my Mother and Father*

Contents

Preface to the Paperback Edition

It has been gratifying that this book has been widely used and praised by thousands of statisticians and engineers concerned with reliability data analysis. Also, it is pleasing that John Wiley and Sons now makes it available at a bargain price in a paperback edition, thanks to the fine efforts of Mr. Steve Quigley, Ms. Susanne Steitz, and the Wiley staff.

Although the field of reliability data analysis continues to advance, this book remains a sound introduction to basic and intermediate topics. This preface briefly discusses a few important advances to help bring the reader up to date.

For certain advances, the reader may wish to consult the following two books:

- *Accelerated Testing: Statistical Models, Test Plans, and Data Analyses,* by Wayne Nelson, Wiley (1990), www.wiley.com. This book is the best single source on accelerated testing.
- *Recurrent Events Data Analysis for Product Repairs, Disease Recurrences, and Other Applications,* by Wayne Nelson, ASA-SIAM (2003), www.siam.org. This new book provides needed innovative models and statistical analyses for such data, going well beyond the simple Poisson model and data analyses appearing in *Applied Life Data Analysis*.

Since 1982 when the cloth edition book was published, there have been great advances in commercially available software for the analysis of reliability data. A brief survey of such software is outlined on the following pages.

WAYNE B. NELSON
Consultant,
WNconsult@aol.com

Schenectady, New York
September 2003

Software for Life Data Analyses

The following reliability data analysis software provides the new likelihood ratio confidence limits, which are more accurate than the normal-approximation limits in this book. All provide probability plots and maximum likelihood fitting.

• AGSS (A Graphical Statistical System), developed at IBM, is a general statistical package written in APL with extensive linear and nonlinear curve fitting, exploratory data analysis, linear and nonlinear regression, design of experiments, quality control, reliability, and accelerated life testing. It fits life distributions and regression models to censored life and accelerated test data (constant- and step-stress) and provides confidence intervals. It runs on PCs, IBM mainframes, and Sun Solaris Systems. Contact Dr. Martin Schatzoff, schatzoff@att.net for information, or order from IBM at 1-800-CALL IBM.

• The SAS RELIABILITY Procedure provides comprehensive statistical modeling and analysis tools for life data, accelerated life test data, recurrence data, and regression models. All standard life distributions can be fitted to data with interval, right, and left censoring. The Procedure creates graphs, including probability plots and percentile plots for survival data, and mean cumulative function plots for recurrence data. The SAS System runs on Microsoft Windows, workstations (UNIX and OpenVMS), OS/2, and the mainframe. For general information, visit www.sas.com/rnd/app/. For questions about the RELIABILITY Procedure, email software@sas.com.

• JMP from SAS Institute, Inc. is a general interactive statistical package for engineers and other researchers. It includes Design of Experiments and Statistical Quality Control. Reliability features include fitting of life distributions, parametric regression and accelerated test relationships, the proportional hazards (Cox) model, and maximum likelihood fitting of user programmed models. JMP has features for recurrence events data analysis and fitting competing risk models. Contact jmp@sas.com, www.jmp.com, SAS Institute, Inc., JMP Sales, SAS Campus Drive, Cary, NC 27513, 1-800-594-6567 for information.

• ReliaSoft's RELIABILITY OFFICE SUITE includes WEIBULL++ for life data analysis, ALTA for accelerated life data analysis, BLOCKSIM and BLOCKSIM FTI for complex system analysis and simulation utilizing reliability block diagrams and/or fault tree diagrams, RGA for reliability growth and repairable systems analysis, XFMEA for failure modes and effects analysis and additional tools for recurrence data analysis and degradation analysis. All have detailed documentation. They can be purchased as a suite or individually. All run under Microsoft Windows (95, 98, NT, 2000, and XP). Free evaluation copies can be downloaded from http://www.reliasoft.com/products.htm. Contact ReliaSoft Corp., 115 S. Sherwood Village Dr., Tucson, AZ 85710, ReliaSoft@ReliaSoft.com, www.ReliaSoft.com, (888)886-0410.

- SPLIDA is an add-on to S-Plus, a general statistical package of Insightful (www.insightful.com), (800)569-0123. SPLIDA provides a wide range of data analysis, modeling, and test planning tools for reliability data. SPLIDA fits all standard life distributions and regression relationships to censored and truncated data. It has special regression tools for accelerated test data analysis. Output (graphical and tabular) includes failure probabilities, distribution quantiles, and hazard rates. SPLIDA also analyzes recurrence data and both repeated measures degradation data and destructive degradation data. Numerous examples of SPLIDA output appear in Meeker and Escobar (1998), *Statistical Methods for Reliability Data*, Wiley, New York. SPLIDA runs on Windows 95/98 (and later) and on Windows NT4.0 (and after) using version 2000 (and later) of S-Plus. An easy-to-install self-executing file containing the SPLIDA functions, their graphical user interface, and documentation can be downloaded from www.public.iastate

- WinSMITH™ is composed of WEIBULL and VISUAL modules bundled together in SuperSMITH™ for probability plotting, growth modeling, and accelerated test analysis. Electronic user guides and DEMO software can be downloaded at www.weibullnews.com. WinSMITH 4.0W A, the latest version, runs on Windows-based systems (3.1, 95, 98, 2000, NT, XT). Contact Wes Fulton, Fulton Findings, 1251 West Sepulveda Blvd. (PMB 800), Torrance, CA 90502, (310)548-6358, wes33@pacbell.com; or Dr. Bob Abernethy, 536 Oyster Road, North Palm Beach, FL 33408-4328, Weibull@worldnet.att.net, (561) 842-4082.

Life Data Features of Statistical Packages

	IBM AGSS	Relia-Soft Suite	SASProc RELIA-BILITY	JMP 5.1	S-PLUS SPLIDA Meeker	Win-SMITH
DATA						
Observed &Right Cens'd	Yes	Yes	Yes	Yes	Yes	Yes
Left Censored	Yes	Yes	Yes	Yes	Yes	Yes
Interval	Yes	Yes	Yes	Yes	Yes	Yes
Transformations	Yes	Yes	In SAS	Yes	Yes	Yes
Subset Selection	Yes	Yes	Yes	Yes	Yes	Yes
Simulation	Yes	Yes	In SAS	Yes	Yes	Yes
DISTRIBUTIONS						
Exponential	Yes	Yes	Yes	Yes	Yes	Yes
Weibull	Yes	Yes	Yes	Yes	Yes	Yes
(Log)Normal	Yes	Yes	Yes	Yes	Yes	Yes
(Log)Logistic	Yes	Yes	Yes	Yes	Yes	No
Gamma	No	Yes	No	Yes	Yes	No
Extreme Value	Yes	No	Yes	Yes	Yes	No
Generalized Gamma	No	Yes	Yes	No	Yes	No
Other	No	Various	No	Yes	Yes	No
User Programmed Dist's	No	No	In SAS	Yes	w Effort	No
RELATIONSHIPS						
Linear for Location Param	Yes	Yes	Yes	Yes	Yes	Yes
Linear without Intercept	Yes	Yes	Yes	Yes	Yes	Yes
Log Linear Scale Param	No	Yes	Yes	No	Yes	Yes
Cox Proportional Hazards	Yes	Yes	In SAS	Yes	In S+	No
User Programmed Relation	No	No	In SAS	Yes	w Effort	No
ML FITTING						
Stepwise	No	Yes	No	No	In S+	No
Hold Coef.s/Params Fixed	No	Yes	Some	No	Yes	Yes
Freq. Count Data, Weights	Yes	Yes	Yes	Yes	Yes	Yes
FIT OUTPUT						
Ests& Normal Conf Limits	Yes	Yes	Yes	Yes	Yes	Yes
Parameters	Yes	Yes	Yes	Yes	Yes	Yes
Percentiles	Yes	Yes	Yes	Yes	Yes	Yes
Fraction Failing	Yes	Yes	Yes	Yes	Yes	Yes
User Function of Param's	No	No	No	Yes	w Effort	No
Covar. Matrix of Est's	Yes	Yes	Yes	No	Yes	No
Max. Log Likelihood	Yes	Yes	Yes	Yes	Yes	Yes
LR Confidence Limits	Yes	Yes	Yes	Yes	Some	Yes
Plot of Fitted Relationship	No	Yes	Yes	Yes	Yes	Yes
Plot of Fined CDF	Yes	Yes	Yes	Yes	Yes	Yes
MODEL EVALUATION						
Residuals	Yes	Yes	Yes	No	Yes	No
Prob. Plots (Right Cens'd)	Yes	Yes	Yes	Yes	Yes	Yes
Exponential	Yes	Yes	Yes	Yes	Yes	Yes
Weibull	Yes	Yes	Yes	Yes	Yes	Yes
Extreme Value	Yes	No	Yes	Yes	Yes	No
Lognormal	Yes	Yes	Yes	Yes	Yes	Yes
Normal	Yes	Yes	Yes	Yes	Yes	Yes
Linear% & Data Scales	Yes	Yes	In SAS	Yes	Yes	Yes
Other	No	Logistic	Logistic	No	Yes	No
Peto-Tumbull cdf est.	No	No	Yes	Yes	Yes	No
LR Tests	Yes	Yes	Yes	Yes	Yes	Yes
Crossplots	Yes	Yes	In SAS	No	Yes	Yes

Preface

Many major companies spend millions of dollars each year on product reliability. Much management and engineering effort goes into evaluating risks and liabilities, predicting warranty costs, evaluating replacement policies, assessing design changes, identifying causes of failure, and comparing alternate designs, vendors, materials, manufacturing methods, and the like. Major decisions are based on product life data, often from a few units. This book presents modern methods for extracting from life test and field data the information needed to make sound decisions. Such methods are successfully used in industry on a great variety of products by many who have modest statistical backgrounds.

This book is directed to engineers and industrial statisticians working on product life data. It will also aid workers in other fields where survival is studied, for example, in medicine, biology, actuarial science, economics, business, and criminology. Also, this book may supplement texts for many statistics and engineering courses, since it gives a wealth of practical examples with real data, emphasizes applied data analysis, employs computer programs, and systematically presents graphical methods, the method of maximum likelihood, censored data analysis, prediction methods, and linear estimation.

Life data generally contain running times on unfailed units, which require special statistical methods. In the past, these rapidly developing methods were associated with aerospace applications, but they are more widely used for consumer and industrial products. This book presents many applications to diverse products ranging from simple dielectrics and small appliances to locomotives and nuclear reactors.

This book draws from my experience teaching courses on life data analysis throughout the General Electric Company and at Rensselaer Polytechnic Institute and Union College. These courses have been popular with practicing engineers and graduate students in engineering, statistics, and operations research.

This book is organized to serve practitioners. The simplest and most widely useful material appears first. The book starts with basic models and simple graphical analyses of data, and it progresses through advanced analytic methods. All preliminary material for a topic is stated, and each topic is self-contained for easy reference, although this results in some repetition. Thus this book serves as a reference as well as a textbook. Derivations are generally omitted unless they help one understand the material. Such derivations appear in advanced sections for those who seek a fundamental understanding and wish to develop new statistical models and data analyses.

Readers of this book need a previous course in statistics and, for some advanced material, facility in calculus or matrix algebra. While many methods employ new and advanced statistical theory, the book emphasizes how to apply them. Certain methods (particularly those in Chapters 8 and 12), while important, are difficult to use unless one has special computer programs, which are now available.

There is much literature on life data analysis. So I have selected topics useful in my consulting. However, I briefly survey other topics in the final chapter.

Chapter 1 describes life data analysis, provides background material, and gives an overview of the book in detail. Chapter 2 presents basic concepts and statistical distributions for product life. Chapters 3 and 4 present graphical methods for estimating a life distribution from complete and censored life data. Chapter 5 explains statistical models and analyses for data on competing failure modes and on series systems. Chapters 6, 7, and 8 provide analytic methods, mainly linear and maximum likelihood methods, for estimating life distributions from complete and censored data. Chapter 9 provides methods for analyzing inspection data (quantal-response and interval data). Chapters 10, 11, and 12 provide methods for comparing samples (hypothesis tests) and for pooling estimates from a number of samples. Chapter 13 surveys other topics.

The real data in all examples come mostly from my consulting for the General Electric Company and other companies. Many of these real data sets are messy. Proprietary data were protected by vaguely naming a product and by multiplying the data by a factor. So engineers are advised not to use examples as typical of any product.

For help on this book I am overwhelmed with a great feeling of gratitude to many. Dr. Gerald J. Hahn, my co-worker, above all others, encouraged me, helped me to obtain support from General Electric, generously contributed much personal time reading the manuscript, and offered many useful suggestions. Gerry is the godfather of this book. I am much indebted for support from management at General Electric Corporate Research and Development—Dr. Art Bueche, Mr. Stu Miller, Mr. Virg Lucke, Dr. Dick Shuey, Mr. E. Lloyd Rivest, Dr. Dave Oliver, Dr. Hal Chestnut, and Mr. Bill Chu. Professor Al Thimm, encouraged by Professor Josef Schmee, both of Union College, kindly provided me with an office, where I worked on this book, and a class that I taught from my manuscript during a leave from GE. Professor John Wilkinson of Rensselaer Polytechnic Institute gave me the original opportunity to teach courses and develop preliminary material for this book.

Colleagues have generously given much time reading the manuscript and offering their suggestions. I am particularly grateful to Paul Feder, Gerry Hahn, Joseph Kuzawinski, Bill Meeker, John McCool, Ron Regal, Josef Schmee, Bob Miller, Bill MacFarland, Leo Aroian, Jim King, Bill Tucker, and Carolyn Morgan.

Many clients generously let me use their data. They also inspired methods (such as hazard plotting) that I developed for their problems. Many students contributed suggestions. There are too many to name, unfortunately.

The illustrations are mostly the superb work of Mr. Dave Miller. The manuscript benefited much from the skillful technical typing of Jean Badalucco, Ceil Crandall, Edith White, and Ruth Dodd.

WAYNE NELSON

Schenectady, New York
November 1981

About the Author

Dr. Wayne Nelson is a leading expert on statistical analysis of reliability and accelerated test data. He currently privately consults with companies on diverse engineering and scientific applications of statistics and develops new statistical methods and computer programs. He presents courses and seminars for companies, universities, and professional societies. He also works as an expert witness. An employee of General Electric Corporation Research and Development for 23 years, he consulted across GE.

For his contributions to reliability, accelerated testing, and reliability education, he was elected a Fellow of the American Statistical Association (1973), the American Society for Quality (1983), and the Institute of Electrical and Electronics Engineers (1988). GE R&D presented him the 1981 Dushman Award for outstanding developments and applications of statistical methods for product reliability and accelerated test data.

He has authored more than 120 literature publications on statistical methods, mostly for engineering applications. For publications, he was awarded the 1969 Brumbaugh Award, the 1970 Youden Prize, and the 1972 Wilcoxon Prize, all of the American Society for Quality. The ASA has awarded him eight Outstanding Presentation Awards for papers at the national Joint Statistical Meetings.

In 1990, he was awarded the first NIST/ASA/NSF Senior Research Fellowship at the National Institute of Standards and Technology to collaborate on modeling electromigration failure of microelectronics. In 2001, he was awarded a Fulbright Award for research and lecturing on reliability data analysis in Argentina.

Dr. Nelson authored the book *Applied Life Data Analysis,* published by Wiley in 1982; it was translated into Japanese in 1988 by the Japanese Union of Scientists and Engineers. In 1990, Wiley published his landmark book *Accelerated Testing: Statistical Models, Test Plans, and Data Analyses*. In 2003, ASA-SIAM published his book *Recurrent Events Data Analysis for Product Repairs, Disease Episodes, and Other Applications*. He has authored various book chapters and tutorial booklets, and he contributed to standards of engineering societies.

He can be reached at WNconsult@aol.com or 739 Huntingdon Dr., Schenectady, NY 12309.

1

Overview
And Background

1. INTRODUCTION

This chapter presents (1) an overview of this book's contents and (2) background information for the rest of the book. To read this chapter and the rest of the book, one needs a basic statistics course. Although addressed mostly to engineering applications, this book applies to many other fields. A key characteristic of life data analysis distinguishes it from other areas in statistics: namely, data are usually censored or incomplete in some way. Like other areas, it is concerned with estimates and confidence limits for population parameters and with predictions and prediction limits for future samples. The following paragraphs describe applications and the history of life data analysis.

Applications. This book presents methods for analysis of product life data and gives many engineering applications. In this book, examples of applications to products include diesel engine fans, transformers, locomotive controls, generator field windings, material strength, generator retaining rings, locomotive reliability, electrical insulating fluids and oils, the strength of electrical connections, Class-B and Class-H motor insulations, appliance cords, fuses, turbine disks, shave dies, alarm clocks, batteries, toasters, capacitors, cryogenic cables, motor coils, engine cylinders and pistons, power lines, large motor warranty costs, turbine wheels, and distribution transformers.

The methods apply to other fields and types of data as the following examples show. Economists and demographers study the length of time people are in the work force (Kpedekpo, 1969). Employers are concerned with the length of time employees work before changing jobs. Mental health officials use tables of length of stay in facilities to predict patient load. Businessmen wish to know the shelf life of goods and the time it takes inventory to turn over; for example, one manufacturer wanted to know the distribution of time from manufacture to installation of a major appliance. Wildlife managers use mortality tables to predict wildlife population sizes and determine hunting seasons. Hoadley (1970) studied the length of time telephones remain disconnected in vacant quarters in order to determine which telephones to remove for use elsewhere and which to leave in for the next customer. Kosambi (1966) proposed that knowledge of the distribution of the time that coins remain in circulation can help the mint plan production. The success of medical treatments for certain diseases is measured by the length of patient survival (Gross and Clark, 1975). The distribution of time from prison release to committing a crime measures the success of prison programs. A trading stamp company estimated the proportion of stamps that would be redeemed; this was used to determine needed cash reserves to cover outstanding stamps. Potency of some pesticides (and chemicals) is bioassayed by observing the times to death (or other reaction) of a sample of insects or animals. Life insurance companies determine premiums from mortality tables. The life of TV programs has been evaluated (Prince, 1967). Jaeger and Pennock (1957) estimated service life of household goods. The Association for the Advancement of Medical Instrumentation (1975) has a proposed standard with methods for estimating the life of heart pacemakers. Zahn (1975) described a psychological experiment on the time a (planted) "stranded" motorist must wait for aid from a passerby. The durability of garments is studied by manufacturers (Goldsmith, 1968). Wagner and Altman (1973) studied the time in the morning when baboons come down from the trees.

This book presents engineering applications and uses mostly engineering and reliability terminology. Biomedical, actuarial, and other fields have their own terminology for many concepts; some of their terminology is mentioned. Differences in terminology may cause initial difficulties to those who read publications in other fields.

History.   Todhunter (1949) describes the early human life table of Halley (Chapter 2) and Bernoulli's work on the effect of smallpox innoculation on the distribution of life. Insurance companies have long used actuarial methods for constructing human life tables. Early in this century, actuarial methods were used to estimate survivorship of (1) medical patients under different treatments and of (2) equipment, particularly on railroads. In the 1950s and 1960s, reliability engineering blossomed; this resulted from demands for more reliable equipment from military and space programs. In this period, engineering design methods for reliable equipment made great strides. However, reliability data analysis mostly employed the oversimple exponential and Poisson distributions. In the 1950s and 1960s, most advances in life data analysis came from biomedical applications. Now methods are widely being developed for engineering applications to many consumer and industrial products. This book brings together recent methods for life data analysis. This field continues to grow, although many important problems remain unsolved, as this book shows.

2. OVERVIEW OF THE BOOK

This section describes this book's contents, organization, and how to use the book. The types of data mentioned here are described in Section 3.

Chapter 1 gives an overview of the book and presents needed background. Chapter 2 describes distributions for life and failure data. Chapter 3 presents simple probability plots for analyzing complete and singly censored data. Chapter 4 presents hazard plots for multiply censored data. Chapter 5 describes models for and graphical analyses of data with a mix of failure modes. Chapters 6 through 9 give analytic methods for (1) estimates and confidence limits for distribution parameters, percentiles, reliabilities, and other quantities and for (2) predictions and prediction limits for future samples. Chapter 6 treats analysis of complete data. Chapter 7 gives linear methods for singly censored data. Chapter 8 gives maximum likelihood methods for multiply censored data. Chapter 9 gives maximum likelihood methods for quantal-response and interval data. Chapter 10 presents various methods for comparing samples of complete data by confidence intervals and hypothesis tests; such methods include 1-, 2-, and K-sample comparisons and estimation by pooling a number of samples. Chapter 11 presents such comparisons based on linear estimates from singly censored samples. Chapter 12 presents such comparisons based on maximum likelihood methods for multiply censored and other types of data. Chapter 13 surveys topics in reliability and life data analysis that are not presented in the book.

Figure 2.1 shows this book's chapters. They are organized by type of data (complete, singly censored, multiply censored, etc.) and by statistical method (elementary, linear, and maximum likelihood). The chapters are in order of difficulty. Early chapters present simple graphical methods, and later ones present advanced analytic methods. The arrows in Figure 2.1 show which chapters are background for later chapters. Also, each chapter introduction

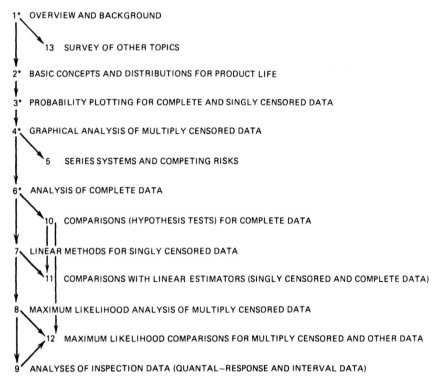

Figure 2.1.   Book organization (asterisk denotes basic material).

refers to needed background and describes the difficulty of the chapter. Some section introductions do the same, and they state if a section is theoretical and can be skipped. Most sections are written for those who just wish to apply the methods. The first four chapters are simple and basic reading for all. The more advanced Chapters 5 through 9 are in order of difficulty. Chapters 10, 11, and 12 can be read after the corresponding Chapters 6 through 9.

Maximum likelihood methods (Chapters 8, 9, and 12) are versatile and apply to most distributions and types of data. Also, they have good statistical properties. If time is limited, one might skip the linear methods (Chapter 7) in favor of maximum likelihood methods.

The book employs the following scheme for numbering sections, equations, figures, and tables. Within each chapter, the sections are numbered simply 1, 2, 3, etc.; subsections are numbered 4.1, 4.2, etc. Equation numbers give the (sub)section number and equation number; for example,

(2.3) is the third numbered equation in Section 2. Figure and table numbers include the section number; Figure 2.3 is the third figure in Section 2. Unless another chapter is stated, any reference to an equation, figure, or table is to one in the same chapter.

There are two types of problems at the end of a chapter. One type involves an analysis of data with the methods in that chapter; the other involves extending the results of the chapter to other problems. An asterisk (*) marks more laborious or difficult problems.

The book cites references by means of the Harvard system. A citation includes the author's name, year of publication, and his publications in that year. For example, "Nelson (1972b)" refers to Nelson's second referenced publication in 1972. All references are listed near the end of the book.

Basic statistical tables are in an appendix near the end of the book. Other tables must be obtained from the literature and are referenced.

The index of the book is detailed. It will be an aid to those who wish to use the book as a reference for selected methods. Also, to aid users, each section is written to be self-contained, thus repeating some material.

The book omits many derivations. Reasons for this are the following: (1) users can properly apply most methods, not knowing derivations, (2) many derivations are easy for a reader or instructor to supply, and (3) more time can be spent on methods useful in practice. Many derivations appear in Mann, Schafer, and Singpurwalla (1974), Gross and Clark (1975), Bain (1978), and Lawless (1982).

3. BACKGROUND MATERIAL

Background material useful for the rest of this book is briefly presented here. The topics are (1) statistical models, (2) population and sample, (3) valid data, (4) failure and exposure, (5) types of data, (6) nature of data analysis, (7) estimates and confidence intervals, (8) hypothesis tests, (9) predictions, (10) practical and statistical significance, (11) numerical calculations, (12) notation.

Statistical models. Supposedly identical units made and used under the same conditions usually have different values of performance, dimensions, life, etc. Variability of such a performance variable is inherent in all products, and it is described by a statistical model or distribution.

Population and sample. A statistical model describes some **population**. A manufacturer of fluorescent lamps is concerned with the future production of a certain lamp—an essentially infinite population. A manufacturer of locomotives is concerned with a small population of locomotives. A metallurgist is concerned with the future production of a new alloy—an essentially infinite population. A generator manufacturer is concerned with the performance of a small population of units to be manufactured next year. To obtain information, we use a **sample** (a set of units) from the population. We analyze the sample data to get information on the underlying population distribution or to predict future data from the population.

Valid data.   There are many practical aspects to the collection of valid and meaningful data. Some are described below. Throughout, this book assumes that such aspects are properly handled.

Most statistical work assumes that the sample is from the population of interest. A sample from another population or a subset of the population can give misleading information. For example, failure data from appliances on a service contract may overestimate failure rates for appliances not on contract. Also, laboratory test data may differ greatly from field data. Data on units made last year may not adequately predict this year's units. In practice, it is often necessary to use such data. Then engineering judgment must determine how well such data represent the population of interest and how much one can rely on the information.

Most statistical work assumes that the data are obtained by simple random sampling from the population of interest. Such sampling gives each possible set of n units the same chance of being the chosen sample; random numbers should be used to ensure random selection. In practice, other statistical sampling methods are sometimes used, the most common methods being stratified sampling and two-stage sampling. Data analyses must take into account the sampling method. This book assumes throughout that simple random sampling is used. Some samples are taken haphazardly, that is, without probability sampling. Such samples may be quite misleading.

In practice, measurements must be meaningful and correct. Also, one needs to avoid blunders in handling data. Bad data can be unknowingly processed by computers and by hand.

Failure and exposure.   Failure must be precisely defined in practice. For dealings between producers and consumers, it is essential that the definition of a failure be agreed upon in advance to minimize disputes. For many products, failure is catastrophic, and it is clear when failure occurs. For some products, performance slowly degrades, and there is no clear end of life. One can then define that a failure occurs when performance degrades below a specified value. Of course, one can analyze data according to each of a number of definitions of failure. One must decide whether time is calendar time or operating hours or some other measure of exposure, for example, the number of start-ups, miles traveled, energy output, cycles of operation, etc. Also, one must decide whether to measure time of exposure starting at time of manufacture, time of installation, or whatever. Engineers define failure and exposure.

Types of data. The proper analysis of data depends on the type of data. The following paragraphs describe the common types of life data from life tests and actual service.

Most nonlife data are **complete**; that is, the value of each sample unit is observed. Such life data consist of the time to failure of each sample unit. Figure 3.1a depicts a complete sample. Chapters 3, 6, and 10 treat such data. Much life data are incomplete. That is, the exact failure times of some units are unknown, and there is only partial information on their failure times. Examples follow.

Sometimes when life data are analyzed, some units are unfailed, and their failure times are known only to be beyond their present running times. Such data are said to be **censored on the right**. Unfailed units are called run-outs, survivors, removals, and suspended units. Similarly, a failure time known only to be before a certain time is said to be **censored on the left**. If all unfailed units have a common running time and all failure times are earlier, the data are said to be **singly censored** on the right. Singly censored data arise when units are started on test together and the data are analyzed before all units fail. Such data are singly **time censored** if the censoring time is fixed; then the number of failures in that fixed time is random. Figure 3.1b depicts such a sample. Time censored data are also called **Type I censored**. Data are singly **failure censored** if the test is stopped when a specified number of failures occurs, the time to that fixed number of failures being random. Figure 3.1c depicts such a sample. Time censoring is more common in practice; failure censoring is more common in the literature, as it is mathematically more tractable. Chapters 3, 7, and 11 treat singly censored data.

Much data censored on the right have differing running times intermixed with the failure times. Such data are called **multiply censored** (also progressively, hyper-, and arbitrarily censored). Figure 3.1d depicts such a sample. Multiply censored data usually come from the field, because units go into service at different times and have different running times when the data are recorded. Such data may be time censored (running times differ from failure times, as shown in Figure 3.1d) or failure censored (running times equal failure times, as shown in Figure 3.1e). Chapters 4, 8, and 12 treat such data.

A mix of **competing failure modes** occurs when sample units fail from different causes. Figure 3.1f depicts such a sample. Data on a particular

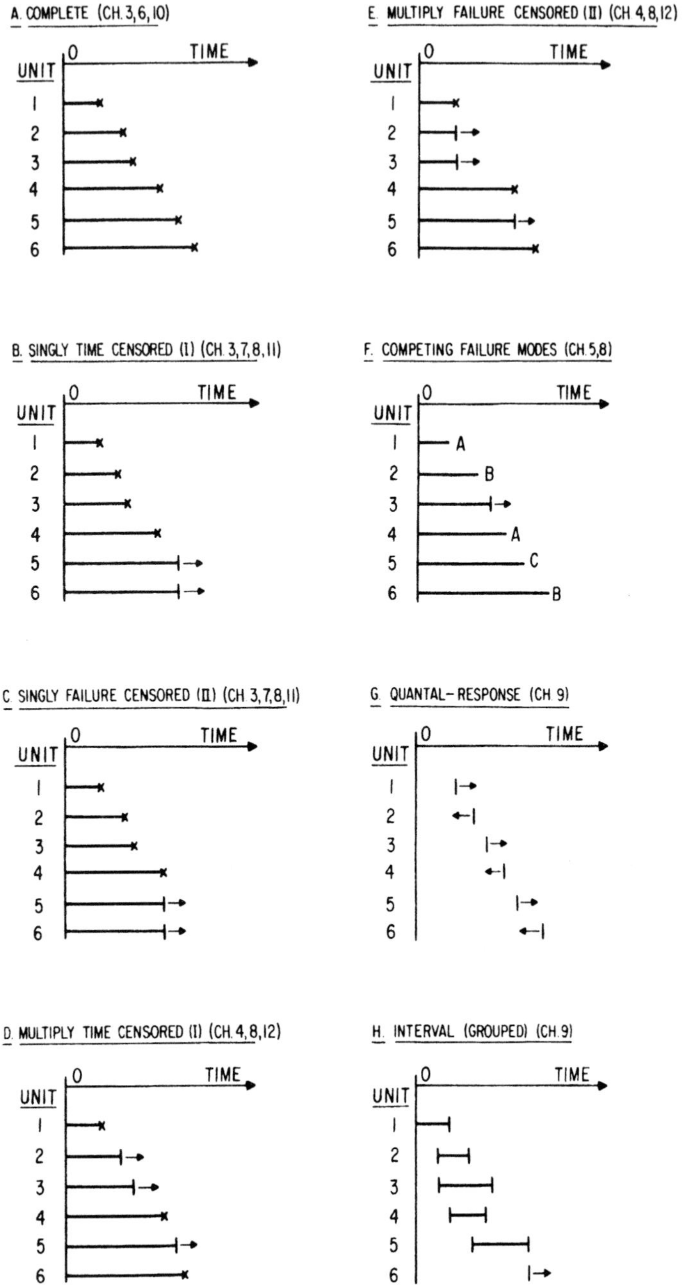
A. COMPLETE (CH. 3,6,10)
E. MULTIPLY FAILURE CENSORED (II) (CH 4,8,12)
B. SINGLY TIME CENSORED (I) (CH. 3,7,8,11)
F. COMPETING FAILURE MODES (CH 5,8)
C. SINGLY FAILURE CENSORED (II) (CH 3,7,8,11)
G. QUANTAL-RESPONSE (CH 9)
D. MULTIPLY TIME CENSORED (I) (CH. 4,8,12)
H. INTERVAL (GROUPED) (CH. 9)
UNIT
0
TIME
1
2
3
4
5
6
A
B
A
C
B

Figure 3.1. Types of data (failure time ×, running time ↦, failure occurred earlier ↤)

failure mode consist of the failure times of units failing by that mode. Such data for a mode are multiply censored. Chapter 5 treats such data in detail; Chapter 8 does so briefly.

Sometimes one knows only whether the failure time of a unit is before or after a certain time. Each observation is either censored on the right or else on the left. Such life data arise if each unit is inspected once to see if it has already failed or not. These data are **quantal-response** data, also called sensitivity, probit, and all-or-nothing response data. Figure 3.1*g* depicts such a sample. Chapter 9 treats such data.

When units are inspected for failure more than once, one knows only that a unit failed in an interval between inspections. So-called **interval** or grouped data are depicted in Figure 3.1*h*. Such data can also contain right and left censored observations. Chapter 9 treats such data.

Data may also consist of a mixture of the above types of data.

Analyses of such censored and interval data have much the same purposes as analyses of complete data, for example, estimation of model parameters and prediction of future observations.

Nature of data analysis.   This section briefly describes the nature of data analysis. It advises how to define a statistical problem, select a mathematical model, fit the model to data, and interpret the results.

The solution of a real problem involving data analysis has seven basic steps.

1.   Clearly state the real problem and the purpose of the data analysis. In particular, specify the numerical information needed in order to draw conclusions and make decisions.

2.   Formulate the problem in terms of a model.

3.   Plan both collection and analyses of data that will yield the desired numerical information.

4.   Obtain appropriate data for estimating the parameters of the model.

5.   Fit the model to the data, and obtain the needed information from the fitted model.

6.   Check the validity of the model and data. As needed, change the model, omit or add data, and redo steps 5 and 6.

7.   Interpret the information provided by the fitted model to provide a basis for drawing conclusions and making decisions for the real problem.

This book gives methods for steps 5 and 6. The other steps involve the judgment of engineers, managers, scientists, etc. Each of the steps is discussed below, but full understanding of these steps comes only with experience. Data analysis is an iterative process, and one usually subjects a data set to many analyses to gain insight. Thus, many examples in this book involve different analyses of the same set of data.

1. A clear statement of a real problem and the purpose of a data analysis is half of the solution. Having that, one can usually specify the numerical information needed to draw practical conclusions and make decisions. Of course, an analysis provides no decisions—only numerical information for people who make them. If one has difficulty specifying the numerical information needed, the following may help. Imagine that any desired amount of data is available (say, the entire population), and then decide what values calculated from the data would be useful. Statistical analysis estimates such values from limited sample data. If such thinking does not clarify the problem, one does not understand the problem. Sometimes there is a place for exploratory data analyses that do not have clear purposes but that may reveal useful information. Data plots are particularly useful for such analyses.

2. To state a problem in terms of a model, one chooses a statistical distribution for performance. Often the model is a simple and obvious one, widely used in practice: for example, a lognormal distribution for time to insulation failure. When a suitable model is not obvious, display the data various ways, say, on different probability papers. Such plots often suggest a suitable model. Indeed, a plot often reveals needed information and can serve as a model itself. Another approach is to use a very general model that is likely to include a suitable one as a special case. After fitting the general model to the data, one often sees which special case is suitable. Still another approach is to try various models and select the one that best fits the data. The chosen model should, of course, provide the desired information. Examples of these approaches appear in later chapters.

3. Ideally a tentative model is chosen before the data are collected, and the data are collected so that the model parameters can be estimated from the data. Sometimes when data are collected before a model and data analyses are determined, it may not be possible to fit a desired model, and a less realistic model must be used.

4. Practical aspects of data collection and handling need much forethought and care. For instance, data may not be collected from the population of interest; for example, data may be from appliances on service contract (self-selection) rather than from the entire population. Many companies go to great expense collecting test and field data but end up with inadequate data owing to lack of forethought.

5. To fit a chosen model to the data, one has a variety of methods. This step is straightforward; it involves using methods described in this book.

Much of the labor can (and often must) be performed by computer programs.

6. Of course, one can mechanically fit an unsuitable model just as readily as a suitable one. An unsuitable model may yield information leading to wrong conclusions and decisions. Before using information from a fitted model, one should check the validity of the model and the data. Such checks usually employ graphical displays that allow one to examine the model and the data for consistency with each other. The model may also be checked against new data. Often different models fit a set of data within the range of the data. However, they can give very different results outside that range.

7. Interpretation of results from the fitted model is easy when the above steps are done properly, as practical conclusions and decisions are usually apparent. A possible difficulty is that the information may not be accurate or conclusive enough for practical purposes. Then more data for the analysis is needed or one must be content with less reliable information. Also, most models and data are inaccurate to some degree. So the uncertainty in any estimate or prediction is greater than is indicated by the corresponding confidence or prediction interval.

Data analysis methods.   Some specific data analysis methods are discussed below—estimates, confidence intervals, hypothesis tests, and predictions. These methods are treated in detail in later chapters.

Estimates and confidence intervals.   Using sample data, the book provides estimates and confidence intervals for the parameters of a model. The estimates approximate the true parameter values. By their width, confidence intervals for parameters indicate the uncertainty in estimates. If an interval is too wide for practical purposes, a larger sample may yield one with the desired width. Chapters 6 through 9 provide such analytical estimates and confidence limits and examples.

Hypothesis tests.   Chapters 10, 11, and 12 provide statistical tests of hypotheses about model parameters. A statistical test compares sample data with a hypothesis about the model. A common hypothesis is that a parameter equals a specified value; for example, the hypothesis that a Weibull shape parameter equals unity implies that the distribution is exponential. Another common hypothesis is that corresponding parameters of two or more populations are equal; for example, the standard two-sample t-test compares two population means for equality. If there is a statistically significant difference between the data and the hypothesized model, then there is convincing evidence that the hypothesis is false. Otherwise, the hypothesis is a satisfactory working assumption. Also, a test of fit or a test for outliers may result in rejection of the model or data.

Predictions. Most statistical methods are concerned with population parameters (including percentiles and reliabilities) when a population is so large that it can be regarded as infinite. One then uses estimates, confidence intervals, and hypothesis tests for **parameters** (or **constants**). However, in many business and engineering problems, the data can be regarded as a sample from a theoretical distribution. Then one usually wishes to predict the **random values** in a **future** sample from the same distribution. For example, one may wish to predict the random warranty costs for the coming year or the random number of product failures in the coming quarter, using past data. Then one wants a **prediction** for the future random value and a **prediction interval** that encloses that future random value with high probability. Many prediction problems go unrecognized and are incorrectly treated with methods for population parameters. Methods for such problems are now being developed, and this book brings together many of them. Chapters 6 through 8 include prediction methods and examples.

Practical and statistical significance. Confidence intervals indicate how (im)precise estimates are and reflect the inherent scatter in the data. Hypothesis tests indicate whether observed differences are statistically significant; that is, whether a difference between a sample of data and a hypothesized model (or whether the difference between a number of samples) is large relative to the inherent random scatter in the data. Statistical significance means that the observed differences are large enough to be convincing. In contrast, practical significance depends on the true differences in the actual populations; they must be big enough to be important in practice. Although results of an analysis may be important in practice, one should not rely on them unless they are also statistically significant. Statistical significance assures that results are real rather than mere random sampling variation.

A confidence interval for such differences is often easier to interpret than a statistical hypothesis test. The interval width allows one to judge whether the results are accurate enough to identify true differences that are important in practice. Chapters 10, 11, and 12 give examples of such confidence intervals and their application to comparisons.

Numerical calculations. Numerical examples are generally calculated with care. That is, extra figures are used in intermediate calculations. This good practice helps assure that answers are accurate to the final number of figures shown. For most practical purposes, two or three significant final figures suffice. A reasonable practice is to give estimates and confidence limits to enough figures so that they differ in just the last two places, for example, $\mu^* = 2.76$ and $\tilde{\mu} = 2.92$. Tabulated values used in examples do not always have the same number of significant figures as the tables in this book.

Many calculations for examples and problems can easily be done with an electronic pocket calculator. However, some calculations, particularly maximum likelihood calculations, will require computer programs. Readers can develop their own programs from the descriptions given here or use standard programs.

Notation. This book mostly follows modern statistical notation. Random variables are usually denoted by capital letters such as $Y, Y_1, \ldots, Y_n, T$, etc. Observed outcomes or possible values of random variables are usually denoted by lower-case letters such as $y, y_1, \ldots, y_n, t$, etc. Population or distribution parameters, also called true values, are usually denoted by Greek letters such as $\mu, \sigma, \alpha, \beta, \theta$, etc. However, Latin letters denote some parameters. Estimates for parameters are usually denoted by the corresponding Latin letter (a estimates α) or by the parameter with "*" or "^" (α^* or $\hat{\alpha}$). Notation often does not distinguish between estimators (random variables) and estimates (the value from a particular sample).

Commonly used symbols and their meanings follow.

$F(P; a, b)$ $100P$ th F percentile with a degrees of freedom in the numerator and b in the denominator (Section 7 of Chapter 2).

K_γ $[100(1+\gamma)/2]$th standard normal percentile two-sided (Section 2 of Chapter 2).

u_P $= \ln[-\ln(1-P)]$, the standard extreme value $100P$ th percentile (Section 5 of Chapter 2).

z_γ 100γ th standard normal percentile, one-sided (Section 2 of Chapter 2).

α Weibull scale parameter (Section 4 of Chapter 2); level of a hypothesis test (Chapter 10).

β Weibull shape parameter (Section 4 of Chapter 2).

δ Extreme value scale parameter (Section 5 of Chapter 2).

θ Exponential mean (Section 1 of Chapter 2); general notation for a parameter.

$\hat{\theta}$ Maximum likelihood estimator for θ (Chapter 8).

θ^* Best linear unbiased estimator for θ (Chapter 7).

$\underset{\sim}{\theta}$ Lower confidence limit for θ.

$\tilde{\theta}$ Upper confidence limit for θ.

λ Exponential failure rate (Section 1 of Chapter 2); Poisson occurrence rate (Section 12 of Chapter 2); extreme value location parameter (Section 5 of Chapter 2).

μ Mean; normal distribution mean (Section 2 of Chapter 2); lognormal parameter (Section 3 of Chapter 2).

σ Standard deviation; normal distribution standard deviation (Section 2 of Chapter 2); lognormal parameter (Section 3 of Chapter 2).

$\varphi(\ )$ Standard normal probability density (Section 2 of Chapter 2).

$\Phi(\ )$ Standard normal cumulative distribution function (Section 2 of Chapter 2).

$\chi^2(P; \nu)$ $100P$ th χ^2 percentile with ν degrees of freedom (Section 6 of Chapter 2).

2

Basic Concepts and Distributions for Product Life

This chapter presents basic concepts and distributions for analyzing life and failure data. It covers continuous and discrete distributions and their properties. This chapter is essential background for all work with product life data. To use this chapter, read the general theory, and then go to specific distributions of interest.

Background. While this chapter is self-contained, readers may find further background on basic probability theory helpful. For example, see Hahn and Shapiro (1967), Parzen (1960), or Hoel (1960).

Continuous distributions. Section 1 on continuous distributions begins with general theory and basic concepts. These are illustrated by a human mortality table and the exponential distribution. The theory is then applied to the other most commonly used life distributions— the normal (Section 2), lognormal (Section 3), Weibull (Section 4), and extreme value (Section 5) distributions. In addition, there is advanced material on the gamma and chi-square distributions (Section 6), other selected distributions (Section 7), shifted distributions (Section 8), conditional and truncated distributions (Section 9), and joint distributions (Section 10). The advanced Sections 6 through 10 may be skipped on a first reading.

Discrete distributions. Section 11 on discrete distributions begins with general theory and basic concepts, illustrated by the geometric distribution.

The theory is then applied to other commonly used discrete distributions—the Poisson (Section 12), binomial (Section 13), hypergeometric (Section 14), and multinomial (Section 15) distributions.

1. GENERAL THEORY ON CONTINUOUS DISTRIBUTIONS (ILLUSTRATED BY THE EXPONENTIAL DISTRIBUTION)

This essential section presents basic concepts and theory for continuous statistical distributions. Such distributions are used as models for the life of products, materials, people, television programs, and many other things. Also, they are used for representing tensile strength, impact resistance, voltage breakdown, and many other properties. Such distributions include the commonly used exponential, normal, lognormal, Weibull, and extreme value distributions. These and other distributions are presented below and are used to analyze data. The basic concepts include:

the probability density
events and their probabilities
the cumulative distribution function
the reliability function
percentiles
the mode, mean, variance, and standard deviation
the hazard function (instantaneous failure rate)

After reading the basic concepts in this section, one may proceed to any distribution of interest.

   A **continuous distribution** with a probability density is a probability model, denoted by Y, that consists of

1. the possible numerical **outcomes** y that are a subset $\mathcal{Y}$ of the line $(-\infty, \infty)$ and

2. a **probability density** $f(y)$, a function of y with the properties

(a) $f(y) \geq 0$ for all y in $\mathcal{Y}$ and

(b) $\int_{\mathcal{Y}} f(y)\,dy = 1$; where the integral runs over the range $\mathcal{Y}$.

The probability density $f(y)$ is the mathematical model for the population histogram (e.g., Figure 1.1a) containing relative frequencies, which (a) must be greater than or equal to zero and (b) must sum up to unity. Any subset $\mathcal{Y}$ of the line $(-\infty, \infty)$ and function with properties (a) and (b) is a continuous distribution with a probability density. The capital letter Y denotes the distribution, which is also loosely called a **random variable**. Of course, Y is

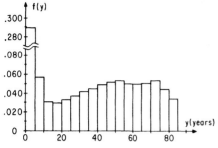

Figure 1.1a. Mortality table histogram.

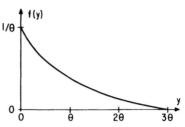

Figure 1.1b. Exponential probability density.

neither random nor a variable; it is a probability model consisting of outcomes and their corresponding probabilities. Y can also be loosely thought of as the potential random value of a randomly selected unit from the population. Lower-case letters y, y_1, y_2,... denote particular outcomes.

Mortality table example.   The basic concepts are illustrated with an early human mortality table published by Halley in 1693 and discussed by Todhunter (1949). This table appears in Table 1.1 and is given in steps of five years. The histogram $f(y)$ appears in Figure 1.1a; it roughly corresponds to a probability density. Table 1.1 and the histogram of Figure 1.1a show, for example, that 0.057 (5.7%) of this population died between the

Table 1.1.   Halley's Mortality Table

y	f(y)	F(y)	R(y)	h(y)
0	–	0	1.000	
0–5	.290	.290	.710	.058
5–10	.057	.347	.653	.016
10–15	.031	.378	.622	.010
15–20	.030	.408	.592	.010
20–25	.032	.440	.560	.011
25–30	.037	.477	.523	.013
30–35	.042	.519	.481	.016
35–40	.045	.564	.436	.019
40–45	.049	.613	.387	.022
45–50	.052	.665	.335	.027
50–55	.053	.718	.282	.032
55–60	.050	.768	.232	.035
60–65	.050	.818	.182	.043
65–70	.051	.869	.131	.056
70–75	.053	.922	.078	.081
75–80	.044	.966	.034	.113
80–85	.034	1.000	0	.200

ages of 5 and 10. In what follows, it is useful to regard Table 1.1 as the life distribution for either (1) a large population or (2) for 1000 people from birth. Then, for example, 57 people died between the ages of 5 and 10. This "discrete" life distribution is used as a simple means of motivating the concepts below, and it is treated as if it were a continuous distribution.

The **exponential distribution** consists of outcomes $y \geq 0$ and the probability density

$$f(y) = (1/\theta)\exp(-y/\theta), \qquad y \geq 0, \tag{1.1}$$

where the parameter θ is called the "distribution mean" and must be positive. θ is expressed in the same units as y, for example, hours, months, cycles, etc. Figure 1.1b depicts this probability density. The exponential distribution is a continuous distribution, since (1) $f(y) \geq 0$ for $y \geq 0$ and (2) $\int_0^\infty (1/\theta)\exp(-y/\theta)\,dy = 1$.

The exponential density is also written as

$$f(y) = \lambda \exp(-\lambda y), \qquad y \geq 0, \tag{1.2}$$

where the parameter $\lambda = 1/\theta$ is called the **failure rate**. The **standard exponential distribution** has $\theta = 1$ or, equivalently, $\lambda = 1$.

The simple exponential life distribution has been widely used (and misused) as a model for the life of many products. As explained later, it describes products subject to "chance failures"; that is, the failure rate is the same at each point in time. This distribution may describe the time between failures for repairable products (Section 12). It is also used to describe product life after units have gone through a burn-in up to the time they start to wear out. As the exponential distribution is easy to fit to data, it has often been misapplied to situations that require a more complex distribution.

An **event** is any subset of outcomes of $\mathcal{Y}$, and the **probability of an event** E is defined by

$$P\{E\} = \int_E f(y)\,dy, \tag{1.3}$$

where the integral runs over the outcomes in the event E. That is, the probability is the area under the probability density function for the points of the event. $P\{Y \leq y_0\}$ denotes the probability of the event consisting of all outcomes of the model Y that satisfy the relationship, that is, are less than or equal to the value y_0. More generally, the notation for the probability of an event is $P\{\ \}$, where the relationship in the braces indicates that the event consists of the outcomes of the model Y that satisfy the relationship, for example $P\{Y^2 > y_0\}$ and $P\{y_0 < Y \leqslant y_0'\}$; also, see (1.6). The probability of an event has two interpretations: (1) it is the proportion of population units that have values in the event, and (2) it is the chance that a random observation is a value from the event.

For example, for Halley's table, the probability of death by age 20 is the sum (corresponding to an integral) of the probabilities up to age 20, namely $0.290 + 0.057 + 0.031 + 0.030 = 0.408$ (40.8% or 408 people out of 1000). Also, for example, the probability of death between the ages of 20 and 50 is the sum of the probabilities between those ages, namely, $0.032 + 0.037 + 0.042 + 0.045 + 0.049 + 0.052 = 0.257$ (25.7% or 257 people). These sums approximate the integral (1.3).

Engine fan example. The exponential distribution with a mean of $\theta = 28{,}700$ hours was used to describe the hours to failure of a fan on diesel engines. The failure rate is $\lambda = 1/28{,}700 = 34.8$ failures per million hours. The probability of a fan failing on an 8,000 hour warranty is

$$P\{Y \leqslant 8000\} = \int_0^{8000} \tfrac{1}{28{,}700} \exp(-y/28{,}700)\, dy = 0.24.$$

That is, 24% of such fans fail on warranty. This information helped management decide to replace all such fans with an improved design.

The cumulative distribution function (cdf) $F(y)$ for a continuous distribution with a probability density $f(y)$ is

$$F(y) = P\{Y \leqslant y\} = \int_{-\infty}^{y} f(u)\, du, \qquad -\infty < y < \infty, \qquad (1.4)$$

where the integral runs over all outcomes less than or equal to y. The cdf $F(y)$ is defined for all y values.

$F(y)$ and $f(y)$ are alternative ways of representing a distribution. They are related as shown in (1.4) and by

$$f(y) = dF(y)/dy. \qquad (1.5)$$

Any such cumulative distribution function $F(y)$ has the following properties:

1. it is a continuous function for all y,
2. $\lim_{y \to -\infty} F(y) = 0$ and $\lim_{y \to \infty} F(y) = 1$, and
3. $F(y) \leqslant F(y')$ for all $y < y'$.

Conversely, any function $F(\ )$ that has properties 1, 2, and 3 is a continuous cumulative distribution function. For product life, $F(y)$ is interpreted as the population fraction failing by age y.

The relationships below express probabilities of common events in terms of $F(y)$, which is conveniently tabled for many distributions:

$$P\{Y \leqslant y\} = \int_{-\infty}^{y} f(y)\,dy = F(y),$$

$$P\{Y > y\} = \int_{y}^{\infty} f(y)\,dy = 1 - F(y), \qquad (1.6)$$

$$P\{y < Y \leqslant y'\} = \int_{y}^{y'} f(y)\,dy = F(y') - F(y).$$

$F(y)$ for Halley's table appears in Table 1.1 and Figure 1.2a. These show, for example, that the probability of death by age 20 is $F(20) = 0.408$; that is, 40.8% of the population died by age 20. Similarly, the probability of death by age 50 is $F(50) = 0.665$.

The exponential cumulative distribution function is

$$F(y) = \int_{0}^{y}(1/\theta)\exp(-y/\theta)\,dy = 1 - \exp(-y/\theta), \qquad y \geqslant 0. \quad (1.7)$$

Figure 1.2b shows this cumulative distribution function. In terms of the failure rate $\lambda = 1/\theta$,

$$F(y) = 1 - \exp(-\lambda y), \qquad y \geqslant 0. \qquad (1.8)$$

For the engine fans, the probability of failure on an 8000 hour warranty is $F(8000) = 1 - \exp(-8000/28{,}700) = 0.24$.

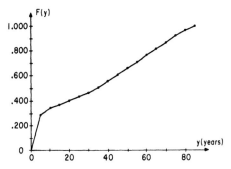

Figure 1.2a. Cumulative distribution function.

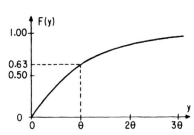

Figure 1.2b. Exponential cumulative distribution.

When y is small compared to $\theta = 1/\lambda$, a useful approximation for the exponential cdf is

$$F(y) \simeq y/\theta \quad \text{or} \quad F(y) \simeq \lambda y. \tag{1.9}$$

In practice this usually is satisfactory for $y/\theta = \lambda y \leqslant 0.01$ or even $\leqslant 0.10$. For example, for the diesel engine fans, the probability of failure in the first 100 hours is $F(100) \simeq 100/28{,}700 = 0.003484$. The exact probability is $F(100) = 1 - \exp(-100/28{,}700) = 0.003478$.

The reliability function $R(y)$ for a life distribution is

$$R(y) = P\{Y > y\} = \int_{y}^{\infty} f(y)\,dy = 1 - F(y). \tag{1.10}$$

This is the probability of (or population fraction) surviving beyond age y. It is also called the **survivorship function**. Statisticians and actuaries have long worked with cumulative distributions for the fraction failing, but public-relations-minded engineers turned them around and called them "reliability functions."

The survivorship (or reliability) function $R(y)$ for Halley's mortality table appears in Table 1.1 and in Figure 1.3a. These show, for example, that the survivorship for 20 years is $R(20) = 0.592$. That is, of 1000 newborn, 592 survive to 20. Similarly, the survivorship for 50 years is $R(50) = 0.335$.

The exponential reliability function is

$$R(y) = \exp(-y/\theta), \qquad y \geqslant 0. \tag{1.11}$$

Figure 1.3b shows this reliability function. For the engine fans, reliability

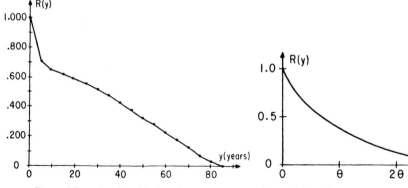

Figure 1.3a. Survivorship function.   **Figure 1.3b.** Exponential reliability function.

for 8000 hours is $R(8000) = \exp(-8000/28{,}700) = 0.76$. That is, 76% of such fans survive at least 8000 hours, the warranty period.

When y is small compared to $\theta = 1/\lambda$, a useful approximation is

$$R(y) \simeq 1 - (y/\theta) \quad \text{or} \quad R(y) \simeq 1 - \lambda y. \qquad (1.12)$$

(1.12) often suffices for $y/\theta = \lambda y \leqslant 0.01$ or even $\leqslant 0.10$. For the diesel engine fans, the reliability for the first 100 hours is $R(100) \simeq 1 - (100/28{,}700) = 0.9965$ or 99.65%.

The **100Pth percentile** y_P of a continuous distribution is the age by which a proportion P of the population has failed. It is the solution of

$$P = F(y_P). \qquad (1.13)$$

It is also called the "100P percent point" or "P fractile." In life data work, one often wants to know low percentiles such as the 1 and 10% points. The 50% point is called the **median** and is commonly used as a "typical" life. It is the middle of a distribution in the sense that 50% of the population fail before that age and 50% survive it.

Percentiles of Halley's human life distribution may be obtained from the cumulative distribution function in Table 1.1 or in Figure 1.2a. To get, say, the 50th percentile, enter the figure on the scale for the fraction failing at 50%, go horizontally to the cumulative distribution function, and then go down to the age scale to read the median life as 33 years. This can be regarded as a typical life. In a recent table for the United States, the median life is 72 years. The median life could also be obtained by backwards interpolation in Table 1.1. The 90th percentile is about 73 years in Halley's table.

The **100Pth exponential percentile** is the solution of $P = 1 - \exp(-y_P/\theta)$, namely,

$$y_P = -\theta \ln(1 - P). \qquad (1.14)$$

For example, the mean θ is roughly the 63rd percentile of the exponential distribution, an often used fact. For the diesel engine fans, median life is $y_{.50} = -28{,}700 \ln(1 - 0.5) = 19{,}900$ hours. The first percentile is $y_{.01} = -28{,}700 \ln(1 - 0.01) = 288$ hours.

The **mode** y_m is the value where the probability density is a maximum. Thus it is the "most likely" time to failure, and it may be regarded as another "typical" life for many distributions. This age may usually be found by the usual calculus method for locating the maximum of a function;

namely, by finding the age where the derivative of the probability density with respect to time equals zero.

The mode of Halley's human life distribution is in the first five years of life (actually peaks at birth). This shows up in the histogram in Figure 1.1a. There is a second mode between the ages of 50 and 75. In a recent table for the United States, the mode is 76 years, and there is a smaller peak in the first year. Medical science has increased human life mainly by reducing infant mortality.

The mode of the exponential distribution is $y_m = 0$ for any value of the mean θ. This may be seen in Figure 1.1b. So the mode is not a useful typical life for the exponential distribution.

The following presents the mean, variance, and standard deviation of a distribution. These values are often used to summarize a distribution. Also, they are used later to obtain certain theoretical results.

The mean $E(Y)$ of a variable Y with a continuous distribution with a probability density $f(y)$ is

$$E(Y) \equiv \int_{-\infty}^{\infty} yf(y)\,dy. \tag{1.15}$$

The integral runs over all possible outcomes y and must exist mathematically. For a theoretical life distribution, the mean is also called the **average** or **expected life**. It corresponds to the arithmetic average of the lives of all units in a population, and it is used as still another "typical" life. An alternative equation for the mean, if the range of Y is positive, is

$$E(Y) = \int_0^{\infty} R(y)\,dy, \tag{1.16}$$

where $R(y)$ is the reliability function (1.10). (1.16) is obtained by integrating (1.15) by parts.

The mean life for Halley's table is approximated by the sum $2.5(0.290) + 7.5(0.057) + \cdots + 82.5(0.034) = 33.7$ years. Here the midpoint of each five-year interval is used as the age at death. This is somewhat crude for the interval from zero to five years, since the deaths are concentrated in the first year. If those deaths are regarded as occuring at birth (an extreme), the mean life is 33.0 years. In a recent table for the United States, the mean life is 70 years.

The mean of the exponential distribution is

$$E(Y) = \int_0^{\infty} y(1/\theta)\exp(-y/\theta)\,dy = \theta. \tag{1.17}$$

This shows why the parameter θ is called the "mean time to failure" (MTTF). In terms of the failure rate λ, $E(Y)=1/\lambda$. For the diesel engine fans, the expected life is $E(Y)=28,700$ hours. Some repairable equipment with many components has exponentially distributed times **between** failures, particularly if most components have been replaced a number of times. Then θ is called the "mean time between failures" (MTBF).

The variance $\mathrm{Var}(Y)$ of a variable Y with a continuous distribution with a probability density $f(y)$ is

$$\mathrm{Var}(Y)\equiv\int_{-\infty}^{\infty}\left[y-E(Y)\right]^{2}f(y)\,dy. \tag{1.18}$$

The integral runs over all possible outcomes y. This is a measure of the spread of the distribution. An alternative formula is

$$\mathrm{Var}(Y)=\int_{-\infty}^{\infty}y^{2}f(y)\,dy-\left[E(Y)\right]^{2}. \tag{1.19}$$

$\mathrm{Var}(Y)$ has the dimensions of Y squared, for example, hours squared.

The variance of life for Halley's table is approximated by the sum $(2.5-33.7)^{2}0.290+(7.5-33.7)^{2}0.057+\cdots+(82.5-33.7)^{2}0.034=75.5$ years squared.

The variance of the exponential distribution is

$$\mathrm{Var}(Y)=\int_{0}^{\infty}y^{2}(1/\theta)\exp(-y/\theta)\,dy-\theta^{2}=\theta^{2}. \tag{1.20}$$

This is the square of the mean. For the diesel engine fans, the variance of the time of failure is $\mathrm{Var}(Y)=(28,700)^{2}=8.24\times10^{8}$ hours^{2}.

The standard deviation $\sigma(Y)$ of a variable Y is

$$\sigma(Y)\equiv\left[\mathrm{Var}(Y)\right]^{1/2}, \tag{1.21}$$

the positive square root. This has the dimensions of Y, for example, hours. The standard deviation is a more commonly used measure of distribution spread than is the variance.

The standard deviation of life for Halley's table is $(75.5)^{1/2}=8.7$ years.

The standard deviation of the exponential distribution is

$$\sigma(Y)=(\theta^{2})^{1/2}=\theta. \tag{1.22}$$

For the exponential distribution, the standard deviation equals the mean.

For the diesel engine fans, the standard deviation of life is $\sigma(Y)=28,700$ hours.

The hazard function $h(y)$ of a continuous distribution of life Y with a probability density is defined for all possible outcomes y as

$$h(y)\equiv f(y)/[1-F(y)]=f(y)/R(y). \qquad (1.23)$$

It is the **(instantaneous) failure rate** at age y. That is, in the short time Δ from y to $y+\Delta$, a proportion $\Delta\cdot h(y)$ of the population that reached age y fails. Thus the hazard function is a measure of proneness to failure as a function of age. The hazard function is also called the **hazard rate**, the **mortality rate**, and the **force of mortality**. In many applications, one wants to know whether the failure rate of a product increases or decreases with product age.

Any function $h(y)$ satisfying

1. $h(y)\geq 0$ for $-\infty<y<\infty$ and
2. $\lim_{y\to-\infty}\int_{-\infty}^{y}h(y)\,dy=0$ and $\lim_{y\to\infty}\int_{-\infty}^{y}h(y)\,dy=\infty$ (1.24)

is a hazard function of a distribution. A hazard function is another way to represent a distribution. Properties (1) and (2) are equivalent to those for cumulative distribution functions and probability densities.

The hazard function (mortality rate) for Halley's mortality table appears in Table 1.1 and in Figure 1.4a. The yearly mortality rate for 20-year-olds (who die before the age of 25) is calculated from the fraction 0.592 surviving to age 20 and the fraction 0.032 that die between the ages of 20 and 25 as $(0.032/0.592)/5=0.011$ or 1.1% per year. A useful point of view is that 592 people per 1000 reach age 20, and 32 of them die before age 25. Thus the

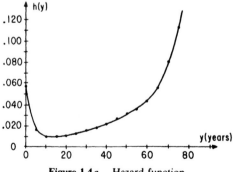

Figure 1.4a. Hazard function.

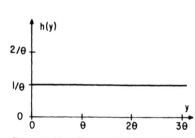

Figure 1.4b. Exponential hazard function.

yearly mortality rates for those reaching 20 is $(32/592)/5$. This is an average value over the five years, and it is tabulated and plotted against the age of 20 years. In a recent table for the United States, the rate is 0.2% per year for those ages.

The exponential hazard function is

$$h(y)=[(1/\theta)\exp(-y/\theta)]/\exp(-y/\theta)=1/\theta, \qquad y\geqslant 0. \quad (1.25)$$

Figures 1.4*b* shows this hazard function. In terms of the failure rate parameter λ,

$$h(y)=\lambda, \qquad y\geqslant 0. \qquad\qquad (1.26)$$

This instantaneous failure rate is constant over time, a key characteristic of the exponential distribution. This means that the chance of failure for an unfailed old unit over a specified short length of time is the same as that for an unfailed new unit over the same length of time. For example, engine fans will continue to fail at a constant rate of $h(y)=34.8$ failures per million hours.

A decreasing hazard function during the early life of a product is said to correspond to **infant mortality**. Such a hazard function appears in Figure 1.4*a*. Such a failure rate often indicates that the product is poorly designed or suffers from manufacturing defects. On the other hand, some products, such as some semiconductor devices, have a decreasing failure rate over the observed time period.

An increasing hazard function during later life of a product is said to correspond to **wear-out** failure. The hazard function in Figure 1.4*a* has this feature. Such failure rate behavior often indicates that failures are due to the product wearing out. Many products have an increasing failure rate over the entire range of life.

The bathtub curve. A few products show a decreasing failure rate in the early life and an increasing failure rate in later life. Figure 1.5 shows such a hazard function. Reliability engineers call such a hazard function a "bathtub curve." In most products, the infant mortality corresponds to a small percentage of the population, and it may go undetected unless there is much data. Some products, such as high-reliability capacitors and semiconductor devices, are subjected to a burn-in to weed out infant mortality before they are put into service, and they are removed from service before wear out starts. Thus they are in service only in the low failure rate portion of their life. This increases their reliability in service. While in service, such products

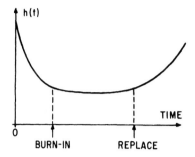

Figure 1.5. "Bathtub curve."

may have an essentially constant failure rate, and the exponential distribution may adequately describe their lives. Jensen and Petersen (1982) comprehensively treat planning and analysis of burn-in procedures, including the economics.

Theoretical distributions with a bathtub hazard function have been presented by many, including Hjorth (1980) and Kao (1959), who present a mixture of two Weibull distributions and a composite Weibull distribution. Section 5 describes distributions with polynomial hazard functions of time, which may have a bathtub shape. Also, a competing failure modes model (Chapter 5) can have a bathtub hazard function. Such distributions typically have three, four, or five parameters. Hence fitting such distributions to data is complicated and requires large samples. Few products require such a complicated distribution.

The cumulative hazard function of a distribution is

$$H(y) \equiv \int_{-\infty}^{y} h(y)\, dy, \qquad -\infty < y < \infty, \qquad (1.27)$$

where the integral runs over all possible outcomes less than or equal to y. For example, the cumulative hazard function for the exponential distribution is

$$H(y) = \int_{0}^{y} (1/\theta)\, dy = y/\theta, \qquad y \geqslant 0. \qquad (1.28)$$

This is a linear function of time. Of course, for $y < 0$, $H(y) = 0$.

Integration of (1.23) yields the **basic relationship**

$$H(y) = \int_{-\infty}^{y} f(y)[1 - F(y)]^{-1}\, dy = -\ln[1 - F(y)]. \qquad (1.29)$$

Equivalently,

$$F(y) = 1 - \exp[-H(y)] \tag{1.30}$$

and

$$R(y) = \exp[-H(y)]. \tag{1.31}$$

Any continuous function $H(y)$ satisfying

1. $H(y) \leq H(y')$ if $y < y'$ (an increasing function),
2. $\lim_{y \to -\infty} H(y) = 0$ and $\lim_{y \to \infty} H(y) = \infty$,   and   (1.32)
3. $H(y)$ is continuous on the right

is a cumulative hazard function of a continuous distribution. Properties 1, 2, and 3 are like those of a cumulative distribution function as a result of the basic relationship (1.29).

For certain purposes, it is more convenient to work with the cumulative hazard function of a distribution than with the cumulative distribution function. For example, for hazard plotting of multiply censored life data, it is easier to work with cumulative hazard paper than with (cumulative) probability paper; see Chapter 4.

2. NORMAL DISTRIBUTION

This basic section presents the normal distribution, which empirically fits many types of data. It often describes dimensions of parts made by automatic equipment, natural physical and biological phenomena, and certain types of life data. The Central Limit Theorem appears in most statistics texts. According to it, the normal distribution may be applicable when each data value is the sum of a (large) number of random contributions. Because its hazard function is strictly increasing, the normal distribution may be appropriate for the life of products with wear-out types of failure. For this reason, the normal (or Gaussian) distribution is only occasionally used for life data.

Many books provide further information on the normal distribution, for example, Hahn and Shapiro (1967) and Johnson and Kotz (1970).

The normal probability density is

$$f(y) = (1/\sigma)(2\pi)^{-1/2} \exp[-(y-\mu)^2/(2\sigma^2)], \qquad -\infty < y < \infty. \tag{2.1}$$

The parameter μ is the mean and may have any value. The parameter σ is the standard deviation and must be positive. μ and σ are in the same units as

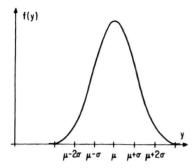

Figure 2.1. Normal probability density.

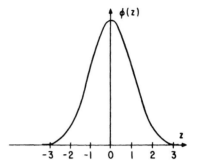

Figure 2.2. Standard normal density.

y, for example, hours, months, cycles, etc. Figure 2.1 depicts this probability density, which is symmetric about the mean μ, which is also the median. This symmetry should be considered when one decides whether the distribution is appropriate for a problem. Figure 2.1 shows that μ is the distribution median and σ determines the spread.

The standard normal probability density is

$$\varphi(z)\equiv(2\pi)^{-1/2}\exp(-z^2/2), \quad -\infty<z<\infty. \tag{2.2}$$

This corresponds to a normal distribution with $\mu=0$ and $\sigma=1$. Figure 2.2 depicts this standard density, which is symmetric about zero. Any normal probability density may be written in terms of the standard one as

$$f(y)=(1/\sigma)\varphi[(y-\mu)/\sigma]. \tag{2.3}$$

The range of possible outcomes y is from $-\infty$ to $+\infty$. Life must, of course, be positive. Thus the fraction of the distribution below zero must be small for the normal distribution to be a satisfactory approximation in practice. For many practical purposes, it may be satisfactory if the mean μ is at least two or three times as great as the standard deviation σ. This rule of thumb is based on the fact that about 2.5% of the distribution is below $\mu-2\sigma$ and 0.14% is below $\mu-3\sigma$.

The normal cumulative distribution function for the population fraction below y is

$$F(y)=P\{Y\leqslant y\}=\int_{-\infty}^{y}(2\pi\sigma^2)^{-1/2}\exp\left[-(y-\mu)^2/(2\sigma^2)\right]dy,$$

$$-\infty<y<\infty. \tag{2.4}$$

This integral cannot be expressed in a simple closed form. Figure 2.3 depicts this function.

The standard normal cumulative distribution function is

$$\Phi(z) = \int_{-\infty}^{z} (2\pi)^{-1/2} \exp(-u^2/2)\, du, \qquad -\infty < z < \infty. \qquad (2.5)$$

$\Phi(\ )$ is tabulated in Appendix A1. More extensive and accurate tables are referenced by Greenwood and Hartley (1962). $F(y)$ of any normal distribution can then be calculated from

$$F(y) = \Phi[(y-\mu)/\sigma], \qquad -\infty < y < \infty. \qquad (2.6)$$

Many tables of $\Phi(z)$ give values only for $z \geqslant 0$. One then uses

$$\Phi(-z) = 1 - \Phi(z). \qquad (2.7)$$

This comes from the symmetry of the probability density.

Transformer example.   A normal life distribution with a mean of $\mu = 6250$ hours and a standard deviation of $\sigma = 2600$ hours was used to represent the life of a transformer in service. The fraction of the distribution with negative life times is $F(0) = \Phi[(0-6250)/2600] = \Phi(-2.52) = 1 - \Phi(2.52) = 0.0059$. This small fraction is ignored hereafter.

The normal reliability function gives the proportion surviving an age of at least y and is

$$R(y) = P\{Y > y\} = 1 - \Phi[(y-\mu)/\sigma], \qquad -\infty < y < \infty. \qquad (2.8)$$

For example, the reliability function for the transformers is $R(y) = 1 - \Phi[(y-6250)/2600]$. Their reliability for $y = 4000$ hours is $R(4000) = 1 - \Phi[(4000-6250)/2600] = 1 - \Phi[-0.865] = 0.81$ or 81%.

The $100P$th normal percentile is obtained from (2.6) as

$$y_P = \mu + z_P \sigma, \qquad (2.9)$$

where $z_P = \Phi^{-1}(P)$ is the $100P$th percentile of the standard normal distribution. That is, z_P satisfies $P = \Phi(z_P)$ and is tabulated in Appendix A2. The **median** (50th percentile) of the normal distribution is

$$y_{.50} = \mu, \qquad (2.10)$$

since $z_{.50} = 0$. Thus the median and mean are equal for a normal distribution. The 16th percentile is $y_{.16} \cong \mu - \sigma$. The relationship $y_{.50} - y_{.16} \cong \sigma$ is often used to estimate σ. The standard normal distribution is symmetric

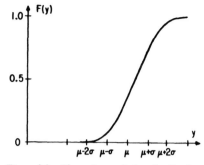

Figure 2.3. Normal cumulative distribution.   **Figure 2.4.**   Normal hazard function.

about the mean. Thus

$$z_P = -z_{1-P}. \tag{2.11}$$

For example, $z_{.84} \simeq 1$, so $z_{.16} \simeq -1$. Some standard percentiles are:

$100P\%$	0.1	1	2.5	5	10	50	90	97.5	99
z_P	-3.090	-2.326	-1.960	-1.645	-1.282	0	1.282	1.960	2.326

For example, the median of transformer life is $y_{.50} = 6250$ hours, and the 10th percentile is $y_{.10} = 6250 + (-1.282)2600 = 2920$ hours.

The normal mode, the most likely value, equals the mean μ, which is also the median. For the normal distribution, these three "typical" values are equal.

For the transformers, the most likely time to failure is 6250 hours.

The mean, variance, and standard deviation are

$$E(Y) = \mu, \tag{2.12}$$

$$\text{Var}(Y) = \sigma^2, \tag{2.13}$$

$$\sigma(Y) = \sigma. \tag{2.14}$$

For the transformers, $E(Y) = 6250$ hours, $\text{Var}(Y) = (2600)^2 = 6.76 \times 10^6$ hours^2, $\sigma(Y) = 2600$ hours.

The normal hazard function at age y is

$$h(y) = (1/\sigma)\varphi[(y-\mu)/\sigma]/\{1-\Phi[(y-\mu)/\sigma]\}, \qquad -\infty < y < \infty. \tag{2.15}$$

Figure 2.4 shows that the normal distribution has an **increasing failure rate** (wear-out) behavior with age.

A key engineering question on the transformers was, "Will their failure rate increase as they age?" If so, a preventative replacement program should replace older units first. The increasing failure rate of the normal distribution assures that older units are more failure prone.

Readers interested in normal data analysis may wish to go directly to Chapter 3 on probability plotting.

3. LOGNORMAL DISTRIBUTION

This basic section presents the lognormal distribution. This distribution empirically fits many types of data adequately, because it has a great variety of shapes. This distribution is often useful if the range of the data is several powers of 10. It is often used for economic data, data on response of biological material to stimulus, and certain types of life data, for example, metal fatigue and electrical insulation life. It is also used for the distribution of repair times of equipment. The lognormal and normal distributions are related as shown below. This fact is used later to analyze data from a lognormal distribution with methods for data from a normal distribution.

Aitchison and Brown (1957), Hahn and Shapiro (1967), and Johnson and Kotz (1970) provide further information on the lognormal distribution.

The lognormal probability density is

$$f(y) = \left\{ 0.4343 / \left[(2\pi)^{1/2} y\sigma \right] \right\} \exp\left\{ -\left[\log(y) - \mu \right]^2 / (2\sigma^2) \right\}, \qquad y > 0; \tag{3.1}$$

the parameter μ is called the **log mean** and may have any value; it is the mean of the **log** of life—not of life. The parameter σ is called the **log standard deviation** and must be positive; it is the standard deviation of the **log** of life—not of life. μ and σ are not "times" like y; instead they are dimensionless pure numbers. In (3.1), $0.4343 = 1/\ln(10)$. Note that log( ) is used throughout for the common (base 10) logarithm. Some authors define the distribution in terms of the natural (base e) logarithm, which is denoted by ln( ) throughout.

Figure 3.1 shows probability densities. It shows that the value of σ determines the shape of the distribution, and the value of μ determines the 50% point and the spread. This variety of shapes makes the lognormal distribution flexible for describing product life. The distribution is defined only for positive y values; so it is suitable as a life distribution.

The lognormal cumulative distribution function for the population fraction below y is

$$F(y) = \Phi\left\{ \left[\log(y) - \mu \right] / \sigma \right\}, \qquad y > 0. \tag{3.2}$$

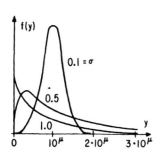

Figure 3.1. Lognormal probability densities.

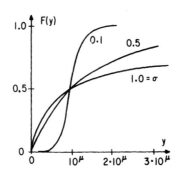

Figure 3.2. Lognormal cumulative distributions.

Here $\Phi(\ )$ is the standard normal cumulative distribution function defined in (2.5) and tabulated in Appendix A1. Figure 3.2 shows lognormal cumulative distribution functions.

Locomotive control example. The life (in thousands of miles) of a certain type of electronic control for locomotives was approximated by a lognormal distribution where $\mu = 2.236$ and $\sigma = 0.320$. The fraction of such controls that would fail on an 80-thousand-mile warranty is $F(80) = \Phi\{[\log(80) - 2.236]/0.320\} = \Phi(-1.04) = 0.15$ from Appendix A1. This percentage was too high, and the control was redesigned.

The lognormal reliability function for life Y gives the proportion surviving an age of at least y and is

$$R(y) = 1 - \Phi\{[\log(y) - \mu]/\sigma\}, \qquad y > 0. \tag{3.3}$$

For the locomotive control, the reliability for the 80-thousand-mile warranty is $R(80) = 1 - \Phi\{[\log(80) - 2.236]/0.320\} = 0.85$ or 85%.

The 100Pth lognormal percentile is obtained from (3.2) as

$$y_P = \text{antilog}(\mu + z_P \sigma), \tag{3.4}$$

where z_P is the 100Pth standard normal percentile; z_P is tabulated in Appendix A2. For example, the **median** (50th percentile) is $y_{.50} = \text{antilog}(\mu)$; this fact is used to estimate $\mu = \log(y_{.50})$ from data.

For the locomotive control, $y_{.50} = \text{antilog}(2.236) = 172$ thousand miles; this can be regarded as a typical life. The 1% life is $y_{.01} = \text{antilog}[2.236 + (-2.326)0.320] = 31$ thousand miles.

The lognormal mode, the most likely value, is

$$y_m = \text{antilog}(\mu - 2.303\ \sigma^2).\tag{3.5}$$

Here $2.303 = \ln(10)$.

For the locomotive control, $y_m = \text{antilog}[2.236 - 2.303(0.320)^2] = 100$ thousand miles.

The lognormal mean is

$$E(Y) = \text{antilog}(\mu + 1.151\ \sigma^2),\tag{3.6}$$

where $1.151 = 0.5 \cdot \ln(10)$. This is the mean life for a population of units.

For the locomotive control, $E(Y) = \text{antilog}[2.236 + 1.151(0.320)^2] = 226$ thousand miles. The mode is 100 thousand miles, and the median is 172 thousand miles.

For a lognormal distribution, mode < median < mean always holds.

The lognormal variance is

$$\text{Var}(Y) = [E(Y)]^2 [\text{antilog}(2.303\ \sigma^2) - 1].\tag{3.7}$$

For the locomotive control, $\text{Var}(Y) = (226)^2 \times \{\text{antilog}[2.303(0.320)^2] - 1\} = 3.68 \times 10^4$ (1000 miles)^2.

The lognormal standard deviation is

$$\sigma(Y) = E(Y)[\text{antilog}(2.303\ \sigma^2) - 1]^{1/2}.\tag{3.8}$$

For the locomotive control, the standard deviation of life is $\sigma(Y) = (3.68 \times 10^4)^{1/2} = 192$ thousand miles.

The lognormal hazard function at age y is

$$h(y) = [0.4343/(y\sigma)]\varphi[(\log(y) - \mu)/\sigma]/\{1 - \Phi[(\log(y) - \mu)/\sigma]\},$$
$$y > 0.\tag{3.9}$$

Figure 3.3 shows lognormal hazard functions. For all lognormal distributions, $h(y)$ is zero at time zero, increases with age to a maximum, and then decreases back down to zero with increasing age. For many products, the failure rate does not go to zero with increasing age. Even so, the lognormal distribution may adequately describe most of the range of life, particularly early life.

For $\sigma \approx 0.5$, $h(y)$ is essentially constant over much of the distribution. For $\sigma \leqslant 0.2$, $h(y)$ increases over most of the distribution and is much like that of

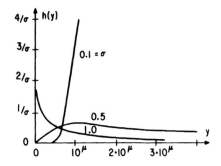

Figure 3.3.   Lognormal hazard functions.

a normal distribution. For $\sigma \geq 0.8$, $h(y)$ decreases over most of the distribution. Thus, the lognormal can describe an increasing, decreasing, or relatively constant failure rate. This flexibility makes the lognormal distribution popular and suitable for many products. Nelson (1972b) describes hazard functions further.

For the locomotive control, $\sigma = 0.320$. So the behavior of $h(y)$ is midway between the increasing and roughly constant hazard functions in Figure 3.3.

Readers interested in lognormal data analysis may wish to go directly to Chapter 3 on probability plotting.

The relationship between the lognormal and normal distributions helps one understand the lognormal distribution in terms of the simpler normal distribution. The (base 10) log of a variable with a lognormal distribution with parameters μ and σ has a normal distribution with mean μ and standard deviation σ. This relationship implies that a lognormal cumulative distribution function plotted against a logarithmic scale for life is identical to the distribution function for the corresponding normal distribution.

The relationship means that the same data analysis methods can be used for both the normal and lognormal distributions. To do this, one takes the logarithms of lognormal data and analyzes them as if they came from a normal distribution.

The relationship comes from the following. Suppose that the variable Y has a lognormal distribution with a log mean μ and a log standard deviation σ. The distribution of the (base 10) log value $V = \log(Y)$ can be seen from

$$F(v) = P\{V \leq v\} = P\{\log(Y) \leq v\} = P\{Y \leq 10^v\}. \qquad (3.10)$$

The final probability is the cumulative distribution function of Y evaluated at 10^v; that is,

$$F(v) = \Phi\big(\{\log[10^v] - \mu\}/\sigma\big) = \Phi\big[(v - \mu)/\sigma\big]. \qquad (3.11)$$

Thus the log value $V = \log(Y)$ has a normal distribution with a mean μ and a standard deviation σ.

For small σ values (say, less than 0.2), the lognormal distribution is close to a normal distribution, and either may then provide an adequate description of product life. For example, Figure 3.1 shows a lognormal probability density that is close to a normal one.

4. WEIBULL DISTRIBUTION

This basic section presents the Weibull distribution. The Weibull distribution is useful in a great variety of applications, particularly as a model for product life. It has also been used as the distribution of strength of certain materials. It is named after Waloddi Weibull (1951), who popularized its use among engineers. One reason for its popularity is that it has a great variety of shapes. This makes it extremely flexible in fitting data, and it empirically fits many kinds of data.

It may be suitable for a "weakest link" type of product. In other words, if a unit consists of many parts, each with a failure time from the same distribution (bounded from below), and if the unit fails with the first part failure, then the Weibull distribution may be suitable for such units. For example, the life of a capacitor is thought to be determined by the weakest (shortest lived) portion of dielectric in it.

If Y has a Weibull distribution, then $\ln(Y)$ has an extreme value distribution, described in Section 5. This fact is used to analyze Weibull data with the simpler extreme value distribution.

Further details on the Weibull distribution are given by Hahn and Shapiro (1967), Johnson and Kotz (1970), and Gumbel (1958). Harter's (1978) bibliography on extreme-value theory gives many references on the Weibull distribution.

The Weibull probability density function is

$$f(y) = (\beta/\alpha^\beta) y^{\beta-1} \exp\left[-(y/\alpha)^\beta\right], \qquad y > 0. \tag{4.1}$$

The parameter β is called the **shape parameter** and is positive. The parameter α is called the **scale parameter** and is also positive; α is called the "characteristic life," since it is always the $100 \times (1 - e^{-1}) \cong 63.2$th percentile. α has the same units as y, for example, hours, months, cycles, etc. β is a dimensionless pure number. Figure 4.1 with Weibull probability densities shows that β determines the shape of the distribution and α determines the spread. The distribution is defined only for positive y; so it is suitable as a life distribution.

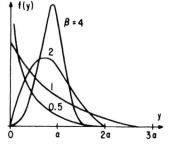

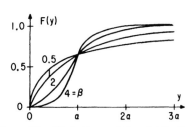

Figure 4.1. Weibull probability densities.   **Figure 4.2.** Weibull cumulative distributions.

For the special case $\beta = 1$, the Weibull distribution is the simple exponential distribution, which was once widely used as a product life distribution but later found to be inadequate for many products. The more flexible Weibull distribution is now widely used, partly because it includes the familiar exponential distribution. For $\beta = 2$, the Weibull distribution is the Rayleigh distribution. For a large value of the shape parameter, say $\beta \geq 10$, the shape of the Weibull distribution is close to that of the smallest extreme value distribution, shown later in Figure 5.1.

Also, for shape parameter values in the range $3 \leq \beta \leq 4$, the shape of the Weibull distribution is close to that of the normal distribution (see Figure 4.1). For much life data, the Weibull distribution is more suitable than the exponential, normal, and extreme value distributions. Try it first.

The Weibull cumulative distribution function is

$$F(y) = 1 - \exp\left[-(y/\alpha)^\beta\right], \qquad y > 0. \tag{4.2}$$

Figure 4.2 shows Weibull cumulative distribution functions. The Weibull distribution parameters are sometimes expressed differently. For example,

$$F(y) = 1 - \exp(-\lambda y^\beta), \tag{4.3}$$

where $\lambda = 1/\alpha^\beta$. Also,

$$F(y) = 1 - \exp(-y^\beta/\theta), \tag{4.4}$$

where $\theta = 1/\lambda = \alpha^\beta$.

Winding example. The life in years of a type of generator field winding was approximated by a Weibull distribution where $\alpha = 13$ years and $\beta = 2$. Thus the fraction of such windings that would fail on a two-year warranty is $F(2.0) = 1 - \exp[-(2.0/13)^2] = 0.023$, or 2.3%.

The Weibull reliability function for the proportion surviving an age of at least y is

$$R(y)=\exp\left[-(y/\alpha)^\beta\right], \qquad y>0. \tag{4.5}$$

For the windings, the reliability for two years (the fraction surviving warranty) is $R(2.0)=\exp[-(2.0/13)^2]=0.977$, or 97.7%.

The $100P$th Weibull percentile is obtained from (4.2) as

$$y_P=\alpha\left[-\ln(1-P)\right]^{1/\beta}, \tag{4.6}$$

where $\ln(\ )$ is the natural logarithm. For example, the $100(1-e^{-1})\cong63$rd percentile is $y_{.63}\cong\alpha$ for any Weibull distribution. This may be seen in Figure 4.2.

For the windings, the 63rd percentile is $y_{.63}=13[-\ln(1-0.63)]^{1/2}=13$ years, the characteristic life. The 10th percentile is $y_{.10}=13[-\ln(1-0.10)]^{1/2}=4.2$ years.

The Weibull mode, the most likely value, is

$$y_m=\begin{cases}\alpha[1-(1/\beta)]^{1/\beta} & \text{for } \beta\geqslant1, \\ 0 & \text{for } 0<\beta\leqslant1.\end{cases} \tag{4.7}$$

That is, the probability density (4.1) is a maximum at y_m.

For the windings, $y_m=13(1-(1/2))^{1/2}=9.2$ years.

The Weibull mean is

$$E(Y)=\alpha\cdot\Gamma\left[1+(1/\beta)\right], \tag{4.8}$$

where $\Gamma(v)=\int_0^\infty z^{v-1}\exp(-z)\,dz$ is the **gamma function**, which is tabulated, for example, by Abramowitz and Stegun (1964). For integer v, $\Gamma(v)=(v-1)!=(v-1)(v-2)\cdots2\cdot1$.

For the windings, $E(Y)=13\Gamma(1+(1/2))=13(0.886)=11.5$ years.

The Weibull variance is

$$\text{Var}(Y)=\alpha^2\left\{\Gamma[1+(2/\beta)]-\{\Gamma[1+(1/\beta)]\}^2\right\}. \tag{4.9}$$

For the windings, $\text{Var}(Y)=13^2\{\Gamma(1+(2/2))-\{\Gamma[1+(1/2)]\}^2\}=36$ years^2.

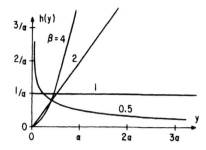

Figure 4.3. Weibull hazard functions.

The Weibull standard deviation is

$$\sigma(Y) = \alpha\left\{\Gamma[1+(2/\beta)] - \left\{\Gamma[1+(1/\beta)]\right\}^2\right\}^{1/2}. \qquad (4.10)$$

For the windings, $\sigma(Y) = 13\{\Gamma(1+(2/2)) - [\Gamma(1+(1/2))]^2\}^{1/2} = 6.0$ years.
The Weibull hazard function is

$$h(y) = (\beta/\alpha)(y/\alpha)^{\beta-1}, \qquad y>0. \qquad (4.11)$$

Figure 4.3 shows Weibull hazard functions. A power function of time, $h(y)$ increases with time for $\beta>1$ and decreases with time for $\beta<1$. For $\beta=1$ (the exponential distribution), the failure rate is constant. The ability to describe increasing or decreasing failure rates contributed to making the Weibull distribution popular for life data analysis.

For the windings, the shape parameter is $\beta=2$. Since β is greater than unity, the failure rate for such windings increases with their age, a wear-out type of behavior. This tells utilities who own such generators that windings get more prone to failure with age. Thus preventive replacement of old windings will avoid costly failures in service.

Readers interested in Weibull life data analysis may wish to go directly to Chapter 3 on probability plotting.

5. SMALLEST EXTREME VALUE DISTRIBUTION

This section presents the smallest extreme value distribution and is essential background for analytic methods (Chapters 6 through 12) for Weibull data. The distribution adequately describes certain extreme phenomena such as temperature minima, rainfall during droughts, electrical strength of materials, and certain types of life data, for example, human mortality of the aged. The distribution is mainly of interest because it is related to the Weibull distribution, and Weibull data are conveniently analyzed in terms of the simpler extreme value distribution. In this book, the extreme value distribution is the **smallest** extreme value distribution. The largest extreme value distribution (Section 7) is seldom used in life and failure data analysis. The extreme value distribution is also closely related to the Gompertz–Makeham distribution, which has been used to describe human life (Gross and Clark, 1975, Sec. 6.5).

Like the Weibull distribution, the smallest extreme value distribution may be suitable for a "weakest link" product. In other words, if a unit consists of identical parts from the same life distribution (unbounded below) and the unit fails with the first part failure, then the smallest extreme value distribution may describe the life of units.

Gumbel (1958), Hahn and Shapiro (1967), and Johnson and Kotz (1970) provide further information on the extreme value distribution. Harter's (1978) bibliography on extreme-value theory gives many references.

The extreme value probability density is

$$f(y) = (1/\delta)\exp[(y-\lambda)/\delta] \cdot \exp\{-\exp[(y-\lambda)/\delta]\}, \qquad -\infty < y < \infty.$$

$$(5.1)$$

The parameter λ is called the **location parameter** and may have any value. The parameter δ is called the **scale parameter** and must be positive. The parameter λ is a characteristic value of the distribution, since it is always the $100(1-e^{-1}) \simeq 63.2\%$ point of the distribution. λ and δ are in the same units as y, for example, hours, months, cycles, etc. Figure 5.1 shows the probability density, which is asymmetric.

The range of possible outcomes y is from $-\infty$ to $+\infty$. Lifetimes must, of course, be positive. Thus the distribution fraction below zero must be small for the extreme value distribution to be satisfactory for life in practice. For most practical purposes, it is satisfactory if the location parameter λ is at least four times as great as the scale parameter δ.

The extreme value cumulative distribution function is

$$F(y) = 1 - \exp\{-\exp[(y-\lambda)/\delta]\}, \qquad -\infty < y < \infty. \qquad (5.2)$$

Figure 5.2 depicts this function. The standard extreme value cumulative distribution function ($\lambda = 0$ and $\delta = 1$) is

$$\Psi(z) = 1 - \exp[-\exp(z)], \qquad -\infty < z < \infty, \qquad (5.3)$$

where $z = (y-\lambda)/\delta$ is called the "standard deviate." $\Psi(z)$ is tabulated by

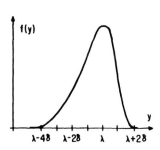

Figure 5.1. Extreme value density.

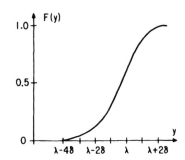

Figure 5.2. Extreme value cumulative distribution.

Meeker and Nelson (1974). Then any extreme value cumulative distribution is given by

$$F(y) = \Psi[(y-\lambda)/\delta], \qquad -\infty < y < \infty. \qquad (5.4)$$

Material strength example. Weibull (1951) resported that the ultimate strength of a certain material can be described by an extreme value distribution with $\lambda = 108$ kg/cm^2 and $\delta = 9.27$ kg/cm^2. The proportion of such specimens with a strength below 80 kg/cm^2 is $F(80) = \Psi[(80 - 108)/9.27] = \Psi(-3.02) = 0.048$ or 4.8%.

The extreme value reliability function is

$$R(y) = \exp\{-\exp[(y-\lambda)/\delta]\}, \qquad -\infty < y < \infty. \qquad (5.5)$$

For the material, the reliability for a stress of 80 kg/cm^2 is $R(80) = 1 - F(80) = 0.952$, or 95.2%. That is, 95.2% of such specimens withstand a stress of 80 kg/cm^2.

The 100Pth extreme value percentile is obtained from (5.2) as

$$y_P = \lambda + u_P \delta, \qquad (5.6)$$

where $u_P = \ln[-\ln(1-P)]$ is the 100Pth standard extreme value percentile ($\lambda = 0$ and $\delta = 1$). Values of u_P are tabulated by Meeker and Nelson (1974). For example, $y_{.632} = \lambda$, the location parameter, since $u_{.632} \cong 0$. Some standard percentiles are

100P%	0.1	1	5	10	50	63.2	90	99
u_P	-6.907	-4.600	-2.970	-2.250	-0.367	0	0.834	1.527

For the material, the median strength is $y_{.50} = 108 + (-0.367)9.27 = 104.6$ kg/cm^2.

The extreme value mode, the most likely value, equals λ, which is also the 63.2nd percentile.

The most likely strength for the material is $\lambda = 108$ kg/cm^2.

The extreme value mean is

$$E(Y) = \lambda - 0.5772\delta, \qquad (5.7)$$

where 0.5772... is Euler's constant. The mean is the 42.8% point of the distribution. For any extreme value distribution, mean < median < λ.

For the material, $E(Y) = 108 - 0.5772(9.27) = 102.6$ kg/cm^2.

The extreme value variance is

$$\mathrm{Var}(Y) = 1.645\,\delta^2, \qquad (5.8)$$

where $1.645 = \pi^2/6$.

For the material, $\mathrm{Var}(Y) = 1.645(9.27)^2 = 141.4$ kg^2/cm^4.

The extreme value standard deviation is

$$\sigma(Y) = 1.283\,\delta. \qquad (5.9)$$

For the material, $\sigma(Y) = 1.283(9.27) = 11.9$ kg/cm^2.

The extreme value hazard function (instantaneous failure rate) is

$$h(y) = (1/\delta)\exp[(y-\lambda)/\delta], \qquad -\infty < y < \infty. \qquad (5.10)$$

Figure 5.3 shows that $h(y)$ is an increasing failure rate with age (wear out). The extreme value distribution adequately describes human mortality in old age. That is, people wear out with an exponentially increasing failure rate.

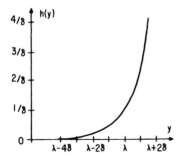

Figure 5.3. Extreme value hazard function.

The relationship between extreme value and Weibull distributions is used in analyzing data. Suppose a Weibull distribution has shape and scale parameters β and α. The distribution of the base e log of a Weibull variable has an extreme value distribution with location and scale parameters

$$\lambda = \ln(\alpha), \tag{5.11}$$

$$\delta = 1/\beta. \tag{5.12}$$

The Weibull parameters expressed in terms of the extreme value parameters are

$$\alpha = \exp(\lambda), \tag{5.13}$$

$$\beta = 1/\delta. \tag{5.14}$$

Similarly, the Weibull parameters can be expressed in terms of the standard deviation $\sigma(Y)$ and mean $E(Y)$ of the corresponding extreme value distribution as

$$\beta = 1.283/\sigma(Y) \tag{5.15}$$

and

$$\alpha = \exp[E(Y) + 0.4501\sigma(Y)]. \tag{5.16}$$

The relationship is derived as follows. Suppose that the variable Y has a Weibull distribution with a scale parameter α and shape parameter β. The distribution of the (base e) log value $W = \ln(Y)$ can be seen from

$$F(w) = P\{W \le w\} = P\{\ln(Y) < w\} = P\{Y \le \exp(w)\}. \tag{5.17}$$

The final probability is the cumulative distribution function of Y evaluated at $\exp(w)$; that is,

$$F(w) = 1 - \exp\left\{-[\exp(w)/\alpha]^{\beta}\right\}$$

$$= 1 - \exp\left\{-\exp[(w - \ln(\alpha))/(1/\beta)]\right\}. \tag{5.18}$$

Thus the log value $W = \ln(Y)$ has an extreme value distribution with a location parameter $\lambda = \ln(\alpha)$ and a scale parameter $\delta = 1/\beta$. This relationship between the Weibull and extreme value distributions is similar to that between the lognormal and normal distributions.

The preceding relationships are used in the analysis of Weibull data. Data analyses are carried out on the base e logs of Weibull data. The resulting extreme value data are simpler to handle, as that distribution has a single shape and simple location and scale parameters, similar to the normal distribution.

Readers interested in life data analysis may wish to go directly to Chapter 3 on probability plotting.

6. GAMMA AND CHI-SQUARE DISTRIBUTIONS

This specialized section presents the gamma and chi-square distributions. Gross and Clark (1975) apply the gamma distribution to biomedical survival data. The gamma hazard function has a behavior that makes it unsuitable for most products. The chi-square distribution is a special case of the gamma distribution and has many uses in data analysis and statistical theory. For example, it arises in queuing theory.

Hahn and Shapiro (1967), Johnson and Kotz (1970), Lancaster (1969), Lawless (1982), and Mann, Schafer, and Singpurwalla (1974) provide further information on the gamma and chi-square distributions.

The gamma probability density is

$$f(y)=[1/\Gamma(\beta)](y^{\beta-1}/\alpha^{\beta})\exp(-y/\alpha), \qquad y>0. \qquad (6.1)$$

The parameter α is a scale parameter and must be positive; the parameter β is a shape parameter and must be positive. Also, $\Gamma(\beta)=\int_0^\infty u^{\beta-1}\exp(-u)\,du$ is the **gamma function**; it is tabulated, for example, by Abramowitz and Stegun (1964). For integer β, $\Gamma(\beta)=(\beta-1)(\beta-2)\cdots2\cdot1=(\beta-1)!$ Figure 6.1 shows gamma probability densities. Figure 6.1 shows that the distribution has a variety of shapes; thus it flexibly describes product life. For the special case $\beta=1$, the gamma distribution is the exponential distribution.

The chi-square distribution is a special case of the gamma distribution with $\alpha=2$ and $\beta=\nu/2$, where ν is an integer. ν is the number of **degrees of freedom** of the distribution. For example, the chi-square distribution with $\nu=2$ degrees of freedom is the exponential distribution with a mean equal to 2.

The gamma cumulative distribution function is

$$F(y)=[1/\Gamma(\beta)]\int_0^y(y^{\beta-1}/\alpha^{\beta})\exp(-y/\alpha)\,dy$$

$$=\Gamma(y/\alpha,\beta), \qquad y>0, \qquad (6.2)$$

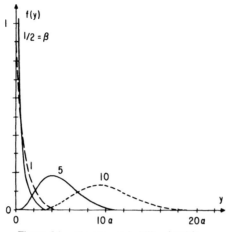

Figure 6.1. Gamma probability densities.

where

$$\Gamma(u,\beta)=[1/\Gamma(\beta)]\int_0^u u^{\beta-1}\exp(-u)\,du \tag{6.3}$$

is the **incomplete gamma function ratio**; it is tabulated by Harter (1964). Figure 6.2 shows gamma cumulative distribution functions. For β an integer, the gamma cumulative distribution function can be expressed as

$$F(y)=\Gamma(y/\alpha,\beta)=1-P(\beta-1,y/\alpha), \tag{6.4}$$

where $P(\beta-1,y/\alpha)$ is the cumulative distribution function for $\beta-1$ or

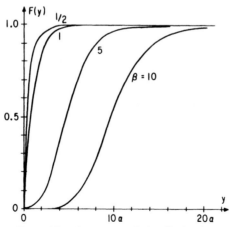

Figure 6.2. Gamma cumulative distributions.

fewer occurrences for a Poisson distribution (Section 12) with a mean equal to (y/α). Similarly, if the degrees of freedom ν is an even integer, the chi-square cumulative distribution function can be expressed as

$$F_{\chi^2}(y)=\Gamma(y/2,\nu/2)=1-P(\nu/2-1,y/2). \qquad (6.5)$$

The 100Pth gamma percentile is the solution y_P of

$$P=\Gamma(y_P/\alpha,\beta). \qquad (6.6)$$

Percentiles of the chi-square distributions are tabulated in Appendix A3. For a χ^2 distribution with ν degrees of freedom, the **Wilson–Hilferty approximation** to the 100Pth percentile $\chi^2(P;\nu)$ is

$$\chi^2(P;\nu)\simeq\nu\left[1-2(9\nu)^{-1}+(z_P/3)(2/\nu)^{1/2}\right]^3, \qquad (6.7)$$

where z_P is the 100Pth standard normal percentile. The approximation (6.7) improves as ν increases, but it is remarkably accurate for ν as small as unity when $0.20\leqslant P\leqslant0.99$.

The mode of the gamma distribution is $\alpha(\beta-1)$ for $\beta>1$, and 0 for $0<\beta\leqslant1$.

The mean, variance, and standard deviation of the gamma distribution are

$$E(Y)=\beta\alpha, \qquad \text{Var}(Y)=\beta\alpha^2, \qquad \sigma(Y)=\beta^{1/2}\alpha. \qquad (6.8)$$

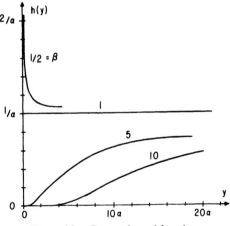

Figure 6.3. Gamma hazard functions.

The hazard function of the gamma distribution is

$$h(y) = [1/\Gamma(\beta)](y^{\beta-1}/\alpha^{\beta})\exp(y/\alpha)/[1-\Gamma(y/\alpha,\beta)]. \quad (6.9)$$

Figure 6.3 shows gamma hazard functions. For $\beta < 1$, $h(y)$ decreases to a constant value $1/\alpha$. For $\beta > 1$, $h(y)$ starts at zero and increases to a constant value $1/\alpha$. Few products have such a failure rate behavior.

The following property of the chi-square distribution is used in the analysis of data from an exponential distribution. Suppose that $Y_1, \ldots, Y_k$ are independent observations from an exponential distribution with a mean of θ. Then the sum $Y = 2(Y_1 + \cdots + Y_k)/\theta$ has a chi-square distribution with $2k$ degrees of freedom.

7. OTHER DISTRIBUTIONS

This specialized section briefly presents a variety of distributions. A general reference for these and many other distributions is Johnson and Kotz (1970). Only the material on sampling distributions of statistics is useful for later chapters, particularly, Chapter 6.

Sampling Distributions of Statistics

Certain distributions from statistical theory are the sampling distributions of statistics (functions of the sample observations) and are not used as life distributions. They include the chi-square distribution of Section 6 and the t and F distributions below. They are used in data analyses later.

The t distribution has a variety of important uses in statistical theory. It is also called **Student's distribution**. The following paragraphs present its basic properties. Johnson and Kotz (1970) give more detail.

The probability density of the t distribution is

$$f(t) = [1 + (t^2/\nu)]^{-(\nu+1)/2}(\pi\nu)^{-1/2}\Gamma[(\nu+1)/2]/\Gamma[\nu/2], \quad (7.1)$$

where $-\infty < t < \infty$, and the positive integer parameter ν is called the **degrees of freedom**. $\Gamma(\ )$ is the gamma function.

The distribution is symmetric about the origin. Thus odd moments are all zero. In particular, the mean (which equals the median) is zero and exists for $\nu > 1$. Even moments of order $2m$ exist for $2m < \nu$; namely,

$$ET^{2m} = 1 \cdot 3 \cdots (2m-1)\nu^m/[(\nu-2)(\nu-4) \cdots (\nu-2m)]. \quad (7.2)$$

In particular, for $m = 1$ and $\nu > 2$, $\mathrm{Var}(T) = ET^2 = \nu/(\nu-2)$.

For large ν, the cumulative t distribution is close to a standard normal distribution. Then standard normal percentiles and probabilities approximate those of the t distribution. Exact percentiles of the t distribution are tabulated in most statistics texts and in Appendix A4.

The t distribution is often used in data analysis. Its basis follows. Suppose that Z has a standard normal distribution and is statistically independent of Q, which has a chi-square distribution with ν degrees of freedom. The statistic

$$T=Z/[Q/\nu]^{1/2} \tag{7.3}$$

has a t distribution with ν degrees of freedom. For example, suppose that a random sample of n observations comes from a normal distribution with mean μ and standard deviation σ. Let $\overline{Y}$ And S^2 denote the sample mean and variance. $Z=n^{1/2}(\overline{Y}-\mu)/\sigma$ has a standard normal distribution, and $Q=(n-1)S^2/\sigma^2$ has a chi-square distribution with $\nu=n-1$ degrees of freedom. So

$$T=n^{1/2}(\overline{Y}-\mu)/S \tag{7.4}$$

has a t distribution with $n-1$ degrees of freedom.

The noncentral t distribution is used to obtain confidence limits for normal distribution percentiles and reliabilities (Chapter 6) and has many other applications. Suppose that Z' has a normal distribution with mean δ and variance 1 and is statistically independent of Q, which has a chi-square distribution with ν degrees of freedom. Then $T'=Z'/(Q/\nu)^{1/2}$ has a noncentral t distribution with ν degrees of freedom. Owen (1968) describes this distribution, its uses, and its tabulations.

The F distribution has a number of important uses in statistical theory. The following paragraphs present its basic properties. For more detail, see Johnson and Kotz (1970) and most statistics texts.

The probability density of the F distribution is

$$g(F)=F^{(\alpha-2)/2}[\beta+(\alpha F/\beta)]^{-(\alpha+\beta)/2}$$
$$\times(\alpha/\beta)^{\alpha/2}\Gamma[(\alpha+\beta)/2]/[\Gamma(\alpha/2)\Gamma(\beta/2)], \tag{7.5}$$

where $0\leqslant F<\infty$, the parameter $\alpha(\beta)$ is called the numerator (denominator) **degrees of freedom** and is positive, and $\Gamma(\ )$ is the gamma function.

The mean exists for $\beta>2$ and is

$$E(F)=\beta/(\beta-2). \tag{7.6}$$

The variance exists for $\beta > 4$ and is

$$\text{Var}(F) = 2\beta^2(\alpha + \beta - 2) / \left[\alpha(\beta - 2)^2(\beta - 4)\right]. \qquad (7.7)$$

F percentiles are tabulated in most statistics texts and in Appendix A5. The tables give only upper percentiles $F(P; \alpha, \beta)$ for $P \geq 0.50$. Lower percentiles are obtained from

$$F(P; \alpha, \beta) = 1 / F(1 - P; \beta, \alpha); \qquad (7.8)$$

the degrees of freedom are reversed on the two sides of this equation.

For large α and β, the cumulative distribution of $\ln(F)$ is close to a normal one. Then

$$F(P; \alpha, \beta) \simeq \exp\left\{z_P[2(\alpha + \beta)/(\alpha\beta)]^{1/2}\right\}, \qquad (7.9)$$

where z_P is the $100P$th standard normal percentile.

The basis of the F statistic follows. Suppose that U and V respectively have chi-square distributions with α and β degrees of freedom and that U and V are statistically independent. Then the statistic

$$F = (U/\alpha)/(V/\beta) \qquad (7.10)$$

has an F distribution with α degrees of freedom in the numerator and β in the denominator. For example, the ratio of the variances of two independent samples from normal distributions with a common true variance has an F distribution. This fact is used in Chapter 10 to compare variances of normal distributions and σ parameters of lognormal distributions. Also, the ratio of the means of two independent samples from exponential distributions with a common true mean has an F distribution. This fact is used in Chapters 10 and 11 to compare means of exponential distributions.

Logistic Distribution

The cumulative distribution function of a logistic distribution is

$$F(y) = 1 / \left\{1 + \exp\left[-(y - \mu)/\sigma\right]\right\}, \qquad -\infty < y < \infty. \qquad (7.11)$$

The parameter μ is called the **location parameter** and may have any value; it is the mean and median of the distribution. The parameter σ is called the **scale parameter** and must be positive. The range of possible outcomes y is from $-\infty$ to $+\infty$. Thus the proportion of the distribution below zero must be small if it is to be a satisfactory life distribution in practice. Like the normal distribution, it is symmetric, but it has more probability in its tails.

Further information on this distribution is provided by Johnson and Kotz (1970).

If Y has a logistic distribution, then the distribution of $U=\exp(Y)$ is called a **log logistic distribution**. The range of possible outcomes of this distribution is from 0 to $+\infty$, and it is thus suitable as a life distribution.

Largest Extreme Value Distribution

The largest extreme value distribution has been used as a model for some types of largest observations, for example, flood heights, extreme wind velocities, and the age of the oldest person dying each year in a community. It is briefly presented here for completeness and comparison with the smallest extreme value distribution.

The cumulative distribution function of a largest extreme value distribution is

$$F(y)=\exp\{-\exp[-(y-\lambda)/\delta]\}, \qquad -\infty<y<\infty. \qquad (7.12)$$

The parameter λ is called the **location parameter** and may have any value. The parameter δ is called the **scale parameter** and must be positive. The parameter λ is always the $100e^{-1}\cong36.8$th percentile of the distribution.

The smallest and largest extreme value distributions have a simple relationship. If Y has a largest extreme value distribution with parameters λ and δ, then $Y'=-Y$ has a smallest extreme value distribution with a location parameter equal to $-\lambda$ and a scale parameter equal to δ. Data analysis methods for the smallest extreme value distribution can be used on data from a largest extreme value distribution after changing the signs of the data values.

Distributions with Linear and Polynomial Hazard Rates

The following paragraphs show how to obtain a distribution from a hazard function. Any function $h(y)$ that has the properties of a hazard function can be used to obtain the corresponding distribution. The following is a simple and natural hazard function to use; it consists of the first two terms in a power series expansion of a general hazard function.

A linear hazard function has the form

$$h(y)=a+by, \qquad y\geqslant0. \qquad (7.13)$$

The parameters a and b must be positive. This distribution has an increasing failure rate with age, a wear-out behavior. For any distribution, the cumulative hazard function satisfies

$$H(y) \equiv \int_{-\infty}^{y} h(y)\,dy = \int_{-\infty}^{y} f(y)[1 - F(y)]^{-1}\,dy = -\ln[1 - F(y)].$$

(7.14)

For the linear hazard function

$$H(y) = \int_{0}^{y} h(y)\,dy = ay + (b/2)y^2, \qquad y \geq 0.$$

(7.15)

Hence, by (7.14),

$$F(y) = 1 - \exp[-ay - (b/2)y^2], \qquad y \geq 0.$$

(7.16)

Other properties of this distribution can be derived from $F(y)$, for example, the density, mean, variance, percentiles, etc. Kodlin (1967) presents this distribution and maximum likelihood methods (Chapter 8) for estimating its parameters from data. Krane (1963) presents general polynomial hazard functions and methods for estimating their parameters; he applies them to estimating survival of vehicles of a public utility. Such a hazard function can have a bathtub shape. Such many-parameter hazard functions are generally useful only if there is much data.

Uniform Distribution

The uniform distribution is used as a theoretical distribution in some data analyses. Also, if a Poisson process (see Section 12) is observed from time 0 to time t, then the times of the occurrences are a random sample from a uniform distribution on the interval from 0 to t. Selected properties of this distribution are presented below.

The probability density of a uniform distribution from 0 to θ is

$$f(y) = \begin{cases} 1/\theta & \text{for } 0 \leq y \leq \theta, \\ 0 & \text{elsewhere.} \end{cases}$$

(7.17)

For $\theta = 1$, the distribution is called the standard uniform distribution; it is defined on the unit interval from 0 to 1.

The cumulative distribution function of a uniform distribution is

$$F(y) = \begin{cases} 0, & \text{for } y < 0, \\ y/\theta, & \text{for } 0 \leq y \leq \theta, \\ 1, & \text{for } y > \theta. \end{cases}$$

(7.18)

The $100P$th percentile of a uniform distribution is

$$y_P = \theta P. \tag{7.19}$$

The mean, variance, and standard deviation of the uniform distribution are

$$E(Y) = \theta/2, \quad \text{Var}(Y) = \theta^2/12, \quad \sigma(Y) = \theta/\sqrt{12}. \tag{7.20}$$

Distributions for Special Situations

Distributions with failure at time zero. A fraction of a population may already be failed at time zero. Many consumers encounter items that do not work when purchased. The model for this consists of the proportion p that fail at time zero and a continuous probability distribution for the rest. Such a cumulative distribution function is depicted in Figure 7.1. The sample proportion failing at time zero is used to estimate p, and the failure times in the remainder of the sample are used to estimate the continuous distribution. Estimation methods are explained in later chapters.

Distributions with eternal survivors. Some units may never fail. This applies to (1) the time to contracting a disease (or dying from it) when some individuals are immune, (2) the time until released prisoners commit a crime where some are completely rehabilitated, (3) the time to redemption of trading stamps, since some stamps are lost and never redeemed, and (4) the time to failure of a product from a particular defect when some units lack that defect. Regal and Larntz (1978) use this model for the time it takes a person to solve a problem when a proportion $1-p$ of the population never solves the problem.

The cumulative distribution function is

$$G(y) = pF(y), \tag{7.21}$$

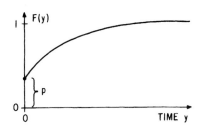

Figure 7.1. A cumulative distribution with a fraction failing at time zero.

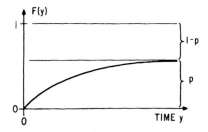

Figure 7.2. A cumulative distribution degenerate at infinity.

where p is the proportion failing and $F(y)$ is a cumulative distribution. A proportion $(1-p)$ survive forever. Such a cumulative distribution is depicted in Figure 7.2. Such a distribution is said to be degenerate at infinity.

Mixtures of distributions.   A population may consist of two or more subpopulations. Units manufactured in different production periods may have different life distributions due to differences in design, raw materials, handling, etc. It is often important to identify such a situation and the production period, customer, environment, etc., that has poor units. Then suitable action may be taken on that portion of the population. The model consists of the proportions $p_1, \ldots, p_K$ of the population in each of the K subpopulations and the corresponding probability densities $f_1(y), \ldots, f_K(y)$. The probability density for the entire population is then

$$f(y) = p_1 f_1(y) + \cdots + p_K f_K(y). \tag{7.22}$$

Figure 7.3 depicts this situation. A mixture of distributions should be distinguished from competing failure modes, described in Chapter 5.

Cox (1959) presents mixtures of exponential distributions and gives maximum likelihood methods for estimating their parameters when the population a failure comes from is (1) identified or else (2) not identified. Proschan (1963) shows that a mixture of exponential distributions has an increasing hazard function. Everett and Hand (1981) survey mixtures.

IFR and IFRA distributions.   Barlow and Proschan (1965, 1975) present theory for increasing failure rate (IFR) and increasing failure rate average (IFRA) distributions. The theory assumes little about the mathematical form of the distribution (nonparametric), and the exponential distribution is

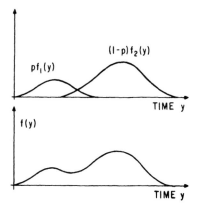

Figure 7.3.   A mixture of distributions.

a limiting case of such distributions. They also present statistical methods for analysis of data from such distributions.

More general distributions.   At times it is useful to work with a general distribution that includes others as special cases. For example, the Weibull distribution includes the exponential distribution. Farewell and Prentice (1979) present the log gamma distribution: it has three parameters and includes the normal (lognormal) and extreme value (Weibull) distributions as special cases. Lawless (1982) presents the generalized gamma distribution.

8.   SHIFTED DISTRIBUTIONS

This highly specialized section presents shifted distributions, which are sometimes used in life and failure data analysis. Such a distribution is obtained by shifting a distribution that is defined for outcomes between 0 and $+\infty$ so that the distribution starts at a value different from 0. The shifted exponential, lognormal, and Weibull distributions are briefly presented below. Also, authors have proposed using a shifted gamma distribution, and one could shift other distributions, such as the log-logistic distribution, that range from 0 to ∞.

This section is not essential background for later material in this book.

The shifted exponential distribution has the cumulative distribution function

$$F(y) = 1 - \exp\left[-(y-\gamma)/\theta\right], \qquad y \geqslant \gamma. \qquad (8.1)$$

The parameter γ is called the **shift** or **threshold parameter** and may have any value, and θ is then called the "scale parameter." The time γ is sometimes called a **minimum life** or **guarantee time**, since all units survive it. Figure 8.1 shows the probability density for such a distribution. This distribution is the exponential distribution shifted by an amount γ. The difference $(Y-\gamma)$ has an exponential distribution with a mean θ. Thus the properties of this

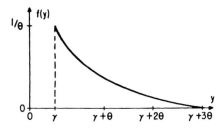

Figure 8.1.   Shifted exponential probability density.

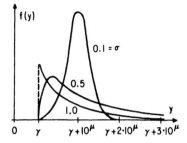

Figure 8.2. Shifted lognormal probability densities.

distribution are like those of the exponential distribution (Section 1). This distribution is also called the **two-parameter exponential distribution**. For $\gamma=0$, it reduces to the exponential distribution of Section 1. Mann, Schafer, and Singpurwalla (1974) and Lawless (1982) describe the distribution in more detail.

The **shifted lognormal distribution** has the cumulative distribution function

$$F(y)=\Phi\{[\log(y-\gamma)-\mu]/\sigma\}, \qquad y>\gamma, \qquad (8.2)$$

here $\Phi(\ )$ is the standard normal cumulative distribution function. The parameter γ is called the shift or location parameter; it may have any value. Figure 8.2 shows probability densities of this distribution. This distribution is also called the **three-parameter lognormal distribution**. The difference $(Y-\gamma)$ has a lognormal distribution with a log mean μ and log standard deviation σ. The properties of this distribution are like those of the lognormal distribution described in Section 3. For $\gamma=0$, it reduces to the lognormal distribution. Aitchison and Brown (1957) provide further information on this distribution.

The **shifted Weibull distribution** has the cumulative distribution function

$$F(y)=1-\exp\{-[(y-\gamma)/\alpha]^{\beta}\}, \qquad y>\gamma. \qquad (8.3)$$

The shift parameter γ may have any value. α and β are the scale and shape parameters. Figure 8.3 shows the probability densities of this distribution. This distribution is the Weibull distribution shifted by an amount γ; that is, the difference $(Y-\gamma)$ has a Weibull distribution. Thus the properties of this distribution are like those of the Weibull distribution described in Section 4. This distribution is also called the **three-parameter Weibull distribution**. For $\gamma=0$, it reduces to the two-parameter Weibull distribution.

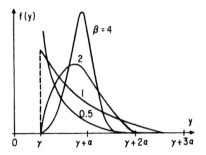

Figure 8.3. Shifted Weibull probability densities.

9. CONDITIONAL AND TRUNCATED DISTRIBUTIONS

This highly specialized section presents conditional and truncated life distributions. It is not essential background for later material in this book. The method of Table 9.2 for predicting a coming number of failures is very useful in practice.

One may be interested in the life distribution of units that have reached a specific age. For example, an insurance company that insures lives of 40-year-olds is interested in the distribution of their ages at death rather than the distribution for newborns. Similarly, an owner of equipment in service will base a replacement policy on the distribution of remaining life for equipment of each age. Conditional distributions represent such situations. Properties of such conditional distributions are presented here.

A distribution that has a range from $-\infty$ to $+\infty$ is not strictly appropriate as a life distribution, since lifetimes must be positive. The corresponding conditional distribution for units that fail above an age of zero has a range from 0 to ∞; so it is suitable as a life distribution. Such distributions are called **truncated distributions**; and they can be viewed as conditional distributions.

General Theory

Motivation. The following example motivates conditional life distributions. It concerns the life distribution of living 40-year-olds in Halley's life table (Table 1.1). Table 1.1 gives the population fractions dying in five-year intervals beyond the age of 40. It is helpful to think of the reliability of 0.436 at age 40 as if 436 people reached age 40; other fractions can be interpreted similarly. Of the "436" people that reached age 40, "49" died between ages 40 and 45. Thus the fraction $f(45|40)$ of 40-year-olds dying between 40 and 45 is $f(45|40)=0.049/0.436=0.112$ (11.2%). In contrast, 4.9% of newborns die between the ages 40 and 45. Similarly, the fraction $f(50|40)$ of 40-year-olds that die between ages 45 and 50 is $f(50|40)=$

Table 9.1.   Conditional Human Life Table

Age y	f(y|40)	F(y|40)	R(y|40)	h(y|40)
40	-	0	1.000	
40-45	.112	.112	.888	.022
45-50	.120	.232	.768	.027
50-55	.121	.353	.647	.032
55-60	.115	.468	.532	.035
60-65	.115	.583	.417	.043
65-70	.117	.700	.300	.056
70-75	.121	.821	.179	.081
75-80	.101	.922	.078	.113
80-85	.078	1.000	0.	.200

$0.052/0.436=0.120$. Such calculations yield the conditional probability density of the distribution of age at death of those that reach age 40. This histogram or probability density $f(y|40)$ is tabulated in Table 9.1 and is depicted in Figure 9.1a. This example motivates the definition below.

The conditional probability density of age Y at failure for units that reach age y' is

$$f(y|y')=f(y)/R(y'), \qquad y \geqslant y', \qquad (9.1)$$

where $f(\ )$ is the probability density and $R(\ )$ is the reliability function of the original distribution. The divisor $R(y')$ "rescales" the density so that the probability under it from y' to ∞ is 1. That is,

$$\int_y^\infty f(y|y')\,dy = \int_y^\infty f(y)[R(y')]^{-1}\,dy = R(y')[R(y')]^{-1}=1.$$

Figure 9.1b depicts this rescaling. $f(y|y')$ has all the properties of a probability density.

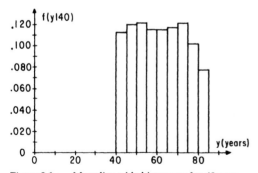

Figure 9.1a.   Mortality table histogram for 40-year-olds.

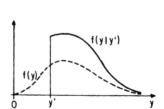

Figure 9.1b.   Original and conditional probability densities.

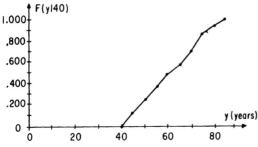

Figure 9.2a. Cumulative distribution function for 40-year-olds.

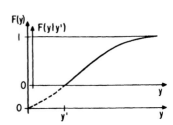

Figure 9.2b. Original and conditional cumulative distributions.

The conditional cumulative distribution function for age Y at failure for units that reach age y' is

$$F(y|y')=\int_{y'}^{y} f(y|y')\,dy=\left[F(y)-F(y')\right]/R(y'), \qquad y\geqslant y', \quad (9.2)$$

where $F(\ )$ is the cumulative distribution function of the original distribution and $R(y')=1-F(y')$. $F(y|y')$ has all the properties of a cumulative distribution function (Section 1).

For the mortality table, the cumulative fraction of 40-year-olds that die before they reach 60 is $F(60|40)=[F(60)-F(40)]/R(40)=(0.768-0.564)/0.436=0.468$. In contrast, the proportion of newborns that die before they reach 60 is $F(60)=0.768$. Figure 9.2a shows the conditional cumulative distribution function for 40-year-olds, and Table 9.1 tabulates it. Figure 9.2b shows the relationship between $F(y|y')$ and $F(y)$.

The conditional reliability for failure at age Y among units that reach age y' is

$$R(y|y')=1-F(y|y')=R(y)/R(y'), \qquad y\geqslant y'. \qquad (9.3)$$

This simple ratio of the reliabilities is easy to remember.

For Halley's life table (Table 1.1), the reliability of 40-year-olds to survive age 60 is $R(60|40)=R(60)/(40)=0.232/0.436=0.532$. In contrast, the proportion of newborns that survive age 60 is $R(60)=0.232$. So one improves one's chances of living to 60 by first living to 40. Figure 9.3a shows the reliability function for 40-year-olds, and Table 9.1 tabulates it. Figure 9.2c shows the relationship between $R(y|y')$ and $R(y)$.

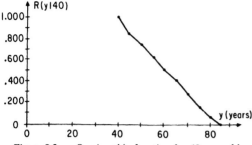

Figure 9.3a. Survivorship function for 40-year-olds.

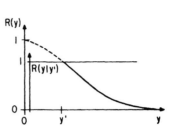

Figure 9.3b. Original and conditional reliability functions.

The 100*P*th percentile y_P of the conditional distribution of age y at failure for units that reach an age y' is the solution of

$$P = F(y_P | y') = [F(y_P) - F(y')] / R(y'). \qquad (9.4)$$

Then y_p is the $100P'$th percentile of the original distribution, where $P' = (1 - P)F(y') + P$.

In the mortality table, the median age at death for 40-year-olds is about 61 years. This is obtained graphically from the conditional cumulative distribution function in Figure 9.2a. Enter the figure on the vertical axis at 0.50, go horizontally to the cumulative distribution function, and then go down to the time scale to read the median life of 61 years.

The conditional mean (or expectation of) age Y at failure among units that reach age y' is

$$E(Y|y') = \int_{y'}^{\infty} yf(y|y')\,dy. \qquad (9.5)$$

Equivalently,

$$E(Y|y') = \left[\int_{y'}^{\infty} yf(y)\,dy \right] \Big/ [1 - F(y')]. \qquad (9.6)$$

For the mortality table, the conditional mean age at death among those who reach 40 is the following sum, which approximates (9.5): $E(Y|40) = (42.5)0.112 + (47.5)0.120 + \cdots + (82.5)0.078 = 61.5$ years. Here the midpoint of a five-year interval is used as the age at death; for example, 42.5 is used for the interval 40 to 45. In comparison, the mean life for newborns is 33.0 years.

The conditional variance of age Y at failure among units that reach age y' is

$$\text{Var}(Y|y') = \int_{y'}^{\infty} [y - E(Y|y')]^2 f(y|y') \, dy. \tag{9.7}$$

Equivalently,

$$\text{Var}(Y|y') = \int_{y'}^{\infty} y^2 f(y|y') \, dy - [E(Y|y')]^2. \tag{9.8}$$

The conditional standard deviation of age Y at failure among units that reach an age y' is

$$\sigma(Y|y') = [\text{Var}(Y|y')]^{1/2}. \tag{9.9}$$

For the mortality table, $\text{Var}(Y|40) \cong (42.5 - 61.5)^2 0.112 + (47.2 - 61.5)^2 0.120 + \cdots + (82.5 - 61.5)^2 0.078 = 154.8$ years^2, and $\sigma(Y|40) = (154.8)^{1/2} = 12.4$ years.

The conditional hazard function for age Y at failure among units that reach age y' is

$$h(y|y') = f(y|y') / [1 - F(y|y')] = f(y) / [1 - F(y)] = h(y), \qquad y \geqslant y'. \tag{9.10}$$

Thus the conditional failure rate at age y is the same as the unconditional failure rate. This fact indicates why burn-in and preventive replacement can reduce the failure rate in service. If units initially have a high failure rate, they can be run in a factory burn-in; then survivors that are put into service have a lower failure rate. Also, such units can be replaced at an age when their failure rate begins to increase appreciably. Units are then used only over the low failure rate portion of their distribution.

For the mortality table, the yearly conditional hazard rate for 40-year-olds in the five-year period between the ages 40 and 45 is $h(45|40) = [f(45|40)/R(40|40)]/5 = 0.112/1.000)/5 = 0.22$, or 2.2% per year. Similarly, $h(50|40) = [f(50|40)/R(45|40)]/5 = (0.120/0.888)/5 = 0.027$. $h(y|40)$ is tabulated in Table 9.1 and graphed in Figure 9.4a. Figure 9.4b shows the relationship between $h(y|y')$ and $h(y)$.

Remaining life for units reaching an age y' is

$$Z = Y - y' \tag{9.11}$$

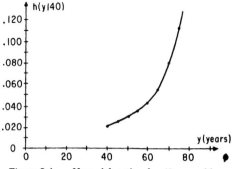

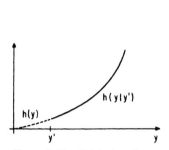

Figure 9.4a. Hazard function for 40-year-olds.

Figure 9.4b. Original and conditional hazard functions.

and is often convenient to use. This is equivalent to moving the origin of the Y time axis to the value y'. All quantities may be expressed in terms of Z as follows. The conditional probability density is

$$g(z|y')=f(z+y')/[1-F(y')], \qquad z\geqslant 0. \tag{9.12}$$

The conditional cumulative distribution is

$$G(z|y')=[F(z+y')-F(y')]/[1-F(y')], \qquad z\geqslant 0. \tag{9.13}$$

The conditional reliability is

$$\overline{G}(z|y')=1-G(z|y'), \qquad z\geqslant 0. \tag{9.14}$$

The $100P$ th percentile of the conditional distribution is

$$z_P=y_{P'}-y', \tag{9.15}$$

where

$$P'=(1-P)F(y')+P. \tag{9.16}$$

The conditional mean or expectation is

$$E(Z|y')=E(Y|y')-y'. \tag{9.17}$$

The conditional variance of the remaining life is

$$\mathrm{Var}(Z|y')=\mathrm{Var}(Y|y'); \tag{9.18}$$

this is the same as the conditional variance of age at death. The conditional standard deviation of remaining life is

$$\sigma(Z|y')=\sigma(Y|y'); \qquad (9.19)$$

this is the same as the conditional standard deviation of age at death. The conditional hazard function is

$$h(z|y')=h(z+y'), \qquad z\geqslant 0. \qquad (9.20)$$

For the mortality table, the preceding quantities for the remaining life of 40-year-olds may readily be calculated from the equations above.

Conditional Exponential Distribution

The cumulative distribution function of a conditional (or truncated) exponential distribution of time Y to failure among units that reach age y' is

$$F(y|y')=1-\exp[-(y-y')/\theta], \qquad y\geqslant y'. \qquad (9.21)$$

Here θ is the mean of the original exponential distribution described in Section 1. For the remaining life $Z=Y-y'$, the cumulative distribution function is

$$G(z|y')=1-\exp(-z/\theta), \qquad z\geqslant 0. \qquad (9.22)$$

That is, for units of any age, the distribution of remaining life is exponential with a mean θ. Thus it is convenient to work with the remaining life. This "lack of memory" is frequently used in the analysis of exponential data.

An example on fans on diesel engines appeared in Section 1. Such fans have an exponential life distribution with a mean of $\theta=28,700$ hours. For such fans in service for 3200 hours, the chance of failure in the next $z=3300$ hours is $G(3300|3200)=1-\exp(-3300/28,700)=11\%$. For fans in service for 6400 hours, the chance of failure in the next 3300 hours is $G(3300|6400) =11\%$, which is the same as the previous value. This is so for any given age. This illustrates that such fans do not become more or less prone to failure as they age.

The reliability of a conditional (or truncated) exponential distribution of time Y to failure among units that reach an age y' is

$$R(y|y')=\exp[-(y-y')/\theta], \qquad y\geqslant y'. \qquad (9.23)$$

For the remaining life $Z = Y - y'$, the reliability function is

$$\bar{G}(z|y') = \exp(-z/\theta), \qquad z \geqslant 0. \tag{9.24}$$

For diesel engine fans 3200 hours old, the conditional reliability for an additional 3300 hours is $\bar{G}(3300|3200) = \exp(-3300/28,700) = 89\%$.

The $100P$th percentile of the conditional exponential distribution of the life Y of units that reach an age y' is

$$y_P = y' + \theta \ln[1/(1-P)]. \tag{9.25}$$

The $100P$th percentile of the conditional distribution of remaining life $Z = Y - y'$ of units that reach an age y' is

$$z_P = \theta \ln[1/(1-P)]; \tag{9.26}$$

this is independent of y' for the exponential distribution.

The median age at failure for fans reaching an age of 3200 hours is $y'_{.50} = 3200 + 28,700 \ln[1/(1-0.50)] = 23,100$ hours. The median remaining life for such fans is $z_{.50} = 28,700 \ln[1/(1-0.50)] = 19,900$ hours.

The conditional probability density for the life Y of units that reach an age y' is

$$f(y|y') = (1/\theta) \exp[-(y-y')/\theta], \qquad y \geqslant y'. \tag{9.27}$$

The conditional probability density for the remaining life $Z = Y - y'$ of units that reach an age y' is

$$g(z|y') = (1/\theta) \exp(-z/\theta), \qquad z \geqslant 0. \tag{9.28}$$

For the fans, the conditional probability density of remaining life for fans with 3200 hours in service is

$$g(z|3200) = (1/28,700) \exp(-z/28,700), \qquad z \geqslant 0.$$

The conditional mean life Y of units that reach an age y' is

$$E(Y|y') = y' + \int_{y'}^{\infty} y(1/\theta) \exp[-(y-y')/\theta] \, dy = y' + \theta. \tag{9.29}$$

This is just the current age y' plus the remaining expected life. The conditional mean of the remaining life $Z = Y - y'$ of units that reach an age

y' is

$$E(Z|y')=\theta=EY. \tag{9.30}$$

For fans that are 3200 hours old, $E(Y|3200)=3200+28,700=31,900$ hours and $E(Z|3200)=28,700$ hours.

The total expected remaining life of a group of units is useful for determining their current value and a replacement policy. The total expected remaining life for a group of units is the sum of the expected remaining lives of the units. For example, the expected remaining life for any fan is 28,700 hours, and the total expected remaining life for 58 such fans in service is $58\times28,700=1.6$ million hours.

The conditional variances of the life Y and of the remaining life $Z=Y-y'$ of units that reach an age y' are

$$\mathrm{Var}(Y|y')=\mathrm{Var}(Z|y')=\theta^2. \tag{9.31}$$

The corresponding **conditional standard deviations** are

$$\sigma(Y|y')=\sigma(Z|y')=\theta. \tag{9.32}$$

For fans that are 3200 hours old, $\sigma(Y|3200)=\sigma(Z|3200)=28,700$ hours.

This completes the discussion of the conditional exponential distribution.

Conditional Normal Distribution

The following paragraphs present the properties of the conditional normal distribution. Also, a method for predicting the number of failures among a population with a mix of ages is given. The method applies to any life distribution.

The cumulative distribution function of a conditional (or truncated) normal distribution of the time Y to failure among units that reach an age y' is

$$F(y|y')=\{\Phi[(y-\mu)/\sigma)]-\Phi[(y'-\mu)/\sigma]\}/\{1-\Phi[(y'-\mu)/\sigma]\},$$
$$y\geqslant y', \tag{9.33}$$

where $\Phi(\ )$ is the standard normal cumulative distribution function that is tabulated in Appendix A1. Here μ and σ are the mean and standard deviation of the original normal distribution.

Section 2 presents an example on transformer life. The life of such transformers is approximated by a distribution with a mean $\mu=6250$ hours and a standard deviation $\sigma=2500$ hours. The chance of failure in the next month (500 hours of service) for units 2000 hours old is $F(2500|2000) = \{\Phi[(2500-6250)/2500] - \Phi[(2000-6250)/2500]\} / \{1 - \Phi[(2000-6250)/2500]\} = 0.023$, or 2.3%.

The **reliability** of a conditional (or truncated) normal distribution of the time Y to failure among units that reach an age y' is

$$R(y|y') = R(y)/R(y') = \{1 - \Phi[(y-\mu)/\sigma]\} / \{1 - \Phi[(y'-\mu)/\sigma]\},$$

$$y \geqslant y'. \tag{9.34}$$

For the transformers, $R(2500|2000) = 1 - 0.023 = 0.977$.

The **100Pth percentile** of the conditional normal distribution of the life Y of units that reach an age y' is

$$y'_P = \mu + z_{P'}\sigma, \tag{9.35}$$

where $z_{P'}$ is the 100P'th standard normal percentile (tabulated in Appendix A2), and $P' = (1-P)\Phi[(y'-\mu)/\sigma] + P$.

The median age at failure for transformers $y' = 2000$ hours old is given by $P' = (1-0.5)\Phi[(2000-6250)/2500] + 0.5 = 0.5223$, $z_{.5223} = 0.0557$, and $y_{.5} = 6250 + 0.0557(2500) = 6390$ hours. For new units, the median life is 6250 hours.

The **probability density** of the conditional or truncated normal distribution of the time Y to failure among units that reach an age y' is

$$f(y|y') = (1/\sigma)\varphi[(y-\mu)/\sigma] / \{1 - \Phi[(y'-\mu)/\sigma]\}, \qquad y \geqslant y', \tag{9.36}$$

where $\varphi(\ )$ is the standard normal probability density.

The **conditional mean** life Y of units that reach an age y' is

$$E(Y|y') = \mu + \sigma\varphi(u)[1 - \Phi(u)]^{-1}, \tag{9.37}$$

where $u = (y'-\mu)/\sigma$.

For transformers that are 2000 hours old, the mean age at failure is $E(Y|2000) = 6250 + 2500\,\varphi[(2000-6250)/2500]\{1 - \Phi[(2000-6250)/2500]\}^{-1} = 6870$ hours. The expected remaining life of such units is $E(Z|2000) = 6870 - 2000 = 4870$ hours.

The **conditional variance** of the life Y of units that reach an age y' is

$$\mathrm{Var}(Y|y') = \sigma^2\{1 + u\varphi(u)[1 - \Phi(u)]^{-1} - \varphi^2(u)[1 - \Phi(u)]^{-2}\}, \tag{9.38}$$

where $u = (y'-\mu)/\sigma$.

The **conditional standard deviation** $\sigma(Y|y')$ is the square root of this variance.

For transformers that are 2000 hours old, the standard deviation of age at failure comes from $u = (2000 - 6250)/2500 = -1.700$ and $\sigma(Y|2000) = 2500\{1 + (-1.700)\varphi(-1.700)[1 - \Phi(-1.700)]^{-1} - \varphi^2(-1.700)[1 - \Phi(-1.700)]^{-2}\}^{1/2} = 2230$ hours.

Prediction of numbers of failures. The following method predicts the expected number of units in service that will fail between the current time and a specified future time—an important practical problem. The prediction is used to plan manufacture of replacements. It uses the conditional failure probabilities for the units that will be in service for that period of time. For each running unit, calculate the conditional probability of failure based on the current age of the unit and the length of the specified period. The sum of these conditional probabilities, expressed as proportions, is an estimate of the expected number of failures in the specified period. If the conditional failure probabilities are all small, then the coming number of failures has a probability distribution that is approximately Poisson (see Section 12), and the Poisson mean is estimated by the sum of the conditional probabilities of failure. This distribution allows one to make probability statements about the coming number of failures in the specified period.

Table 9.2 shows a calculation of an estimate of the expected number of transformer failures over the next 500 service hours. The calculation involves the current ages of the 158 unfailed transformers. To reduce the labor in calculating the 158 conditional failure probabilities, the transformers are grouped by age. Then, for each group, a nominal conditional probability of failure is calculated for a nominal unit age, as shown in Table 9.2. The sum of the conditional probabilities for a group is approximated by the nominal probability for the group times the number of transformers in the group. The expected number of transformer failures over the next 500 hours is obtained by the calculations shown in Table 9.2, and it is 7.8 failures. Using a table of Poisson probabilities, one finds that the number of failures in 500

Table 9.2.   Calculation of Expected Number of Failures in 500 Hours

Age Range of Group	Nominal Age	Conditional Probability		Number of Transformers		Expected No. of Failures
1751-2250	2000	0.023	x	17	=	0.39
2251-2750	2500	0.034	x	54	=	1.84
2751-3250	3000	0.044	x	27	=	1.19
3251-3750	3500	0.056	x	17	=	0.95
3751-4250	4000	0.070	x	19	=	1.33
4251-4750	4500	0.086	x	24	=	2.06
				Total Expected		7.8

hours will be 12 or fewer, with 95% probability. This probability statement does not take into account the statistical uncertainty in the estimate of the life distribution of transformers.

Retaining ring example. For steel specimens taken from generator retaining rings, the distribution of 0.2% yield strength is assumed to be normal with a mean $\mu = 150.23$ ksi and a standard deviation $\sigma = 4.73$ ksi. The ring manufacturer ships only those rings testing above the specification of 145 ksi. The strength distribution of rings received by the customer is a truncated normal distribution, given that the strengths are above 145 ksi. Before knowing the distribution of yield strength, the customer wanted to raise the specification to 150 ksi in order to increase generator efficiency. The proportion of rings scrapped under the old specification is $F(145) = \Phi[(145 - 150.23)/4.73] = 0.13$, or 13%, and the proportion scrapped under the new specification would be $F(150) = \Phi\{(150 - 150.23)/4.73\} = 0.48$, or 48%. Upon learning that the new specification would result in rejection of about 48% of production, the customer decided that it would be better to work on developing a stronger alloy.

Conditional Lognormal Distribution

The following paragraphs present the properties of the conditional lognormal distribution.

The cumulative distribution function of a conditional (or truncated) lognormal distribution of the time Y to failure among units that reach an age y' is

$$F(y|y') = \left(\Phi\{ [\log(y) - \mu]/\sigma \} - \Phi\{ [\log(y') - \mu]/\sigma \} \right) /$$

$$\left(1 - \Phi\{ [\log(y') - \mu]/\sigma \} \right), \qquad y \geq y'. \tag{9.39}$$

The standard normal cumulative distribution function $\Phi(\ )$ is tabulated in Appendix A1. Also, μ is the log mean and σ is the log standard deviation of the original lognormal distribution described in Section 3.

For the locomotive controls described in Section 3, the distribution of miles (in thousands) to failure is taken to be lognormal with $\mu = 2.236$ and $\sigma = 0.320$. For controls with 240 thousand miles on them, the conditional probability of failure in the next 20 thousand miles is

$$F(260|240) = \left(\Phi\{ [\log(260) - 2.236]/0.320 \} - \Phi\{ [\log(240) - 2.236]/ \right.$$

$$\left. 0.320 \} \right) / \left(1 - \Phi\{ [\log(240) - 2.236]/0.320 \} \right) = 0.118.$$

Similarly, $F(140|120) = 0.112$ and $F(20|0) = 0.0018$.

The reliability of a conditional lognormal distribution of the time Y to failure among units that reach an age y' is

$$R(y|y') = R(y)/R(y')$$

$$= \left(1 - \Phi\left\{[\log(y) - \mu]/\sigma\right\}\right) / \left(1 - \Phi\left\{[\log(y') - \mu]/\sigma\right\}\right),$$

$$y \geqslant y'. \tag{9.40}$$

For such controls, $R(260|240) = 0.882$, $R(140|120) = 0.888$, and $R(20) = 0.9982$.

The $100P$th percentile of the conditional lognormal distribution of the life Y of units that reach an age y' is

$$y_P = \text{antilog}(\mu + z_{P'}\sigma), \tag{9.41}$$

where $z_{P'}$ is the $100P'$th standard normal percentile, tabulated in Appendix A2, and

$$P' = \left(1 - \Phi\left\{[\log(y') - \mu]/\sigma\right\}\right)P + \Phi\left\{[\log(y') - \mu]/\sigma\right\}.$$

The median mileage at failure for controls with 240 thousand miles is given by $P' = (1 - \Phi\{[\log(240) - 2.236]/0.320\})0.50 + \Phi\{[\log(240) - 2.236]/0.320\} = 0.837$, $z_{.837} = 0.982$, $y_{.50} = \text{antilog}[2.236 + 0.982(0.320)] = 355$ thousand miles. For comparison, $y_{.50} = 172$ thousand miles for new controls.

The probability density of the conditional lognormal distribution of time Y to failure among units that reach an age y' is

$$f(y|y') = [0.4343/(y\sigma)]\varphi\left\{[\log(y) - \mu]/\sigma\right\} / \left(1 - \Phi\left\{[\log(y') - \mu]/\sigma\right\}\right),$$

$$y \geqslant y', \tag{9.42}$$

where $\varphi(\ )$ is the standard normal probability density.

The conditional mean life Y of units that reach an age y' is

$$E(Y|y') = E(Y)\left(1 - \Phi\left\{[\log(y') - (\mu + 2.303\sigma^2)]/\sigma\right\}\right) /$$

$$\left(1 - \Phi\left\{[\log(y') - \mu]/\sigma\right\}\right) \tag{9.43}$$

where $E(Y) = \text{antilog}(\mu + 1.151\sigma^2)$ is the mean of the (unconditional) lognormal distribution.

For controls with 240 thousand miles, the conditional expected mileage at failure is

$$E(Y|240) = \text{antilog}\left[2.236 + 2.303(0.320)^2\right]$$
$$\times \left[1 - \Phi\left(\left\{\log(240) - \left[2.236 + 2.303(0.320)^2\right]\right\}/0.320\right)\right]/$$
$$\left(1 - \Phi\left\{\left[\log(240) - 2.236\right]/0.320\right\}\right) = 423 \text{ thousand miles.}$$

For comparison, $E(Y) = 225$ thousand miles for new controls.

Aitchison and Brown (1957) provide further details on the conditional lognormal distribution, which they call the truncated distribution. For example, they provide an expression for the moments of the distribution, and the second moment can be used to obtain the conditional variance and standard deviation.

Conditional Weibull Distribution

The following paragraphs present the properties of the conditional Weibull distribution.

The cumulative distribution function of a conditional (or truncated) Weibull distribution of the time Y to failure among units that reach an age y' is

$$F(y|y') = 1 - \exp\left[(y'/\alpha)^\beta - (y/\alpha)^\beta\right], \qquad y \geq y', \qquad (9.44)$$

where α and β are the scale and shape parameters of the original Weibull distribution described in Section 4.

For the generator field windings described in Section 4, time to failure is described with a Weibull distribution with a characteristic life $\alpha = 13$ years and a shape parameter $\beta = 2.0$. For windings 6.5 years old the chance of failure in the next two years (by age 8.5 years) is $R(8.5|6.5) = 1 - \exp[(6.5/13)^{2.0} - (8.5/13)^{2.0}] = 0.163$. Electric utilities use such information in determining when to replace such windings.

The reliability of a conditional Weibull distribution of the time Y to failure among units that reach an age y' is

$$R(y|y') = \exp\left[(y'/\alpha)^\beta - (y/\alpha)^\beta\right], \qquad y \geq y'. \qquad (9.45)$$

For the windings, $R(8.5|6.5) = 1 - 0.163 = 0.837$.

The 100Pth percentile of the conditional Weibull distribution of the time Y to failure among units that reach an age y' is

$$y_P = \alpha \left\{ \ln\left[1/(1-P') \right] \right\}^{1/\beta}, \qquad (9.46)$$

where $P' = 1 - (1-P)\exp[-(y'/\alpha)^\beta]$.

For such windings at an age of 6.5 years, the conditional median age at failure is given by $P' = 1 - (1-0.50)\exp[-(6.5/13)^{2.0}] = 0.611$, $y_{.50} = 13\{\ln[1/(1-0.611)]\}^{1/2.0} = 12.6$ years. In comparison, the median for new windings is 10.8 years.

The probability density of the conditional Weibull distribution of the time Y to failure among units that reach an age y' is

$$f(y|y') = (\beta y^{\beta-1}/\alpha^\beta)\exp\left[(y'/\alpha)^\beta - (y/\alpha)^\beta \right], \qquad y \geq y'. \quad (9.47)$$

The conditional mean life Y of units that reach an age y' is

$$E(Y|y') = \int_{y'}^{\infty} y \exp\left[(y'/\alpha)^\beta \right] \alpha^{-\beta}\beta y^{\beta-1}\exp\left[-(y/\alpha)^\beta \right] dy \qquad (9.48)$$

$$= \alpha \exp\left[(y'/\alpha)^\beta \right] \left\{ \Gamma[1 + (1/\beta)] - \Gamma^*\left[(y'/\alpha)^\beta; 1 + (1/\beta) \right] \right\},$$

where $\Gamma^*(u; v) \equiv \int_0^u z^{v-1}e^{-z}\,dz$ is the incomplete gamma function. Note that $\Gamma^*(\infty, v) = \Gamma(v)$, the gamma function. The incomplete gamma function ratio is $\Gamma(u; v) \equiv \Gamma^*(u; v)/\Gamma(v)$; this is a gamma cumulative distribution. The incomplete gamma function ratio is tabulated by Harter (1964). In terms of the incomplete gamma function ratio, the conditional Weibull mean is

$$E(Y|y') = \alpha\Gamma[1 + (1/\beta)]\exp\left[(y'/\alpha)^\beta \right] \left\{ 1 - \Gamma\left[(y'/\alpha)^\beta; 1 + (1/\beta) \right] \right\}$$

$$= E(Y)\exp\left[(y'/\alpha)^\beta \right] \left\{ 1 - \Gamma\left[(y'/\alpha)^\beta; 1 + (1/\beta) \right] \right\}. \qquad (9.49)$$

For windings at an age of 6.5 years, the conditional mean age at failure is $E(Y|6.5) = 13\Gamma[1 + (1/2)]\exp[(6.5/13)^{2.0}]\{1 - \Gamma[(6.5/13)^{2.0}; 1 + (1/2)]\} = 13.6$ years. The remaining median life of 6.5-year-old windings is $13.6 - 6.5 = 7.1$ years. For comparison, $E(Y) = 11.5$ years for new windings.

The expected remaining life $E(Z|y') = E(Y|y') - y'$ increases with age y' for $\beta < 1$ and decreases with age for $\beta > 1$. For $\beta = 1$, the distribution is exponential, and the expected remaining life is constant and equal to α.

The conditional variance and standard deviation can be expressed in terms of the incomplete gamma function ratio. They are seldom needed and are not given here.

Conditional or Truncated Extreme Value Distribution

The cumulative distribution function of a conditional distribution (or truncated) smallest extreme value distribution of time Y to failure among units that reach an age y' is

$$F(y|y') = \left(1 - \exp\{-\exp[(y-\lambda)/\delta]\}\right)/\exp\{-\exp[(y'-\lambda)/\delta]\},$$

$$y \geqslant y'. \tag{9.50}$$

This truncated distribution is also known as the Gompertz–Makeham distribution. It has been used as a model for human life in old age. Its properties are given by Gross and Clark (1975, Sec. 6.5). Bailey, Homer, and Summe (1977) present the related modified Makeham distribution.

10. JOINT DISTRIBUTIONS AND THE JOINT NORMAL DISTRIBUTION

This highly specialized section is **advanced**; it is useful (but not essential) background for applications using Chapters 6, 7, 8, and 9 on analytic estimates, which approximately have a joint normal distribution for large samples. Some material involves matrix notation, but this can be skipped. For more information on multivariate distributions, particularly the multivariate normal distribution, see Morrison (1976).

Often one works with K numerical variables $Y_1, \ldots, Y_K$. They can be K observations in a sample, K statistics or estimates calculated from a sample, K order statistics of a sample (Chapter 7), K properties of a sample unit, times to a unit's failure from K possible causes (Chapter 5), etc. This section presents theory for models for K variables, that is, for their joint (multivariate) distributions. The theory here is for continuous joint distributions, but it applies to discrete distributions after obvious modifications. The multinomial distribution (Section 15) is a joint discrete distribution. The joint normal distribution appears here as an example of a joint continuous distribution.

The contents of this section are

Joint Probability Density,

Events and Their Probabilities,

Joint Cumulative Distribution,

Marginal Distributions,
Independent Random Variables,
Moments,
Linear Sums of Variables, and
Other Joint Distributions.

Joint Probability Density

General. A **joint continuous distribution** with a probability density is a probability model for K variables $Y_1, \ldots, Y_K$. The model consists of (1) the possible joint numerical outcomes $(y_1, \ldots, y_K)$ that are points in K-dimensional Euclidean space E_K and (2) a **joint probability density** $f(y_1, \ldots, y_K)$ that is a function with the properties

1. $f(y_1, \ldots, y_K) \geq 0$ for all $(y_1, \ldots, y_K)$ and   (10.1)

2. $\displaystyle\int_{-\infty}^{\infty} \cdots \int_{-\infty}^{\infty} f(y_1, \ldots, y_K)\, dy_1 \cdots dy_K = 1.$

As before, capital letters $Y_1, \ldots, Y_K$ denote the probability model or distribution, and small letters $(y_1, \ldots, y_K)$ denote a particular outcome or point in E_K. Figure 10.1 depicts a joint (bivariate) probability density for two variables.

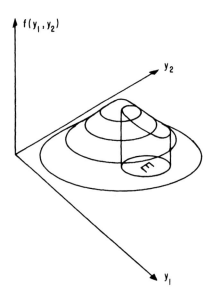

Figure 10.1. Bivariate probability density.

Joint normal density. A bivariate normal probability density for two variables is

$$f(y_1, y_2) = \left\{2\pi\left[v_{11}v_{22}(1-\rho^2)\right]^{1/2}\right\}^{-1}\exp\left\{-\tfrac{1}{2}(1-\rho^2)^{-1}\left\{\left[(y_1-\mu_1)^2/v_{11}\right]\right.\right.$$

$$\left.\left. -2\rho\left[(y_1-\mu_1)(y_2-\mu_2)/(v_{11}v_{22})^{1/2}\right]+\left[(y_2-\mu_2)^2/v_{22}\right]\right\}\right\},$$

$$(10.2)$$

where the parameter μ_k is the mean, v_{kk} is the variance of variable $k=1,2$, $v_{12}=v_{21}$ is their covariance, $\rho\equiv v_{12}/(v_{11}v_{22})^{1/2}$ is the **correlation coefficient** of Y_1 and Y_2, $-\infty<y_1<\infty$, and $-\infty<y_2<\infty$. Figure 10.1 shows such a "bell-shaped" density. Such a probability density has a constant value on ellipses centered on (μ_1,μ_2), as shown in Figure 10.2. This figure also shows how the shape of the ellipses depends on ρ, which must be in the range $(-1,1)$. When ρ is positive (negative), the contours slope positively (negatively), as in Figure 10.2. Data from such a bivariate distribution could be plotted in the figure; the points tend to cluster near (μ_1,μ_2), thin out with distance from (μ_1,μ_2), and have an elliptical pattern like the contours.

A K-variate normal probability density for K variables is most simply expressed in matrix notation as

$$f(y_1,\ldots,y_K)=(2\pi)^{-K/2}|\mathbf{V}|^{-1/2}\exp\left[-\tfrac{1}{2}(\mathbf{y}-\boldsymbol{\mu})'\mathbf{V}^{-1}(\mathbf{y}-\boldsymbol{\mu})\right], \quad (10.3)$$

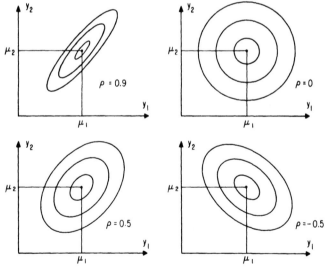

Figure 10.2. Bivariate normal contours of constant $f(y_1, y_2)$.

where $'$ denotes a transpose, $\mathbf{y} = (y_1, \ldots, y_K)'$ is the column vector for the observation, $\boldsymbol{\mu} = (\mu_1, \ldots, \mu_K)'$ is the column vector of means, $\mathbf{V}$ is the K-by-K covariance matrix of variances v_{kk} and covariances $v_{kk'}$, $k, k' = 1, \ldots, K$, and $|\mathbf{V}| \neq 0$ is the determinant of $\mathbf{V}$. $\mathbf{V}$ is symmetric and positive definite. The **correlation coefficient** for Y_k and $Y_{k'}$ is

$$\rho_{kk'} \equiv v_{kk'} / (v_{kk} v_{k'k'})^{1/2}; \qquad (10.4)$$

it is always in the range -1 to 1. For $K = 3$, the probability density is constant on ellipsoids centered on (μ_1, μ_2, μ_3). Data from such a trivariate distribution could be plotted in three dimensions; the points tend to cluster near (μ_1, μ_2, μ_3), thin out with distance from (μ_1, μ_2, μ_3), and have an ellipsoidal pattern like the contours.

Events and Their Probabilities

General.   An **event** is any subset of points of E_K. The **probability of an event** E is

$$P\{E\} \equiv \int_E \cdots \int f(y_1, \ldots, y_K) \, dy_1, \ldots, dy_K, \qquad (10.5)$$

where the multiple integral runs over the points of E. For example, for $K = 2$, the probability of an event is the volume between the event E and the probability density as shown in Figure 10.1. As before, notation such as $P\{Y_1 \leqslant 6 \text{ and } Y_2^2 > 3\}$ denotes the probability of the event that consists of the outcomes satisfying the relationships inside the braces. For example, $P\{Y_1 \leqslant y_1' \text{ and } \cdots \text{ and } Y_K \leqslant y_K'\}$ consists of all points $(y_1, \ldots, y_K)$, where $y_1 \leqslant y_1'$ and $\cdots$ and $y_K \leqslant y_K'$.

Joint normal.   There are tables of probabilities of various shaped regions for joint normal distributions. Greenwood and Hartley (1962) list such tables.

Cumulative Distribution Function

General.   The joint cumulative distribution function for $Y_1, \ldots, Y_K$, which have a joint density $f(y_1, \ldots, y_K)$, is

$$F(y_1, \ldots, y_K) \equiv P\{Y_1 \leqslant y_1 \text{ and } \cdots \text{ and } Y_K \leqslant y_K\}$$

$$= \int_{-\infty}^{y_1} \cdots \int_{-\infty}^{y_K} f(y_1', \ldots, y_K') \, dy_1' \cdots dy_K'. \qquad (10.6)$$

Such a $F(y_1,\ldots, y_K)$ has the following properties:

1. It is a continuous function for all $(y_1,\ldots, y_K)$,
2. $\lim_{\text{any } y_k \to -\infty} F(y_1,\ldots, y_K) = 0$ and $\lim_{\text{all } y_k \to \infty} F(y_1,\ldots, y_K) = 1$, and
3. $F(y_1,\ldots, y_K) \leqslant F(y_1',\ldots, y_K')$ for all $y_1 \leqslant y_1',\ldots, y_K \leqslant y_K'$.

Any function that has properties 1, 2, and 3 is a continuous joint cumulative distribution function. The cumulative distribution function and the probability density also have the relationship

$$f(y_1,\ldots, y_K) = \partial^K F(y_1,\ldots, y_K)/\partial y_1 \cdots \partial y_K. \qquad (10.7)$$

Joint normal.   Greenwood and Hartley (1962) list tables of joint normal cumulative distribution functions.

Marginal Distributions

Single variable.   The marginal distribution of a single variable, say Y_1, has probability density

$$f_1(y_1) \equiv \int_{-\infty}^{\infty} \cdots \int_{-\infty}^{\infty} f(y_1, y_2,\ldots, y_K)\, dy_2 \cdots dy_K. \qquad (10.8)$$

The marginal cumulative distribution of Y_1 is

$$F_1(y_1) \equiv P\{Y_1 \leqslant y_1\} \qquad (10.9)$$

$$= \int_{-\infty}^{y_1}\int_{-\infty}^{\infty} \cdots \int_{-\infty}^{\infty} f(y_1',\ldots, y_K')\, dy_1' dy_2' \cdots dy_K' = \int_{-\infty}^{y_1} f_1(y_1)\, dy_1.$$

The marginal distribution of Y_1 is the distribution of just Y_1 when the values of all other variables are ignored.

Joint normal.   The marginal distribution of a variable Y_k of a joint normal distribution is a normal distribution with mean μ_k and variance v_{kk}, $k = 1,\ldots, K$.

A number of variables.   The marginal joint distribution of K' variables, say $Y_1,\ldots, Y_{K'}$, has joint probability density

$$f_{1\ldots K'}(y_1,\ldots, y_{K'}) \equiv \int_{-\infty}^{\infty} \cdots \int_{-\infty}^{\infty} f(y_1,\ldots, y_K)\, dy_{K'+1} dy_{K'+2} \cdots dy_K.$$

$$(10.10)$$

Their joint cumulative distribution function is

$$F_{1\ldots K'}(y_1,\ldots,y_{K'})=F(y_1,\ldots,y_{K'},\infty,\ldots,\infty). \qquad (10.11)$$

Joint normal. The marginal joint distribution of K' variables, say $Y_1,\ldots,Y_{K'}$, of a joint normal distribution is a joint normal distribution with corresponding means μ_k, variances v_{kk}, and covariances $v_{kk'}$, $k,k'=1,\ldots,K'$.

Independent Random Variables

General. Variables $Y_1,\ldots,Y_{K'}$ are **statistically independent** if and only if their joint density satisfies

$$f_{1\ldots K'}(y_1,\ldots,y_{K'})=f_1(y_1)\times\cdots\times f_{K'}(y_{K'}). \qquad (10.12)$$

For example, the bivariate normal density for independent variables is

$$f_{12}(y_1,y_2)=(2\pi v_1)^{-1/2}\exp\left[-(y_1-\mu_1)^2/(2v_1)\right]$$
$$\times(2\pi v_2)^{-1/2}\exp\left[-(y_2-\mu_2)^2/(2v_2)\right].$$

Equivalently, they are statistically independent if and only if

$$F_{1\ldots K'}(y_1,\ldots,y_{K'})=F_1(y_1)\times\cdots\times F_{K'}(y_{K'}). \qquad (10.13)$$

If $Y_1,\ldots,Y_{K'}$ are independent and all have the same marginal distribution, then $Y_1,\ldots,Y_{K'}$ is said to be a **random sample** from that distribution. The preceding formulas allow one to construct a joint distribution from separate distributions of the individual variables, provided that the variables are statistically independent. Joint distributions of dependent variables can be obtained by other ways; for example, Chapter 7 gives the joint distribution of dependent order statistics of a random sample.

Joint normal. Joint normal variables are statistically independent if and only if their covariances are zero or, equivalently, if their correlation coefficients are zero. Variables with other joint distributions with zero covariances need not be statistically independent. However, any statistically independent variables always have zero covariances and correlations.

Moments

Mean. The mean (or expected value) of a variable, say Y_1, is

$$EY_1\equiv\int_{-\infty}^{\infty}\cdots\int_{-\infty}^{\infty}y_1f(y_1,\ldots,y_K)\,dy_1\cdots dy_K=\int_{-\infty}^{\infty}y_1f_1(y_1)\,dy_1.$$
$$(10.14)$$

Joint normal mean. The mean of a joint normal variable Y_k is the parameter μ_k, $k = 1, \ldots, K$.

Variance and standard deviation. The variance of a variable, say Y_1, is

$$\mathrm{Var}(Y_1) = \int_{-\infty}^{\infty} \cdots \int_{-\infty}^{\infty} (y_1 - EY_1)^2 f(y_1, \ldots, y_K) \, dy_1 \cdots dy_K$$

$$= \int_{-\infty}^{\infty} (y_1 - EY_1)^2 f_1(y_1) \, dy_1. \tag{10.15}$$

The corresponding standard deviation is the square root of the variance.

Joint normal variance and standard deviation. The variance of a joint normal variable Y_k is the parameter v_{kk}, and the standard deviation is $v_{kk}^{1/2}$, $k = 1, \ldots, K$.

Covariance and correlation. The covariance of two variables, say Y_1 and Y_2, is

$$\mathrm{Cov}(Y_1, Y_2) \equiv \int_{-\infty}^{\infty} \cdots \int_{-\infty}^{\infty} (y_1 - EY_1)(y_2 - EY_2) f(y_1, \ldots, y_K) \, dy_1 \cdots dy_K$$

$$= \int_{-\infty}^{\infty} \int_{-\infty}^{\infty} (y_1 - EY_1)(y_2 - EY_2) f_{12}(y_1, y_2) \, dy_1 \, dy_2. \tag{10.16}$$

Always $\mathrm{Cov}(Y_1, Y_2) = \mathrm{Cov}(Y_2, Y_1)$. The corresponding correlation coefficient is

$$\rho_{12} = \rho_{21} \equiv \mathrm{Cov}(Y_1, Y_2) / [\mathrm{Var}(Y_1)\mathrm{Var}(Y_2)]^{1/2}. \tag{10.17}$$

Joint normal covariance. The covariance of joint normal variables Y_k and $Y_{k'}$ is the parameter $v_{kk'}$, and their correlation is $\rho_{kk'} = v_{kk'} / (v_{kk} v_{k'k'})^{1/2}$.

Linear Sums

A linear sum of joint variables $Y_1, \ldots, Y_K$ has the form

$$Z = c_0 + c_1 Y_1 + \cdots + c_K Y_K, \tag{10.18}$$

where the coefficients are constants. A sample average of K observations is such a sum where $c_0 = 0$ and $c_k = 1/K$, $k = 1, \ldots, K$. Such sums occur frequently in data analysis, and they are important. In particular, they occur in large-sample theory for maximum likelihood estimation, linear estimation, and estimation by the method of moments. Basic properties of such sums are presented below.

The mean and variance of such a sum are

$$EZ = c_0 + c_1 EY_1 + \cdots + c_K EY_K, \tag{10.19}$$

$$\mathrm{Var}(Z) = \sum_{k=1}^{K} c_k^2 \mathrm{Var}(Y_k) + 2 \sum\sum_{k<k'} c_k c_{k'} \mathrm{Cov}(Y_k, Y_{k'}). \tag{10.20}$$

The covariance of Z with another sum $Z' = c_0' + c_1' Y_1 + \cdots + c_K' Y_K$ is

$$\mathrm{Cov}(Z, Z') = \sum_{k=1}^{K} c_k c_k' \mathrm{Var}(Y_k) + \sum\sum_{k \neq k'} c_k c_{k'}' \mathrm{Cov}(Y_k, Y_{k'}). \tag{10.21}$$

The joint distribution of such sums of joint normal variables is joint normal. The joint distribution of the sums has the means, variances, and covariances given by the formulas above. In fact, a distribution is joint normal if and only if any linear sum of the variables has a normal distribution.

Other Joint Distributions

The joint normal distribution is the most widely known and used joint distribution. Others used in life data analysis are surveyed by Block and Savits (1981) and include

The joint lognormal distribution.   The logs of the variables have a joint normal distribution.

Joint exponential distributions.   There are various joint exponential distributions; for example, Barlow and Proschan (1975) present a shock model, and Block (1975) presents others.

Joint Weibull distributions.   Such distributions are given by Lee and Thompson (1974) and by Moeschberger and David (1971).

Consult the bibliographies and journal indices (Chapter 13) for other joint distributions.

11.   GENERAL THEORY ON DISCRETE DISTRIBUTIONS (GEOMETRIC DISTRIBUTION)

This essential section presents basic concepts and theory for discrete statistical distributions. Such distributions are models for counts of the number of failures of a product or other occurrences. Such distributions include the geometric, Poisson, binomial, hypergeometric, and multinomial distributions. These distributions and their uses are illustrated with examples on a variety of products. Chapter 6 gives methods for analysis of data from such distributions. Readers interested in analyses of such data may go directly to Chapter 6 after these sections.

The basic concepts include:

the probability function and outcomes of a discrete distribution,

events and their probabilities, and

the mean, variance, and standard deviation of a discrete distribution.

These concepts are explained below with the geometric distribution.

A **discrete distribution** is a probability model Y that consists of a list of distinct possible **outcomes** y_1, y_2, y_3, etc., each with a corresponding **probability** $f(y_1)$, $f(y_2)$, $f(y_3)$, etc., Here $f(\ )$ is called the **probability function**. The probabilities $f(y_i)$ must (1) be zero or greater and (2) their sum must equal 1. Any list of outcomes and corresponding probabilities that satisfy (1) and (2) is a discrete distribution. The capital Y denotes the probability model, which is also loosely called a **discrete random variable** if it has numerical outcomes. Small letters y, y_1, y_2, etc., denote particular outcomes.

The outcomes may be numerical (for example, $0, 1, 2, 3$, etc.) or categories (for example, dead, alive, undecided). For a population, $f(y_i)$ is the population fraction with the value y_i. It is also the chance that a randomly taken observation has the value y_i; that is, $f(y_i)$ is the fraction of the time that y_i would be observed in an infinitely large number of observations.

The **geometric distribution,** for example, consists of the outcomes $y = 1, 2, 3, \ldots$ and corresponding probabilities

$$f(y) = p(1-p)^{y-1}, \tag{11.1}$$

where $0 < p < 1$. Clearly, (1) all $f(y)$ are positive and (2) their sum equals unity. Figure 11.1 depicts this distribution. The geometric probability function is tabulated by Williamson and Bretherton (1963).

Assumptions.   (11.1) is the distribution of the number y of trials (years, months, etc.) to "failure," where each trial is statistically independent of all

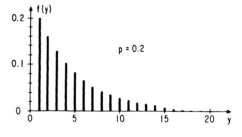

Figure 11.1.   Geometric probability function.

other trials and each trial has the same chance p of failure. Such a trial with two outcomes (failure or survival here) is called a Bernoulli trial. This distribution with $p=\frac{1}{2}$ is a model for the number of flips of a coin until the first heads occurs. Product failure in the yth year of service for a certain type of distribution transformer is described with a geometric distribution with $p=0.00459$.

An event is any set of outcomes of a probability model Y. The **probability of an event** is the sum of the probabilities of the outcomes in the event. Notation for the probability of an event is $P\{\ \}$, where the relationship in the braces indicates that the event consists of the outcomes of Y that satisfy the relationship. For example, $P\{Y=y\}=f(y)$, and $P\{Y\leqslant y\}=\Sigma_{y_i\leqslant y}f(y_i)$ denotes the probability of all Y outcomes that are less than or equal to the value y. For the geometric distribution for transformer life, $P\{Y\leqslant y\}$ would be interpreted as the probability of failure by age y, and $P\{Y>y\}$ is the probability of surviving beyond the age y. An event is said to occur if the observed outcome is in the event. The probability of an event has two possible interpretations: (1) it is the proportion of population units that have values in the event, and (2) it is the chance of an observation being an outcome in the event.

The cumulative distribution function $F(y)$ of a discrete distribution with numerical outcomes is

$$F(y)=P\{Y\leqslant y\}= \sum_{y_i\leqslant y} f(y_i), \qquad -\infty<y<\infty, \qquad (11.2)$$

where the sum runs over all outcomes y_i that are less than or equal to y. For a population, $F(y)$ is the fraction of the population with a value of y or less. This function is defined for all y values between $-\infty$ and $+\infty$, not just for the values y_i. $F(y)$ for a discrete distribution is a staircase function, as in Figure 11.2, and the jump in the function at an outcome y_i equals the probability $f(y_i)$. Thus, $F(y)$ and $f(y)$ may be obtained from each other.

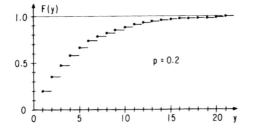

Figure 11.2. Geometric cumulative distribution.

The function $F(y)$ has the properties

1. $\lim_{y \to -\infty} F(y) = 0$,
2. $\lim_{y \to +\infty} F(y) = 1$, and
3. $F(y) \leqslant F(y')$ for all $y < y'$.

The geometric cumulative distribution function is

$$F(y) = p + p(1-p) + p(1-p)^2 + \cdots + p(1-p)^{y-1}$$

$$= 1 - (1-p)^y, \quad \text{for} \quad y = 1, 2, 3, \ldots . \tag{11.3}$$

For noninteger values of y, this $F(y)$ is defined as depicted in Figure 11.2. This function satisfies 1, 2, and 3 and is therefore a cumulative distribution function.

Many discrete distributions have only integer outcomes, for example, the geometric distribution. Then the following relationships express probabilities of some common events in terms of $F(y)$, which is tabulated for many distributions:

$$P\{Y \leqslant y\} = F(y),$$

$$P\{Y < y\} = P\{Y \leqslant y-1\} = F(y-1),$$

$$P\{Y = y\} = P\{Y \leqslant y\} - P\{Y \leqslant y-1\} = F(y) - F(y-1),$$

$$P\{Y > y\} = 1 - P\{Y \leqslant y\} = 1 - F(y), \tag{11.4}$$

$$P\{Y \geqslant y\} = 1 - P\{Y < y\} = 1 - F(y-1),$$

$$P\{y' < Y \leqslant y\} = P\{Y \leqslant y\} - P\{Y \leqslant y'\} = F(y) - F(y'),$$

$$P\{y' \leqslant Y \leqslant y\} = P\{Y \leqslant y\} - P\{Y < y'\} = F(y) - F(y'-1), \quad \text{etc.}$$

It is important to note when inequalities are strict and when they include equality.

Transformer example. For example, the probability of a distribution transformer failure after a five-year warranty is $P\{Y > 5\} = 1 - F(5) = 1 - [1 - (1 - 0.00459)^5] = 0.977$, or 97.7%. This is the expected proportion of a manufacturer's production surviving warranty. Similarly, the probability of failure in the 6th through 10th years in service is

$$P\{6 \leqslant Y \leqslant 10\} = F(10) - F(5)$$

$$= 1 - (1 - 0.00459)^{10} - \left[1 - (1 - 0.00459)^{5} \right]$$

$$= 0.022, \quad \text{or} \quad 2.2\%.$$

The mean, variance, and standard deviation are used (1) to summarize a discrete distribution and (2) to calculate approximate probabilities with the normal distribution. They are defined next.

The mean of a discrete random variable Y is

$$E(Y) = y_1 f(y_1) + y_2 f(y_2) + y_3 f(y_3) + \cdots, \qquad (11.5)$$

provided that the sum exists. The sum runs over all outcomes y_i. The mean can be loosely regarded as a middle value of a distribution. It corresponds to a population average, and the average of a large sample tends to be close to the distribution mean. It is also called the **expected value** of Y, and it need not equal a possible y_i value.

The mean of the geometric distribution is

$$E(Y) = 1p + 2p(1-p) + 3p(1-p)^2 + \cdots = 1/p. \qquad (11.6)$$

For the distribution transformers, the mean life is $E(Y) = 1/0.00459 = 218$ years. This assumes (probably incorrectly) that the geometric distribution adequately describes the entire range of transformer life, whereas it is satisfactory for just the lower tail.

The variance of a discrete random variable Y is

$$\text{Var}(Y) = \left[y_1 - E(Y) \right]^2 f(y_1) + \left[y_2 - E(Y) \right]^2 f(y_2)$$

$$+ \left[y_3 - E(Y) \right]^2 f(y_3) + \cdots, \qquad (11.7)$$

where the sum runs over all possible outcomes y_i. Equivalently,

$$\text{Var}(Y) = \left[y_1^2 f(y_1) + y_2^2 f(y_2) + y_3^2 f(y_3) + \cdots \right] - \left[E(Y) \right]^2. \quad (11.8)$$

The variance is a measure of the spread of a distribution about its mean. It has the dimensions of Y^2; for example, if Y is in years, $\text{Var}(Y)$ is in years squared.

The variance of the geometric distribution, for example, is

$$\text{Var}(Y)=\left[1^2p+2^2p(1-p)+3^2p(1-p)^2+\cdots\right]-\left[1/p\right]^2=(1-p)/p^2.$$

$$(11.9)$$

For the distribution transformers, the variance of life is $\text{Var}(Y)=(1-0.00459)/(0.00459)^2 = 47{,}247$ years^2.

The standard deviation of a discrete random variable Y is

$$\sigma(Y)=\left[\text{Var}(Y)\right]^{1/2}.\qquad(11.10)$$

This is another measure of the spread of a distribution about its mean. It has the dimensions of Y; for example, if Y is in years, $\sigma(Y)$ is in years.

The standard deviation of the geometric distribution, for example, is

$$\sigma(Y)=\left[(1-p)/p^2\right]^{1/2}.\qquad(11.11)$$

For the distribution transformers, the standard deviation of life is $\sigma(Y)=(47{,}247)^{1/2}=217$ years.

Normal approximation. Approximate probabilities for many discrete distributions may be calculated using a normal distribution. This reduces computing labor and is often useful outside the range of existing tables of discrete distributions. Suppose that a discrete random variable Y has **integer** outcomes, mean $E(Y)$, and standard deviation $\sigma(Y)$. The basic approximation is then

$$P\{Y\leqslant y\}=F(y)\simeq\Phi\{\left[y+0.5-E(Y)\right]/\sigma(Y)\},\qquad(11.12)$$

where $\Phi(\ )$ is the standard normal cumulative distribution tabulated in Appendix A1. Related approximations include:

$$P\{Y>y\}=1-F(y)\simeq1-\Phi\{\left[y+0.5-E(Y)\right]/\sigma(Y)\}.$$

$$P\{y'<Y\leqslant y\}=F(y)-F(y')\simeq\Phi\{\left[y+0.5-E(Y)\right]/\qquad(11.13)$$

$$\sigma(Y)\}-\Phi\{\left[y'+0.5-E(Y)\right]/\sigma(Y)\}.$$

In using such formulas, note which inequalities are strict and which are not. These formulas apply to the discrete distributions below. The "0.5" in the equations is called the continuity correction; it is intended to bring the continuous approximate cumulative distribution closer to the leading edge of the steps of the exact discrete cumulative distribution (see Figures 12.4 and 13.2).

A property of the geometric distribution.   If a unit from a geometric life distribution has survived y trials, then its chance of failing on the next trial is p for any value of y. Thus, such a unit does not become more or less prone to failure as it ages. For example, no matter how many times tails has turned up in flips of a fair coin, the probability of heads turning up on the next flip is $p = \frac{1}{2}$. Such a life distribution is suitable for drinking glasses, atomic particles, coins in circulation, and other items that fail from a chance event. This discrete distribution is analogous to the continuous exponential distribution.

Following sections use the basic concepts above to present commonly used discrete distributions for life and failure data. These distributions include the Poisson, binomial, multinomial, and other distributions. Johnson and Kotz (1969) and Patil and Joshi (1968) describe these and other discrete distributions in detail.

12. POISSON DISTRIBUTION

This basic section presents the Poisson distribution. It is a widely used model for the number of occurrences of some event within some observed time, area, volume, etc. For example, it has been used to describe the number of soldiers of a Prussian regiment kicked to death yearly by horses, the number of flaws in a length of wire or computer tape, the number of defects in a sheet of material, the number of failures of a repairable product over a certain period, the number of atomic particles emitted by a sample in a specified time, and many other phenomena. This model describes situations where (1) the occurrences occur independently of each other over time (area, volume, etc.), (2) the chance of an occurrence is the same for each point in time (area, volume, etc.), and (3) the potential number of occurrences is essentially unlimited.

The Poisson probability function is

$$f(y) = (1/y!)(\lambda t)^y \exp(-\lambda t), \qquad (12.1)$$

where the number of occurrences is $y = 0, 1, 2, \ldots$. Here the quantity t is the "length" or exposure of the observation; it may be a time, length, area, volume, etc. For example, for failures of a power line, the length of observation includes the length of the line and the length of time; so exposure t is their product in 1000 ft-years. Also, the parameter λ must be positive and is the **occurrence rate**; it is expressed as the number of occurrences per unit "length." For example, for power line failures, λ would be a **failure rate** and expressed in failures per 1000 ft per year. Many authors present the Poisson distribution in terms of the single parameter $\mu=\lambda t$. Figure 12.1 depicts the distribution probability function. Poisson probabilities $f(y)$ are conveniently tabulated by Molina (1949) and General Electric (1962).

Power line example. For a particular power line wire, the yearly number of failures is assumed to have a Poisson distribution with $\lambda=0.0256$ failures per year per 1000 ft. For $t=515.8$ 1000 feet of such wire in service, the probability of no failures in a year is $f(0)=(1/0!)(0.256\times515.8)^0\exp(-0.0256\times515.8)=\exp(-13.20)=1.8\times10^{-6}$, about two in a million.

The Poisson cumulative distribution function for the probability of y or fewer occurrences is

$$F(y)=P(Y\leqslant y)=\sum_{i=0}^{y}(1/i!)(\lambda t)^i\exp(-\lambda t). \qquad (12.2)$$

Figure 12.2 shows cumulative Poisson distribution functions. The chart in Figure 12.3 provides a Poisson probability $F(y)$ as follows. Enter the chart on the horizontal axis at the value of $\mu=\lambda t$. Go up to the curve labeled y. Then go horizontally to the vertical scale to read $F(y)$. For example, for the

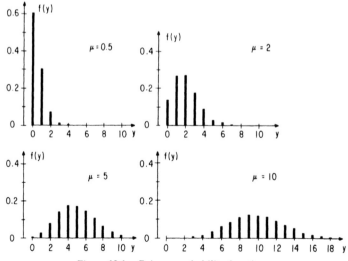

Figure 12.1. Poisson probability functions.

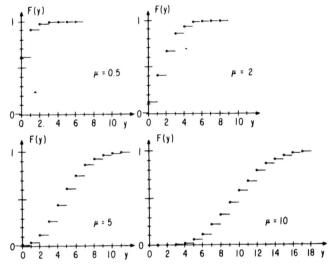

Figure 12.2.   Poisson cumulative distributions.

power line wire, $\lambda t = 0.0256 \times 515.8 = 13.2$, and the probability of 15 or fewer failures is $F(15) = 0.75$, or 75%, from the chart. $F(y)$ is tabulated by Molina (1949), General Electric (1962), and briefly in Appendix A6.

$F(y)$ can also be expressed as

$$F(y) = 1 - G(2\mu; 2y+2), \tag{12.3}$$

where

$$G(2\mu; 2y+2) = (1/y!)2^{-y-1}\int_0^{2\mu} e^{-z/2}z^y\, dz$$

is the chi-square cumulative distribution function with $(2y+2)$ degrees of freedom evaluated at 2μ. Thus the chi-square distributions are used to analyze Poisson data, and their percentiles appear in Appendix A3.

The Poisson mean of the number Y of occurrences is

$$E(Y) = 0(1/0!)(\lambda t)^0 \exp(-\lambda t) + 1(1/1!)(\lambda t)^1$$

$$\times \exp(-\lambda t) + 2(1/2!)(\lambda t)^2 \exp(-\lambda t) + \cdots$$

$$= \lambda t. \tag{12.4}$$

This is simply the occurrence rate λ (occurrences per unit "length") times

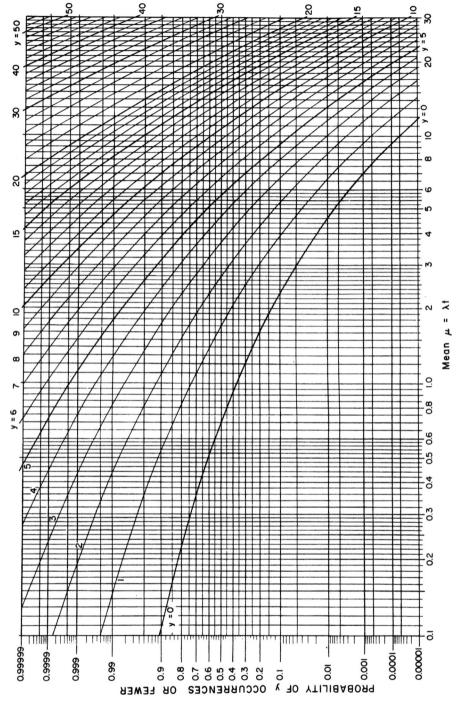

Figure 12.3. Poisson cumulative probabilities. From H. F. Dodge and H. G. Romig, *Sampling Inspection Tables*, Wiley, New York, 1944, Fig. 6.

87

the "length" t of observation. $E(Y)$ is also called the "expected number of occurrences." For the power line, the expected number of failures in a year is $\lambda t = 0.0256(515.8) = 13.2$ failures; this number is useful in maintenance planning. Equation (12.4) can be written as

$$\lambda = E(Y)/t. \qquad (12.5)$$

This shows why λ is called the occurrence rate; it is the expected number of occurrences per unit time or "length."

The Poisson variance of the number Y of occurrences is

$$\text{Var}(Y) = \left[0^2(1/0!)(\lambda t)^0 \exp(-\lambda t) + 1^2(1/1!)(\lambda t)^1 \exp(-\lambda t) \right.$$
$$\left. + 2^2(1/2!)(\lambda t)^2 \exp(-\lambda t) + \cdots \right] - (\lambda t)^2$$
$$= \lambda t. \qquad (12.6)$$

The Poisson variance and mean both equal λt. For the power line, the variance of the number of failures in a year is $\text{Var}(Y) = 0.0256(515.8) = 13.2$.

The Poisson standard deviation of the number Y of occurrences is

$$\sigma(Y) = (\lambda t)^{1/2}. \qquad (12.7)$$

For the power line, $\sigma(Y) = (13.2)^{1/2} = 3.63$ failures per year.

A normal approximation to the Poisson $F(y)$ is

$$F(y) \cong \Phi\left\{ \left[y + 0.5 - E(Y) \right]/\sigma(Y) \right\} = \Phi\left[(y + 0.5 - \lambda t)/(\lambda t)^{1/2} \right], \qquad (12.8)$$

where $\Phi(\ )$ is the standard normal cumulative distribution function; it is tabulated in Appendix A1. This approximation is more exact the larger λt and the closer y to λt. It is satisfactory for many practical purposes if $\lambda t \geqslant 10$. Figure 12.4 shows the normal approximation as a straight line on normal probability paper. Each exact Poisson cumulative distribution is a staircase function. The normal approximation should be close to the exact function at integer y values.

For example, for the power line ($\lambda t = 13.2$), the approximate probability of 15 or fewer failures in a year is $F(15) \cong \Phi[(15 + 0.5 - 13.2)/3.63] = \Phi(0.63) = 0.74$, or 74%. The exact probability is 75%.

Sum of Poisson counts. In some applications, Poisson counts are summed. The following result is useful for such sums of K counts. Suppose

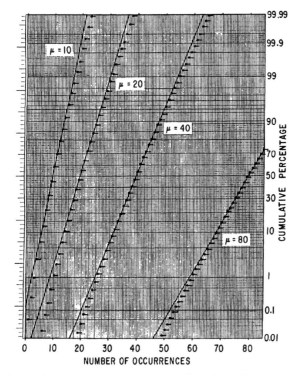

Figure 12.4. Normal approximation to the Poisson distribution.

that the kth count is Y_k, the corresponding occurrence rate is λ_k, and the length of observation is t_k, $k=1,\ldots,K$. If the counts are statistically independent, then the sum $Y=Y_1+\cdots+Y_K$ has a Poisson distribution with a mean of $\mu=\lambda_1 t_1+\cdots+\lambda_K t_K$.

For example, the numbers of failures of the power line in two successive years may be regarded as statistically independent. Then the total number of failures in the two years has a Poisson distribution with a mean of $\mu=0.0256(515.8)+0.0256(515.8)=26.4$ failures.

However, the expected numbers of failures on a power line in two different years may not be statistically independent, since the line sees different weather, which affects the mean number of failures. Thus the total number of failures for the two years may not have a Poisson distribution, but the distribution may serve as a first approximation to the true one.

Demonstration testing commonly involves the Poisson distribution. Often a manufacturer of military hardware must demonstrate its reliability. This entails testing a number of units for a specified combined time t. Units that fail are repaired and kept on test. For example, an electronic system was to be tested for $t=10,000$ hours. It was agreed that the hardware would pass the test if there were no more than a specified number y of failures. For the electronic system $y=2$ failures were allowed.

A manufacturer needs to know the desired λ value to design into the hardware to assure that it pass the test with a desired high probability $100(1-\alpha)\%$ (90% chosen for the electronic system). The hardware will then fail the test with $100\alpha\%$ probability, which is called the **producer's risk.** For the electronic system, the producer's risk was chosen as 10%. Suppose that the observed number Y of failures has a Poisson distribution with a mean $\mu=\lambda t$. Then, to obtain the desired design failure rate, one must find the value of $\mu=\lambda t$ such that the Poisson probability $F_{\mu}(y)=1-\alpha$. To do this, enter Figure 12.3 on the vertical axis at $1-\alpha$, go horizontally to the curve for y or fewer failures, and then go down to the horizontal axis to read the appropriate μ value. Then the desired design failure rate is $\lambda=\mu/t$. For the electronic system, $\lambda=1.15/10,000=0.115$ failures per 1000 hours.

Poisson process. In some applications, occurrences are observed at random points in time. The Poisson process model describes many such situations. These include, for example, (1) failures in a stable fleet of repairable items, (2) phone calls coming into an exchange, (3) atomic particles registering on a counter, and (4) power line failures. The model is defined by the following properties: (1) the number of occurrences in any period of length t has a Poisson distribution with mean $\mu=\lambda t$, where λ is the (positive) occurrence rate, and (2) the numbers of occurrences in any number of separate intervals are all statistically independent. More formal definitions are given by Parzen (1962) and by Cox and Miller (1965).

In particular, the number Y of occurrences from time 0 to time t has a Poisson distribution with mean λt, the expected number of occurrences by time t.

A consequence of the definition is that the times between successive occurrences are statistically independent and have an exponential distribution with failure rate λ. That is, if the first, second, third, etc., occurrences occur at times $0 \leqslant T_1 \leqslant T_2 \leqslant T_3 \leqslant \cdots$, then the differences $D_1=T_1-0$, $D_2=T_2-T_1$, $D_3=T_3-T_2,\ldots$ have an exponential distribution with failure rate λ and are statistically independent. That is, $P\{D_i \leqslant t\}=1-\exp(-\lambda t)$, $t \geqslant 0$, $i=1,2,3,\ldots$.

In certain reliability demonstration tests of units from an exponential life distribution, failed units are replaced immediately. This is called **testing with replacement.** Then the number of failures in a fixed total test time has a Poisson distribution. Its failure rate is the same as that of the exponential distribution and the exposure is the total test time summed over all units. Section 1 of Chapter 10 gives an example of this.

The Poisson process is often used as a model for failures in a fleet of repairable units that have a stable mix of part ages. Consequently, for a population that is aging, a "nonhomogeneous" Poisson process may be appropriate. For such a process, the failure rate is a function of time $\lambda(t)$. Such nonhomogenous processes are described by Parzen (1962) and Cox and Miller (1965).

The numbers of occurrences in independent Poisson processes may be summed, and the sum is a Poisson process. The occurrence rate of the sum is the sum of the occurrence rates. For example, if the number of failures on each of a number of power lines is a Poisson process, then the total number of failures on all lines is a Poisson process. Also, for example, if the number of failures on each of a number of parts in a product is a Poisson process, then the total number of failures of the product is a Poisson process. This is also true for nonhomogeneous Poisson processes.

Analyses of Poisson data are presented in Chapters 6 and 10. Readers may wish to go directly to the Poisson analyses there.

13. BINOMIAL DISTRIBUTION

This basic section presents the binomial distribution. It is widely used as a model for the number of sample units that are in a given category. The distribution is used if each unit is classified as in the category or else not in the category, a dichotomy. For example, it is used for the number of defective units in samples from shipments and production, the number of units that fail on warranty, the number of one-shot devices (used once) that work properly, the number of sample items that fall outside of specifications, the number of sample people that respond affirmatively to a question, the number of heads in a number of coin tosses, and many other situations.

Chapters 6 and 10 present methods for analysis of binomial data. After reading this section, one may go directly to that material.

Assumptions of the model are (1) each of n sample items has the same chance p of being in the category and (2) the outcomes of the n sample items are statistically independent. Of course, the number of sample items in the category can range from zero to n. The binomial distribution is suitable for small samples taken from large populations. When the sample is a large fraction of the population (say, 10% or greater), the hypergeometric distribution is appropriate (Section 14), since the sample outcomes are not independent.

The binomial probability function for a sample of n units is

$$f(y) = \frac{n!}{y!(n-y)!} p^y (1-p)^{n-y}. \tag{13.1}$$

where the possible number of units in the category is $y = 0, 1, 2, \ldots, n$, and p is the population proportion in the category ($0 \leqslant p \leqslant 1$). As required, the $f(y)$ are positive and sum to unity. Figure 13.1 depicts binomial probability functions.

In reliability work, if the category is "failure" of a device, the proportion p is also called a failure rate and is often expressed as a percentage. Note that this failure rate differs from the Poisson failure rate λ, which has the dimensions of failures per unit time. If the category is "successful operation" of a device, the proportion p is called the *reliability* of the device.

A locomotive control under development was assumed to fail on warranty with probability $p = 0.156$. A sample of $n = 96$ such controls were field tested on different locomotives so that their failures should be statistically independent. There occurred 15 failures on warranty. The binomial probability of $y = 15$ failures is $f(15) = 96![15!(96-15)!]^{-1}(0.156)^{15}(1-0.156)^{96-15} = 0.111$.

The Poisson approximation to binomial probabilities simplifies their calculation; namely,

$$f(y) \simeq (1/y!)(np)^y \exp(-np). \qquad (13.2)$$

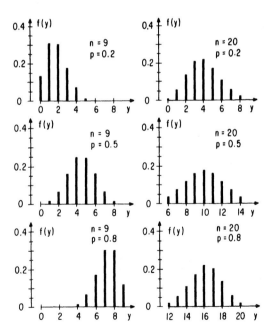

Figure 13.1. Binomial probability functions.

This is the Poisson probability function where the mean is $\mu=np$. Such probabilities may be calculated or taken from a Poisson table. This approximation is more exact the larger n and the small (np), and it suffices for most purposes if $n\geqslant50$ and $(np)\leqslant10$. For example, for the locomotive control, $f(15)\cong(1/15!)(96\times0.156)\exp(-96\times0.156)=0.102$ (0.111 exact).

The binomial cumulative distribution function for the probability of y or fewer sample items being in the category is

$$F(y)=\sum_{i=0}^{y}\frac{n!}{i!(n-i)!}p^{i}(1-p)^{n-i}, \qquad y=0,1,2,\ldots,n. \qquad (13.3)$$

$F(y)$ is laborious to calculate, but it is tabulated by the Harvard Computation Laboratory (1955), the National Bureau of Standards (1950), Romig (1953), Weintraub (1963) for small p values, and briefly in Appendix A7. There are computer routines and personal calculators that calculate $F(y)$.

For example, the probability of 15 or fewer warranty failures of the 96 locomotive controls occurring is $F(15)=0.571$. This was found by interpolation in a binomial table.

A normal approximation to the binomial $F(y)$ is

$$F(y)\cong\Phi\{[y+0.5-E(Y)]/\sigma(Y)\}$$

$$=\Phi\{(y+0.5-np)/[np(1-p)]^{1/2}\}, \qquad (13.4)$$

where $\Phi(\ )$ is the standard normal cumulative distribution function and is tabulated in Appendix A1. This approximation is more exact the larger the sample size n and the closer p to $\frac{1}{2}$ and the closer y to (np). It is satisfactory for many practical purposes if $(np)\geqslant10$ and $n(1-p)\geqslant10$. Figure 13.2 shows the normal approximation to the binomial cumulative distribution as a straight line on normal probability paper. Each exact binomial cumulative distribution is a staircase function. The approximation should be close to the exact function at the integer y values.

For example, for the sample of locomotive controls, the approximate probability of 15 or fewer failures occurring is $F(15)\simeq\Phi\{(15+0.5-96\times0.156)/[96\times0.156(1-0.156)]^{1/2}\}=0.556$. Similarly, the approximate probability of 15 failures occurring is $P\{Y=15\}=F(15)-F(14)\simeq0.556-\Phi\{(14+0.5-96\times0.156)/[96\times0.156(1-0.156)]^{1/2}\}=0.112$ (0.111 exact).

The binomial mean of the number Y of sample items in the category is

$$E(Y)=np. \qquad (13.5)$$

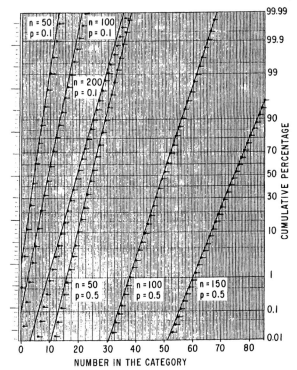

Figure 13.2.   Normal approximation to the binomial distribution.

This is simply the number n in the sample times the population proportion p in the category. For example, the expected number of failures in a sample of 96 locomotive controls is $E(Y)=96\times0.156=15.0$ failures.

The binomial variance is

$$\mathrm{Var}(Y)=np(1-p). \qquad (13.6)$$

For a sample of 96 locomotive controls, the variance of the number of failed controls is $\mathrm{Var}(Y)=96\times0.156\times(1-0.156)=12.7$.

The binomial standard deviation is

$$\sigma(Y)=\left[np(1-p)\right]^{1/2}. \qquad (13.7)$$

For a sample of 96 locomotive controls, $\sigma(Y)=[96\times0.156\times(1-0.156)]^{1/2}=3.56$ failures.

Acceptance sampling. The following material briefly describes another binomial example. Acceptance sampling plans are treated in detail in many quality control books, for example, Grant and Leavenworth (1980), Schilling (1982), and Dodge and Romig (1959). Also, Chapter 10 provides more detail on sampling plans.

The binomial distribution is commonly used for the number of defective units in a random sample from a large lot. An **acceptance sampling plan** specifies the number n of units in the sample and the acceptable number y of defectives in the sample. If there are more than y defectives in a sample, the consumer is entitled to reject the shipment. Such a plan should accept most good lots and reject most poor ones. A plan had $n=20$ and $y=1$. If a shipment has a proportion defective of $p=0.01$, the chance of it passing inspection is

$$F(1) = \frac{20!}{0!(20-0)!}0.01^0 0.99^{20} + \frac{20!}{1!(20-1)!}0.01^1 0.99^{19} = 0.983,$$

which could also be read from a binomial table. The chance of passing as a function of p is called the **operating characteristic (OC) curve** of the plan (n, y). The OC curve for the sampling plan above appears in Figure 13.3. Plans with different n and y have different OC curves.

Train example. A certain freight train requires three locomotives that must all complete the run without breakdown to avoid a costly delay. Experience shows that each locomotive successfully completes the run with probability $p=0.9$. Then a train with three randomly chosen locomotives successfully completes a run with binomial probability $P(Y=3) = [3!/(3!0!)](0.9)^3(0.1)^0 = 0.729$. A fourth locomotive increases the chance of at least three completing the run to

$$P(Y \geqslant 3) = P(Y=4) + P(Y=3) = [4!/(4!0!)](0.9)^4(0.1)^0$$

$$+ [4!/(3!1!)](0.9)^3(0.1)^1 = 0.945.$$

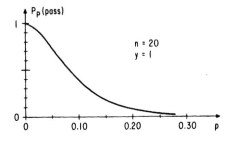

Figure 13.3. OC curve of acceptance sampling plan.

This probability is called the reliability of the train. The extra locomotive improves the reliability of the train. Such redundancy through extra components is commonly used for designing reliable systems with less reliable parts; redundancy is described in many reliability texts, for example, Shooman (1968).

Analyses of binomial data appear in Chapters 6 and 10. Readers may wish to go directly to the binomial analyses there.

14. HYPERGEOMETRIC DISTRIBUTION

This specialized section presents the hypergeometric distribution. It is used as a model for the number of sample units that are in a given category where the sampled population has a **finite** number of units. The distribution is used if each unit is classified as in the category or else not in the category, a dichotomy. For example, it used to describe the number of defective items in samples from shipments and production lots, the number of people that respond affirmatively to a question, etc.

This model describes the number Y of sample items in a given category when samples of n items are randomly taken from a population of N items where D of the population items are in the category. The population proportion in the category is $p = D/N$. The binomial distribution describes the same type of situation, but with an infinite population; the simpler binomial distribution is usually accurate enough if $n/N \leqslant 0.10$.

The probability function of the hypergeometric distribution for the number Y of sample items in the category is

$$f(y) = \binom{D}{y}\binom{N-D}{n-y} \bigg/ \binom{N}{n}, \tag{14.1}$$

where $\binom{a}{b} \equiv a!/[b!(a-b)!]$ is the binomial coefficient, and the possible numbers y are the integers from $\max(0, D-N+n)$ through $\min(D, n)$. Hypergeometric probabilities $f(y)$ are tabulated by Lieberman and Owen (1961).

An acceptance sampling plan for lots of $N = 500$ units of a product required a random sample of $n = 20$ from each lot. The lot is accepted if it contains $y = 1$ or fewer defectives. If a lot contains $D = 50$ defectives, the probability of no defectives in the sample is $f(0) = [50!(500-50)!20!(500-20)!]/[0!(50-0)!(20-0)!(500-50-20+0)!500!] = 0.115$. Similarly, the probability of 1 defective is $f(1) = [50!(500-50)!20!(500-20)!]/[1!(50-1)!(20-1)!(500-50-20+1)!500!] = 0.270$. The probability of $y = 1$ or fewer defectives is $f(0) + f(1) = 0.385$; the corresponding binomial probability was found in Section 13 to be 0.392. The values of factorials are tabulated, for example, by Lieberman and Owen (1961).

The binomial approximation to a hypergeometric probability is

$$f(y) \cong \frac{n!}{y!(n-y)!} p^y (1-p)^{n-y}, \tag{14.2}$$

where $p = D/N$ is the population proportion in the category. This is the binomial probability for samples of n items where is $p = D/N$. Such probabilities may be calculated or taken from a binomial table, which is more widely available. This approximation is more exact the smaller the sampling fraction n/N. It is often adequate if the sample contains less than 10 or even 20% of the population. The approximation for the preceding example is $f(0) \cong 20![0!(20-0)!]^{-1} 0.1^0 (1-0.1)^{20-0} = 0.122$, where $p = 50/500 = 0.1$.

Other approximations come from the Poisson and normal approximations to the binomial distribution. The Poisson approximation ($\mu \cong nD/N$) is

$$f(y) \cong (1/y!)(nD/N)^y \exp(-nD/N). \tag{14.3}$$

This is the Poisson distribution with a mean $\mu = nD/N$. Such probabilities may be taken from a Poisson table. As before, this approximation is more exact the larger n and the smaller $p = D/N$, and it is usually satisfactory if $n \geqslant 50$, $(np) \leqslant 10$, and $n/N \leqslant 0.10$. For example, the approximate Poisson probability for the example above is $f(0) \cong (1/0!)(20 \times 50/500)\exp(-20 \times 50/500) = 0.135$.

The hypergeometric cumulative distribution function for the probability of y or fewer sample items in the category is

$$F(y) = \sum_{i \leqslant y} f(i), \tag{14.4}$$

where the sum runs over all possible numbers i that are less than or equal to y. This function is laborious to calculate, but it is tabulated by Lieberman and Owen (1961). The hypergeometric $F(y)$ can be approximated by the binomial one with $p = D/N$ if n is less than, say, $0.10N$.

A normal approximation to the hypergeometric cumulative distribution function $F(y)$ is

$$F(y) \cong \Phi\{[y + 0.5 - E(Y)]/\sigma(Y)\}$$
$$= \Phi\Big([y + 0.5 - n(D/N)]/\{n(D/N)[1 - (D/N)]$$
$$\times (N-n)/(N-1)\}^{1/2}\Big), \tag{14.5}$$

where $\Phi(\ )$ is the standard normal cumulative distribution function and is tabulated in Appendix A1. This approximation has the same properties as the normal approximation to the binomial distribution where $p=D/N$.

Thy hypergeometric mean of the number Y of sample items in the category is

$$E(Y)=nD/N. \tag{14.6}$$

This expected number is simply the number n in the sample times the population proportion $p=D/N$ in the category. For the example above, the expected number of defectives in such a sample is $E(Y)=20\times50/500=2$.

The hypergeometric variance of the number Y of sample items in the category is

$$\mathrm{Var}(Y)=[(N-n)/(N-1)]n(D/N)[1-(D/N)]. \tag{14.7}$$

This is like the binomial variance (13.6) where $p=D/N$. The quantity in the first pair of square brackets is called the finite population correction. It is 1 if N is "infinitely large" compared to n; then (14.7) is the same as the binomial variance. For the example above, the variance of the number of defectives in such a sample is $\mathrm{Var}(Y)=[(500-20)/(500-1)]\times20\times(50/500)[1-(50/500)]=1.731$. The corresponding binomial variance is $\mathrm{Var}(Y)=20(50/500)[1-(50/500)]=1.800$.

The hypergeometric standard deviation of the number Y of sample items with the characteristic is

$$\sigma(Y)=\{[(N-n)/(N-1)](D/N)[1-(D/N)]\}^{1/2}. \tag{14.8}$$

This is similar to the binomial standard deviation where $p=D/N$. The comments above on the hypergeometric variance also apply to the standard deviation. For the example above, $\sigma(Y)=(1.731)^{1/2}=1.32$ defectives.

15. MULTINOMIAL DISTRIBUTION

This specialized section presents the multinomial distribution. It is a model for the numbers of n sample observations that fall into each of M categories. The binomial distribution is a special case where $M=2$. The multinomial distribution is a model, for example, for (1) the numbers of sample units below specification, in specification, and above specification, (2) the numbers of sample people who give each of the M responses to a certain question, and (3) the numbers of units failing from each of M possible causes. Its assumptions are (1) each sample unit has the same chance of being in a particular category and (2) the outcomes of the sample units are statistically independent. The number of sample items falling in a particular category may range from zero to the entire number n of items in the sample.

The **multinomial probability function** for the numbers $Y_1, Y_2, \ldots, Y_M$ of sample items in categories $1, 2, \ldots, M$, respectively, is

$$f(y_1, y_2, \ldots, y_M) = \frac{n!}{y_1! y_2! \cdots y_M!} \pi_1^{y_1} \pi_2^{y_2} \cdots \pi_M^{y_M}. \tag{15.1}$$

Here the possible numbers $y_1, y_2, \ldots, y_M$ each have integer values from 0 to n and must satisfy $y_1 + y_2 + \cdots + y_M = n$; π_m is the population proportion in category m, $m = 1, \ldots, M$. The π_m must be between 0 and 1 and satisfy $\pi_1 + \pi_2 + \cdots + \pi_M = 1$.

The **mean** of the number Y_m of sample items from category m is

$$E(Y_m) = n\pi_m, \qquad m = 1, \ldots, M. \tag{15.2}$$

This is the same as the binomial mean (13.5) for just category m when all other categories are regarded as a single category.

The **variance** of the number Y_m of sample items from category m, $m = 1, \ldots, M$, is

$$\mathrm{Var}(Y_m) = n\pi_m(1 - \pi_m). \tag{15.3}$$

This is the same as the binomial variance (13.6) for just category m where all other categories are combined and called "not category m."

The **standard deviation** of the number Y_m of sample items from category m, $m = 1, \ldots, M$, is

$$\sigma(Y_m) = \left[n\pi_m(1 - \pi_m) \right]^{1/2}. \tag{15.4}$$

This is the same as the binomial standard deviation (13.7) for just category m.

The **covariance** of the numbers Y_m and $Y_{m'}$ of sample items from the categories m and m' is

$$\mathrm{Cov}(Y_m, Y_{m'}) = -n\pi_m\pi_{m'}. \tag{15.5}$$

A multivariate normal distribution (Section 10) with the means, variances, and covariances above can be used to calculate approximate multinomial probabilities of events for large n.

PROBLEMS

2.1.   Mortality table.   The following American experience mortality table gives the proportion living as a function of age, starting from age 10 in increments of 10 years.

Age	10	20	30	40	50	60	70	80	90	100
Living	1.000	.926	.854	.781	.698	.579	.386	.145	.008	.000

(a)   Calculate the percentage dying in each 10-year interval and plot the histogram.

(b)   Calculate the average lifespan of 10-year-olds. Use the midpoints of the intervals as the age at death.

(c)   Calculate the hazard function for each 10-year interval (as a percent **per year**) and plot it.

(d)   Draw the survivorship curve for 10-year-olds and obtain the 10, 50, and 90% points of the distribution of their lifespans. What proportion of 10-year-olds reach age 65?

(e)   Calculate and draw the survivorship curve for 30-year-olds and obtain the 10, 50, and 90% points of the distribution of their lifespans and their expected lifespan.

(f)   Repeat (e) for 50-year-olds. What proportion of 50-year-olds reach age 65?

(g)   Calculate and plot the hazard function for 50-year-olds and compare it with that for 10-year-olds. What general conclusion does this indicate?

2.2.   Weibull.   For the Weibull cumulative distribution function, $F(y) = 1 - \exp[-(y/\alpha)^\beta]$, $y > 0$, derive the following.

(a)   Probability density.
(b)   Hazard function.
(c)   Cumulative hazard function.
(d) $100F$th percentile.
(e)   Mean.
(f)   Standard deviation.

A Weibull distribution for engine fan life has $\alpha = 26,710$ hours and $\beta = 1.053$. Calculate the following.

(g)   Median life.
(h)   Mean life.

(i)   Most likely (modal) life.

(j)   Standard deviation of life.

(k)   Fraction failing on an 8000-hour warranty.

2.3. Log$_{10}$ normal.   For the lognormal cumulative distribution function, $F(y)=\Phi\{[\log_{10}(y)-\mu]/\sigma\}$, $y>0$, derive (a) through (f) in Problem 2.2.

A lognormal life distribution for a Class B electrical insulation at 170°C has $\mu=3.6381$ and $\sigma=0.2265$. Calculate (g) through (j) in Problem 2.2.

2.4. Log$_e$ normal.   The lognormal distribution with base e logs has the cumulative distribution function

$$F(y)=\Phi\{[\ln(y)-\mu']/\sigma'\}, \qquad 0<y<\infty,$$

where μ' is the mean and σ' is the standard deviation of $\log_e$ life. Obtain the formula for the following.

(a) $100P$th percentile.

(b)   Probability density in terms of the standard normal probability density $\varphi(\ )$.

(c)   Mode.

(d)   Mean.

(e)   Variance.

(f)   Hazard function.

(g)   Relationship between μ' and σ' and the μ and σ of the same lognormal distribution expressed with base 10 logs.

2.5. Logistic.   The logistic cumulative distribution function is

$$F(y)=1/\{1+\exp[-(y-\mu)/\sigma]\}, \qquad -\infty<y<\infty.$$

(a)   Determine the range of allowable values of the location and scale parameters μ and σ.

(b)   Verify that the function is a cumulative distribution function.

(c)   Plot the cumulative distribution function as a function of the standardized variable $z=(y-\mu)/\sigma$.

(d)   Give the formula for the probability density.

(e)   Plot the probability density as a function of the standardized variable z.

(f)   Give the formula for the hazard function.

(g)   Plot the hazard function as a function of the standardized variable z.

(h)   Give the expression for the $100F$th percentile of this distribution.

(i)   Give an expression for the distribution median in terms of μ and σ.

(j) . Give the probability of a unit with age equal to μ surviving an age $\mu+\sigma$.

(k)* Calculate the distribution mean.

(l)* Calculate the distribution variance.

(m)* Calculate the distribution standard deviation.

(n)   Find the cumulative distribution of $w=\exp(y)$, the log-logistic distribution.

(o)   Find the probability density of w.

2.6.*   Mixture of exponentials.   Suppose that a population contains a proportion p of units from an exponential life distribution with mean θ_1 and the remaining proportion $1-p$ from an exponential life distribution with mean θ_0. Proschan (1963) treats such a problem.

(a)   Derive the hazard function of the mixture distribution.

(b)   Show that the failure rate of the mixture distribution decreases with age.

2.7.   Exponential and Poisson prediction.   Fifty-eight fans in service come from an exponential distribution with a mean of 28,700 hours.

(a)   Predict the number of such fans that will fail in the next 2000 hours of service on each fan; assume that failed fans are replaced by a fan with a new design that does not fail.

(b)   Do (a), but assume that each failed fan is immediately replaced by a fan of the old design.

(c)   For (a), calculate a limit that is above the observed number of failures with 90% probability.

(d)   For (b), do (c).

2.8.   Binomial acceptance sampling.   For a binomial acceptance sampling plan with sample size $n=20$ and acceptance number $y=0$, calculate and plot the OC function. Do the same for $n=10$ and $y=0$. Which OC curve is preferable from the viewpoint of (1) the supplier and (2) the customer?

*Asterisk denotes laborious or difficult.

3

Probability Plotting of Complete and Singly Censored Data

Those who analyze data know that probability plots are very useful for getting information from data. This chapter explains how to make and use such plots for complete data (all units failed) and singly censored data (a common running time for unfailed units). To read this chapter, one needs to know the basic concepts and distributions in Sections 1 through 5 of Chapter 2. This introduction briefly states the advantages and disadvantages of probability plots and outlines this chapter.

Advantages

Probability plots are often preferred over the numerical analyses in later chapters because plots serve many purposes, which no single numerical method can. A plot has many advantages.

1. It is fast and simple to use. In contrast, numerical methods may be tedious to compute and may require analytic know-how or an expensive statistical consultant. Moreover, the added accuracy of numerical methods over plots often does not warrant the effort.

2. It presents data in an easy-to-grasp form. This helps one draw conclusions from data and also to present data to others.

3. It provides simple estimates for a distribution—its percentiles, parameters, nominal life, percentage failing on warranty, failure rate, and many other quantities.

4. It helps one assess how well a given theoretical distribution fits the data.

5. It applies to both complete and censored data.

6. It helps one spot unusual data. The peculiar appearance of a data plot or certain plotted points may reveal bad data or yield important insight when the cause is determined.

7. It lets one assess the assumptions of analytic methods applied to the data.

Limitations

Some limitations of a data plot in comparison to analytic methods are the following.

1. It is not objective. Two people using the same plot may obtain somewhat different estimates. But they usually come to the same conclusion, of course.

2. It does not provide confidence intervals (Chapter 6) or a statistical hypothesis test (Chapter 10). However, a plot is often conclusive, and leaves little need for such analytic results.

Usually a thorough analysis combines graphical and analytical methods.

Chapter Overview

Section 1 motivates probability plots. Sections 2 and 3 explain how to make and interpret such plots for complete data (each sample unit has a failure time). Sections 4, 5, 6, 7, and 8 explain plots for the exponential, normal, lognormal, Weibull, and extreme value distributions. Section 9 gives practical aids for all such plots. Section 10 presents such plots for singly censored data. Section 11 briefly discusses theory for constructing probability papers.

King (1971) comprehensively presents probability plotting for many distributions.

1. MOTIVATION

Data can usually be regarded as a sample from a population, as described in Chapter 1. For example, times to breakdown of an insulating fluid in a test are regarded as a random sample from the entire production of units containing the fluid. Similarly, breaking strengths of a sample of wire connections are regarded as a random sample from the entire production of wire connections. As described here, the sample cumulative distribution function is used to estimate the population cumulative distribution function.

Insulating fluid example. The motivation uses the data in Table 1.1, which contains times to breakdown of an insulating fluid between electrodes recorded at seven different voltages. The plotting positions in Table 1.1 are explained later. A test purpose was to assess whether time to breakdown at each voltage has an exponential distribution as predicted by theory. If appropriate, the distribution can be used to estimate the probability of fluid breakdown in actual use.

Table 1.1.   Times to Breakdown of an Insulating Fluid

26 kV		28 kV		30 kV		32 kV	
Min-utes	Plotting Position	Min-utes	Plotting Position	Min-utes	Plotting Position	Min-utes	Plotting Position
5.79	16.3	68.85	10.0	7.74	4.5	0.27	3.3
1579.52	50.0	108.29	30.0	17.05	13.6	0.40	10.0
2323.70	83.3	110.29	50.0	20.46	22.7	0.69	16.7
		426.07	70.0	21.02	31.8	0.79	23.3
		1067.60	90.0	22.66	40.9	2.75	30.0
				43.40	50.0	3.91	36.7
				47.30	59.1	9.88	43.3
				139.07	68.2	13.95	50.0
				144.12	77.3	15.93	56.7
				175.88	86.4	27.80	63.3
				194.90	95.5	53.24	70.0
						82.85	76.7
						89.29	83.3
						100.58	90.0
						215.10	96.7

34 kV		36 kV		38 kV	
Min-utes	Plotting Position	Min-utes	Plotting Position	Min-utes	Plotting Position
0.19	2.6	0.35	3.3	0.09	6.2
0.78	7.9	0.59	10.0	0.39	18.7
0.96	13.2	0.96	16.7	0.47	31.2
1.31	18.4	0.99	23.3	0.73	43.7
2.78	23.7	1.69	30.0	0.74	56.2
3.16	28.9	1.97	36.7	1.13	68.7
4.15	34.2	2.07	43.3	1.40	81.2
4.67	39.5	2.58	50.0	2.38	93.7
4.85	44.7	2.71	56.7		
6.50	50.0	2.90	63.3		
7.35	55.3	3.67	70.0		
8.01	60.5	3.99	76.7		
8.27	65.8	5.35	83.3		
12.06	71.1	13.77	90.0		
31.75	76.3	25.50	96.7		
32.52	81.6				
33.91	86.8				
36.71	92.1				
72.89	97.4				

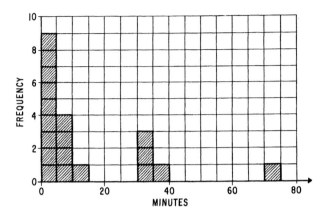

Figure 1.1.   Histogram of the 34-kV data.

Histogram.   Data are commonly plotted in a histogram, as shown in Figure 1.1 with the insulating fluid data at 34 kV. To make a histogram, divide a data axis into intervals (usually of equal length), and tally the number of observations in each interval with squares, X's, 1's, etc. A sample histogram corresponds to the probability density of a theoretical distribution. A histogram is satisfactory for moderate- to large-size samples of complete data. It is not as informative as a probability plot and does not apply to censored data. So probability plots are used here.

Cumulative distributions.   The value of the **population** cumulative distribution function (cdf) at a given time is the population fraction failing by that time. Similarly, the value of the **sample** cdf at a time is the sample fraction failing by that time. That is, if a sample has i of n observations below a particular time, then the sample cdf at that time is i/n, or $100(i/n)\%$. Figure 1.2 shows this sample staircase function for the 34-kV data. The sample cdf is an estimate of the population cdf.

Sample cdf.   Construct the sample cdf as follows.

1.   Order the n data values from smallest to largest, as in Table 1.1. The smallest has rank 1, the next larger has rank 2,...., and the largest has rank n.

2.   For the ith ranked data value, calculate its sample cdf as $100(i/n)\%$, $i=1,...,n$. This percentage is used to plot the sample cdf. Better plotting percentages for probability paper are given later.

3.   Plot each observation against its cdf plotting percentage, as shown in Figure 1.2, with a dot.

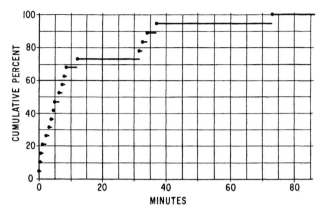

Figure 1.2. Sample cumulative distribution function of the 34-kV data.

4. Draw in the staircase for the sample cdf, if desired, as in Figure 1.2. For example, Figure 1.2 shows the 10th time (6.50 minutes) plotted against $100 \times 10/19 = 53\%$.

Estimate of the cdf. For most large populations, the population cdf contains many closely spaced steps, and it can be well approximated by a smooth curve. In contrast, a sample cdf usually has few points and a somewhat ragged appearance, as in Figure 1.2. To approximate the population cdf, one can draw a smooth curve through the sample cdf points by eye. This curve estimates the population cdf and provides information on the population.

Fit a theoretical distribution. Instead of a smooth curve, one can fit a particular theoretical cdf to a sample cdf. Then the theoretical curve estimates the population cdf. Chapter 2 describes such distributions and their properties. Such fitting is easy to do with a plot of a sample cdf on probability paper as described in this chapter. Chapters 6, 7, 8, and 9 present analytic methods for fitting a theoretical distribution to data.

Use of probability paper. On probability paper,* the data and cumulative probability scales are constructed so that any such theoretical cdf plots as a straight line as shown in Figure 2.1. One plots the sample cdf on the probability paper for a distribution, and draws a straight line through the

---

*Many probability, hazard, and other data analysis papers are offered by TEAM (Technical and Engineering Aids for Management), Box 25, Tamworth, NH 03886. Some probability papers are offered by the CODEX Book Co., Norwood, MA 02062 and Keuffel & Esser Co., Hoboken, NJ.

data. Like a smooth curve, the line estimates the population cdf. In addition to estimates of percentiles and percentages failed, a probability plot provides estimates of the parameters of the theoretical distribution fitted to the data. Details of how to make and interpret such plots follow.

2. HOW TO MAKE A PROBABILITY PLOT OF COMPLETE DATA

A probability plot of a **complete** sample of life data provides a variety of information. In particular, it provides graphical estimates of distribution parameters, percentiles, etc. Also, it provides a check on the validity of the data and an assumed distribution. To make a probability plot, do the following seven steps.

1. **Order the *n* failure times from smallest to largest.** Table 1.1 shows this for the 34-kV data.

2. **Assign a rank to each failure.** Give the earliest failure rank 1, the second failure rank 2, etc., and the last failure rank *n*.

3. **Calculate probability plotting positions F_i.** For the failure with rank *i*, the "midpoint" plotting position (corresponding sample cumulative percentage failed) is

$$F_i = 100(i - 0.5)/n, \qquad i = 1, \ldots, n. \tag{2.1}$$

Table 1.1 shows these for the 34-kV data. Appendix A9 is a table of these plotting positions. Section 9 motivates these plotting positions and describes other commonly used plotting positions.

4. **Plot the failure times on probability paper.** There are probability papers for many distributions, including the exponential, normal, lognormal, Weibull, extreme value, and chi-square distributions. The distribution should be chosen from experience or an understanding of the physical phenomena. For example, The Weibull distribution often describes the life and breakdown voltage of capacitors and cables, and the lognormal distribution often describes the life of electrical insulation. Label the data scale (vertical scale in Figure 2.1) to include the range of the data. Then plot each failure against its time on the data scale and against its plotting position on the cumulative probability scale (horizontal scale in Figure 2.1). Figure 2.1 shows the 34-kV data on exponential probability paper. The probability scale appears on the vertical axis on some papers, and on the horizontal axis on other papers.

5. **Assess the data and assumed distribution.** If the plotted points tend to follow a straight line, then the chosen distribution appears to be adequate. For the data in Figure 2.1, the exponential distribution might be

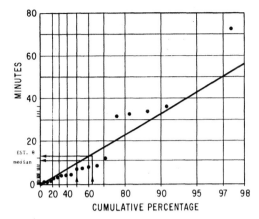

Figure 2.1.   Exponential probability plot of the 34-kV data.

questioned. Such an assessment is subjective, and two people analyzing the same data may get somewhat different results. Moreover, an adequate fit for one application may not be adequate for another. Analytic tests of fit for adequacy of distribution may be useful. Moderate deviations from linearity occur because of random sampling. A valid plot with many observations tends to follow a straight line. A curved plot indicates that the distribution does not fit the data; one should then try plotting the data on other paper.

An outlier point extremely out of line with the rest is suspect and might be discarded. But **its cause should be sought**, since that information may help improve the product or the data collection. Often plots with and without suspect points yield the same results for practical purposes. If not, one must make a difficult choice and may choose the more conservative results.

If the sample is small, the plot may be erratic and not follow a straight line very well, particularly in the tails. Only pronounced peculiarities should be interpreted as inadequate fit or data. Inexperienced analysts expect plots to be too orderly and straight.

6.   Draw a straight line through the plotted data.   Determine the line by eye to minimize the deviations between the line and the plotted points. For exponential paper, the fitted line must pass through the origin, as in Figure 2.1. The line estimates the cumulative distribution function—the relationship between the percentage failing and time. Sometimes a straight line does not fit the data well enough on any probability paper. Then one might fit a curve and use it as explained later. Depending on how the line will be used, it can be fitted to the whole sample, the center of the data, the lower tail, or whatever is appropriate, ignoring or taking into account outliers. Also, the line can be fitted to yield conservative results.

7. Obtain the desired information. Methods for obtaining information from such plots follow. Section 3 below explains how to estimate percentiles and probabilities of failure. Methods for estimating distribution parameters depend on the distribution; Sections 4 through 8 give such methods for each basic distribution. Section 9 gives aids for making and interpreting plots. The examples give insights on how to use and interpret such plots.

3. PERCENTILES AND PROBABILITIES

The following estimates of percentiles and percentages failing apply whether a straight line or a curve is fitted to a plot.

Percentiles. A graphical estimate of a percentile is obtained as follows. Enter the probability scale at the desired percentage. Go (vertically in Figure 2.1) to the fitted line, and then go (horizontally in Figure 2.1) to the corresponding point on the data scale to read the percentile estimate. For example, from Figure 2.1, this estimate of the 50th percentile is 10 minutes.

Percentage failing. An estimate of the percentage failing by a given time is obtained as follows. Enter the plot on the data scale at that time. Go (horizontally in Figure 2.1) to the fitted line, and then go (vertically in Figure 2.1) to the corresponding point on the probability scale to read the estimate. For example, from Figure 2.1, 20% fail by five minutes.

4. EXPONENTIAL PROBABILITY PLOT

This section explains exponential probability plots and, particularly, how to estimate the distribution parameters. The insulating fluid data are the example.

The mean of an exponential distribution is the 63rd percentile. For example, the graphical estimate of the mean (63rd percentile) from Figure 2.1 is 14 minutes.

The failure rate of an exponential distribution is the reciprocal of the mean. For example, this estimate from Figure 2.1 is $1/14 = 0.07$ failures per minute.

The plotted points do not follow a straight line very well. So the exponential fit is crude. The sample is small (19 observations), so it reveals only large departures from an exponential distribution. A more sensitive evaluation of an exponential fit is obtained from a Weibull plot, as described in Section 7. A Weibull plot generally shows more than an exponential plot, particularly in the lower tail of the distribution, which is usually of greatest interest.

5.   NORMAL PROBABILITY PLOT

This section describes normal probability plots. It presents an example and methods for estimating the distribution mean and standard deviation.

Connection strength example.   Table 5.1 shows breaking strengths of 23 wire connections from King (1971). The wires are bonded at one end to a semiconductor wafer and at the other end to a terminal post. Table 5.1 shows whether the wire or a bond failed. Engineering wanted to know if such connections meet the specification that no more than 1% of the strengths be below 500 mg.

Figure 5.1 is a normal probability plot of the data made by a computer. The computer program uses hazard plotting positions, described in Chapter 4. The plot suggests that over 10% of the strengths are below 500 mg — much over 1%.

The plot shows that the 3150 value and the two 0 values are out of line with the others and are therefore suspect. The suspect 3150 value was discussed in detail by Nelson (1972b). The 0 values are bond failures, presumably from bonds that were not made. This information suggested that engineering should find the reason for such bonds and eliminate them.

Table 5.1.   Connection Strength Data

Breaking Strength	Type of Break	Breaking Strength	Type of Break
0	Bond	1250	Bond
0	Bond	1350	Wire
550	Bond	1450	Bond
750	Wire	1450	Bond
950	Bond	1450	Wire
950	Wire	1550	Bond
1150	Wire	1550	Wire
1150	Bond	1550	Wire
1150	Bond	1850	Wire
1150	Wire	2050	Bond
1150	Wire	3150	Bond
1250	Bond		

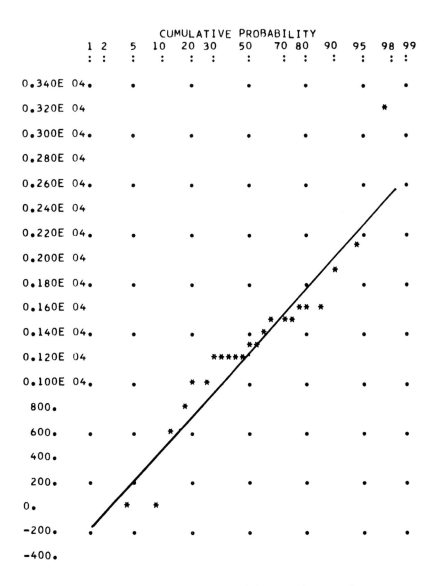

Figure 5.1. Normal probability plot of the connection strengths.

A new plot without the two 0 values indicates that the specification is still not met, even when the no-bonding problem is eliminated. So one must separately look at the wire and bond strength distributions to determine which needs improvement. Chapter 5 describes such analyses. A plot without the 3150 value would also be useful. Because the sample is small, excluding two or three values changes the plot appreciably for practical purposes.

Although the range of the normal distribution includes negative values, normal paper usefully displays the strength data. The distribution fits the data well enough for practical purposes.

Normal mean.   The mean μ of a normal distribution equals the median (50th percentile). From Figure 5.1, this estimate is 1250 mg.

Normal standard deviation.   The slope of the fitted line in a normal plot corresponds to the standard deviation σ. To estimate σ, estimate the mean and the 16th percentile. The estimate of σ is the difference between the mean and the 16th percentile. From Figure 5.1, the estimate of the 16th percentile is 610 mg. The estimate of σ is $1250 - 610 = 640$ mg.

6. LOGNORMAL PROBABILITY PLOT

This section describes lognormal probability plots. It presents an example and estimates of distribution parameters.

Class H insulation example.   The data in Table 6.1 consist of the times to failure of specimens of a new Class H electrical insulation at temperatures

Table 6.1.   Class-H Insulation Life Data

Hours to Failure				Plotting Position
190°	220°	240°	260°	
7228	1764	1175	600	5%
7228	2436	1175	744	15
7228	2436	1521	744	25
8448	2436	1569	744	35
9167	2436	1617	912	45
9167	2436	1665	1228	55
9167	3108	1665	1320	65
9167	3108	1713	1464	75
10511	3108	1761	1608	85
10511	3108	1953	1896	95

of 190, 220, 240, and 260°C. Some failure times are equal, because specimens were inspected periodically and a given time is the midpoint of the period when the failure occurred. The purpose of the experiment was to estimate insulation life at 180°C where a nominal life of 20,000 hours was desired. The distribution line for 180°C in Figure 6.1 was estimated as described by Nelson (1971). The purpose of plotting the data is to assess the lognormal fit to the data and to check for odd data.

The plotting positions appear in Table 6.1. The data for each temperature are plotted on lognormal paper in Figure 6.1. Experience has shown that this distribution usually fits such data. The plotted points for each temperature tend to follow a straight line, so the lognormal distribution appears adequate. For small samples, such plots may appear erratic.

Relation to normal paper. Lognormal and normal probability paper have the same cumulative probability scale. However, lognormal paper has a log data scale, whereas normal paper has a linear one. Thus a normal probability plot of the logs of lognormal data has the same appearance as a log normal plot of the original data. In this way, one can make a lognormal plot when lognormal paper is not available.

Parameter μ. The parameter μ is the log of the median (50th percentile). From Figure 6.1, the estimate of median insulation life at 180°C is 11,500 hours, well below the desired 20,000 hours. The estimate of μ is $\log(11,500) = 4.061$. One usually uses the median, a typical life, rather than μ.

Parameter σ. The slope of the fitted line in a lognormal plot corresponds to the log standard deviation σ. Estimate it as follows. After

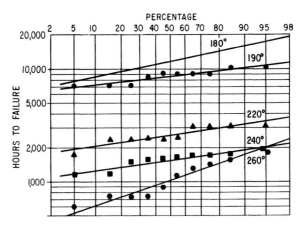

Figure 6.1.   Lognormal plot of Class-H data.

estimating the log mean, estimate the 16th percentile. The estimate of σ equals the difference between the μ estimate and the log of the 16th percentile. From Figure 6.1, the estimate of the 16th percentile at 180°C is 9000 hours. The estimate of σ is $4.061 - \log(9000) = 0.11$. This small value indicates that the distribution is close to normal; so the insulation failure rate increases with age (wear-out behavior).

Arrhenius model. The Arrhenius model (Nelson, 1971) for such insulation data assumes that (1) life has a lognormal distributions at each temperature, and (2) each distribution has the same σ, i.e., the true distribution lines should be parallel in a lognormal plot. The fitted lines in Figure 6.1 are not parallel, owing to random variation or unequal true σ's. In particular, the slope of the 260°C data is greater than the other slopes. This suggests that the failure mode at 260°C may differ from that at the other test temperatures. See problem 5.2 of Chapter 5.

7. WEIBULL PROBABILITY PLOT

This section describes Weibull probability plots—how to estimate the Weibull parameters and how to fit an exponential distribution to a Weibull plot. Nelson and Thompson (1971) describe many available Weibull probability papers.

Insulating fluid example. Table 1.1 shows data on time to breakdown of an insulating fluid. Such data from selected voltages are plotted on Weibull paper in Figure 7.1.

Scale parameter. The scale parameter α of a Weibull distribution is the 63rd percentile. For example, from Figure 7.1, this estimate for 34 kV is eight minutes.

Shape parameter. The slope of the fitted line in a Weibull plot corresponds to the shape parameter β. To estimate β, use the point labeled "origin" and the shape parameter scale. Draw a line passing through the origin and parallel to the fitted line, as in Figure 7.1. The value where this line intersects the shape parameter scale is the estimate of β. For the 34-kV data, the estimate is 0.88. Some Weibull papers require that the line through the origin be drawn perpendicular to the fitted line.

Exponential fit. The Weibull distribution with a shape parameter of 1 is the exponential distribution. Weibull paper displays data in the lower tail better than exponential paper does. Interest usually focuses on this tail (early failures). If the exponential distribution does not fit early data, a Weibull plot shows this better than does an exponential plot. In Figure 7.1,

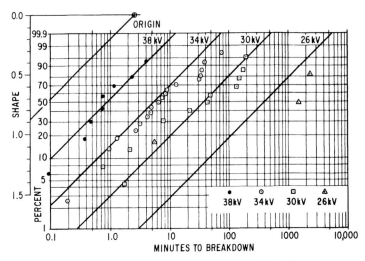

Figure 7.1. Weibull probability plot of the insulating fluid data.

each plot is relatively straight and has a shape parameter near unity. Thus the exponential distribution appears to describe the data at each voltage. There is one suspect data point—the smallest value at 26 kV.

Theory for such fluid assumes that the shape parameter has the same value at each voltage, and thus the slopes of the plots should be equal. The slopes look about equal when random variation in the data is taken into account. Parallel lines were fitted to the data in Figure 7.1. This can be done by eye or by the analytic methods of Chapters 11 and 12.

An exponential distribution can be fitted to data on a Weibull plot as follows. Draw a line from the origin to pass through the shape parameter scale at unity. Then fit through the data a line that is parallel to the first line. This fitted line is an exponential distribution. The estimate of the exponential mean is the 63rd percentile from the fitted line.

8. EXTREME VALUE PROBABILITY PLOT

This section describes extreme value probability plots. It presents an example and estimates of the distribution parameters.

Residuals example. After fitting a statistical model to data, one can assess the validity of the model and data by examining the log residuals. For the insulating fluid data, theory assumes that such residuals come from an extreme value distribution as described by Nelson (1970). As a check on this, the residuals are plotted on extreme value paper in Figure 8.1. The

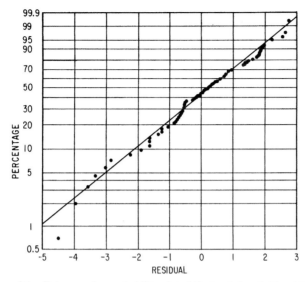

Figure 8.1.   Extreme value probability plot of the insulating fluid residuals.

points follow a straight line well, and there are no peculiar points. So the distribution and data look satisfactory.

Location parameter.   The location parameter λ of an extreme value distribution is the 63rd percentile. For the example, this estimate from Figure 8.1 is 0.7.

Scale parameter.   The slope of the fitted line in an extreme value plot corresponds to the scale parameter, δ. The estimate of δ is the difference between the estimates of the location parameter and the 31st percentile. For the residuals, the estimate of the 31st percentile is -0.5, and the estimate of δ is $0.7-(-0.5)=1.2$.

9.   AIDS FOR DATA PLOTTING

The following discussion of data plotting aids includes references, plotting positions, computer plots, choice of distribution, a shortcut, nonparametric fit, fitting the line, failure rate behavior, coarse data, plotting of selected points, and extended data scales.

References.   Probability plotting appears in few statistical texts. Many texts emphasize mathematical theory and overlook plotting, a simple and valuable tool for data analysis. King (1971) explains a great variety of plotting methods for practical applications.

Plotting positions. The F_i are motivated as follows. The smallest of n observations represents the first $(100/n)\%$ of the population, that is, the population between 0 and $(100/n)\%$. The midpoint of this interval is the first plotting position $[100(1-0.5)/n]\%$. Similarly, the second smallest of n observations represents the second $(100/n)\%$ of the population, that is, the population between $(100/n)\%$ and $[100(2/n)]\%$. The midpoint of this interval is the second plotting position $[100(2-0.5)/n]\%$. This pattern continues through the nth (the largest) observation, which represents the last $(100/n)\%$ of the population, that is, that between $[100(n-1)/n]\%$ and $[100(n/n)]\%$. The midpoint of this interval is the nth plotting position $[100(n-0.5)/n]\%$.

Different plotting positions have been zealously advanced. In general, the ith plotting position is a "typical" population percentage near to which the ith ordered observation falls. The "mean" plotting position is popular and is

$$F_i' = 100i/(n+1), \qquad i=1,\ldots,n.$$

King (1971) tabulates F_i'; they are the expected (mean) percentage of the sample below the ith ordered observation. Johnson (1964) advocates and tabulates median plotting positions, well approximated by

$$F_i'' \approx 100(i-0.3)/(n+0.4),$$

as shown by Benard and Bos-Levenbach (1953). Also, some authors advocate the expected values of the order statistics of the standardized distribution as plotting positions. Some advocate plotting positions that yield "best" estimates of the distribution parameters when a straight line is fitted to the plotted data by least squares, for example, Chernoff and Lieberman (1954); however, in life data analysis, one is usually more interested in estimating low percentiles rather than parameters. Section 5 of Chapter 7 provides theory for plotting positions.

In practice, plotting positions differ little compared with the randomness of the data. One could use F_i or F_i', depending on which is easier to calculate mentally. One should consistently use one kind in comparing different samples.

Computer plots. Most computer packages now offer probability plots for complete data (STATSYSTEM, BMDP, SAS, and OMNITAB). Some plot singly and multiply censored data, for example, STATPAC. Such packages reduce the labor of plotting. Hahn, Nelson, and Cillay (1975) describe STATSYSTEM. Nelson and others (1978) describe STATPAC. Dixon and Brown (1977) describe BMDP.

Choice of distribution.   If experience does not suggest a distribution for a set of data, try different plotting papers and determine the one that gives the straightest plot, particularly in the region of interest (usually the lower tail).

If a probability plot of data significantly bows up or down from a straight line, replot the data on other paper. Curvature can best be judged by laying a transparent straight edge along the points. In an exponential plot, a failure rate that increases and then decreases suggests the use of lognormal paper. If that lognormal plot is not straight, the units may have two or more failure modes (Chapter 5). Further interpretation of curved plots is given by King (1971) and Nelson (1979).

A smooth curve through an exponential plot may clearly pass through the time axis above zero. A possible explanation is that the distribution may have no failures before a minimum time. For example, time to product failure is figured from the data of manufacture, and there may be a minimum time to get a unit into service. An estimate of this time is the point where the smooth curve passes through the time axis. Subtract the minimum time from each time to failure, and plot the differences on probability paper. The distribution is a shifted one (Section 8 of Chapter 2).

Shortcut.   Ordering the data from smallest to largest consumes time, particularly for large samples. A shortcut avoids this. In any order, read each observation and put a tick mark on the data scale as shown in Figure 2.1. When done, move over from the tick mark for the smallest observation and plot it as a dot at its plotting position (read from Appendix A9). Do this for the second smallest observation, third smallest, etc., while working through the table of plotting positions.

Nonparametric fit.   A set of data may not plot as a straight line on available papers. Then one may draw a smooth curve through a plot and use the curve to estimate distribution percentiles and probabilities of failure. Such a curve is usually adequate within the data. Extrapolation of the curve beyond the data is subject to (possibly much) error. For such a fit that emphasizes early times to failure, plot the data on Weibull paper and draw a smooth curve through the data. Such a fit is called **nonparametric** because no particular mathematical form of the fitted distribution is assumed.

Fitting the line.   To estimate distribution parameters, fit the straight line to the data points near the center of the sample. This improves estimates if there are peculiar extreme smallest or largest sample values. Such peculiar values fall far from the true distribution line, and the middle values tend to lie close to it.

On the other hand, if interested in the lower tail of a distribution, fit a straight line to the data in the lower tail, particularly if that line differs from one through the rest of the data.

Failure rate behavior.   Whether the failure rate increases or decreases with age is often a key concern. For an exponential plot, the reciprocal of the slope of a curve through the plotted data is the instantaneous failure rate. Chapter 4 gives the basis for this.

Coarse data.   Failure times are typically recorded to the nearest hour, day, month, 100 miles, etc. That is, the data are rounded. For example, if a unit is inspected periodically, then one knows only the period in which the unit failed. For plotting such data, the intervals should be small, say, less than one-fifth of the standard deviation. If some observations have the same value, the data are coarse; a plot has a flat spot as in Figure 6.1. Then estimates of parameters, percentiles, and failure probabilities can be somewhat crude.

One can plot equal failure times at equally spaced times over the corresponding time interval. For example, suppose there are five failures at 1000 hours, all between 950 and 1050 hours. The equally spaced plotting times for the five failures are 960, 980, 1000, 1020, and 1040 hours. This smooths the steps out of the plot and tends to make the plot and estimates more accurate.

Chapter 9 gives numerical analyses for such coarse (or **interval**) data.

Plot selected points.   When a data set has many failure times, one need not plot them all. This reduces the work and the clutter of data points on the plot. One might plot every kth ordered point, all points in a tail of the sample, or only some points near the center of the sample. Choose the points to be plotted according to how the plot will be used. For example, if interested in the lower tail of the distribution, plot all the early failures. Often there are so few failures that all should be plotted. When plotting selected points, use the entire sample to calculate plotting positions.

Extended data scales.   For some data, the ranges of the scales of the available papers are not large enough. To extend the data scale of any paper, join two or more pages together.

10.   PROBABILITY PLOTS OF SINGLY CENSORED DATA

Introduction.   Often life data are censored, because life data are analyzed before all sample units run to failure. Data are **singly censored** if all failure times are before a single censoring time.

The method. The method for probability plotting singly censored life data is like that for complete data. In particular, plot the ith ordered observation against the plotting position $F_i = 100(i-0.5)/n$ or $F_i' = 100i/(n+1)$. Here n is the total sample size **including the nonfailures**. Nonfailures are not plotted, since their failure times are unknown. Only the early failure times are observed, and they estimate the lower part of the life distribution, usually of greatest interest.

Sometimes one estimates the lower or upper tail of a distribution from a singly censored sample by extending a straight line beyond the plotted points. The accuracy of such extrapolation depends on how well the theoretical distribution describes the true one into the extrapolated tail.

The following examples illustrate how to plot and interpret singly censored samples. Plotting and interpreting are the same as for complete samples.

Appliance cord example. Electric cords for a small appliance are flexed by a test machine until failure. The test simulates actual use, but highly accelerated. Each week, 12 cords go on the machine and run a week. After a week, the unfailed cords come off test to make room for a new sample of cords. Table 10.1 shows data on (1) the standard cord and (2) a new cheaper

Table 10.1. Appliance Cord Data

TYPE B6			TYPE B7		
Hours	Rank i	100i/(n+1)	Hours	Rank i	100i/(n+1)
57.5	1	4	72.4	1	7.7
77.8	2	8	78.6	2	15.4
88.0	3	12	81.2	3	23.4
96.9	4	16	94.0	4	30.8
98.4	5	20	120.1	5	38.4
100.3	6	24	126.3	6	46.1
100.8	7	28	127.2	7	53.8
102.1	8	32	128.7	8	61.5
103.3	9	36	141.9	9	69.2
103.4	10	40	164.1+	10	X
105.3	11	44	164.1+	11	X
105.4	12	48	164.1+	12	X
122.6	13	52			
139.3	14	56			
143.9	15	60			
148.0	16	64			
151.3	17	68			
161.1+	18	X			
161.2+	19	X			
161.2+	20	X			
162.4+	21	X			
162.7+	22	X			
163.1+	23	X			
176.8+	24	X			

X REMOVED FROM TEST BEFORE FAILURE.

cord. A "+" marks the running time on each unfailed cord. The basic question is, "How do the lives of the two types of cords compare on test?" For cord type B6, 17 of the 24 cords failed. Plotting positions appear in Table 10.1. The two samples are plotted on normal paper in Figure 10.1. Normal paper was chosen after the data were plotted on several papers. It yields a reasonably straight plot, and the distribution was familiar to the engineers. Plots for the two cords roughly coincide, whatever the paper used; thus the life of new cord is comparable to that of standard cord for engineering purposes. The B6 data show a gap between 105 and 140 hours, roughly over the weekend. No reason for this gap was found, but it does not affect the conclusion. Straight lines through the two samples would estimate the distributions, but they are not needed to answer the basic question.

Class B insulation example.   To test a new Class B electrical insulation for electric motors, 10 motorettes were run at each of four temperatures (150, 170, 190, and 220°C). The main purpose of the test was to estimate the median life at the design temperature of 130°C. When the data were analyzed, there were seven failures at 170°C, five each at 190 and 220°C, and none at 150°C. Such motorettes are periodically inspected for failure, and Table 10.2 records the midpoint of the period in which a failure occurred (Crawford, 1970).

Experience indicates that the lognormal distribution describes such insulation life. The plots provide (1) a check on the assumption of a lognormal

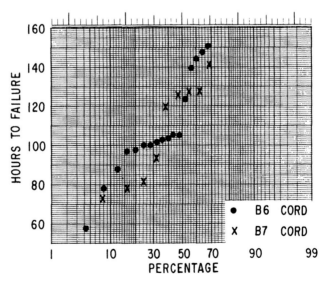

Figure 10.1.   Appliance cord data.

Table 10.2. Class-B Insulation Life Data

$\underline{150^{o}C}$ all 10 motorettes still on test without failure at 8064 hours.

$\underline{170^{o}C}$ Hours to Failure	Plotting Position $\underline{100(1-0.5)/n}$	$\underline{190^{o}C}$ Hours to Failure	Plotting Position $\underline{100(1-0.5)/n}$	$\underline{220^{o}C}$ Hours to Failure	Plotting Position $\underline{100(1-0.5)/n}$
1764	5	408	5	408	5
2772	15	408	15	408	15
3444	25	1344	25	504	25
3542	35	1344	35	504	35
3780	45	1440	45	504	45
4860	55	1680 +	—	528 +	—
5196	65	1680 +	—	528 +	—
5448 +	-	1680 +	—	528 +	—
5448 +	-	1680 +	—	528 +	—
5448 +	-	1680 +	—	528 +	—

distribution, (2) a check for suspect data, and (3) estimates of the median lives at the test temperatures. The medians are used as explained by Hahn and Nelson (1971) to estimate the line for 130°C, the design temperature.

Table 10.2 shows the plotting positions, and Figure 10.2 shows lognormal plots with lines fitted to the data at each temperature. The two earliest failures at 190°C appear early compared to the other data. Otherwise the plots are reasonably straight; so the lognormal distribution appears satisfactory. The experiment was reviewed to seek a cause of the early failures, but none was found. Analyses yield the same conclusions whether or not those failures are included. The estimates of the medians are X's in the plots.

Left censored data. Table 10.3 shows data on time to breakdown of an insulating fluid tested at constant voltage stresses. These data are similar to those in Section 1. The test purpose was to estimate the relationship between time to breakdown and voltage. A probability plot of the data checks the assumed Weibull distribution and the validity of the data. Some times to breakdown at 45 kV occurred too early to be recorded—before the times labeled " − "; such data are said to be **singly censored on the left**.

Include the left censored data in determining the sample size and the ranks of the failures and their plotting positions. For 45 kV there are three times censored on the left among the sample of 12. For example, the three-second time has rank 6 and its plotting position is $100(6-0.5)/12=$

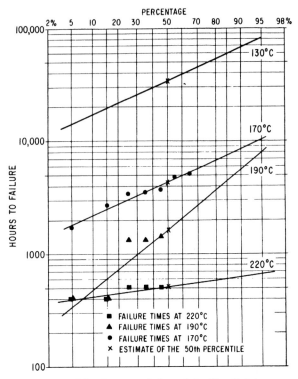

Figure 10.2. Lognormal plots of the Class-B data.

Table 10.3. Insulating Fluid Times to Breakdown with Censoring

TIME TO BREAKDOWN (SECONDS)					Plotting
45 kV	40 kV	35 kV	30 kV	25 kV	Position
1-	1	30	50	521	4.2
1-	1	33	134	2,517	12.5
1-	2	41	187	4,056	20.8
2	3	87	882	12,553	29.1
2	12	93	1,448	40,290	37.4
3	25	98	1,468	50,560+	45.7
9	46	116	2,290	52,900+	54.3
13	56	258	2,932	67,270+	62.6
47	68	461	4,138	83,990 *	70.9
50	109	1182	15,750	85,500+	79.2
55	323	1350	29,180+	85,700+	87.5
71	417	1495	86,100+	86,420+	95.8

- denotes left censored (failure occurred earlier).

+ denotes right censored (unfailed).

* unplotted failure.

124

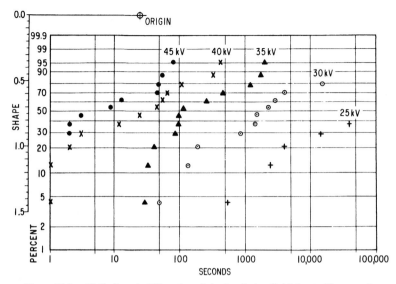

Figure 10.3. Weibull probability plot of the insulating fluid data with censoring.

45.7%. The data are plotted on Weibull paper in Figure 10.3. One can examine the plots to check the data and the fit of the Weibull distribution, using the methods for complete samples. The censored times are not plotted, since the failure times are not known, but the censored times are used to determine the plotting positions of the failures. Also, samples may contain data censored on the left and data censored on the right. Then only the observed failures in the middle of the sample are plotted. For 25 kV, the failure at 83,990 seconds requires a special plotting position, as described in Section 2 of Chapter 4.

11. THEORY FOR PROBABILITY PAPERS

This **advanced** section presents theory for probability papers for the exponential, normal, lognormal, Weibull, and smallest extreme value distributions. This section explains how to determine the probability and data scales so that a theoretical cumulative distribution function (cdf) is a straight line on the paper. Such theory is useful only to those who desire a deeper understanding or want to develop new probability papers.

Exponential distribution.   The exponential cdf is

$$F(y) = 1 - \exp(-y/\theta), \qquad y \geq 0, \tag{11.1}$$

where θ is the mean. Exponential paper is based on (11.1) rewritten as

$$y_F = -\theta \ln(1-F). \tag{11.2}$$

This is a linear relationship between the $100F$th percentile y_F and the function $-\ln(1-F)$. Therefore the data scale is linear, and the probability scale has the value F located at $-\ln(1-F)$. For example, $F=0$ is at $-\ln(1-0)=0$, $F=0.10$ at $-\ln(1-0.10)=0.105$, $F=0.632$ at $-\ln(1-0.632) \simeq 1$, and $F=0.99$ at $-\ln(1-0.99)=4.60$. The probability scale for F is a reversed log scale. Also, (11.2) shows that θ determines the slope of that linear relationship and θ is the $F=100(1-e^{-1}) \simeq 63$rd percentile. Exponential paper appears in Figure 2.1.

Normal distribution.   The normal cdf is

$$F(y)=\Phi[(y-\mu)/\sigma], \qquad -\infty<y<\infty, \tag{11.3}$$

where $\Phi(\ )$ is the standard normal cdf, μ is the mean, and σ is the standard deviation. Normal paper is based on (11.3) rewritten as

$$y_F=\mu+\sigma\Phi^{-1}(F), \tag{11.4}$$

where $\Phi^{-1}(F)$ is the inverse of the standard normal cdf and is its $100F$th percentile z_F. This is a linear relationship between the $100F$th percentile y_F and the function $z_F=\Phi^{-1}(F)$. Thus the data scale is linear, and the probability scale has the value F located at $z_F=\Phi^{-1}(F)$. For example, the 50% point is at $z_{.50}=0$, the 84% point at $z_{.84} \approx 1$, and the 97.5% point at $z_{.975} \approx 2$. Also, (11.4) shows that σ determines the slope of that linear relationship. Normal paper appears in Figure 5.1. Some normal papers have a linear scale labeled "normal deviates" or "probits" (top of Figure 10.1). The $100F$th percentage point is located at $z_F=\Phi^{-1}(F)$ on the probit scale.

When normal paper is not available, one can plot the ith ordered observation against $\Phi^{-1}(F_i)$ on a linear scale where $100F_i$ is its plotting position. Employed by computer programs, this method also applies to many other distributions when probability paper is not available.

Lognormal distribution.   The lognormal cdf is

$$F(y)=\Phi[(\log(y)-\mu)/\sigma], \qquad y>0, \tag{11.5}$$

where $\Phi(\ )$ is the standard normal cdf, μ is the log mean, σ is the log standard deviation, and $\log(\ )$ is a base 10 log. Lognormal paper is based on (11.5) rewritten as

$$\log(y_F) = \mu + \sigma \Phi^{-1}(F), \tag{11.6}$$

where $\Phi^{-1}(F)$ is the inverse of the standard normal cdf and is its $100F$th percentile z_F. This is a linear relationship between the $\log(y_F)$ of the $100F$th percentile and the function $z_F = \Phi^{-1}(F)$. The data scale is logarithmic, and the probability scale has the value F located at $z_F = \Phi^{-1}(F)$. Also, (11.6) shows that σ determines the slope of that linear relationship. Lognormal paper appears in Figure 6.1.

Normal and lognormal probability papers have the same probability scale. However, normal paper has a linear data scale, and lognormal paper has a log data scale. This means that the logs of lognormal data can be plotted on normal paper.

Weibull distribution. The Weibull cdf is

$$F(y) = 1 - \exp\left[-(y/\alpha)^\beta\right], \qquad y > 0, \tag{11.7}$$

where β is the shape parameter and α the scale parameter. Weibull paper is based on (11.7) rewritten as

$$\log(y_F) = \log(\alpha) + (1/\beta)\log\left[-\ln(1-F)\right]. \tag{11.8}$$

This is a linear relationship between $\log(y_F)$ of the $100F$th percentile and the function $\log[-\ln(1-F)]$. Therefore, the data scale is logarithmic, and the probability scale has the value F located at $\log[-\ln(1-F)]$. For example, $F = 0.01$ is at $\log[-\ln(1-0.01)] \cong -2$, $F = 0.632$ at $\log[-\ln(1-0.632)] \cong 0$, and $F = 0.99$ at $\log[-\ln(1-0.99)] \cong 0.66$. Also (11.8) shows that β determines the slope of that linear relationship. Weibull paper appears in Figure 7.1.

Extreme value distribution. The smallest extreme value cdf is

$$F(y) = 1 - \exp\left\{-\exp\left[(y-\lambda)/\delta\right]\right\}, \qquad -\infty < y < \infty, \tag{11.9}$$

where λ is the location parameter and δ the scale parameter. Extreme value paper is based on (11.9) rewritten as

$$y_F = \lambda + \delta \ln\left[-\ln(1-F)\right]. \tag{11.10}$$

This is a linear relationship between the $100F$th percentile y_F and the function $\ln[-\ln(1-F)]$. Thus, on extreme value paper, the data scale is linear, and the probability scale has the value F located at $\ln[-\ln(1-F)]$. Also (11.10) shows that δ determines the slope of that linear relationship. Extreme value paper appears in Figure 8.1.

Extreme value and Weibull papers have the same probability scale aside from a multiplicative factor that converts from natural to common logs. Extreme value paper has a linear data scale, and Weibull paper has a log data scale. Thus, the Weibull distribution could be called the log extreme value distribution.

PROBLEMS

3.1. Insulations. Specimen lives (in hours) of three electrical insulations at three test temperatures appear below.

(a) On separate lognormal probability paper for each insulation, plot the data from the three test temperatures.

(b) Are there any pronounced peculiarities in the data?

(c) How do the three insulations compare at 200°C, the usual operating temperature, with respect to the median and spread in life?

(d) How do the three compare at 225 and 250°C, occasional operating temperatures?

(e) How do the three compare overall, and are any differences convincing to you?

Insulation 1			Insulation 2			Insulation 3		
200°C	225°C	250°C	200°C	225°C	250°C	200°C	225°C	250°C
1176	624	204	2520	816	300	3528	720	252
1512	624	228	2856	912	324	3528	1296	300
1512	624	252	3192	1296	372	3528	1488	324
1512	816	300	3192	1392	372			
3528	1296	324	3528	1488	444			

(f) On a lognormal probability paper make three plots of the data for insulation 2 at 225°C, using the (1) "midpoint", (2) "mean", and (3) median plotting positions. (1) How do the graphical estimates of the distribution median, 1% point, and σ compare for the three plotting positions? (2) Which yields the most conservative (pessimistic) and optimistic estimates for reliability purposes? (3) Do the differences in the estimates look large compared to the uncertainties in the estimates?

3.2. Alarm clock.   12 alarm clocks yielded 11 failure times (in months), 30.5, 33, 33, 36, 42, 55, 55.5, 76, 76, 106, 106, and one survival time 107.5. Make a Weibull probability plot.

(a)   Does the Weibull distribution adequately fit the data?

(b)   Fit a line to the data and obtain estimates of the shape and scale parameters.

(c)   Comment on the nature of the failure rate (increasing or decreasing with age).

(d)   Estimate the median life. Is this estimate sensitive to the Weibull assumption?

3.3. Insulating fluid.   Table 2.1 of Chapter 7 shows samples of times to breakdown of an insulating fluid at five test conditions. According to engineering theory, these distributions are exponential.

(a)   Separately plot each of the five samples on the same sheet of Weibull probability paper.

(b)   Graphically estimate the five shape parameters.

(c)   Do the shape parameters appear comparable, subjectively taking into account the randomness in the data?

(d)   Do exponential distributions adequately fit the data?

(e)   Are there any peculiarities or other noteworthy features of the data?

3.4. Circuit breaker.   A mechanical life test of 18 circuit breakers of a new design was run to estimate the percentage failed by 10,000 cycles of operation. Breakers were inspected on a schedule, and it is known only that a failure occurred between certain inspections as shown.

1000 cycles	10–15	15–17.5	17.5–20	20–25	25–30	30+
Number of failures	2	3	1	1	2	9 survive

(a)   Make a Weibull plot with each failure as a separate point.

(b)   How well does the Weibull distribution appear to fit the data?

(c)   Graphically estimate the percentage failing by 10,000 cycles.

(d)   What is your subjective estimate of the uncertainty in the estimate?

(e)   Graphically estimate the Weibull shape parameter. Does the plot convince you that the true shape parameter differs from unity?

(f) The old breaker design had about 50% failure by 10,000 cycles. Is the new design clearly better?

(g) Another sample of 18 breakers of the new design was assembled under different conditions. These breakers were run on test for 15,000 cycles without failure and removed from test. Do you subjectively judge the two samples consistent?

3.5.* The logistic cumulative distribution function is

$$F(y)=1/\{1+\exp[-(y-\mu)/\sigma]\}, \qquad -\infty<y<\infty.$$

(a) Give the expression for its $100F$th percentile.

(b) Make probability paper for this distribution. Show lines on the probability scale for 1, 2, 5, 10, 20, 50, 80, 90, 95, 98, and 99% and show the calculations.

(c) Explain how to estimate the parameters μ and σ from a plot.

(d) Plot the data from Problem 3.2 on the paper.

---

*Asterisk denotes laborious or difficult.

4

Graphical Analysis of Multiply Censored Data

Data plots are used for display and interpretation of data because they are simple and effective. Such plots are widely used to analyze field and life test data on products consisting of electronic and mechanical parts, ranging from small electrical appliances through heavy industrial equipment. This chapter presents hazard and probability plotting for analysis of multiply censored life data. The methods of Chapter 3 do not apply to such data.

Multiply censored data consist of failure times intermixed with running times, called censoring times, as depicted in Figure 1.1. Such life data are common and can result from (1) removal of units from use before failure, (2) loss or failure of units due to extraneous causes, and (3) collection of data while units are still running (common for field data and much test data). Note that this chapter does not apply to failures found on inspection where it is known only that the failure actually occurred earlier; methods of Chapter 9 apply to such inspection data.

Section 1 shows step by step how to make a hazard plot of such data. Hazard plots look like probability plots and are interpreted in the same way. They can also be used for complete and singly censored data, but hazard plotting positions are slightly different from probability plotting positions. Section 2 shows how to make a probability plot of such data. Hazard plots give the same information as the probability plots, but with less labor. Section 3 presents the theory for hazard plotting papers. Chapter 3 is useful background for this chapter.

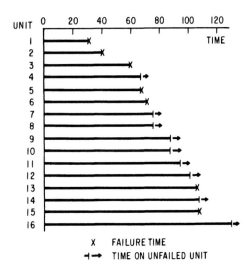

Figure 1.1. Field winding life data.

1. HAZARD PLOTTING

This section gives step-by-step instructions for making a hazard plot to estimate a life distribution from multiply censored data.

Fan failure data. Data that illustrate the hazard plotting method appear in Table 1.1. These data are the hours to fan failure on 12 diesel generators and the running hours on 58 generators without a fan failure. Each running time is marked with a "+" to indicate that the failure time for such a unit is beyond the running time. Failure times are unmarked. One problem was to estimate the percentage failing on warranty. Another was to determine if the failure rate of the fans decreased or increased with age; that is, would the problem get better or worse as the remaining fans aged? This information helped management decide whether to replace the unfailed fans with a better fan.

Steps to Make a Hazard Plot

1. The data on n units (70 fans here) consist of the failure times and the running (censoring) times. Order the n times from smallest to largest as shown in Table 1.1 without regard as to which are censoring or failure times. Label the times with reverse ranks; that is, label the first time with n, the second with $n-1,\dots$, and the nth with 1. Running times are marked with a "+," and failure times are unmarked.

Table 1.1. Fan Data and Hazard Calculations

Hours	Reverse Rank k	Hazard 100/k	Cum. Hazard	Hours	Reverse Rank k	Hazard 100/k	Cum. Hazard
450	70	1.4	1.4	6100+	24		
460+	69			6100+	23		
1150	68	1.5	2.9	6300+	22		
1150	67	1.5	4.4	6450+	21		
1560+	66			6450+	20		
1600	65	1.5	5.9	6700+	19		
1660+	64			7450+	18		
1850+	63			7800+	17		
1850+	62			7800+	16		
1850+	61			8100+	15		
1850+	60			8100+	14		
1850+	59			8200+	13		
2030+	58			8500+	12		
2030+	57			8500+	11		
2030+	56			8500+	10		
2070	55	1.8	7.7	8750+	9		
2070	54	1.9	9.6	8750	8	12.5	35.2
2080	53	1.9	11.5	8750+	7		
2200+	52			9400+	6		
3000+	51			9900+	5		
3000+	50			10100+	4		
3000+	49			10100+	3		
3000+	48			10100+	2		
3100	47	2.1	13.6	11500+	1		
3200+	46						
3450	45	2.2	15.8				
3750+	44						
3750+	43						
4150+	42						
4150+	41						
4150+	40						
4150+	39						
4300+	38						
4300+	37						
4300+	36						
4300+	35						
4600	34	2.9	18.7				
4850+	33						
4850+	32						
4850+	31						
4850+	30						
5000+	29						
5000+	28						
5000+	27						
6100+	26						
6100	25	4.0	22.7				

+Denotes running time.

2. Calculate a hazard value for each **failure** as $100/k$, where k is its reverse rank, as shown in Table 1.1. For example, the fan failure at 1600 hours has reverse rank 65, and its hazard value is $100/65 = 1.5\%$. This hazard value is the observed instantaneous failure rate at the age of 1600 hours, since 1 out of the 65 units that reached that age failed at that age.

3. Calculate the cumulative hazard value for each **failure** as the sum of its hazard value and the cumulative hazard value of the preceding failure. For example, for the failure at 1600 hours, the cumulative hazard value of 5.9 is the hazard value 1.5 plus the cumulative hazard value 4.4 of the preceding failure. The cumulative hazard values (for the fan failures) appear in Table 1.1. Cumulative hazard values may exceed 100% and have no physical meaning

4. Choose the hazard paper of a theoretical distribution. There are hazard papers* for the exponential, Weibull, extreme value, normal, and lognormal distributions. These distributions are described in Chapter 2. The distribution should be chosen on the basis of engineering knowledge of the product life distribution. Otherwise, different distributions can be tried, and one that fits the data well (a straight plot) could be used.

5. On the vertical axis of the hazard paper, mark a time scale that brackets the data. For the fan data, exponential hazard paper was chosen, and the vertical scale was marked off from 0 to 10,000 hours, as shown in Figure 1.2. The time scale must start with zero on exponential paper.

6. On the hazard paper, plot each failure time vertically against its cumulative hazard value on the horizontal axis, as shown in Figure 1.2. Running times are **not** plotted; hazard and cumulative hazard values are not calculated for them. However, the running times do determine the plotting positions of the failure times through the reverse ranks.

7. If the plot of failure times is roughly straight, one may conclude that the distribution adequately fits the data. Then, by eye, fit a straight line through the data points, as shown in Figure 1.2.

The line estimates the cumulative percentage failing, read from the horizontal probability scale as a function of age. The straight line, as explained below, yields information on the life distribution. If the data do not follow a straight line, then plot the data on another hazard paper. If no theoretical distribution fits adequately, draw a smooth curve through the plotted data. Then use the curve in the same way as a straight line to get information on the distribution. The sample cumulative hazard function is a nonparametric (distribution-free) estimate of the true cumulative hazard

---

*Exclusively offered in the catalog of TEAM (Technical and Engineering Aids for Management), Box 25, Tamworth, NH 03886.

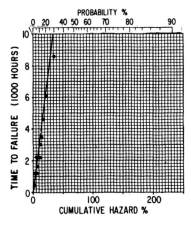

Figure 1.2.   Exponential hazard plot of fan data.

function. By virtue of the basic relationship (3.3), it also provides a nonparametric estimate of the cumulative distribution function.

Modified positions.   The above hazard plotting positions for a complete sample are close to the probability plotting positions $100i/(n+1)$. Hazard plotting positions can be modified to be closer to the probability plotting positions $100(i-0.5)/n$. Such a modified position for a failure is the average of its regular cumulative hazard value and that of the preceding failure. The modified position of the earliest failure is half its regular cumulative hazard value. The modified positions agree better with a distribution fitted by maximum likelihood (Chapter 8).

The basic assumption.   An assumption must be satisfied if the hazard plotting method is to be reliable. It is assumed that the life distribution of units censored at a given age must be the same as the life distribution of units that run beyond that age. For example, this assumption is not satisfied if units are removed from service when they look like they are about to fail. Lagakos (1979) discusses this and alternate assumptions in detail.

Hazard plotting can also be used for some types of data that are multiply censored on the right and on the left. Nelson (1972b) describes this.

How to Use a Hazard Plot

A hazard plot provides information on

The percentage of units failing by a given age,
Percentiles of the distribution,
The failure rate as a function of age,

Distribution parameters,
Conditional failure probabilities for units of any age,
Expected number of failures in a future period,
Life distributions of individual failure modes (Chapter 5), and
The life distribution that would result if certain failure modes were eliminated (Chapter 5).

The probability and data scales on a hazard paper for a distribution are exactly the same as those on the probability paper. Thus, a hazard plot is interpreted in the same way as is a probability plot, and the scales on hazard paper are used like those on probability paper. The cumulative hazard scale is only a convenience for plotting multiply censored data. The discussions in Chapter 3 on how to use and interpret probability plots apply to hazard plots.

Estimates of the percentage failing and percentiles. The population percentage failing by a given age is estimated from the fitted line with the method for probability paper. Enter the plot on the time scale at the given age, go to the fitted line, and then go to the corresponding point on the probability scale to read the percentage. For example, the estimate of the percentage of fans failing by 8000 hours (generator warranty) is 24%; this answers a basic question. Similarly, a percentile is estimated with the method for probability paper. Enter the plot on the probability scale at the given percentage, go to the fitted line, and then go to the corresponding point on the time scale to read the percentile. For example, the estimate of the 50th percentile, nominal fan life, is 14,000 hours (off scale).

Nature of the failure rate. The curvature of the plot of a sample on exponential hazard paper indicates the nature of the failure rate. Figure 1.3 shows curves for increasing, decreasing, and constant failure rates on exponential hazard paper or square grid paper. Figure 1.2 shows an essentially constant failure rate. This indicates that the fans will continue to fail at the same rate. Management decided to replace the unfailed fans with a better fan.

Exponential Hazard Plot

The following explains how to estimate the parameters of the exponential distribution from a hazard plot. This distribution is described in Chapter 2. Other information on how to interpret an exponential plot is in Chapter 3.

Exponential parameter estimates. The mean θ of an exponential distribution is the 63rd percentile, which has a cumulative hazard value of 100%.

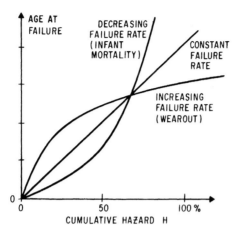

Figure 1.3. Cumulative hazard functions.

To estimate the mean graphically, enter the hazard plot at 100% on the cumulative hazard scale. Go to the fitted line and then to the corresponding point on the data scale to read the estimate. For the fan data in Figure 1.2, the estimate of the mean is 29,000 hours, off the paper. Exponential hazard paper has a square grid.

The failure rate λ of an exponential distribution equals the reciprocal of the mean. For fans, the estimate is $1/29,000 = 0.000035$, or 35 fan failures per million hours.

Normal Hazard Plot

The following explains how to estimate the parameters of a normal distribution from a hazard plot. Also, it shows how to estimate the conditional failure probability of a unit in service and the expected number of failures among units in service. The normal distribution is described in Chapter 2.

Transformer. Table 1.2 shows operating hours to failure on 22 transformers. Hours on 158 unfailed transformers are not shown, but the reverse ranks of the failures take them into account. As in this example, sometimes running times are not known, and they must be estimated from manufacturing or installation dates. Engineering wanted to know if the transformer failure rate increased or decreased with age; this might help identify the cause of failure. Also, management wanted a prediction of the number of failures in the coming 500 operating hours; this would aid in planning manufacture of replacements. The failures are plotted on normal hazard paper in Figure 1.4.

Table 1.2. Transformer Failures and Hazard Calculations

Hours	Reverse Rank k	Hazard	Cum. Hazard
10	180	0.56	0.56
314	179	0.56	1.12
730	178	0.56	1.68
740	177	0.56	2.24
990	176	0.57	2.81
1046	175	0.57	3.38
1570	174	0.57	3.95
1870	169	0.59	4.54
2020	164	0.61	5.15
2040	161	0.62	5.77
2096	157	0.64	6.41
2110	156	0.64	7.05
2177	155	0.65	7.70
2306	146	0.68	8.38
2690	103	0.97	9.35
3200	67	1.49	10.84
3360	66	1.52	12.36
3444	63	1.59	13.95
3508	62	1.61	15.56
3770	46	2.17	17.73
4042	32	3.13	20.86
4186	26	3.85	24.71

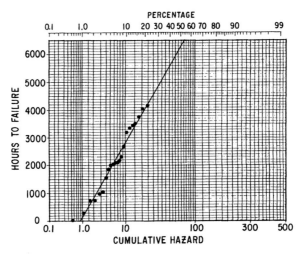

Figure 1.4. Normal hazard plot of transformer data.

Normal parameter estimates.   The following method provides graphical estimates of the normal mean and standard deviation from a hazard plot. It is the same as the method for a probability plot. To estimate the mean, enter the plot at the 50% point on the probability scale. Go to the fitted line and then to the corresponding point on the time scale to read the estimate (6250 hours here). To estimate the standard deviation enter the plot at the 16% point on the probability scale. Go directly to the fitted line and then to the corresponding point on the time scale to read the time as 3650 hours from Figure 1.4. The difference between the estimates of the mean and the 16% point, 6250 − 3650 = 2600 hours, is the estimate of the standard deviation.

Failure rate.   The data give a straight plot on normal paper, shown in Figure 1.4. This indicates that such transformers have an increasing failure rate—a wear-out behavior. The plot of the data on a square grid in Figure 1.5 shows the curvature characteristic of an increasing failure rate.

Conditional failure probabilities.   The prediction of the coming number of failures involves conditional failure probabilities. To estimate the conditional probability of a unit of a given current age failing by a certain future age, use the following. Suppose one wants to estimate the probability of a transformer 2000 hours old failing before it has 2500 hours in service. Enter the hazard plot on the time scale at the current age, 2000 hours. Go to the fitted line and then to the cumulative hazard scale to read the corresponding

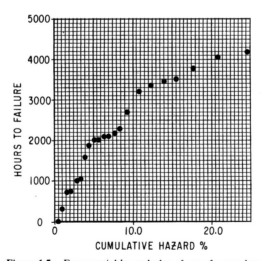

Figure 1.5.   Exponential hazard plot of transformer data.

value as a percentage; it is 5.7% here. Similarly, get the cumulative hazard value, 8.0%, for the future age, 2500 hours. Take the difference, $8.0 - 5.7 = 2.3\%$, between the two values. Then enter the plot on the cumulative hazard scale at the value of the difference, 2.3%, and go up to the probability scale to read the estimate of the conditional failure probability as 2.3%. This is 0.023, expressed as a fraction in Table 1.3.

Prediction of the number of failures. An important practical problem is to predict the number of units that will fail between the current time and a specified future time. Do this as follows. Calculate the conditional failure probability for each unit in or going into service for that period. The sum of these conditional probabilities, expressed as proportions, is a prediction of the number of failures. If the conditional failure probabilities are all small, then the random future number of failures has a probability distribution that is approximately Poisson with a mean equal to the expected number of failures. This distribution allows one to make probability statements about the number of future failures.

A calculation of an estimate of the expected number of transformer failures in a future 500 hours of service appears in Table 1.3. The estimate for the next 500 hours in service is based on the current hours of use of the 158 unfailed transformers. To reduce the calculation, the 158 transformers are grouped by age, and, for each group, a conditional probability of failure is obtained, as explained above. Then the sum of the conditional probabilities for a group is approximated by the nominal probability for the group times the number of transformers in the group as shown in the table. The estimate of the expected number of transformer failures over the next 500 hours is calculated as shown in Table 1.3 and is 7.8 failures. A Poisson table shows that the number of failures will be 12 or fewer with 95% probability.

Table 1.3. Calculation of Expected Number of Failures in 500 Hours

Group Ages	Nom'l Age	Cond'l Prob.	Trans- formers	Exp'd No. Failed
1750–2250	2000	0.023 x	17 =	0.39
2250–2750	2500	0.034 x	54 =	1.84
2750–3250	3000	0.044 x	27 =	1.19
3250–3750	3500	0.056 x	17 =	0.95
3750–4250	4000	0.070 x	19 =	1.33
4250–4750	4500	0.086 x	24 =	2.06
		Total Expected:	=	7.8

This probability statement does not take into account the statistical uncertainty in the estimate 7.8 failures. The calculation could also include the unconditional failure probabilities of new or replacement units going into service during a period.

The calculation in Table 1.3 assumes that all units receive the same amount of use over the period. When the units will receive different amounts of use, a conditional failure probability must be calculated for each unit and be based on its age and projected use. As before, the expected number of failures is the sum of the conditional failure probabilities of all units.

Lognormal Hazard Plot

The following explains how to estimate the parameters of the lognormal distribution from a hazard plot. The lognormal distribution is described in Chapter 2.

Turn failures. Table 1.4 shows data for turn failures of a Class H insulation for electric motors. The insulation was run in 10 motorettes at each test temperature (190, 220, 240, and 260°C) and periodically tested for electrical failure of the insulation. A time to turn failure in Table 1.4 is assigned midway between the test time when the failure was found and the

Table 1.4. Turn Failure Data and Hazard Calculations

190°C Hours	Reverse Rank	Hazard	Cum. Hazard	240°C Hours	Reverse Rank	Hazard	Cum. Hazard
7228	10	10.0	10.0	1175	10	10.0	10.0
7228	9	11.1	21.1	1521	9	11.1	21.1
7228	8	12.5	33.6	1569	8	12.5	33.6
8448	7	14.3	47.9	1617	7	14.3	47.9
9167	6	16.7	64.6	1665	6	16.7	64.6
9167	5	20.0	84.6	1665	5	20.0	84.6
9167	4	25.0	109.6	1713	4	25.0	109.6
9157	3	33.3	142.9	1761	3	33.3	142.9
10511	2	50.0	192.9	1881 +	2		
10511	1	100.0	292.9	1953	1	100.0	242.9

220°C Hours	Reverse Rank	Hazard	Cum. Hazard	260°C Hours	Reverse Rank	Hazard	Cum. Hazard
1764	10	10.0	10.0	1128	10	10.0	10.0
2436	9	11.1	21.1	1464	9	11.1	21.1
2436	8	12.5	33.6	1512	8	12.5	33.6
2436 +	7			1608	7	14.3	47.9
2436	6	16.7	50.3	1632 +	6		
2436	5	20.0	70.3	1632 +	5		
3108	4	25.0	95.3	1632 +	4		
3108	3	33.3	128.6	1632 +	3		
3108	2	50.0	178.6	1632 +	2		
3108	1	100.0	278.6	1896	1	100.0	147.9

time of the previous electrical test. In Table 1.4, each unit without a turn failure has its censoring time marked with a "+." Standard practice uses shorter periods between electrical tests at the higher temperatures. This makes the nominal number of inspections to failure roughly the same for all test temperatures. For the Class H test temperatures, the times between inspections were 28, 7, 2, and 2 days, respectively.

These data are analyzed to verify theory for the Arrhenius model. The theory says that time to such failure has a lognormal distribution at each test temperature and that the log standard deviation has the same value for all temperatures. Nelson (1975) shows further analyses of these data to estimate the distribution of time to turn failure at the design temperature of 180°C.

Figure 1.6 shows a lognormal plot of the turn failures for each test temperature. The plot shows some interesting features.

The plots for the four test temperatures are parallel. This indicates that the log standard deviation for turn failures has the same value at all test temperatures. This is consistent with the Arrhenius model. In contrast, the lognormal plots of the Class H data in Section 6 of Chapter 3 are not parallel.

The 260°C data coincide with the 240°C data. Clearly, insulation life should be less at 260°C than at 240°C, and the spacing between the 260° and 240°C distributions should be about the same as the spacing between the 240° and 220° distributions. This shows the value of a plot for spotting

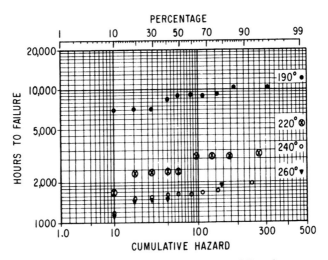

Figure 1.6. Lognormal hazard plot of turn failure data.

the unusual. There are two possible reasons for this. First, the 260°C motorettes were not made when the other 30 motorettes were; so they may differ with respect to materials and handling and thus life. Second, the inspection period at 260°C is two days (less severe handling), whereas it should be one day (more severe) to agree with the periods at the other test temperatures.

Lognormal parameters. Graphical estimates of the log mean and the log standard deviation are obtained from a fitted distribution line as follows. The log mean for a distribution is just the (base 10) logarithm of the median. For example, the median time to turn failure at 220°C is estimated from Figure 1.6 as 2600 hours. The log mean is log (2600)=3.462. The log standard deviation corresponds to the slope of the plotted data. One can fit the distribution lines for the data from the test temperatures so as to be parallel and have a compromise slope. The estimate of the log standard deviation is the difference between the logarithm of the 84th percentile and the log mean. For 220°C, the estimate of the 84th percentile is 3200 hours from Figure 1.6. The estimate of the log standard deviation is log(3200) − log(2600)=0.09. Such a small log standard deviation indicates that turn failures have an increasing failure rate, that is, wear-out behavior.

Weibull Hazard Plot

The following explains how to estimate the Weibull parameters from a hazard plot. It also shows how to use the Weibull distribution to assess whether the failure rate of a product increases or decreases with age. The Weibull distribution is described in Chapter 2.

Field windings. Figure 1.1 and Table 1.5 show data on field windings of 16 generators: months in service on failed windings and months on windings still running. The running and failure times are intermixed because the units were put into service at different times. There were two engineering questions. First, does the failure rate increase with winding age? If so, preventive replacement of old windings is called for. Second, what is the failure probability of a generator's windings before its next scheduled maintenance? This information helps a utility assess the risk of failure if replacement is deferred.

Shape parameter. For data on Weibull hazard paper, the following assesses whether the failure rate increases or decreases with age. A Weibull distribution has failure rate that increases (decreases) if the shape parameter is greater (less) than 1, and a value of 1 corresponds to a constant failure rate. Draw a straight line parallel to the fitted line so that it passes both through the circled dot in the upper left of the Weibull hazard paper and

Table 1.5. Winding Data and Cumulative Hazards

Time	Cum. Hazard
31.7	6.2
39.2	12.9
57.5	20.1
65.0+	
65.8	28.4
70.0	37.5
75.0+	
75.0+	
87.5+	
88.3+	
94.2+	
101.7+	
105.8	62.5
109.2+	
110.0	112.5
130.0+	

+ running time.

through the shape parameter scale, as in Figure 1.7. The value on the shape scale is the estimate; it is 1.94 for the winding data. The value 1.94 indicates that such windings have a failure rate that increases with age; that is, they wear out and should be replaced at some age. Note that Weibull hazard paper is log-log paper.

Scale parameter. To estimate the Weibull scale parameter, enter the hazard plot on the cumulative hazard scale at 100% (63% on the probability scale). Go up to the fitted line and then sideways to the time scale to read the estimate of the scale parameter; this is 116 months in Figure 1.7.

Exponential fit. An exponential distribution can be fitted to data plotted on Weibull paper; see Section 7 of Chapter 3. Force the fitted line to have a slope corresponding to a shape parameter of 1. This line helps one assess whether the simpler exponential distribution adequately fits the data.

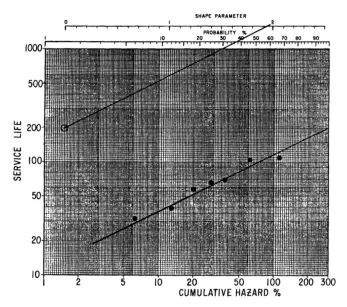

Figure 1.7. Weibull hazard plot of winding life data.

Extreme Value Hazard Plot

The following shows how to estimate the parameters of an extreme value distribution from a hazard plot. The extreme value distribution is described in Chapter 2.

Fuse residuals. After fitting a statistical model to data, one can assess the validity of the model and data by examining (log) residuals calculated from the data as described by Nelson (1973). Such residuals were obtained from fuse life data. According to the model, these multiply censored residuals come from an extreme value distribution. The validity of the assumed distribution and the data can be assessed from the computer plot of the residuals on extreme value paper in Figure 1.8. The plotted points follow a straight line fairly well, and there are no peculiar points. Thus the distribution and data seem satisfactory. The program uses hazard calculations for plotting positions.

Extreme value parameters. The location parameter λ of an extreme value distribution is the 63rd percentile. For the residuals, the estimate of λ from Figure 1.8 is 0.0. The slope of the fitted line in an extreme value plot

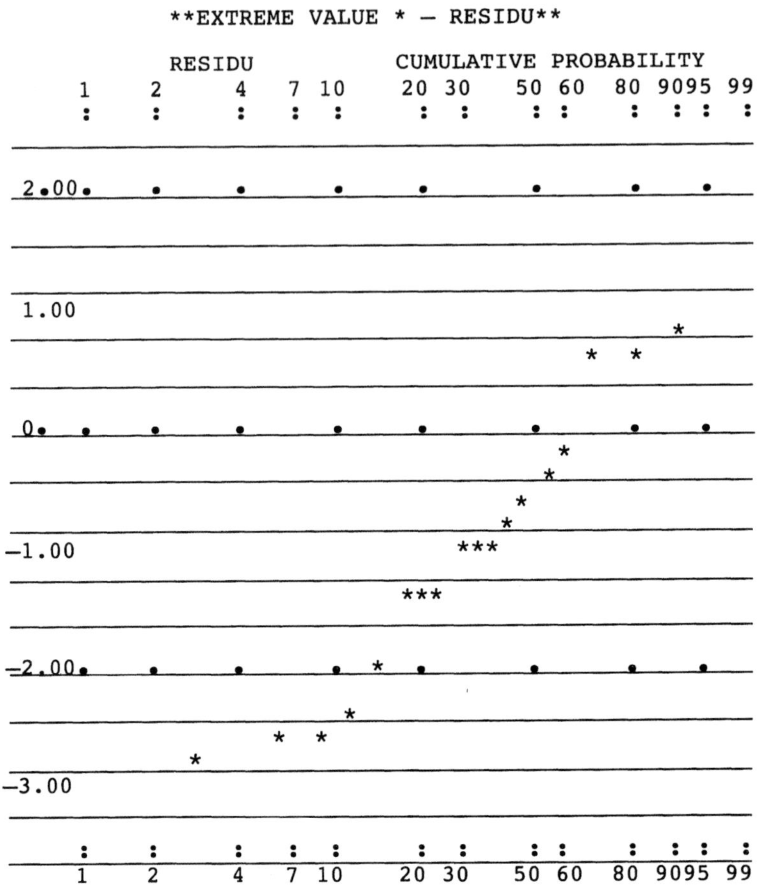

```
              **EXTREME VALUE  *  —  RESIDU**

              RESIDU            CUMULATIVE PROBABILITY
         1    2    4  7 10     20 30   50 60   80 9095 99
         :    :    :  :  :      :  :    :  :    :  :  :   :
    ─────────────────────────────────────────────────────
     2.00•   •    •      •      •      •      •       •
    ─────────────────────────────────────────────────────

    ─────────────────────────────────────────────────────
     1.00                                            *
    ─────────────────────────────────────────────────────
                                          *     *
    ─────────────────────────────────────────────────────
     0•   •    •      •      •      •      •      •       •
                                          *
    ─────────────────────────────────────────────────────
                                        *
                                      *
    ─────────────────────────────────────────────────────
    −1.00                             ***
    ─────────────────────────────────────────────────────
                                    ***
    ─────────────────────────────────────────────────────
    −2.00•   •    •      •   *  •      •      •       •
    ─────────────────────────────────────────────────────
                            *
                        *   *
    ─────────────────────────────────────────────────────
                      *
    −3.00
    ─────────────────────────────────────────────────────

         :    :    :  :  :      :  :    :  :    :  :  :   :
         1    2    4  7 10     20 30   50 60   80 9095 99
```

Figure 1.8. Extreme value hazard plot of fuse residuals.

corresponds to the scale parameter δ. δ is estimated as follows. After estimating the location parameter, estimate the 31st percentile. From Figure 1.8, this estimate is -1.0. The estimate of the scale parameter is the difference between the 63rd and 31st percentiles. So the estimate of δ is $0.0 - (-1.0) = 1.0$.

2. PROBABILITY PLOTTING OF MULTIPLY CENSORED DATA

Probability plots of multiply censored data can be used in place of hazard plots. Hazard plots are less work but require special plotting papers.

Probability and hazard plots of the same data are the same for practical purposes. Probability plotting involves calculating the sample reliability function. This function or the distribution function (the cumulative fraction failing) is plotted against age on probability paper. The Herd–Johnson, Kaplan–Meier (product-limit), and actuarial methods are presented. Each provides a different nonparametric estimate of the reliability function.

Assumption. A basic assumption of each plotting method must be satisfied if it is to yield reliable results. Namely, units censored at a given age must have the same life distribution as units that run beyond that age. For example, this assumption fails to hold if units are removed from service when they look like they are about to fail.

Herd–Johnson Method

Herd (1960) suggested the following probability plotting method, which was popularized by Johnson (1964), who presents it more complexly.

This section first presents an example that illustrates the plotting method. Then it explains how to make a probability plot.

Field winding example. Life data on 16 generators illustrate this method. These data in Table 2.1 and Figure 1.1 are the months in service on failed windings and on windings still running. The running and failure times are intermixed, because the units were put into service at different times. A basic engineering question was, "Does the failure rate increase with winding age?" If so, preventive replacement of old windings is called for. These data were previously analyzed with a hazard plot.

Plotting position calculations. Suppose that there are n times in a multiply censored sample. Order them from smallest to largest, as shown in Table 2.1, and mark each running time "$+$." Number the times backwards with reverse ranks—the smallest time is labeled n, the second smallest is labeled $(n-1)$, etc., and the largest is labeled 1. For the ith failure with reverse rank r_i, recursively calculate the reliability

$$R_i = [r_i/(r_i+1)] R_{i-1} \tag{2.1}$$

where $R_0 = 1$ is the reliability at time 0. The corresponding failure probability is the plotting position $F_i = 100(1 - R_i)$. The F_i are calculated only for failure times, but the running times determine the plotting positions of the failures. These calculations for the field winding data appear in Table 2.1; such calculations are easy to carry out with a pocket calculator. For a complete sample (all units failed), $F_i = 100i/(n+1)$, the mean plotting position. Johnson (1964) gives the plotting positions (2.1) in terms of

Table 2.1.   Winding Data and Herd–Johnson Calculations

TIME	REVERSE RANK r_i	COND'L $r_i/(r_i+1)$ x	RELIABILITY PREVIOUS R_{i-1}	CURRENT = R_i	FAILURE PROB. $100(1-R_i)\%$
31.7	16	(16/17) x	1.000	= 0.941	5.9
39.2	15	(15/16) x	0.941	= 0.883	11.7
57.5	14	(14/15) x	0.883	= 0.824	17.6
65.0 +	13				
65.8	12	(12/13) x	0.824	= 0.761	23.9
70.0	11	(11/12) x	0.761	= 0.697	30.3
75.0 +	10				
75.0 +	9				
87.5 +	8				
88.3 +	7				
94.2 +	6				
101.7 +	5				
105.8	4	(4/5) x	0.697	= 0.557	44.3
109.2 +	3				
110.0	2	(2/3) x	0.557	= 0.372	62.8
130.0 +	1				

+ Running

expected ranks, and he shows how to convert them to median plotting positions, a small and laborious refinement.

Plot and interpret.   Plot each failure time against its plotting position F_i, as shown in Figure 2.1. Only failure times are plotted. This plot is essentially identical to the hazard plot in Figure 1.7. Use the probability plot as described in Chapter 3 to obtain information.

From Figure 2.1, the shape parameter estimate for the winding data is 1.85. This value suggests that the winding failure rate increases with age,

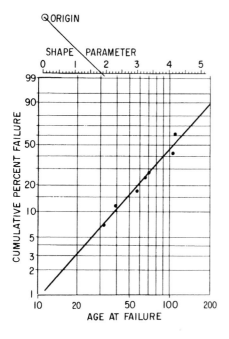

Figure 2.1. Weibull probability plot of winding data.

that is, a wear-out pattern. So windings should be replaced at some age when they are too prone to failure.

Kaplan–Meier Method

Kaplan and Meier (1958) give the *product-limit* estimate, much the same as the Herd–Johnson one. In place of (2.1), Kaplan and Meier use the recursion formula for reliability after failure i:

$$R_i' = [(r_i - 1)/r_i] R_{i-1}' \tag{2.2}$$

where $R_0' = 1$ is the reliability at time 0. For a complete sample, $F_i' = 1 - R_i' = i/n$, the sample fraction failing; this is the usual nonparametric estimate. If the largest time in a sample is a failure, the corresponding F_i' is unity and the failure cannot be plotted on most probability papers.

The BMDP routine PL1 of Dixon and Brown (1977) and SURVREG of Preston and Clarkson (1980) calculate the product-limit estimate of survivorship (reliability) and failure rate. Also, they calculate the standard errors of the reliability estimates at each failure time. For moderate-size samples, a survivorship calculation is easily performed with a pocket calculator. Also, most statistical computer packages for biomedical applications do all calculations.

The Kaplan–Meier method is widely used in biomedical applications to obtain a nonparametric estimate of the survivorship function. Tutorial presentations of the method appear in Gross and Clark (1975) and Kalbfleisch and Prentice (1980).

Actuarial Method

Various actuarial methods for estimating life distributions have been used for 200 years. They were developed and used to construct human life tables. Chiang (1968) describes such methods and biomedical applications of them. The simplest such method is presented here. Such methods are suited to large samples; the Herd–Johnson and Kaplan–Meier methods are better for small samples, since they plot individual failures times.

This section first presents an example that illustrates the actuarial method. Then it explains how to do the actuarial calculations and make a probability plot. Finally, it explains further how to use and interpret a probability plot.

Turbine disk example.   Table 2.2 shows life data on 206 turbine disks of which 13 had failed. The 100-hour intervals with failures are shown. A replacement disk had been designed so it would not fail.

Management needed to decide among three policies: (1) replace disks only when they fail (suitable if most disks last the life of the turbine), (2) replace disks at the scheduled overhaul at 5000 hours, or (3) replace all disks

Table 2.2.   Turbine Disk Data and Actuarial Calculations

y	Hours	In r_y	Fail f_y	Censor c_y	Out r_{y+1}	Reliability $R^*(y-1)[1-f_y(r_y-0.5c_y)^{-1}] = R^*(y)$	% Failure $100[1-R^*(y)]$
1&2	0– 200	206	0	4	202	1.000	0.
3	200– 300	202	1	2	199	$1.000[1-(202-0.5x2)^{-1}] = 0.9950$	0.50
4	300– 400	199	1	11	187	$0.9950[1-(199-0.5x11)^{-1}]= 0.9898$	1.02
5	400– 500	187	3	10	174	$0.9898[1-3(187-0.5x10)^{-1}]=0.9735$	2.65
8	700– 800	142	1	10	131	$0.9735[1-(142-0.5x10)^{-1}]= 0.9664$	3.36
10	900–1000	120	1	9	110	$0.9664[1-(120-0.5x9)^{-1}] = 0.9580$	4.20
13	1200–1300	92	2	5	85	$0.9580[1-2(92-0.5x5)^{-1}] = 0.9366$	6.34
14	1300–1400	85	1	13	71	$0.9366[1-(85-0.5x13)^{-1}] = 0.9248$	7.52
16	1500–1600	57	1	14	42	$0.9248[1-(57-0.5x14)^{-1}] = 0.9063$	9.37
17	1600–1700	42	1	14	27	$0.9063[1-(42-0.5x14)^{-1}] = 0.8804$	11.96
21	2000–2100	9	1	2	6	$0.8804[1-(9-0.5x2)^{-1}] = 0.7704$	22.96

immediately. Replacement of each unfailed disk would cost $3000; replacement of each failed disk would cost $6000 owing to related damage to the turbine.

Interval data. The actuarial method estimates the reliability or survivorship function from multiply censored data. The method is usually used for large samples where the data are grouped into time **intervals**. For example, human mortality data are usually grouped into one-year intervals. The method uses the number of units that enter, r_y, fail, f_y, are censored, c_y, and survive, r_{y+1}, in interval y, as shown in Table 2.2. Of course, $r_{y+1} = r_y - f_y - c_y$. The intervals need not have the same length, and intervals with no failures need not be tabulated. In contrast, preceding methods use and plot individual failure times.

Chain rule. Let $R(y)$ denote the product reliability (survival probability) at the end of interval y; this is the probability of a new unit surviving through interval y. The estimate of $R(y)$ uses the chain rule for conditional probabilities

$$R(y) = R(y|y-1)R(y-1|y-2) \cdots R(2|1)R(1), \qquad (2.3)$$

where $R(i|i-1)$ denotes the conditional reliability that units that have survived interval $(i-1)$ also survive interval i. It is convenient to work with the recursion relationship

$$R(y) = R(y-1)R(y|y-1). \qquad (2.4)$$

That is, the fraction $R(y)$ surviving interval y is the fraction $R(y-1)$ surviving interval $(y-1)$ times the fraction $R(y|y-1)$ of those that reach interval y and survive it. The probability of a new unit failing by the end of interval y is $F(y) = 1 - R(y)$.

Reliability estimates. The actuarial estimate of $R(y)$ employs estimates of the conditional reliabilities $R(i|i-1)$ and the preceding relationships. A simple estimate of $R(i|i-1)$ is

$$R^*(i|i-1) = 1 - [f_i/(r_i - 0.5c_i)]. \qquad (2.5)$$

The $-0.5c_i$ adjusts for the censored units, which run about half the interval. Other censoring adjustments appear in the literature (Chiang, 1968). Then the estimate of $R(y)$ is

$$R^*(y) = R^*(y-1)R^*(y|y-1), \qquad (2.6)$$

where $R^*(0) = 1$.

The recursive calculations of the $R^*(y)$ described above can readily be carried out in the tabular form in Table 2.2 This method provides an estimate $R^*(y)$ of the reliability at the end of each period.

These $R^*(y)$ may be plotted against the interval upper end points on square grid paper. Also, the sample (cdf) fraction failing $F^*(y)=1-R^*(y)$ can be plotted against the interval upper end points on probability paper.

Turbine disk. Table 2.2 shows turbine disk data and the actuarial calculations; the calculations are easy to carry with a pocket calculator. The given times are end points of 100-hour intervals. Only intervals with failures are tabulated, since the sample reliability function decreases only at the ends of such intervals. The sample reliability function is often plotted on square grid paper, as in Figure 2.2; it is difficult to use such a plot to extrapolate and estimate the upper tail of the distribution. The sample cumulative distribution function is plotted on normal probability paper in Figure 2.3; a straight line through this plot easily estimates the upper tail of the distribution if the normal distribution fits there.

Advantages and disadvantages. The actuarial method gives a plot that does not show individual failure times, which help one to assess how accurate the sample cdf is. Kaplan and Meier (1958) extend the method to get the "product-limit" estimate of the cumulative distribution function,

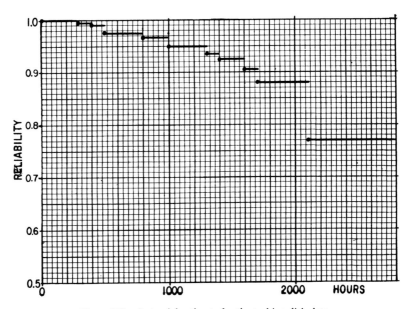

Figure 2.2. Actuarial estimate for the turbine disk data.

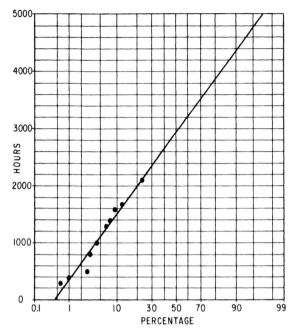

Figure 2.3. Normal probability plot of the actuarial estimate for the turbine disk data.

which shows the individual failure times. Chiang (1968) gives approximate confidence limits for the cumulative distribution, based on the actuarial estimate.

Interpret plot. Interpret such a plot like any other probability plot as described in Chapter 3. The following paragraphs answer the question on disk life.

Disk replacement cost. Figure 2.3 shows that almost all disks would fail by overhaul at 5000 hours. Thus the replacement cost would be about $193 \times \$6000 = \$1,158,000$ with policies (1) and (2), and it would be about $193 \times \$3000 = \$579,000$ with policy (3). The savings of about \$579,000 with policy (3) had to be weighed against the value of the remaining life of the disks and the business implications of 193 more such failures with policies (1) and (2). Plots on other probability papers also indicate policy (3).

Managements' choice of policy (3) raised another question: which disks are most prone to failure and should be replaced first?

Nature of the failure rate. The question was answered by plotting the data on various papers. The straightest plot was on normal paper. The normal distribution has an increasing failure rate with age. Thus the oldest disks are most prone to failure and should be replaced first.

Computer routines.   The BMDP routine PL1 of Dixon and Brown (1977) calculates the actuarial estimate of survivorship (reliability) and failure rate. Also, it calculates the standard error of the survivorship estimate for each interval. Most statistical computer packages for biomedical applications do such calculations.

3.   THEORY FOR HAZARD PAPERS

This **advanced** section presents theory for hazard papers. It first presents general theory on hazard and cumulative hazard functions. Then it applies the theory to various theoretical distributions to obtain their hazard papers. Readers interested only in applications may wish to skip this section.

Hazard and Cumulative Hazard Functions

Hazard plotting is based on the hazard function of a distribution. The basic properties of hazard functions are briefly presented here and then used as a theoretical basis for hazard papers.

The **hazard function** $h(y)$ for a distribution of time y to failure is defined in terms of the cumulative distribution function $F(y)$ and the probability density $f(y)$ as

$$h(y) \equiv f(y)/[1-F(y)]. \tag{3.1}$$

This is also called the instantaneous failure rate and force of mortality. It is a measure of proneness to failure as a function of the age of units in the sense that $\Delta \cdot h(y)$ is the expected proportion of units of age y that fail in a short time Δ. For this reason, it plays a central role in life data analysis. More information on it is in Chapter 2.

The **cumulative hazard function** $H(y)$ of a distribution is

$$H(y) \equiv \int_{-\infty}^{y} h(y)\,dy = -\ln[1-F(y)]. \tag{3.2}$$

As explained below, the scales on the paper for a theoretical distribution are constructed so the relationship between $H(y)$ and time y is linear. (3.2) can be rewritten as

$$F(y) = 1 - \exp[-H(y)]. \tag{3.3}$$

This "basic relationship" between the cumulative probability F and the cumulative hazard H is in all hazard papers. The probability scale on a hazard paper is exactly the same as that on the corresponding probability paper. The cumulative hazard scale is equivalent to the cumulative probability scale and is a convenient alternative scale in the plotting of multiply censored data.

The heuristic basis for hazard plotting is essentially that for plotting on probability paper. For a complete sample of n failures plotted on probability paper, the increase in the sample cumulative distribution function (cdf) at a failure time is $1/n$, the probability for the failure. Then the sample cdf, which is the **sum** of the failure probabilities, approximates the theoretical cdf, which is the **integral** of the probability density. The sample cdf is then plotted on probability paper on which the theoretical cdf is a straight line. Similarly, for a sample plotted on hazard paper, the increase in the sample cumulative hazard function at a failure time is equal to its conditional failure probability $1/k$, where k is its reverse rank. The number k of units in operation at the time of the failure includes the failed unit. Then the sample cumulative hazard function, based on the sum of the conditional probabilities of failure, approximates the theoretical cumulative hazard function, which is the integral of the conditional probability of failure. The sample cumulative hazard function is then plotted on hazard paper, on which the theoretical cumulative hazard function is a straight line. Nelson (1972b) gives a more formal justification for the hazard plotting positions in terms of the properties of order statistics of multiply censored samples.

Hazard Papers

Hazard papers have been developed for the basic theoretical distributions: the exponential, normal, lognormal, Weibull, and extreme value distributions. Their hazard plotting papers are obtained from the general theory above. The data and cumulative hazard scales on hazard paper for a theoretical distribution are chosen so that such a cumulative hazard function is a straight line on the paper.

Exponential distribution. The exponential cumulative distribution function is

$$F(y)=1-\exp(-y/\theta), \qquad y\geqslant 0, \tag{3.4}$$

where θ is the mean time to failure. The cumulative hazard function is

$$H(y)=\int_0^y (1/\theta)\,dy=y/\theta, \qquad y\geqslant 0; \tag{3.5}$$

it is a linear function of time. Then time y to failure as a function of H is

$$y(H) = \theta H. \tag{3.6}$$

Thus time to failure y is a linear function of H that passes through the origin. So exponential hazard paper has a square grid. Its probability scale is given by the basic relationship (3.3). Exponential hazard paper appears in Figure 1.2. The value of θ is the time for which $H=1$ (i.e., 100%); this is used to estimate θ from an exponential hazard plot.

Normal distribution.   The normal cumulative distribution function is

$$F(y) = \Phi[(y - \mu)/\sigma], \qquad -\infty < y < \infty, \tag{3.7}$$

where μ is the mean, σ is the standard deviation, and $\Phi(\ )$ is the standard normal cumulative distribution function. The cumulative hazard function is

$$H(y) = -\ln\{1 - \Phi[(y - \mu)/\sigma]\}, \qquad -\infty < y < \infty. \tag{3.8}$$

Then time y as a function of H is

$$y(H) = \mu + \sigma \Phi^{-1}(1 - e^{-H}) = \mu + z_{(1 - e^{-H})}\sigma, \tag{3.9}$$

where $\Phi^{-1}(P)$ is the inverse of $\Phi(\ )$ and z_p is the $100P$th standard normal percentile. By (3.9), time y is a linear function of $\Phi^{-1}(1 - e^{-H})$. Thus, on normal hazard paper, a cumulative hazard value H is located on the cumulative hazard scale at the position $\Phi^{-1}(1 - e^{-H})$, and the time scale is linear. Normal hazard paper appears in Figure 1.4. The probability scale is used to estimate μ and σ by the methods for normal probability paper.

Lognormal distribution.   The lognormal cumulative distribution function is

$$F(y) = \Phi[(\log(y) - \mu)/\sigma], \qquad y > 0, \tag{3.10}$$

where $\log(y)$ is the base 10 logarithm, $\Phi(\ )$ is the standard normal cumulative distribution function, μ is the log mean, and σ is the log standard deviation. The cumulative hazard function is

$$H(y) = -\ln\{1 - \Phi[(\log(y) - \mu)/\sigma]\}, \qquad y > 0. \tag{3.11}$$

Then time y as a function of H is

$$\log[y(H)] = \mu + \sigma \Phi^{-1}(1 - e^{-H}), \tag{3.12}$$

where $\Phi^{-1}(\ )$ is the inverse of the standard normal distribution function. By (3.12), $\log(y)$ is a linear function of $\Phi^{-1}(1-e^{-H})$. Lognormal hazard paper appears in Figure 1.6. Lognormal and normal hazard papers have the same cumulative hazard and probability scales. On lognormal paper the time scale is logarithmic, whereas it is linear on normal paper. The probability scale is used to estimate μ and σ by the methods for lognormal probability paper.

Weibull distribution. The Weibull cumulative distribution function is

$$F(y)=1-\exp\left[-(y/\alpha)^{\beta}\right], \qquad y>0, \tag{3.13}$$

where β is the shape parameter and α is the scale parameter. The cumulative hazard function is

$$H(y)=\int_0^y (\beta/\alpha)(y/\alpha)^{\beta-1}dy=(y/\alpha)^{\beta}, \qquad y>0; \tag{3.14}$$

this is a power function of time y. Then time y as a function of H is

$$\log(y)=(1/\beta)\log(H)+\log(\alpha). \tag{3.15}$$

This shows that $\log(y)$ is a linear function of $\log(H)$. Thus, Weibull hazard paper is log-log graph paper. The probability scale is given by the basic relationship (3.3). Weibull hazard paper appears in Figure 1.7. The slope of the straight line equals $1/\beta$; this is used to estimate β with the aid of the shape parameter scale. For $H=1(100\%)$, the corresponding time y equals α; this is used to estimate α graphically.

Extreme value distribution. The cumulative distribution function of the smallest extreme value distribution is

$$F(y)=1-\exp\left\{-\exp\left[(y-\lambda)/\delta\right]\right\}, \qquad -\infty<y<\infty, \tag{3.16}$$

where λ is the location parameter and δ is the scale parameter. The cumulative hazard function is

$$H(y)=\exp\left[(y-\lambda)/\delta\right], \qquad -\infty<y<\infty; \tag{3.17}$$

it is an exponential function of time. Then time y as a function of H is

$$y(H)=\lambda+\delta\ln(H). \tag{3.18}$$

This shows that time y is a linear function of $\ln(H)$. Thus, the extreme value hazard paper is semilog paper. The probability scale is given by the basic relationship (3.3). Extreme value hazard paper appears in Figure 1.8. For $H = 1$ (i.e., 100%), the corresponding time y equals λ; this is used to estimate λ from an extreme value plot. The probability scale is used to estimate δ by the method for extreme value probability paper. Weibull and extreme value hazard papers have the same cumulative hazard and probability scales. On Weibull paper the time scale is logarithmic, and on extreme value paper it is linear.

PROBLEMS

4.1. Shave die.   The following are life data (in hours) on a shave die used in a manufacturing process. Operating hours for shave dies that wore out: 13.2   67.8   76   59   30   26.7   26   68   30.1   76.3   43.5. Operating hours on shave dies that were replaced when the production process was stopped for other reasons: 45.1   27   49.5   62.8   75.3   13   58   34   48   49   31.5   18.1   34   41.5   62   66.5   52   60.1   31.3   39   28.6   7.3   40.3   22.1   66.3   55.1

(a)   Calculate hazard (or probability) plotting positions for dies that wore out.

(b)   Plot the data on Weibull hazard (or probability) paper.

(c)   Assess the validity of the data and the fit of the Weibull distribution.

(d)   Graphically estimate the Weibull shape parameter. Does it suggest that the instantaneous failure rate increases or decreases with die life? An increasing one indicates that dies over a certain age should be replaced when the production process is stopped for other reasons, as die wear out stops production and is therefore costly.

4.2. Turbine disk.   Table 2.2 shows turbine disk data (rounded to 100 hours).

(a)   Do the hazard calculations.

(b)   Plot the data on lognormal hazard paper.

(c)   How does the lognormal fit compare with the normal fit in Figure 2.3?

(d)   Estimate the median life. Compare it with the estimate from Figure 2.3.

(e)   Make a Weibull hazard plot and determine whether the failure rate increases or decreases with age.

(f)   Compare the normal, lognormal, and Weibull fits with respect to the upper tails; that is, which is most pessimistic and which is most optimistic?

4.3.*   Censored on right and left.   Hazard plotting also applies to data censored on the left. Just multiply all data values by −1 to get reversed data censored on the right, and plot that data. Data below from Sampford and Taylor (1959) are multiply censored on the right and the left. Mice were paired, and one of each pair got injection A and the other got injection B. For each pair, the difference of their log days of survival appears below. However, observation terminated after 16 days. When one mouse of a pair survives, one knows only that the difference is greater (less) than that observed. There are three such differences in the data: −0.25− and −0.18− censored on the left and 0.30+ censored on the right. The differences for the 17 pairs are −0.83, −0.57, −0.49, −0.25−, −0.18−, −0.12, −0.11, −0.05, −0.04, −0.03, 0.11, 0.14, 0.30, 0.30+, 0.33, 0.43, 0.45.

(a)   Calculate hazard plotting positions for all observed differences above −0.18−. Note that the sample size is 17, not 12. Convert these cumulative hazard values to cumulative probabilities by means of the basic relationship (3.3).

(b)   Reverse the data and calculate hazard plotting positions for all reversed observed differences above −0.30−.

(c)   Reversing the data reverses the probability scale. Convert the two sets of hazard plotting positions to consistent probability plotting positions.

(d)   Plot all observed differences on normal probability paper. Normal paper is used because the distribution of log differences is symmetric about zero if injections A and B have the same effect.

(e)   Does the mean difference differ significantly from zero, corresponding to no difference between the injections? Assess this subjectively.

(f)   Graphically estimate the mean and standard deviation of the distribution of differences.

4.4.   Table 10.3 of Chapter 3 shows a time to breakdown of 83,990 in the 25-kV data.

(a)   Use the Herd–Johnson method to calculate probability plotting positions for the 25-kV data including 83,990.

*Asterisk denotes laborious or difficult.

(b) Plot the data and those from the other voltages on Weibull probability paper.

(c) Is the 83,990 point consistent with the others?

(d) Do (a), (b), and (c) using the Kaplan–Meier method.

4.5. Battery. Service data on the life of a type of battery appear below.

(a) Do the actuarial calculations for the fraction surviving each month.

(b) Plot the sample cumulative distribution function on Weibull probability paper. Does the Weibull distribution adequately fit the data?

(c) Estimate the Weibull shape parameter and comment on the nature of the failure rate (increasing or decreasing?).

(d) Estimate the proportion failing by 24 months, design life.

Interval (Month)	Units Entering	Failures in Interval	Censored in Interval
1	1926	9	217
2	1700	2	180
3	1518	0	178
4	1340	3	172
5	1165	3	125
6	1037	5	134
7	898	5	97
8	796	2	81
9	716	7	92
10	617	2	84
11	531	3	99
12	429	5	84
13	340	0	59
14	281	3	60
15	218	1	69
16	148	1	35
17	112	1	24
18	87	2	25
19	60	0	16
20	44	0	16
21	28	1	9
22	18	0	5
23	13	0	10
24	3	0	

4.6.* The logistic cumulative distribution function is

$$F(y)=1/\{1+\exp[-(y-\mu)/\sigma]\},\qquad -\infty<y<\infty,$$

where the scale parameter σ is positive and the location parameter μ has any value.

(a) Find the relationship between life y and cumulative hazard H.

(b) To make hazard paper, calculate the positions corresponding to cumulative hazards of 1, 2, 5, 10, 20, 50, 100, 200, and 500% on the cumulative hazard scale.

(c) Calculate the positions corresponding to cumulative probabilities of 1, 2, 5, 10, 20, 50, 80, 90, 95, 98, and 99% on the probability scale.

(d) Carefully draw the logistic hazard plotting paper and include the probability scale.

(e) Explain how to use the hazard scale to estimate μ and σ.

5

Series Systems and
Competing Risks

Overview. Many products fail from more than one cause. For example, any part in a household appliance may fail and cause the appliance to fail. Also, humans may die from accidents, various diseases, etc. Sections 1 and 2 present the series-system model for such products, and Section 3 presents graphical analyses of such data. The models give the relationship between the product life distribution and those of its parts. The theory also provides models for the life of a product as a function of its size, for example, for the life of cable as a function of length. The data analyses in Section 3 provide estimates of the life distribution of the product and of its parts or causes of failure. Needed background includes Chapters 2 and 4. Section 4 of Chapter 8 gives maximum likelihood methods for analysis of such data.

Background. Literature on competing failure modes began two hundred years ago with Daniel Bernoulli and d'Alembert. They developed theory (models and data analyses) to evaluate the effect of smallpox inoculation on human mortality (Todhunter, 1949). Much literature is motivated by medical studies, actuarial applications, and bioassay. In these applications, authors refer to competing risks, irrelevant causes of death, and multiple decrement analyses. An early bibliography is given by Chiang (1968). More recently David and Moeschberger (1979), Kalbfleisch and Prentice (1980), and Birnbaum (1979) surveyed the history, models, and statistical methods for such competing risks, emphasizing nonparametric methods and biomedical applications. Such theory applies to product life and other phenomena. For example, Harter (1977a) reviews the use of such theory for the effect of size on material strength. Regal and Larntz (1978) apply such theory to data on the time that individuals and groups take to solve problems. Mayer (1970) studies life of ball bearing assemblies.

1. SERIES SYSTEMS OF DIFFERENT COMPONENTS

This section presents series systems, the product rule for reliability, the addition law for failure rates, and the resulting distribution when some failure modes are eliminated. These are the basic ideas for series systems and products with a number of failure modes.

Series systems and the product rule. Suppose that a product has a potential time to failure from each of M causes (also called competing risks or failure modes). Such a product is called a **series system** if its life is the smallest of those M potential times to failure; that is, the system fails with the first "part" failure. In other words, if $Y_1, \ldots, Y_M$ are the potential times to failure for the M causes (or parts), then the system time Y to failure is

$$Y = \min(Y_1, \ldots, Y_M). \tag{1.1}$$

Let $R(y)$ denote the reliability function of the system, and let $R_1(y), \ldots, R_M(y)$ denote the reliability functions of the M causes (each in the absence of all other causes). Suppose that the times to failure for the different causes (components or risks) are statistically **independent**. Such units are said to have **independent competing risks** or to be **series systems with independent components**. For such systems,

$$R(y) = P\{Y > y\} = P\{Y_1 > y \text{ and } Y_2 > y \text{ and } \cdots \text{ and } Y_M > y\}$$

$$= P\{Y_1 > y\} P\{Y_2 > y\} \cdots P\{Y_M > y\},$$

since $Y_1, \ldots, Y_M$ are statistically independent. Thus

$$R(y) = R_1(y) R_2(y) \cdots R_M(y). \tag{1.2}$$

This key result is the **product rule** for reliability of series systems (with independent components). In contrast, for a mixture of distributions (Section 7 of Chapter 2), each unit has only one possible mode of failure corresponding to the subpopulation to which the unit belongs. Cox (1959) makes this distinction.

Lamp assembly example. Two incandescent lamps (light bulbs) are in an assembly that fails if either lamp fails. The first type of lamp is taken to

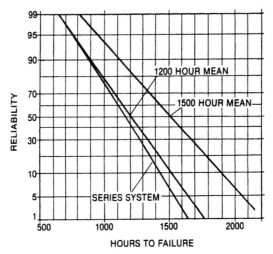

Figure 1.1.   Reliability function of a series system of two lamps.

have a normal life distribution with a mean of 1500 hours and a standard deviation of 300 hours. The second type is taken to have a normal life distribution with a mean of 1200 hours and a standard deviation of 240 hours. Reliability functions of these distributions are depicted as straight lines on normal probability paper in Figure 1.1. The life distribution of such assemblies was needed, in particular, the median life. The lamp lives are assumed to be independent; so the assembly reliability is

$$R(y) = \{1 - \Phi[(y - 1500)/300]\}\{1 - \Phi[(y - 1200)/240]\},$$

where $\Phi(\ )$ is the standard normal cumulative distribution. For example, $R(1200) = \{1 - \Phi[(1200 - 1500)/300]\} \times \{1 - \Phi[(1200 - 1200)/240]\} = 0.421$. $R(y)$ is plotted in Figure 1.1 and is not a straight line. So the life distribution of such assemblies is not normal. The median life is obtained by solving $R(y) = 0.50$ to get $y_{.50} = 1160$ hours; this also can be obtained from the plot.

Addition law for failure rates.   For series systems, it is useful to look at the hazard function (failure rate) and cumulative hazard functions. The product rule (1.2) in terms of the cumulative hazard functions $H(y)$ of the system and $H_1(y), \ldots, H_M(y)$ of each of the M causes is

$$\exp[-H(y)] = \exp[-H_1(y)]\exp[-H_2(y)] \cdots \exp[-H_M(y)]$$

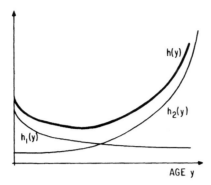

Figure 1.2.   Hazard functions of a series system and its components.

or

$$H(y) = H_1(y) + H_2(y) + \cdots + H_M(y). \tag{1.3}$$

Differentiate this to get the hazard function

$$h(y) = h_1(y) + h_2(y) + \cdots + h_M(y); \tag{1.4}$$

this is called the **addition law for failure rates** for **independent** failure modes (or competing risks). So for series systems, failure rates add. This law is depicted in Figure 1.2, which shows the hazard functions of the two components of a series system and the system hazard function.

A great increase in the failure rate of a product may occur at some age. This may indicate that a new failure cause with an increasing failure rate is becoming dominant at that age. This shows up in hazard or probability plots as depicted in Figure 1.3. There the lower tail of the distribution is spread over a wide age range, and the upper portion of the distribution

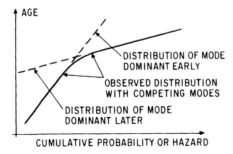

Figure 1.3.   Cumulative distribution of data with competing failure modes.

is confined to a much narrower age range. Such an appearance of a data plot may indicate two or more competing failure modes. Figure 3.3 in Section 3 shows such a plot.

Galambos (1978) presents (asymptotic) theory for systems with many components that may differ and be statistically dependent.

Exponential components.   Suppose that M independent components have exponential life distributions with failure rates $\lambda_1, \ldots, \lambda_M$. Then series systems consisting of such components have an exponential life distribution with a constant failure rate

$$\lambda = \lambda_1 + \cdots + \lambda_M. \tag{1.5}$$

This simple relationship is used for reliability analysis of many systems. Often misused, (1.5) is correct only if all component life distributions are exponential. Similarly, the mean time to failure for such series systems is

$$\theta = \left[(1/\theta_1) + \cdots + (1/\theta_M) \right]^{-1}, \tag{1.6}$$

where the component mean lives are $\theta_1 = 1/\lambda_1, \ldots, \theta_M = 1/\lambda_M$. (1.5) comes from the addition law for failure rates, namely,

$$h(y) = \lambda_1 + \cdots + \lambda_M, \qquad y \geq 0, \tag{1.7}$$

which is a constant. Thus the distribution of time to system failure is exponential.

Freight train example.   A high-priority freight train required three locomotives for a one-day run. If such a train arrived late, the railroad had to pay a large penalty. To assess its risks, the railroad needed to know the reliability of such trains. Experience indicated that times to failure for such locomotives could be approximated with an exponential distribution with a mean of $\theta_0 = 43.3$ days. It was assumed that the three locomotives in a train fail independently. So such a train is a series system, and time to train delay has an exponential distribution with a mean of $\theta = [(1/43.3) + (1/43.3) + (1/43.3)]^{-1} = 14.4$ days. The reliability of the train on a one-day run is $R(1) = \exp(-1/14.4) = 0.933$. The unreliability is $1 - 0.933 = 0.067$; that is, 1 out of $1/0.067 \cong 15$ such trains is delayed.

To reduce the chance of delay, the railroad used trains with four locomotives. Then the train was delayed only if two or more locomotives failed. The reliability of such a train on a one-day run is the binomial probability of one or fewer failures, namely,

$$R(1)=\binom{4}{0}p^0(1-p)^4+\binom{4}{1}p^1(1-p)^3,$$

where $p=1-\exp(-1/43.3)=0.0234$ is the daily probability of locomotive failure. So the reliability for one day is

$$R(1)=(1-0.0234)^4+4\cdot0.0234(1-0.0234)^3=0.9968.$$

The unreliability is $1-0.9968=0.0032$; that is, 1 out of $1/0.0032\cong313$ such trains is delayed. The extra locomotive provides what is called parallel redundancy and increases the reliability.

Design methods for evaluating the reliability of systems with more complex redundancy are described by Shooman (1968), who calls systems such as the four-locomotive train an "r-out-of-n structure" (3 out of 4 here). Such design analyses were developed in the 1950s and 1960s for aerospace and military hardware. They are now used for many consumer and industrial products ranging from heart pacemakers through nuclear reactors.

Weibull components. Suppose that M independent components have Weibull life distributions with scale parameters $\alpha_1,\dots,\alpha_M$ and the **same shape parameter** β. Then series systems consisting of M such components have a Weibull distribution with a scale parameter

$$\alpha=\left[(1/\alpha_1^\beta)+\cdots+(1/\alpha_M^\beta)\right]^{-1/\beta} \tag{1.8}$$

and the same shape parameter β. This can be seen from the system hazard function

$$h(y)=\beta(y^{\beta-1}/\alpha_1^\beta)+\cdots+\beta(y^{\beta-1}/\alpha_M^\beta)=\beta(y^{\beta-1}/\alpha^\beta), \tag{1.9}$$

which is the hazard function for a Weibull distribution. Section 2 presents an example of such a system—a four-slice toaster.

Other component distributions. For a system whose component distributions are all normal, the system distribution is **not** normal. This is true of the lognormal and most other distributions. Similarly, (1.8) does not hold for Weibull distributions with different shape parameters.

Elimination of failure modes. Often it is important to know how elimination of causes of failure will affect the life distribution of a product. As before, suppose those causes are independent and

$$R(y)=R_1(y)R_2(y)\cdots R_M(y), \quad h(y)=h_1(y)+h_2(y)+\cdots +h_M(y),$$

$$H(y)=H_1(y)+H_2(y)+\cdots +H_M(y),$$

$$F(y)=1-[1-F_1(y)]\times \cdots \times [1-F_M(y)]. \tag{1.10}$$

Suppose that cause 1 is eliminated. (This may be a collection of causes.) Then $R_1(y)=1$, $h_1(y)=0$, $H_1(y)=0$, and $F_1(y)=0$, and the life distribution for failure by all other causes is

$$R^*(y)=R_2(y)\cdots R_M(y), \quad h^*(y)=h_2(y)+\cdots +h_M(y),$$

$$H^*(y)=H_2(y)+\cdots +H_M(y),$$

$$F^*(y)=1-[1-F_2(y)]\cdots [1-F_M(y)]. \tag{1.11}$$

Lamp assembly example. If the 1500-hour lamp were replaced by one with essentially unlimited life, the assembly of two lamps would have the life distribution of the 1200-hour lamp.

2. SERIES SYSTEMS OF IDENTICAL PARTS

Some series systems consist of identical parts, each from the same life distribution. For example, tandem specimens are sometimes used in creep-rupture studies of an alloy. Pairs of specimens are linked end to end and stressed until one ruptures. Use of pairs hastens the test and provides more information on the lower tail of the distribution of time to rupture. Also, for example, a power cable might be regarded as a series connection of a large number of small segments of cable. The cable life is the life of its first segment to fail. Similarly, the life of a battery is the life of its first cell to fail. An assembly of ball bearings fails when the first ball fails. Tests of such systems are called "sudden death tests." The following theory for such systems is a special case of the general theory in Section 1.

Failure Rate Proportional to the Number of Components

System life distribution. Suppose that a series system consists of M statistically **independent, identical** components with a component reliability function $R_0(y)$, a cumulative distribution function $F_0(y)$, a hazard function $h_0(y)$, and a cumulative hazard function $H_0(y)$. For a series system of M such independent, identical components, let $R(y)$ denote the system reliability function, $F(y)$ the system cumulative distribution function, $h(y)$ the system hazard function, and $H(y)$ the system cumulative hazard function. Then

$$R(y)=[R_0(y)]^M, \qquad F(y)=1-R(y)=1-[1-F_0(y)]^M,$$
$$h(y)=Mh_0(y), \qquad H(y)=MH_0(y). \tag{2.1}$$

These are special cases of (1.2), (1.3), and (1.4). The life of such a system is the smallest of M (component) lives from the same distribution.

Toaster example.   A proposed four-slice toaster has the same components as standard (two-slice) toasters. However, there are twice as many of most parts in the four-slice toaster. So its failure rate is roughly twice as great as that of the two-slice toaster and its small percentage failing on warranty is about twice as great. More details are in a following paragraph.

Exponential components.   Suppose **independent** components have the same exponential life distribution with failure rate λ_0. Then series systems of M such components have an exponential life distribution with a failure rate

$$\lambda=M\lambda_0. \tag{2.2}$$

This is a special case of (1.5). Similarly, the mean time to failure θ for such series systems is

$$\theta=\theta_0/M, \tag{2.3}$$

where $\theta_0=1/\lambda_0$ is the mean time to failure of the component. The freight train of Section 1 is such a system.

Weibull components.   Suppose a type of component has a Weibull life distribution with shape parameter β and scale parameter α_0. Then series systems of M such independent components have a Weibull distribution with the same shape parameter β and a scale parameter

$$\alpha=\alpha_0/M^{1/\beta}. \tag{2.4}$$

This is a special case of (1.8).

Toaster example.   Life of two-slice toasters is approximated with a Weibull distribution with $\beta=4/3$ and $\alpha_0=15$ years. A proposed four-slice toaster would have essentially $M=2$ of each component in the two-slice toaster; so it would have twice as great a failure rate. Thus four-slice toasters have a Weibull life distribution with $\beta = 4/3$ and $\alpha = 15/2^{1/(4/3)} \cong 9$ years.

Extreme value components. Suppose **independent** components have a smallest extreme value distribution with a scale parameter δ_0 and a location parameter λ_0. Then series systems of M such components have an extreme value distribution with a scale parameter δ_0 and a location parameter

$$\lambda = \lambda_0 - \delta_0 \ln(M). \qquad (2.5)$$

For example, this distribution describes the yearly minimum temperature. The lowest temperature in the design life of M years of a heating system comes from a smallest extreme value distribution with the same scale parameter and a smaller location parameter, λ, given by (2.5). This is used to design system capacity.

A similar result holds for the maximum observation in a sample from a largest extreme value distribution.

Extreme value residuals example. Nelson and Hendrickson (1972) give an example of $M = 360$ standardized residuals from a smallest extreme value distribution with $\lambda_0 = 0$ and $\delta_0 = 1$. The smallest residual is -8.90 and seems too small. The smallest observation comes from a smallest extreme value distribution with $\lambda = 0 - 1 \cdot \ln(360) = -5.886$ and $\delta = 1$. The probability that the smallest sample value is -8.90 or less is $F(-8.90) = 1 - \exp(-\exp\{[-8.90 - (-5.886)]/1\}) = 0.049$. This probability is small enough for the observation to be slightly suspect and possibly regarded as an outlier and investigated.

Other component distributions. For (log) normal component life, the life distribution of series systems of M such components is not (log) normal and depends on M. However, for large samples from any distribution (satisfying mild restrictions), the distribution of the smallest observation is approximately Weibull or extreme value, according to whether the original distribution is respectively bounded below or not. This is an important limit theorem for extreme values (Gumbel, 1958). The result indicates that life of some large series systems may be adequately described by the Weibull or extreme value distribution.

Failure Rate Proportional to Size

The model. Some products that come in various sizes have failure rates that are proportional to product size. For example, the failure rate of a capacitor dielectric is often assumed to be proportional to the area of the dielectric. Also, for example, the failure rate of cable insulation is often assumed to be proportional to cable length.

In general, if $h_0(y)$ is the failure rate for an amount A_0 of such product, then the failure rate $h(y)$ of an amount A of the product is

$$h(y)=(A/A_0)h_0(y). \tag{2.6}$$

That is, such products are regarded as **series systems** of $M=(A/A_0)$ identical components from the same life distribution. A/A_0 need not be an integer. (2.6) is based on the assumption that adjoining portions of the product have statistically **independent** lifetimes.

Formulas for an amount A of product are

$$R(y)=[R_0(y)]^{A/A_0},$$
$$F(y)=1-R(y)=1-[1-F_0(y)]^{A/A_0}, \tag{2.7}$$
$$H(y)=(A/A_0)H_0(y),$$

where the zero subscript denotes the distribution for an amount A_0.

Exponential life. If an amount A_0 of product has an exponential life distribution with failure rate λ_0, then an amount A has an exponential life distribution with failure rate

$$\lambda=(A/A_0)\lambda_0. \tag{2.8}$$

The product mean time to failure is

$$\theta=\theta_0/(A/A_0), \tag{2.9}$$

where $\theta_0=1/\lambda_0$ is the mean time to failure for an amount A_0 of product. These results follow from (2.2) and (2.3).

Motor insulation example. Test specimens of motor insulation were assumed to have an exponential life distribution with a failure rate of $\lambda_0=3.0$ failures per million hours. The insulation area of specimens is $A_0=6$ in.^{2}, and the area of motors is $A=500$ in.^{2}. The life distribution of motor insulation is exponential, with a failure rate of $\lambda=(500/6)3.0=250$ failures per million hours.

Weibull life. Suppose an amount A_0 of product has a Weibull life distribution with shape parameter β and scale parameter α_0. Then an amount A has a Weibull distribution with the same shape parameter β and scale parameter

$$\alpha = \alpha_0 / (A/A_0)^{1/\beta}. \tag{2.10}$$

This follows from (2.4) and (2.6).

Capacitor example. Time to dielectric breakdown of a type of 100-pf capacitor is taken to have a Weibull distribution with $\beta = 0.5$ and $\alpha_0 = 100,000$ hours. The 500-pf capacitor has the same design but $A/A_0 = 5$ times the dielectric area. The dielectric life of 500-pf capacitors is Weibull with $\beta = 0.5$ and $\alpha = 100,000/5^{1/0.5} = 4000$ hours.

Series systems with dependence. Some series-system products contain identical parts with statistically **dependent** lifetimes. For example, adjoining segments of a cable may have positively correlated lives, that is, have similar lives. Figure 10.2 of Chapter 2 depicts positively correlated lives ($\rho = 0.5, 0.9$) for a series system with two parts. Models for dependent part lives are complicated; see Galambos (1978), David and Moeschberger (1979), Harter (1977a), and Moeschberger (1974). There is work on multivariate exponential distributions for such systems; Proschan and Sullo (1976) give some data analyses and references on a fatal shock model. Barlow and Proschan (1975, Chap. 5) present some previous work on multivariate life distributions. Harter (1977a) comprehensively surveys models for the effect of size. Block and Savits (1981) comprehensively survey multivariate distributions.

However, there are simple upper and lower limits for the system life distribution when component lives are positively correlated. The lower limit is the life distribution for a series system of independent parts, and the upper limit is the life distribution for a single part. These crude limits may bracket the true distribution accurately enough for practical purposes.

Cryogenic cable example. Accelerated tests of cryogenic cable specimens indicated that specimen life can be approximated with a Weibull life distribution with $\alpha_0 = 1.05 \times 10^{11}$ years and $\beta = 0.95$ at design conditions. The volume of dielectric of such specimens is 0.12 in.^{3}, and the volume of dielectric of a particular cable is 1.984×10^7 in.^{3}. Suppose that the cable can be regarded as a **series system** of **independent** specimens. Then its (approximate) life distribution is Weibull, with $\alpha = 1.05 \times 10^{11}/(1.984 \times 10^7/0.12)^{1/0.95} = 234$ years and $\beta = 0.95$. The cable engineers wanted to know the 1% point of this distribution; it is $y_{.01} = 234 \cdot [-\ln(1-0.01)]^{1/0.95} = 1.8$ years. If adjoining "specimen lengths" of the cable have positively correlated lives, the true life distribution of such cable cannot exceed the distribution for the life of a single specimen. These two distributions differ appreciably because (A/A_0) is large, but even the pessimistic distribution showed that the design was adequate.

3. HAZARD PLOTS FOR DATA WITH COMPETING FAILURE MODES

This section presents hazard plotting methods for analyzing data with a number of different failure modes:

1. A method for estimating the life distribution that would result if certain failure modes were eliminated from a product. This method uses available data to estimate how product design changes would affect the life distribution.

2. A method for estimating the distribution of time to failure for a given failure mode. This method can reveal the nature of the failure mode and the effect of design changes on the failure mode.

3. A method for combining different sets of data to estimate a life distribution.

The methods apply to field data and life test data. Also, they apply to certain strength and failure data, for example, King (1971, p. 196).

The methods employ either hazard or probability plotting, described in Chapter 4. The methods assume that the product is a **series system** with **independent** competing failure modes. All analyses below require that the cause of each failure be **identified**.

The Appliance Data

The methods are illustrated with life test data on a small electrical appliance. These data in Table 3.1 come from the life testing of 407 units at various stages in a development program. For each unit, the data consist of the number of cycles it ran to failure or to removal from test and its failure code (one of 18 causes). Units with the code 0 were removed from test before failure. Interest centered mostly on early failures. Replacement of unfailed units with new ones allowed more testing of the early life. Notes indicate when design changes were made and the failure modes they were to eliminate.

Two life tests were used: the automatic test (units cycled by automatic test equipment) and the manual test (units cycled manually). For analyses, the units are divided by date of manufacture into five groups. The units in a group have roughly the same design. Figure 3.1 depicts part of the data. The

Table 3.1. Appliance Data Tabulation

GROUP 1

Automatic Test		Manual Test	
Cycles	Code	Cycles	Code
927	1		
2084	2		
241	2		
8	3		
71	4		
97	2		
1529	5		
482	6		
2330	1		
121	6		
1488	6		
1834	7		
10175	0		
49	6		
2397	7		
197	6		
91	6		
10011	0		
11110	0		
10008	0		
IMPROVE MODE 2			
5234			
5212	0	46	6
5974	0	46	3
5320	0	2145	0
4328	0	2145	0
4860	0	270	6
19	8	98	6
89	6	1467	11
65	6	495	11
7	6	2145	0
604	8	16	10
4	8	16	12
244	8	2145	0
250	0	692	11
2494	0	2145	0
2140	0	52	6
4428	0	98	11
		2145	0
		2145	0
		495	6
		1961	0
		1107	12
		1937	11
		12	13
		1961	0
		1467	12
		616	6
		1453	0
		1453	0
		1453	0
		413	11
		1056	11
		1453	0
		1453	0
		1453	0
		557	14
		1193	11

GROUP 2

Automatic Test		Manual Test	
Cycles	Code	Cycles	Code
IMPROVE MODE 6			
11	1	311	11
2223	9	571	0
4329	9	73	11
3112	9	571	11
13403	0	136	0
6367	0	136	6
2451	5	136	0
381	6	136	0
1062	5	136	0
1594	2	1300	0
329	6	1300	0
2327	6	1198	9
958	10	670	11
7846	9	575	11
170	6	1300	0
3059	6	1198	11
3504	9	569	1
2568	9	417	12
2471	9	1164	0
3214	9	1164	0
3034	9	1164	0
2694	9	1164	0
49	15	1164	0
6976	9	608	0
35	15	608	0
2400	9	608	0
1167	9	608	0
2831	2	490	11
2702	10	608	11
708	6	608	0
1925	9	608	0
1990	9	608	0
2551	9	47	11
2761	6	608	0
2565	0	608	0
3478	9	608	0
		608	11
		45	1
		608	0
		608	0
		964	11
		281	12
		964	11
		670	11
		1164	0
		1164	0
		838	11
		731	0
		630	11
		485	0
		485	0
		145	11
		190	0
		190	0

GROUP 3

Manual Test	
Cycles	Code
658	8
90	1
190	1
241	1
349	6
410	12
90	11
90	11
268	11
410	11
410	11
485	11
508	11
631	11
631	11
631	11
635	11
658	11
731	11
739	11
790	11
855	11
980	11
980	11
218	0
218	0
378	0
378	0
739	0
739	0
739	0
739	0
790	0
790	0
790	0
790	0
790	0
790	0
790	0
790	0
790	0
790	0
980	0
980	0
980	0
980	0
980	0
600	0
600	0
600	0
600	0

Table 3.1 *(Continued)*

	GROUP 4				GROUP 5			
	Automatic Test		Manual Test		Automatic Test		(cont'd)	
	Cycles	Code	Cycles	Code	Cycles	Code	Cycles	Code
	CHANGE MODES 1,5,6,9,11,12				45	8	3380	5
	15277	0	2	8	4395	0	470	0
	9042	17	11	8	881	5	4017	5
	11687	0	21	8	3412	5	3985	0
	8519	10	303	8	1117	5	4056	0
	13146	0	762	10	1789	5	NEW PART DESIGN	
	3720	2	984	12	4137	0	AND LUBRICANT	
	8421	17	1126	1	4129	8	FOR MODE 5.	
	4105	2	1128	12	1585	6		
	3765	9	657	0	4368	0	2615	0
	2029	10	681	0	3901	0	4596	0
	3133	0	1072	0	2432	5	4275	0
	4294	0	1096	0	4006	0	3984	0
	5760	0	1098	0	4816	0	4800	0
	5619	0	1098	0	2070	0	4436	0
	5535	0	1098	0	1588	5	4132	0
	5585	18	1098	0	546	5	3597	0
	137	8	1098	0	2157	5	1680	5
	5733	18	1098	0	3008	2	3508	0
	1945	10	1098	0	3164	5	3009	5
	5070	0	1098	0	3688	0	4112	0
	5324	0	1098	0	1312	2	4323	0
	4894	0	1098	0	687	17	3462	5
	2191	5	1098	0	1312	2	3550	0
	1307	5	1098	0	1103	0	4441	0
	738	5	1098	0	535	5	3244	5
	4734	0	1098	0	1342	5	4317	0
	2832	0	1098	0	496	5	2974	5
	364	10	1098	0	2433	0	3264	0
	2748	0	1098	0	3319	17	4349	0
	515	5	1330	0	4319	0	2961	5
	2622	0	1446	0	2203	17	4103	0
	872	0	1447	0	2899	5	4246	5
			1449	0	713	5	3075	10
			1449	0	1321	5	1493	5
			1449	0	816	5	3978	0
			1449	0	2762	5	1624	1
			1449	0	1489	5	1430	0
			1449	0	4413	0	1877	6
			1449	0	649	5	3174	0
			1449	0	1238	5	4256	0
			1451	0	378	0	4252	0
			1451	0	1738	0	3904	0
			1451	0	148	17	4398	0
			1451	0	703	5	3719	0
			1451	0	487	5	3424	0
			1451	0	437	17	3532	0
			1451	0	3892	0	3995	0
			1451	0	3937	0	4384	0
			1462	0	2113	5	4161	0
			1462	0	3552	0	3773	0
					3484	0	3876	0
					3855	0	1015	5
					2757	8	3727	0
					5448	0	4318	0
					4693	0		
					412	5		
					4004	5		
					3830	5		

Failure Codes

0 denotes removal from test before failure.

1 - 18 denote various failure modes.

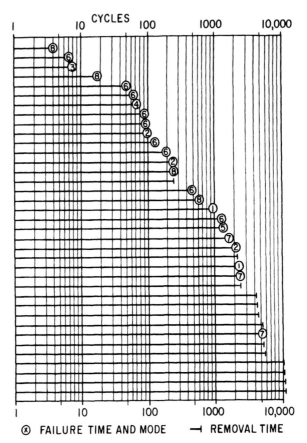

Figure 3.1.   Automatic test data from Group 1.

objective of the development program was to improve the early- and long-life portions of the life distribution. This reflects concern over warranty costs and long-term customer satisfaction.

Life Distributions When Failure Modes Will be Eliminated

Hazard plots can be used to estimate product life distributions when certain failure modes are eliminated. The appliance example illustrates the method. Table 3.2 shows hazard calculations for the Group 1 automatic test data with all modes acting, and Figure 3.2 shows a Weibull hazard plot.

Product life may be improved through engineering changes in design or operation to eliminate one or more failure modes. Often it is costly or time consuming to make changes and to then collect and analyze data to

Table 3.2.   Hazard Calculations for Automatic Test Data from Group 1

CYCLES	FAILURE CODE	REVERSE RANK	HAZARD	CUM-HAZARD
4	8	37	2.70	2.70
7	6	36	2.78	5.48
8	3	35	2.86	8.34
19	8	34	2.94	11.28
49	6	33	3.03	14.31
65	6	32	3.13	17.44
71	4	31	3.23	20.67
89	6	30	3.33	24.00
91	6	29	3.45	27.45
97	2	28	3.57	31.02
121	6	27	3.70	34.72
197	6	26	3.85	38.57
241	2	25	4.00	42.57
244	8	24	4.17	46.74
250+	0	23		
482	6	22	4.55	51.29
604	8	21	4.76	56.05
927	1	20	5.00	61.05
1488	6	19	5.26	66.31
1529	5	18	5.56	71.87
1834	7	17	5.88	77.75
2084	2	16	6.25	84.00
2140+	0	15		
2330	1	14	7.14	91.14
2397	7	13	7.69	98.83
2494+	0	12		
4328+	0	11		
4428+	0	10		
4860+	0	9		
5212+	0	8		
5234	7	7	14.29	113.12
5320+	0	6		
5974+	0	5		
10008+	0	4		
10011+	0	3		
10175+	0	2		
11110+	0	1		

determine the effect of the changes. The following method estimates the life distribution that would result if selected failure modes are **completely eliminated**. Such an estimate allows one to assess whether the elimination of specific failure modes would be worthwhile. This method employs past data with a mixture of all failure modes.

Major design changes were made on the appliance in Group 4 to eliminate failure modes 1, 6, 9, 11, and 12, which affect early and long life. The Group 2 automatic test data were collected before the design changes were made. They are used to predict the life distribution that would result if those failure modes were eliminated. Cumulative hazard values for the predicted life distribution are shown in Table 3.3. These calculations assume that failure modes 1, 6, 9, 11, and 12 are completely eliminated. Each failure

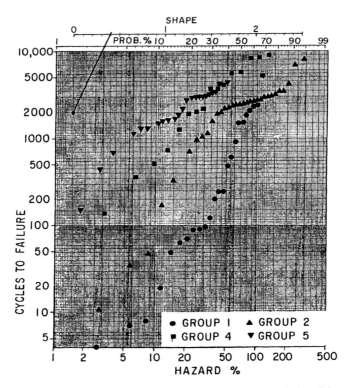

Figure 3.2.   Plots of automatic test data for Groups 1, 2, 4, and 5.

time for mode 1, 6, 9, 11, or 12 is treated as a **censoring time**, that is, as if the appliance were removed from test at the time of such a failure. These are censoring times, since "redesigned" appliances would not run to failure by one of the removed failure modes. The failure times for the remaining modes are plotted against their cumulative hazard values in Figure 3.3 as triangles.

For comparison, Figure 3.3 also shows the plot of all the automatic test failures for Group 2 and the plot for Group 4. The design changes to eliminate these modes are in Group 4. So, the three plots in Figure 3.3 correspond to (1) an estimate of the life distribution before the design changes, (2) a prediction of the life distribution with the changes, and (3) an estimate of the life distribution actually achieved with the changes. Plots (2) and (3) agree well.

Table 3.3.   Hazard Calculations for the Predicted Life Distribution

CYCLES	FAILURE CODE	REVERSE RANK	CUM-HAZARD
11 +		36	
35	15	35	2.86
49	15	34	5.80
170 +		33	
329 +		32	
381 +		31	
708 +		30	
958	10	29	9.25
1062	5	28	12.82
1167 +		27	
1594	2	26	16.67
1925 +		25	
1990 +		24	
2223 +		23	
2327 +		22	
2400 +		21	
2451	5	20	21.67
2471 +		19	
2551 +		18	
2565 +		17	
2568 +		16	
2694 +		15	
2702	10	14	28.81
2761 +		13	
2831	2	12	37.14
3034 +		11	
3059 +		10	
3112 +		9	
3214 +		8	
3478 +		7	
3504 +		6	
4329 +		5	
6367 +		4	
6976 +		3	
7846 +		2	
13403 +		1	

If a design change does not completely eliminate a failure mode, then the resulting life distribution is between the distribution before the change and the prediction. An example of this appears later. Also, a design change may affect failure modes other than intended ones, and it may even introduce new failure modes.

Life Distribution of a Single Failure Mode

Sometimes one desires information on the distribution of time to failure of a particular mode, that is, the life distribution if there were only a single mode. An estimate of this distribution indicates the nature of the failure mode and the effect of design changes on the failure mode. The estimate employs the previous method carried to the extreme, where all modes but one are eliminated. That is, failure times for all other modes are treated as

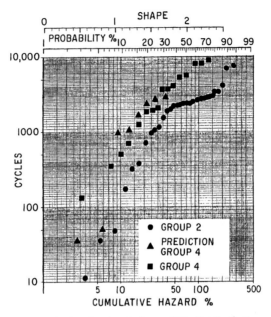

Figure 3.3. Predicted and observed life distributions.

censoring times, since such units were not run long enough to fail by that mode.

The following example illustrates the method. A new design of a part was tried to overcome mode 11 failures. Units on the manual test in Group 2 contained the old design of the part. Units on the manual test in Group 3 contained the new design.

Table 3.4a shows hazard calculations for mode 11 with the old design. Each failure time for all modes except 11 is treated as a censoring time, since such units did not run to failure by mode 11. The failure times for mode 11 with the old part are plotted in Figure 3.4. Table 3.4b shows hazard calculations for mode 11 with the new part, and the hazard plot is shown in Figure 3.4. The normal distribution empirically provides the best fit.

Figure 3.4 indicates that the two designs have the same distribution of time to failure. Mode 11 was later eliminated by another design change. The straight data plot on normal paper indicates that the part has an increasing failure rate—wear-out behavior, confirming engineering expectation.

Table 3.4a. Cumulative Hazard for Mode 11, Group 2

CYCLES	FAILURE MODE	CUM. HAZARD
45 +		
47	11	1.89
73	11	3.81
136 +		
136 +		
136 +		
136 +		
136 +		
145	11	5.98
190 +		
190 +		
281 +		
311	11	8.36
417 +		
485 +		
485 +		
490	11	10.99
569 +		
571 +		
571	11	13.85
575 +		
608 +		
608 +		
608 +		
608 +		
608	11	20.24
608 +		
608 +		
608 +		
608 +		
608 +		
608 +		
608	11	24.79
608 +		
608 +		
630	11	30.05
670	11	35.61
670	11	41.49
731 +		
838	11	48.16
964	11	55.30
964	11	62.99
1164 +		
1164 +		
1164 +		
1164 +		
1164 +		
1164 +		
1164 +		
1198 +		
1198	11	87.99
1300 +		
1300 +		
1300 +		

Table 3.4b. Cumulative Hazard for Mode 11, Group 3

CYCLES	FAILURE MODE	CUM. HAZARD
90 +		
90	11	1.96
90	11	3.96
190 +		
218 +		
218 +		
241 +		
268	11	6.18
349 +		
378 +		
378 +		
410 +		
410	11	8.68
410	11	11.24
485	11	13.87
508	11	16.57
600 +		
600 +		
600 +		
600 +		
631	11	19.70
631	11	22.93
631	11	26.26
635	11	29.71
658 +		
658	11	33.41
731	11	37.26
739	11	41.26
739 +		
739 +		
739 +		
739 +		
790	11	46.26
790 +		
790 +		
790 +		
790 +		
790 +		
790 +		
790 +		
790 +		
790 +		
790 +		
790 +		
855	11	58.76
980	11	73.05
980	11	89.72
980 +		
980 +		
980 +		
980 +		
980 +		

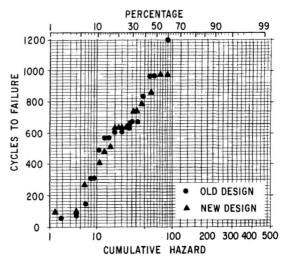

Figure 3.4.   Normal hazard plot of old and new designs.

Life Distributions for a Combination of Failure Modes

Sometimes it is useful to combine data on different failure modes to estimate the distribution of time to failure when all of these modes act. The following method utilizes all data to estimate the life distribution and employs the previous two methods. This method is highly specialized.

A design modification for mode 5 was made partway through Group 5, producing an increase in time to failure for mode 5. So only the subsequent data represent the distribution of time to failure for all causes. This small sample does not accurately estimate the life distribution. The following method uses the mode 5 data (after the modification) to estimate the distribution for mode 5 and all data from Group 5 to estimate the distribution for all other modes. These two distributions are then combined to estimate the distribution of time to failure for all modes. This method assumes that the distribution for all modes other than 5 is the same throughout Group 5. Engineering thought that no design changes should affect other failure modes.

Table 3.5a shows hazard values for the estimate of the life distribution for mode 5; they are based on the automatic test data after the modification. Each time for a failure mode other than mode 5 is treated as a censoring time. The failure times for mode 5 are plotted against their cumulative hazard values in Figure 3.5. This plot merely displays the data and is not part of the method being presented here.

Table 3.5a.   Hazard Values for Mode 5 in Group 5 after Design Change

CYCLES	FAILURE CODE	REVERSE RANK	HAZARD
1015	5	46	2.17
1493	5	44	2.27
1680	5	42	2.38
2961	5	39	2.56
2974	5	38	2.63
3009	5	37	2.70
3244	5	34	2.94
3462	5	31	3.23
4246	5	14	7.14

Table 3.5b shows hazard values for the estimate of the life distribution for all modes other than 5; they are based on all automatic test data from Group 5. Each time for mode 5 is treated as a censoring time. The failure times for all other modes are plotted against their cumulative hazard values in Figure 3.5. This plot merely displays the data and is not part of the method being presented here. The plot estimates the life distribution if failure mode 5 were eliminated.

The cumulative hazard function of a series system is the sum of the cumulative hazard functions of the different modes. So, the remaining calculation consists of summing the cumulative hazard function for mode 5

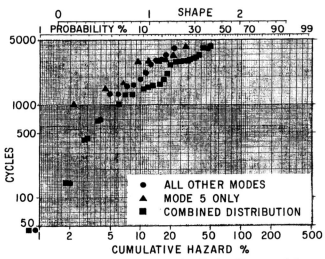

Figure 3.5. Distribution based on Mode 5 data after Design Change and data on all other modes from Group 5.

Table 3.5b. Hazard Values for Group 5 with Mode 5 Eliminated

CYCLES	FAILURE CODE	REVERSE RANK	HAZARD
45	8	110	0.91
148	17	109	0.92
437	17	106	0.94
687	17	99	1.01
1312	2	90	1.11
1312	2	89	1.12
1585	6	83	1.20
1624	1	81	1.23
1877	6	77	1.30
2203	17	73	1.37
2757	8	69	1.45
3008	2	64	1.56
3075	10	62	1.61
3319	17	57	1.75
4129	8	25	4.00

and the cumulative hazard function of all other modes. This provides an estimate of the cumulative hazard function of the distribution of time to failure for all causes. A method for doing this follows.

The failure times in Table 3.5a for mode 5 and the failure times in Table 3.5b for all other modes are arranged from smallest to largest, as shown in Table 3.5c, with their hazard values (not in increasing order). The cumulative hazard value for a failure time is the sum of its hazard value and hazard values for all previous failure times. The failure times are then plotted against their cumulative hazard values, as shown in Figure 3.5. This plot estimates the distribution of time to failure for all causes.

Life Distributions for Systems of Identical Components

Sometimes one must analyze life data from series systems of identical components or from systems with failure rate proportional to size. One may wish to estimate the life distribution of (1) the individual components or (2) the system or (3) systems of different sizes. An example of (1) involves turn failure of electrical insulation, which can occur in either of two coils in a motorette, and failure of one coil terminates the test on the other coil. One can estimate the distribution of time to failure of individual coils from such data. McCool (1970b) called this situation "sudden death" testing in ball bearing life tests. An example of (2) involves a proposed design of a four-slice toaster, which is equivalent to a series system of two two-slice

Table 3.5c. Combination of Mode 5 after Design Change and All Other Modes in Group 5

CYCLES	FAILURE CODE	HAZARD	CUM. HAZARD
45	8	0.91	0.91
148	17	0.92	1.83
437	17	0.94	2.77
687	17	1.01	3.78
1015	5	2.17	5.95
1312	2	1.11	7.06
1312	2	1.12	8.18
1493	5	2.27	10.45
1585	6	1.20	11.65
1624	1	1.23	12.88
1680	5	2.38	15.26
1877	6	1.30	16.56
2203	17	1.37	17.93
2757	8	1.45	19.38
2961	5	2.56	21.94
2974	5	2.63	24.57
3008	2	1.56	26.13
3009	5	2.70	28.83
3075	10	1.61	30.44
3244	5	2.94	33.38
3319	17	1.75	35.13
3462	5	3.23	38.36
4129	8	4.00	42.36
4246	5	7.14	49.50

toasters. Existing data on two-slice toasters can be used to estimate the life distribution of four-slice toasters. An example of (3) involves failures of power cables of different lengths. The data on cables of various lengths can be used to estimate the life distribution of cables of any given length.

Component life may be estimated from system data, as illustrated with the turn failure data on the 20 coils in 10 motorettes in Table 3.6. If a motorette has a turn failure, the coil data consist of the failure time of one coil followed by an equal running time for the other coil. If a motorette is removed from test without a turn failure, the data consist of a running time for each coil. In Table 3.6, running times are labeled " + ." The table also shows the hazard calculations for the 20 coils. Figure 3.6 shows a normal hazard plot of the coil failure times. The earliest failure looks inconsistent with the others.

Table 3.6. Hazard Calculations for Individual Coils

Hours	Reverse Rank	Hazard	Cumulative Hazard
1175	20	5.0	5.0
1175+	19		
1521	18	5.6	10.6
1521+	17		
1569	16	6.2	16.8
1569+	15		
1617	14	7.1	23.9
1617+	13		
1665	12	8.3	32.2
1665+	11		
1665	10	10.0	42.2
1665+	9		
1713	8	12.5	54.7
1713+	7		
1761	6	16.7	71.4
1761+	5		
1881+	4		
1881+	3		
1953	2	50.0	121.4
1953+	1		

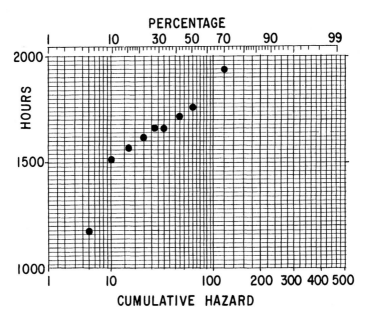

Figure 3.6. Normal hazard plot of coil failures.

186

System life may be estimated from component data by the following method. Suppose such systems consist of M identical, independent components. One must first carry out the hazard calculations for the component data, then multiply the cumulative hazard value of each failure by M, and plot the component failure times against the new cumulative hazard value on hazard paper. This plot provides an estimate of the system life distribution. This was done for the four-slice toaster.

Lives of systems of a given size may be estimated from systems of various sizes (for example, cables of different lengths) by the following method. The running and failure times in the data are tabulated from smallest to largest, as usual for hazard calculations. The size of each system is then expressed as a multiple of the given system size. The multiple need not be an integer and may be less than or greater than unity. The reverse rank of each system is then obtained by summing the value of its multiple and those of all older units—failed and censored. The hazard calculations and plotting are then carried out with these "reverse ranks" of the failures.

PROBLEMS

5.1. Shave die. Use the shave die data of Problem 4.1. Other events that stop the process are thought to occur at random and independently. If so, the times to when the process is stopped for other reasons have an exponential distribution.

(a) Calculate hazard (or probability) plotting positions for times to stopping production for other reasons, treating times to die wear out as censoring times.

(b) Plot the data on Weibull hazard (or probability) paper.

(c) Assess the validity of the data and a Weibull fit to the data.

(d) Graphically estimate the Weibull shape parameter and assess whether the exponential distribution fits adequately.

(e) Make a separate Weibull hazard plot of time to replacement from any cause (wear out or process stopped).

5.2. Class H insulation. Below are times to failure on 10 motors with a new Class H insulation at 220°C. Motors were periodically inspected for failure. A failure time is midway between the time when the failure was found and the previous inspection time. Rounding to the midpoint affects the data little. The table shows a failure or running time for each cause (turn, phase, or ground, each a separate part of the insulation). Each failed part was isolated electrically and could not fail again, and most motors stayed on test to a second or third failure cause. Most other data contain only one failure cause per unit. In use, the first failure from any cause determines the motor life.

Hours to Failure

Motor	Turn	Phase	Ground
1	1764	2436	2436
2	2436	2436	2490
3	2436	2436	2436
4	2436+	2772+	2772
5	2436	2436+	2436
6	2436	4116+	4116+
7	3108	4116+	4116+
8	3108	4116+	4116+
9	3108	3108	3108+
10	3108	4116+	4116+

+ denotes time without failure.

(a) Use the first failure on each motor. Do hazard calculations and plot the failures on lognormal hazard paper. This estimates motor life with all modes acting.

(b) Do the hazard calculations for the turn failures. Plot the data on lognormal hazard paper.

(c) Do (b) for phase failures.

(d) Do (b) for ground failures.

(e) Which failures should be eliminated to improve life most?

(f) Combine the hazard calculations for separate modes to get an estimate of the cumulative hazard function when all modes act. Plot each failure against its cumulative hazard on lognormal hazard paper.

(g) How do the estimates from (a) and (f) compare?

(h) Why do (a) and (f) differ in the upper tail? Which is a better estimate in the upper tail?

5.3. Cylinder life. For certain engines, each piston failure causes its cylinder to fail. Also, there are cylinder failures due to other causes. The table below shows an actuarial estimate of the reliability function of current pistons and a separate actuarial estimate of the reliability function of cylinders. A proposed new piston design would eliminate cylinder damage from piston failure. The following shows how much cylinder life (mileage) would be improved with the new piston.

Thousand Miles	Reliability Estimate	
	Cylinders	Pistons
0	1.00000	1.00000
25	0.99582	0.99907
50	0.99195	0.99907
75	0.99195	0.99761
100	0.96544	0.99528
125	0.94068	0.99372
150	0.90731	0.99082
175	0.84526	0.98738
200	0.80437	0.98186
225	0.77076	0.97216
250	0.71639	0.96290
275	0.66887	0.95554

(a)   Make a Weibull probability plot of the estimate of the reliability function of current pistons. Relabel the probability scale with reliability percentages.

(b)   On the same paper, plot the cylinder reliability function estimate.

(c)   Use the reliability function estimates to calculate an estimate of the reliability function for cylinders with the new piston design. This estimate of reliability increases at some mileages.

(d)   Plot the estimate from part (c) on the same Weibull paper.

(e)   Compare the failure rates of the three distributions—increasing or decreasing with mileage?

(f)   Discuss whether you think customers could be induced to pay a premium price for the new piston design. What important economic considerations determine whether cylinder life has been improved significantly?

5.4.   Oil breakdown voltage.   An insulating oil was tested between a pair of parallel disk electrodes under a voltage that increased linearly with time. The oil breakdown voltage was measured 60 times each, with two sizes of electrodes. According to theory, a pair of 3-in.-diameter electrodes act like a series system of nine independent pairs of 1-in.-diameter electrodes. The following analyses check this.

1-in. Diameter

57	59	56	56	58	64	58	55	58	54
65	61	64	65	65	52	53	60	58	63
60	62	54	63	60	52	62	50	60	57
68	57	57	58	52	67	52	62	56	59
55	65	63	57	67	64	62	58	66	60
57	64	66	52	65	57	58	62	60	59

3-in. Diameter

57	49	49	41	52	40	48	48	43	45
57	54	49	49	52	53	51	46	55	54
49	51	50	49	51	49	47	55	49	51
51	50	50	55	46	55	57	53	54	54
54	41	60	50	55	54	53	54	53	46
55	50	59	58	60	55	55	56	59	51

(a) Make a Weibull hazard plot of the voltage data from the 1-in.-diameter electrodes.

(b) Carry out the hazard calculations for the data from the 3-in.-diameter electrodes, treating each failure as the first among nine 1-in. electrodes. The eight survivor times immediately follow their corresponding failure time.

(c) Plot the results of (b) on the paper with the 1-in. data.

(d) Are the two sets of data comparable?

5.5. Metal fatigue. A low-cycle fatigue test of 384 specimens of a superalloy at a particular strain range and high temperature yielded the following data. Data are grouped by the log (base 10) of the number of cycles to failure into intervals of 0.05. The data show the number of specimens failing from each cause. S denotes failure due to surface defect, I denotes failure due to interior defect. The following analyses indicate the potential improvement in life that would result if a research program yields a new manufacturing method that eliminates surface defects.

(a) On normal paper (for log cycles), make a hazard or probability plot of the distribution when both failure modes act. Take the grouping of the data into account.

(b) Is the usual assumption of lognormal fatigue life of metal specimens adequate?

(c)   Estimate σ and the 50 and 1% points of this fatigue life distribution. Usually one is interested only in low percentage points of a fatigue life distribution. Do you recommend fitting a line to the entire sample to estimate the 1% point? Why?

(d)   On normal paper, make a plot of the distribution of log cycles to failure from interior defect, assuming that failure due to surface defect is eliminated.

(e)   Are failures due to interior defect adequately described by a lognormal fatigue distribution?

(f)   Estimate σ and the 50 and 1% points of this "potential" fatigue life distribution.

(g)   Do (d), (e), and (f) for surface defects.

(h)   Some specimens may not have surface defects and would be prone only to failure from interior defects. How would this affect preceding results?

(i)   The assumption of independent competing failure modes may not hold. How does dependence affect the preceding results?

Log Lower Endpoint	Defect		Log Lower Endpoint	Defect	
	I	S		I	S
3.55		3	4.25	2	3
3.60		6	4.30	3	4
3.65		18	4.35	7	7
3.70	1	23	4.40	12	4
3.75	3	47	4.45	7	2
3.80	1	44	4.50	4	3
3.85		57	4.55	2	1
3.90	1	37	4.60	1	
3.95		20	4.65	1	
4.00		16	4.70		
4.05	3	12	4.75	1	
4.10	3	4	4.80		
4.15	2	3	4.85	1	
4.20	5	10	Total	60	324

(j)   A common *mistake* in estimating the distribution for a particular mode is to treat the failures from that mode as if they were a complete sample, ignoring failures from any other modes. Plot the interior failures as if they were a complete sample of 60 failure times. Compare this plot with the correct plot, especially the 1% and 50% points.

(k)   Repeat (j) for the 324 surface failures.

5.6.*   Probability of failure from a particular cause. Suppose that a series system can fail from one of K independent competing causes with cumulative life distributions $F_1(y),\ldots, F_K(y)$.

(a)   Derive an expression for the proportion $P_k(y)$ of the population that fail from cause k by age y.

(b)   Show $P_1(y)+ \cdots +P_K(y)=F(y)$, the cumulative distribution of time to failure when all causes act.

(c)   Give the special result from (a) when the distributions are all exponential with means θ_k, $k=1,\ldots, K$.

(d)   For (c), what proportion of the failures by age y have come from cause k?

(e)   Give the special result from (a) when all distributions are Weibull distributions with differing scale parameters and a common shape parameter.

(f)   For (e), what proportion of the failures by age y have come from cause k?

---

*Asterisk denotes laborious or difficult.

6

Analysis of Complete Data

This chapter presents standard analytic methods for complete data. Such life data consist of the failure times of all units in the sample. Chapter 3 presents simple graphical methods for analyzing such data. Before using analytic methods, it is important to check with a probability plot that a chosen distribution fits the data (Chapter 3). A combination of graphical and analytic methods is often most informative. Later chapters present methods for analysis of censored data having running times on unfailed units; those methods also apply to complete data.

Section 1 explains the basic ideas for data analysis. Subsequent sections present data analyses for the basic distributions: (1) Poisson, (2) binomial, (3) exponential, (4) normal and lognormal, and (5) Weibull and extreme value distributions. To use these methods, one needs to know the distribution (Chapter 2). Section 7 presents distribution-free (nonparametric) methods. The nonparametric methods apply to data from any distribution, and they may be used even if the distribution is a known parametric one such as those in Sections 2 through 6.

For each basic distribution, this chapter presents estimates and confidence limits for parameters, percentiles, and reliabilities. Such estimates and confidence limits are covered in a first course. Thus they are presented here as handbook formulas, with little space devoted to motivation and derivation. Many statistical computer programs calculate such estimates and confidence limits. Readers lacking previous acquaintance with such basics can refer to Section 1 for a brief general technical introduction to such ideas.

For each basic distribution, this chapter also presents predictions and prediction limits for a **random** value of a **future** sample. For example, one may wish to use past data to obtain a prediction for and limits that enclose the total warranty cost for the coming year **or** the number of product failures in the coming quarter. In contrast, estimates and confidence limits are for **constant** values of a **population**, which is generally assumed large compared to the sample. Relatively new, prediction problems often go unrecognized and are incorrectly treated with estimates and confidence intervals for population values. Nevertheless, prediction methods are presented here with little motivation and no derivation. Readers can refer to Section 1 for a brief general technical introduction to prediction.

Other distributions are sometimes used to analyze life data. These include the negative binomial, logistic, gamma, and shifted exponential, Weibull, and lognormal distributions. Books that survey methods for these and other distributions have been written by Johnson and Kotz (1970), Patil and Joshi (1968), Bain (1978), Mann, Schafer, and Singpurwalla (1974), and Lawless (1982). Also, journal articles on specific distributions are listed in indexes of journals and in Joiner, et al. (1970) and Joiner (1975).

1. BASIC IDEAS FOR DATA ANALYSIS

This section briefly presents basic ideas for data analysis. These ideas include a statistic and its sampling distribution, estimator, confidence intervals, sample size, and prediction (predictors and limits). Later sections apply these ideas to analyses of complete data from the basic distributions. As this section is advanced, readers may first wish to read later sections.

In this chapter, the n observations in a sample are denoted by $Y_1, \ldots, Y_n$. Generally, the population is assumed to be infinite and random sampling is used; that is, the observations are statistically independent and from the same distribution.

A Statistic and Its Sampling Distribution

A **statistic** is a numerical value determined from a sample by some procedure and denoted by $\theta^*(Y_1, \ldots, Y_n)$, a function of the data values. Sample statistics include the sample average, standard deviation, smallest observation, largest observation, median, etc. Suppose one could repeatedly take random samples of given size n from a distribution and determine the sample value of a particular statistic. The values of the statistic would differ from sample to sample and would have a distribution, which is called the **sampling distribution** of the statistic. The form of a sampling distribution depends on the statistic, the sample size, and the parent distribution. For example, for samples from a normal distribution, the sample average has a sampling distribution that is normal. The term **statistic** can refer to (1) the procedure used to determine the statistic value, (2) the value determined for a given sample, or (3) the sampling distribution.

Estimator

A statistic $\theta^*(Y_1, \ldots, Y_n)$ that approximates an unknown population value θ is called an **estimate** when one refers to its value for a particular sample; it is called an **estimator** when one refers to the procedure or the sampling distribution. In practice, one does not know how close a sample estimate is to the population value (called the true value). Instead one judges an estimation method by the sampling distribution. Ideally the distribution should be centered on the true value and have a small spread.

An estimator θ^* for θ is called **unbiased** if the mean $E\theta^*$ of its sampling distribution equals θ. If different from zero, $E\theta^* - \theta$ is called the **bias** of θ^*. Unbiasedness is a reasonable property, but not an essential one. Unbiased estimators receive much attention in the mathematical literature, partly because they are mathematically tractable. Figure 1.1 shows two sampling distributions where the biased one tends to yield estimates closer to the population value and so is better. Both biased and unbiased estimators are presented later.

The spread in the sampling distribution of an estimator θ^* should be small. The usual measure of spread is the distribution variance $\text{Var}(\theta^*)$ or, equivalently, its positive square root, the standard deviation $\sigma(\theta^*)$, called the **standard error** of the estimator. One usually wants estimators that are (almost) unbiased and have small standard errors. A useful biased estimator has a small bias compared to its standard error. If an estimator θ^* for θ is biased, then the **mean square error** (MSE) is a measure of the uncertainty in the estimator:

$$\text{MSE}(\theta^*) \equiv E\left[(\theta^* - \theta)^2\right] = \int_{-\infty}^{\infty} (\theta^* - \theta)^2 f(\theta^*) \, d\theta^*,$$

where $f(\ )$ is the probability density of θ^*. The mean square error, variance,

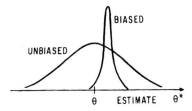

Figure 1.1.   Biased and unbiased estimators.

and bias of an estimator are related by

$$MSE(\theta^*) = Var(\theta^*) + (E\theta^* - \theta)^2.$$

The MSE combines the spread and miscentering of the sampling distribution.

The sampling distributions of most statistics are complex. However, for large sample sizes, many have a cumulative distribution that is close to a normal one. The statistic's mean and standard error completely specify the approximating normal distributions. For example, for large samples of size n from a parent distribution with a mean μ and standard deviation σ, the sample average has a cumulative distribution that is approximately normal with a mean μ and a standard error $\sigma/\sqrt{n}$. This well-known result comes from the central limit theorem for independent, identically distributed random variables. The normal distribution often yields useful approximate results when exact results are unknown or too difficult to obtain. If one selects among unbiased estimators that are approximately normally distributed, the one with the smallest variance is best. Also, then an estimate falls within two standard errors of the true value with about 95% probability, and it is within three standard errors of the true value, with about 99.7% probability.

Confidence Intervals

Estimation provides a single estimate of a population value. The uncertainty in such an estimate may be judged from its standard error. A confidence interval also indicates the uncertainty in an estimate. Calculated from sample data, it encloses the population value with a specified high probability. The length of such an interval indicates if the corresponding estimate is accurate enough for practical purposes. Confidence intervals are generally wider than inexperienced data analysts expect; so confidence intervals help one avoid thinking that estimates are closer to the true value than they really are. Also, the real uncertainty is usually greater than a confidence interval indicates because the interval is based on certain assumptions about the data and model; for example, (1) the sample is a random one (2) from the population of interest, and (3) the assumed theoretical distribution is close enough to the true one for practical purposes. Departures from the assumptions add to the uncertainties in the results.

The following defines a confidence interval for a population value θ. Suppose $\underset{\sim}{\theta} = \underset{\sim}{\theta}(Y_1, \ldots, Y_n)$ and $\bar{\theta} = \bar{\theta}(Y_1, \ldots, Y_n)$ are functions of the sample data $Y_1, \ldots, Y_n$ such that

$$P_\theta\{\underset{\sim}{\theta} \leqslant \theta \leqslant \bar{\theta}\} = \gamma, \tag{1.1}$$

no matter what the value of θ or any distribution parameter. Then the interval $[\underset{\sim}{\theta}, \tilde{\theta}]$ is called a **two-sided $100\gamma\%$ confidence interval** for θ. $\underset{\sim}{\theta}$ and $\tilde{\theta}$ are the lower and upper **confidence limits**, or bounds. The random limits $\underset{\sim}{\theta}$, and $\tilde{\theta}$ enclose θ with probability γ.

Similarly, suppose that $\tilde{\theta} = \tilde{\theta}(Y_1, \ldots, Y_n)$ satisfies

$$P_\theta\{\theta \leqslant \tilde{\theta}\} = \gamma, \tag{1.2}$$

no matter what the value of θ or any distribution parameter. Then the interval $(-\infty, \tilde{\theta})$ is called a **"one-sided $100\gamma\%$ upper confidence interval** for θ," and $\tilde{\theta}$ is called a "one-sided **$100\gamma\%$ upper confidence limit** for θ." One-sided lower confidence intervals and limits are defined similarly. In most applications, it is clear whether to use a one-sided or a two-sided interval. For example, one usually wants a one-sided upper limit for a fraction failing on warranty or a one-sided lower limit for reliability, but two-sided limits for a Weibull shape parameter.

One- and two-sided confidence intervals are related. The upper (lower) limit of a two-sided $100\gamma\%$ confidence interval for a value θ is a one-sided upper (lower) $[100(1+\gamma)/2]\%$ confidence limit for θ. For example, the upper limit of a two-sided 90% interval is the upper limit of a one-sided 95% confidence interval.

Most people would like an interval with high confidence. However, the width of a confidence interval increases with the confidence level, and a 99.9% confidence interval, for example, may be so wide that it has little value in an application. Most data analysts use 90, 95, and 99% confidence intervals. Physicists often use short 50% confidence intervals (for the "probable error"), as they flatter the data. One can calculate and present a number of intervals for different confidence levels.

Approximate confidence limits for a population value θ can be obtained from an (almost) unbiased estimator θ^* that is approximately normally distributed. For example, then

$$P\{(\theta^* - \theta)/\sigma(\theta^*) \leqslant z_\gamma\} \simeq \gamma, \tag{1.3}$$

no matter what the value of θ, where z_γ is the 100γth standard normal percentile. (1.3) becomes

$$P_\theta\{\theta \leqslant \theta^* + z_\gamma\sigma(\theta^*)\} \simeq \gamma. \tag{1.4}$$

This means that

$$\tilde{\theta} = \theta^* + z_\gamma\sigma(\theta^*) \tag{1.5}$$

is a one-sided approximate $100\gamma\%$ upper confidence limit for θ. Often $\sigma(\theta^*)$ is not known and must be estimated from the data. Similarly, a one-sided approximate $100\gamma\%$ lower confidence limit for θ is

$$\underline{\theta} = \theta^* - z_\gamma \sigma(\theta^*), \tag{1.6}$$

and two-sided approximate $100\gamma\%$ confidence limits for θ are

$$\underline{\theta} = \theta^* - K_\gamma \sigma(\theta^*), \qquad \tilde{\theta} = \theta^* + K_\gamma \sigma(\theta^*), \tag{1.7}$$

where K_γ is the $[100(1+\gamma)/2]$th standard normal percentile. Such approximate intervals tend to be narrower than exact ones. Also, a two-sided interval tends to have a confidence closer to $100\gamma\%$ than does a one-sided interval if the sampling distribution of θ^* is unsymmetrical.

Sample Size

Often one must determine a suitable sample size. To do this, one specifies how precise the chosen estimator θ^* for θ must be. Usually one wants the estimate to be within a specified $\pm w$ of θ, with $100\gamma\%$ probability. The following is an approximate sample size n when θ^* is approximately normally distributed with variance V^*/n; V^* may depend on θ and other distribution parameters. Then

$$n \simeq V^* \left(K_\gamma/w \right)^2. \tag{1.8}$$

In practice, the value of V^* is usually unknown, and one must estimate it from experience, a preliminary sample, or similar data. Mace (1964) gives similar but more refined methods for determining sample size to achieve specified expected length of confidence intervals. Gross and Clark (1975, Chap. 8) give guidance on the choice of sample size.

If θ is positive, one may specify that θ^* be within a factor f of θ with probability $100\gamma\%$. That is, θ^* is between θ/f and θf, with $100\gamma\%$ probability. Suppose that $\mathrm{Var}(\theta^*) = V/n$, where n is the sample size and V is a known factor. The appropriate sample size n is

$$n \simeq K_\gamma^2 (V/\theta^2) / [\ln(f)]^2, \tag{1.9}$$

where K_γ is the $[100(1+\gamma)/2]$th standard normal percentile. V/θ^2 usually depends on unknown distribution parameters that must be estimated from experience, a preliminary sample, or similar data. The formula assumes that n is large enough that the distribution of $\ln(\theta^*)$ is approximately normal.

Prediction

The statistical methods above use data from a sample to get information on a distribution, usually its parameters, percentiles, and reliabilities. However, in business and engineering, one is often concerned about a future sample from the same distribution. In particular, one usually wants to use a past sample to predict the future sample value of some statistic and to enclose it with prediction limits. The following presents such two-sample prediction theory.

Predictor. Suppose that $Y_1, \ldots, Y_n$ and $X_1, \ldots, X_m$ are all independent observations in a past and future sample from the same distribution F_θ, where θ denotes one or more unknown distribution parameters. Suppose one wishes to predict the random value $U = u(X_1, \ldots, X_m)$ of some statistic of the future sample. To do this, one uses a **predictor** $U^* = u^*(Y_1, \ldots, Y_n)$ that is a function of the past sample.

The difference $U^* - U$ is the **prediction error**; its sampling distribution depends on the one or more parameters θ. If its mean is 0 for all values of θ, U^* is called an **unbiased predictor** for U. The variance $\text{Var}_\theta(U^* - U)$ of the prediction error or its square root, the standard prediction error, $\sigma_\theta(U^* - U)$ should be small.

Prediction interval. Often one wants an interval that encloses a single future statistic U with high probability. The interval width indicates the statistical uncertainty in the prediction. Statistics $\underset{\sim}{U} = \underset{\sim}{u}(Y_1, \ldots, Y_n)$ and $\tilde{U} = \tilde{u}(Y_1, \ldots, Y_n)$ of the past sample that satisfy

$$P_\theta\{\underset{\sim}{U} < U < \tilde{U}\} = \gamma \tag{1.10}$$

for any value of θ are called "two-sided $100\gamma\%$ **prediction limits** for U." One-sided prediction limits are defined in the obvious way. One can think about such an interval in terms of a large number of pairs of past and future samples. An expected proportion γ of such intervals will enclose U in the corresponding future sample. Prediction limits for a random sample value (say, a future sample mean) are wider than confidence limits for the corresponding constant population value (say, a population mean).

Suppose that the sampling distribution of the prediction error is approximately normal. Then a two-sided approximate $100\gamma\%$ prediction interval for U has limits

$$\underset{\sim}{U} = U^* - K_\gamma \sigma_\theta(U^* - U), \qquad \tilde{U} = U^* + K_\gamma \sigma_\theta(U^* - U), \tag{1.11}$$

where K_γ is the $[100(1+\gamma)/2]$th standard normal percentile. One-sided approximate prediction limits use the 100γth percentile z_γ in place of K_γ. In practice, the standard prediction error $\sigma_\theta(U^* - U)$ is usually unknown and is estimated from the data.

There are also prediction intervals to enclose simultaneously a number of future statistics, for example, to enclose all observations in a future sample.

Tests for Outliers

An outlier is a data value that is far from the rest of the sample. An explanation for it should be sought to reveal faulty or improved products or poor test methods. Such outliers stand out in probability plots. This chapter does not give formal methods for identifying outliers. Such methods are surveyed by Grubbs (1969), David (1970), Hawkins (1980), and Barnett and Lewis (1978).

2. POISSON DATA

The Poisson distribution is a model for the number of occurrences in a given length of observation. This distribution is described in detail in Chapter 2. Poisson data consist of the number Y of occurrences in a "length" t of observation.

This section gives estimates and confidence intervals for the Poisson occurrence rate λ and guidance on sample size. Also, it gives predictions and prediction limits for the number of occurrences in a future sample. Haight (1967) gives justification for the results below.

Estimate of λ

Estimate. The estimate for the true occurrence rate λ is the sample occurrence rate

$$\hat{\lambda} = Y/t. \tag{2.1}$$

It is unbiased, and

$$\mathrm{Var}(\hat{\lambda}) = \lambda/t. \tag{2.2}$$

The sampling distribution of $Y = \hat{\lambda}t$ is the Poisson distribution with mean λt. Suppose the expected number of occurrences λt is large, say, greater than 10. Then the distribution of $\hat{\lambda}$ is approximately normal with a mean of λ and variance (2.2).

Power line example. Two types of power wire were used in a region. A new tree wire had $Y_1 = 12$ failures in $t_1 = 467.9$ 1000 ft·years of exposure, and the standard bare wire had $Y_2 = 69$ and $t_2 = 1079.6$. For tree wire, $\hat{\lambda}_1 = 12/467.9 = 0.0256$ failures per 1000 ft·years; for bare wire, $\hat{\lambda}_2 = 69/1079.6 = 0.0639$.

Confidence Interval for λ

Poisson limits. Conservative two-sided $100\gamma\%$ confidence limits for λ are

$$\underline{\lambda}=0.5\chi^2\big[(1-\gamma)/2;2Y\big]/t, \qquad \tilde{\lambda}=0.5\chi^2\big[(1+\gamma)/2;2Y+2\big]/t, \quad (2.3)$$

where $\chi^2(\delta;\nu)$ is the 100δth percentile of the chi-square distribution with ν degrees of freedom. Such a one-sided lower (upper) limit is

$$\underline{\lambda}=0.5\chi^2(1-\gamma;2Y)/t, \tag{2.4}$$

$$\tilde{\lambda}=0.5\chi^2(\gamma;2Y+2)/t. \tag{2.5}$$

$\tilde{\lambda}$ is often used in reliability work. The limits above are conservative in that the confidence level is at least $100\gamma\%$; exact confidence levels cannot be achieved conveniently for discrete distributions. Nelson (1972a) gives simple charts for the limits (2.3), (2.4), and (2.5).

For tree wire, two-sided 95% confidence limits are $\underline{\lambda}_1=0.5\chi^2[(1-0.95)/2;2\cdot12]/467.9=0.0133$ and $\tilde{\lambda}_1=0.5\chi^2[(1+0.95)/2;2\cdot12+2]/467.9=0.0444$ failures per 1000 ft·years. Each limit is a one-sided 97.5% confidence limit.

Normal approximation. When Y is large, two-sided approximate $100\gamma\%$ confidence limits for λ are

$$\underline{\lambda}\cong\hat{\lambda}-K_\gamma(\hat{\lambda}/t)^{1/2}, \qquad \tilde{\lambda}\cong\hat{\lambda}+K_\gamma(\hat{\lambda}/t)^{1/2}, \tag{2.6}$$

where K_γ is the $[100(1+\gamma)/2]$th standard normal percentile. Such a one-sided lower (upper) limit is

$$\underline{\lambda}\simeq\hat{\lambda}-z_\gamma(\hat{\lambda}/t)^{1/2}, \qquad \tilde{\lambda}=\hat{\lambda}+z_\gamma(\hat{\lambda}/t)^{1/2}, \tag{2.7}$$

where z_γ is the 100γth standard normal percentile. These limits employ the approximate normal distribution of $\hat{\lambda}$; so Y should exceed, say, 10 for most practical purposes.

For bare wire, the two-sided approximate 95% confidence limits are $\underline{\lambda}_2 = 0.0639 - 1.960(0.0639/1079.6)^{1/2} = 0.0488$ and $\tilde{\lambda}_2 = 0.0639 + 1.960(0.0639/1079.6)^{1/2}=0.0790$ failures per 1000 ft·years. Each limit is a one-sided approximate 97.5% confidence limit.

Choice of t for Estimating λ

The estimate $\hat{\lambda}$ is within $\pm w$ of λ with approximate probability $100\gamma\%$ if

$$t = \lambda (K_\gamma/w)^2, \tag{2.8}$$

where K_γ is the $[100(1+\gamma)/2]$th standard normal percentile. In (2.8), one must approximate the unknown λ. This formula employs the normal approximation for the distribution of $\hat{\lambda}$ and is more accurate the larger λt is (say $\lambda t > 10$ for most practical purposes).

Suppose that λ_1 of tree wire is to be estimated within ± 0.0050 failures per 1000 ft·years with 95% probability. One needs approximately $t = 0.0256(1.960/0.0050)^2 = 3,940$ 1000 ft·years. $\lambda_1 t \cong 0.0256(3940) = 100.9 > 10$; so the normal approximation is satisfactory.

Estimate and Confidence Limits for a Function of λ

Suppose $h = h(\lambda)$ is a function of λ. For example, the Poisson probability of y or fewer failures is the function $F_\lambda(y) = \sum_{i=0}^{y} \exp(-\lambda t)(\lambda t)^i/i!$. The usual estimate of h is $\hat{h} = h(\hat{\lambda})$. $\hat{h}$ may be a biased estimator for h, but its bias decreases as t increases. The approximate variance of $\hat{h}$ is

$$\text{Var}(\hat{h}) \simeq [\partial h(\lambda)/\partial\lambda]^2 \lambda/t, \tag{2.9}$$

where the partial derivative is evaluated at the true λ value.

Confidence limits for h are simple when h is a monotone function of λ. For example, $F_\lambda(y)$ decreases with increasing λ. Suppose that $\utilde{\lambda}$ and $\tilde{\lambda}$ are two-sided $100\gamma\%$ confidence limits for λ. Then $\utilde{h} = h(\utilde{\lambda})$ and $\tilde{h} = h(\tilde{\lambda})$ are such limits for an increasing function of λ, and $\utilde{h} = h(\tilde{\lambda})$ and $\tilde{h} = h(\utilde{\lambda})$ are such limits for a decreasing function. This method also yields one-sided limits for h.

Prediction

Often one seeks information on the random number of occurrences in a future sample. For example, those who maintain power lines need to predict the number of line failures in order to plan the number of repair crews.

The following gives a prediction and prediction limits for the number X of occurrences in a future observation of length s with true rate λ. Suppose the past data consist of Y occurrences in an observation of length t with the same true rate λ. The random variations in X and Y are assumed to be statistically independent. Nelson (1970) justifies the results below.

Prediction.   The observed rate of occurrence is $\hat{\lambda} = Y/t$. So the prediction of the future number is

$$\hat{X} = \hat{\lambda}s = (Y/t)s. \tag{2.10}$$

This is unbiased, and the prediction error $(\hat{X} - X)$ has variance

$$\mathrm{Var}(\hat{X} - X) = \lambda s(t+s)/t. \tag{2.11}$$

If λt and λs are large (say, both exceed 10), the distribution of $(\hat{X} - X)$ is approximately normal with mean 0 and variance (2.11).

If all the power line in the region were tree wire, there would be $s = 515.8$ 1000-ft. Then the prediction of the number of failures in the coming year would be $\hat{X} = (12/467.9)515.8 = 13.2$ failures, better rounded to the nearest integer, 13.

Prediction limits.   Two-sided $100\gamma\%$ prediction limits for the future number X of occurrences are the closest integer solutions $\underset{\sim}{X}$ and $\tilde{X}$ of

$$\tilde{X}/s = [(Y+1)/t]\, F[(1+\gamma)/2; 2Y+2, 2\tilde{X}],$$
$$s/(\underset{\sim}{X}+1) = (t/Y)F[(1+\gamma)/2; 2\underset{\sim}{X}+2, 2Y], \tag{2.12}$$

where $F(\delta; a, b)$ is the 100δth F percentile with a degrees of freedom in the numerator and b in the denominator. Such a one-sided lower (upper) limit for X is the closest integer solution $\underset{\sim}{X}$ ($\tilde{X}$) of

$$(\tilde{X}/s) = [(Y+1)/t]\, F(\gamma; 2Y+2, 2\tilde{X}), \tag{2.13}$$
$$s/(\underset{\sim}{X}+1) = (t/Y)F(\gamma; 2\underset{\sim}{X}+2, 2Y). \tag{2.14}$$

Nelson (1970) gives simple charts for $\underset{\sim}{X}$ and $\tilde{X}$. Prediction limits for the random X/s are wider than confidence limits (2.3) for the constant λ.

Two-sided 95% prediction limits for the number of tree wire failures next year are $\underset{\sim}{X} = 4.0$ and $\tilde{X} = 26.0$. Each limit is a one-sided approximate 97.5% prediction limit.

Normal approximation.   When Y and X are large, two-sided approximate 100% prediction limits for X are

$$\underset{\sim}{X} \cong \hat{X} - K_\gamma \big[\hat{\lambda}s(t+s)/t\big]^{1/2}, \qquad \tilde{X} \cong \hat{X} + K_\gamma \big[\hat{\lambda}s(t+s)/t\big]^{1/2}, \tag{2.15}$$

where K_γ is the $[100(1+\gamma)/2]$th standard normal percentile. These limits employ the approximate normal distribution of $(\hat{X}-X)$; so Y (or λt) and λs should be large, say, both exceeding 10. For such a one-sided limit, replace K_γ by z_γ, the 100γth standard normal percentile.

Two-sided approximate 95% prediction limits for the number of tree wire failures next year are $\underset{\sim}{X} = 13.2 - 1.960[(0.0256)515.8(467.9 + 515.8)/467.9]^{1/2} = 2.9$ failures and $\tilde{X} = 13.2 + 1.960[(0.0256)515.8(467.9 + 515.8)/467.9]^{1/2}=23.5$ failures, that is, 3 to 24 failures. Here $Y=12>10$ and $\hat{\lambda}s=0.0256(515.8)=13.2>10$; and the approximation is adequate.

3. BINOMIAL DATA

This section presents estimates and confidence intervals for the binomial probability p and guidance on sample size. Also, it presents prediction and prediction limits for the number of failures in a future sample. Binomial data consist of the number Y of "category" units among n statistically independent sample units, which each have the same probability p of being a category unit.

Estimate of p

The estimate of the population proportion p is the sample proportion

$$\hat{p}=Y/n. \tag{3.1}$$

This is unbiased, and

$$\text{Var}(\hat{p})=p(1-p)/n. \tag{3.2}$$

The sampling distribution of $n\hat{p}$ is the binomial distribution of Y. If np and $n(1-p)$ are large (say, exceed 10), the distribution of $\hat{p}$ is approximately normal, with a mean of p and variance (3.2).

Locomotive control example. Of $n=96$ locomotive controls on test, $Y=15$ failed on warranty. The estimate of the population proportion failing on warranty is $\hat{p}=15/96=0.156$ or 15.6%.

Confidence Interval for p

Standard limits. Two-sided $100\gamma\%$ confidence limits for p are

$$\underset{\sim}{p} = \{1+(n-Y+1)Y^{-1}F[(1+\gamma)/2;2n-2Y+2,2Y]\}^{-1},$$

$$\tilde{p} = \left(1+(n-Y)\{(Y+1)F[(1+\gamma)/2;2Y+2,2n-2Y]\}^{-1}\right)^{-1}, \tag{3.3}$$

where $F(\delta; a, b)$ is the 100δ th F percentile with a degrees of freedom in the numerator and b in the denominator. Such a lower (upper) one-sided limit for p is

$$\underline{p} = \left[1 + (n - Y + 1) Y^{-1} F(\gamma; 2n - 2Y + 2, 2Y)\right]^{-1}, \tag{3.4}$$

$$\tilde{p} = \left\{1 + (n - Y)\left[(Y + 1) F(\gamma; 2Y + 2, 2n - 2Y)\right]^{-1}\right\}^{-1}. \tag{3.5}$$

Obtaining such limits is easier with the Clopper–Pearson charts in Appendix A8. Enter the appropriate chart on the horizontal axis at $\hat{p} = Y/n$; go vertically to the curve labeled with the sample size n (separate curves for the lower and upper limits); and go horizontally to the vertical scale to read the confidence limit. Also, there are tables of such limits in, for example, Owen (1962) and Natrella (1963).

Such limits are conservative in that the confidence level is at least $100\gamma\%$; exact confidence levels cannot conveniently be achieved for discrete distributions.

For a 95% confidence interval for the locomotive control, enter the 95% chart in Appendix A8 on the horizontal axis at $\hat{p} = 15/96 = 0.156$; go vertically to a curve for $n = 96$ (interpolate); and go horizontally to the vertical scale to read $\underline{p} = 9\%$ and $\tilde{p} = 24\%$. Each limit is a one-sided 97.5% confidence limit for p.

Normal approximation. If Y and $n - Y$ are large, two-sided approximate $100\gamma\%$ confidence limits for p are

$$\underline{p} \cong \hat{p} - K_\gamma\left[\hat{p}(1 - \hat{p})/n\right]^{1/2}, \qquad \tilde{p} \cong \hat{p} + K_\gamma\left[\hat{p}(1 - \hat{p})/n\right]^{1/2}, \tag{3.6}$$

where K_γ is the $[100(1 + \gamma)/2]$th standard normal percentile. Such a one-sided lower (upper) $100\gamma\%$ confidence limit is

$$\underline{p} \cong \hat{p} - z_\gamma\left[\hat{p}(1 - \hat{p})/n\right]^{1/2}, \tag{3.7}$$

$$\tilde{p} \cong \hat{p} + z_\gamma\left[\hat{p}(1 - \hat{p})/n\right]^{1/2}, \tag{3.8}$$

where z_γ is the 100γ th standard normal percentile. These limits employ the approximate normal distribution of $\hat{p}$; so $n\hat{p} = Y$ and $n(1 - \hat{p}) = n - Y$ should exceed, say, 10 for most purposes.

For the locomotive control, such 95% limits are

$$\underset{\sim}{p} \cong 0.156 - 1.960[0.156(1-0.156)/96]^{1/2} = 0.083,$$

$$\tilde{p} \cong 0.156 + 1.960[0.156(1-0.156)/96]^{1/2} = 0.229,$$

that is, 8.3 to 22.9%. $Y=15>10$ and $n-Y=81>10$ are large enough. Each limit is a one-sided 97.5% confidence limit for p.

Poisson approximation. For n large and Y small, two-sided approximate $100\gamma\%$ confidence limits for p are

$$\underset{\sim}{p} \cong 0.5\chi^2[(1-\gamma)/2;2Y]/n, \qquad \tilde{p} \cong 0.5\chi^2[(1+\gamma)/2;2Y+2]/n, \quad (3.9)$$

where $\chi^2(\delta; \nu)$ is the 100δth percentile of the chi-square distribution with ν degrees of freedom. Such a one-sided lower (upper) limit is

$$\underset{\sim}{p} \cong 0.5\chi^2(1-\gamma;2Y)/n, \tag{3.10}$$

$$\tilde{p} \cong 0.5\chi^2(\gamma;2Y+2)/n. \tag{3.11}$$

These limits employ the Poisson approximation to the distribution of Y; so Y should be small, say, under $n/10$.

For the locomotive control, two-sided approximate 95% confidence limits are

$$\underset{\sim}{p} \cong 0.5\chi^2(0.025;30)/96 = 0.5(16.8)/96 = 0.087,$$

$$\tilde{p} \cong 0.5\chi^2(0.975;32)/96 = 0.5(49.5)/96 = 0.258,$$

that is, 8.7% to 25.8%. Each limit is a one-sided 97.5% confidence limit. Here $Y=15>n/10=9.6$; so the approximate limits are a bit crude.

Sample Size n to Estimate p

The estimate $\hat{p}$ is within $\pm w$ of p with approximately $100\gamma\%$ probability if

$$n \cong p(1-p)(K_\gamma/w)^2, \tag{3.12}$$

where K_γ is the $[100(1+\gamma)/2]$th standard normal percentile. To use (3.12), one must approximate the unknown p. If one does not want to approximate

p, the largest sample size results from $p=1/2$. Then

$$n \cong 0.25\left(K_\gamma/w \right)^2 \qquad (3.13)$$

is sure to be large enough. For $0.3<p<0.7$, (3.12) is close to (3.13), which is conservative and thus often preferable. The formulas employ the normal approximation for the distribution of $\hat{p}$ and are usually adequate if $np>10$ and $n(1-p)>10$.

Suppose that the percentage of locomotive controls failing on warranty is to be estimated within $\pm 5\%$ with 95% probability. Then $n \cong 0.156(1-0.156)(1.960/0.05)^2=202$ controls; $\hat{p}$ was used in place of the unknown p. The upper limit is $n \cong 0.25(1.960/0.05)^2=384$ controls in all. Here $n\hat{p}=202(0.156)=31.5>10$, and $n(1-\hat{p})=202(1-0.156)=170.5>10$. So the approximation is satisfactory.

Prediction

The following provides a prediction and prediction limits for the number X of category units in a future sample of m units. Suppose the past sample consists of Y category units in a sample of n units from the same population.

The previously observed proportion is $\hat{p}=Y/n$. So the prediction of the number X of future category units is

$$\hat{X}=m\hat{p}=m(Y/n). \qquad (3.14)$$

This is unbiased. Its prediction error variance is

$$\text{Var}(\hat{X}-X)=mp(1-p)(m+n)/n, \qquad (3.15)$$

where p is the true unknown population proportion. If np, $n(1-p)$, mp, and $m(1-p)$ are all large (say, all exceed 10), then the distribution of the prediction error $(\hat{X}-X)$ is approximately normal, with mean zero and variance (3.15).

Locomotive control example. Suppose that the proposed production is $m=900$ controls. The previous sample consisted of $Y=15$ failures among a sample of $n=96$ controls; thus, $\hat{p}=15/96=0.156$. The prediction of the number of failures in the proposed production is $\hat{X}=900(15/96)=141$ controls.

Prediction Limits

Exact limits. Exact prediction limits involve the hypergeometric distribution and are laborious to calculate (Lieberman and Owen, 1961). The following approximate limits will serve for most applications.

Normal approximation. For Y, $n-Y$, X, and $m-X$ all large (say, greater than 10), two-sided approximate $100\gamma\%$ prediction limits for the future number X of category units are

$$\underset{\sim}{X} \cong \hat{X} - K_\gamma \left[m\hat{p}(1-\hat{p})(m+n)/n \right]^{1/2},$$

$$\tilde{X} \cong \hat{X} + K_\gamma \left[m\hat{p}(1-\hat{p})(m+n)/n \right]^{1/2},$$

(3.16)

where K_γ is the $[100(1+\gamma)/2]$th standard normal percentile. Such a one-sided lower (upper) limit for X is

$$\underset{\sim}{X} \cong \hat{X} - z_\gamma \left[m\hat{p}(1-\hat{p})(m+n)/n \right]^{1/2},$$ (3.17)

$$\tilde{X} \cong \hat{X} + z_\gamma \left[m\hat{p}(1-\hat{p})(m+n)/n \right]^{1/2},$$ (3.18)

where z_γ is the 100γth standard normal percentile. These limits employ the approximate normal distribution of $\hat{X} - X$; so Y, $n-Y$, $\hat{X}$, and $m-\hat{X}$ should all be large, say, over 10.

For the locomotive control, such two-sided 95% prediction limits are

$$\underset{\sim}{X} \cong 141 - 1.960 \left[(900)0.156(1-0.156)(900+96)/96 \right]^{1/2} = 72,$$

$$\tilde{X} \cong 141 + 1.960 \left[(900)0.156(1-0.156)(900+96)/96 \right]^{1/2} = 210.$$

Each limit is a one-sided 97.5% prediction limit. $Y=15$, $n-Y=81$, $\hat{X}=141$, and $m-\hat{X}=759$ exceed 10; so the approximate limits are accurate enough for practical purposes.

Poisson approximation. For n large and Y small (say $Y/n<0.1$), two-sided approximate $100\gamma\%$ prediction limits for X are the solutions $\underset{\sim}{X}$ and $\tilde{X}$ of

$$(\tilde{X}/m) = \left[(Y+1)/n \right] F \left[(1+\gamma)/2; 2Y+2, 2\tilde{X} \right],$$

$$m/(\underset{\sim}{X}+1) = (n/Y) F \left[(1+\gamma)/2; 2\underset{\sim}{X}+2, 2Y \right],$$ (3.19)

where $F(\delta; a, b)$ is the 100δth F percentile with a degrees of freedom in the numerator and b in the denominator. Such a one-sided lower (upper) prediction limit for X is the solution of the corresponding equation above, where $(1+\gamma)/2$ is replaced by γ. Nelson (1970) provides simple charts for X and $\tilde{X}$. The limits (3.19) employ the Poisson approximation of the distributions of Y and X; so Y should be small (say, $Y<n/10$) and n large.

For the locomotive control, $Y=15$, $n=96$, and $m=900$. Two-sided approximate 95% "Poisson" prediction limits are $X=76$ and $\tilde{X}=237$. Here $Y=15>n/10=9.6$, and the approximate limits are a bit crude. Each limit is a one-sided approximate 97.5% limit.

4. EXPONENTIAL DATA

The exponential distribution of Chapter 2 is a model for the life of products with a constant failure rate λ and mean time to failure $\theta=1/\lambda$. Complete exponential data consist of n observations $Y_1,\ldots,Y_n$.

This section provides estimates and confidence intervals for θ, λ, distribution percentiles, and reliabilities. Also, it provides predictions and prediction intervals for a future observation, sample average (or total), and smallest sample observation. Bain (1978) and Mann, Schafer, and Singpurwalla (1974) give justification for the following results. Chapter 3 gives graphical methods to check how well the exponential distribution fits data. Check this before using the analytic methods below, as the exponential distribution is often inappropriate.

ESTIMATES AND CONFIDENCE LIMITS

Mean θ

Estimate.   The usual estimate of the mean θ is the sample mean

$$\bar{Y}=(Y_1+ \cdots + Y_n)/n. \tag{4.1}$$

It is unbiased, and

$$\mathrm{Var}(\bar{Y})=\theta^2/n. \tag{4.2}$$

The sampling distribution of $2n\bar{Y}/\theta$ is chi square with $2n$ degrees of freedom. If n is large (say, over 15), then the distributions of $\bar{Y}$ and $\ln(\bar{Y})$ are approximately normal.

Insulating fluid example.   Table 1.1 of Chapter 3 shows times to breakdown of an insulating fluid at a number of voltages. The 34-kV data are

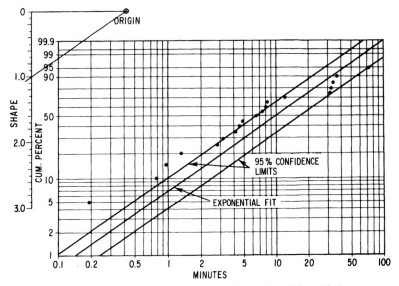

Figure 4.1. Weibull plot with exponential fit and confidence limits.

plotted on Weibull probability paper in Figure 4.1. For 34 kV, the estimate of the mean time to failure is $\overline{Y} = (0.19 + 0.78 + \cdots + 72.89)/19 = 14.3$ minutes. The exponential distribution with this mean is drawn as a straight line in Figure 4.1. Such a plot is a useful way of displaying data and a fitted distribution. Contrary to engineering theory, the exponential distribution does not pass through the data well; the Weibull distribution (Section 6) is an alternative.

Exact confidence limits for θ. Two-sided $100\gamma\%$ confidence limits for θ are

$$\underline{\theta} = 2n\overline{Y}/\chi^2\big[(1+\gamma)/2; 2n\big], \qquad \tilde{\theta} = 2n\overline{Y}/\chi^2\big[(1-\gamma)/2; 2n\big], \quad (4.3)$$

where $\chi^2(\delta; 2n)$ is the 100δth chi-square percentile with $2n$ degrees of freedom and is tabulated in Appendix A3. Lower and upper one-sided $100\gamma\%$ confidence limits for θ are

$$\underline{\theta} = 2n\overline{Y}/\chi^2(\gamma; 2n), \qquad (4.4)$$

$$\tilde{\theta} = 2n\overline{Y}/\chi^2(1-\gamma; 2n). \qquad (4.5)$$

For 34 kV, two-sided 95% confidence limits for the mean time to breakdown are

$$\underset{\sim}{\theta} = 2(19)14.3/\chi^2[(1+0.95)/2; 2\times 19] = 9.6 \text{ minutes,}$$

$$\tilde{\theta} = 2(19)14.3/\chi^2[(1-0.95)/2; 2\times 19] = 23.7 \text{ minutes.}$$

In Figure 4.1, these values are the limits for the 63rd percentile.

Approximate limits. Two-sided approximate $100\gamma\%$ confidence limits for θ are

$$\underset{\sim}{\theta} \cong \overline{Y}/\exp(K_\gamma/n^{1/2}), \qquad \tilde{\theta} \cong \overline{Y}\cdot\exp(K_\gamma/n^{1/2}), \tag{4.6}$$

where K_γ is the $[100(1+\gamma)/2]$th standard normal percentile. Lower and upper one-sided approximate $100\gamma\%$ confidence limits for θ are

$$\underset{\sim}{\theta} \cong \overline{Y}/\exp(z_\gamma/n^{1/2}), \tag{4.7}$$

$$\tilde{\theta} \cong \overline{Y}\cdot\exp(z_\gamma/n^{1/2}), \tag{4.8}$$

where z_γ is the 100γth standard normal percentile.

These limits employ the approximate normal distribution of $\ln(\overline{Y})$; so n should exceed, say, 15.

For 34 kV, the two-sided approximate 95% confidence limits for θ are

$$\underset{\sim}{\theta} \cong 14.3/\exp(1.960/19^{1/2}) = 9.1 \text{ minutes,}$$

$$\tilde{\theta} \cong 14.3/\exp(1.960/19^{1/2}) = 22.4 \text{ minutes.}$$

Sample Size for Estimating the Mean θ

Suppose the estimate $\overline{Y}$ is to be between θ/f and θf, with $100\gamma\%$ probability. The approximate sample size that does this is

$$n \cong [K_\gamma/\ln(f)]^2, \tag{4.9}$$

where K_γ is the $[100(1+\gamma)/2]$th standard normal percentile. This formula employs the normal approximation to the distribution of $\ln(\overline{Y})$. So n should be large, say, $n > 15$.

Suppose the mean time to breakdown of insulating fluid at 34 kV is to be estimated within 20%, with 95% probability. Then $f=1.20$, and the sample size is $n \approx [1.960/\ln(1.2)]^2 = 116$. When the required sample is prohibitively large, then one may have to be content with less precision.

Failure Rate λ

The following results come directly from the relationship $\lambda = 1/\theta$ and the corresponding results for θ.

Estimate.   The usual estimate of the failure rate $\lambda = 1/\theta$ is

$$\hat{\lambda} = 1/\overline{Y} = n/(Y_1 + \cdots + Y_n). \qquad (4.10)$$

This is the number n of failures divided by the total running time, that is, the observed failure rate. This estimator is biased; the bias decreases with increasing sample size. The sampling distribution of $2n\lambda/\hat{\lambda}$ is chi square with $2n$ degrees of freedom.

For example, for 34 kV, $\hat{\lambda} = 1/14.3 = 0.070$ failures per minute.

Limits.   Two-sided $100\gamma\%$ confidence limits for λ are

$$\begin{aligned}
\lambda &= 1/\tilde{\theta} = \chi^2[(1-\gamma)/2; 2n]/(2n\overline{Y}), \\
\tilde{\lambda} &= 1/\underline{\theta} = \chi^2[(1+\gamma)/2; 2n]/(2n\overline{Y}),
\end{aligned} \qquad (4.11)$$

where $\chi^2(\delta; 2n)$ is the 100δth chi-square percentile with $2n$ degrees of freedom. One-sided limits for λ employ (4.4) and (4.5). In practice, only an upper limit is usually of interest.

For 34 kV, two-sided 95% confidence limits are $\lambda = 1/23.7 = 0.042$ and $\tilde{\lambda} = 1/9.6 = 0.104$ failures per minute.

Percentile

Estimate.   The usual estimate of the $100P$th percentile $y_P = -\theta \ln(1-P)$ is

$$Y_P = -\overline{Y} \ln(1-P). \qquad (4.12)$$

This is unbiased, and

$$\mathrm{Var}(Y_P) = [\ln(1-P)]^2 \theta^2/n. \qquad (4.13)$$

For 34 kV, the estimate of the 10th percentile is $Y_{10} = -14.3 \ln(1-0.10) = 1.5$ minutes. This estimate can also be obtained from the fitted distribution line in Figure 4.1.

Limits. Two-sided $100\gamma\%$ confidence limits for the $100P$th percentile are

$$\underline{y}_P = -\underline{\theta}\ln(1-P) = -2n\overline{Y}\big[\ln(1-P)\big]/\chi^2\big[(1+\gamma)/2;2n\big],$$

$$\tilde{y}_P = -\tilde{\theta}\ln(1-P) = -2n\overline{Y}\big[\ln(1-P)\big]/\chi^2\big[(1-\gamma)/2;2n\big], \tag{4.14}$$

where $\chi^2(\delta;2n)$ is the 100δth chi-square percentile with $2n$ degrees of freedom. One-sided limits for y_P employ the one-sided limits for θ in (4.4) and (4.5).

For 34 kV, two-sided 95% confidence limits for $y_{.10}$ are $\underline{y}_{.10} = -9.6\ln(1-0.10) = 1.0$ and $\tilde{y}_{.10} = -23.7\ln(1-0.10) = 2.5$ minutes. These limits are plotted in Figure 4.1. On Weibull paper, the limits for exponential percentiles are straight lines that are parallel to the distribution line. Such lines show the statistical uncertainty in the fitted distribution.

Reliability

Estimate. The usual estimate of the reliability $R(y) = \exp(-y/\theta)$ for age y is

$$R^*(y) = \exp(-y/\overline{Y}). \tag{4.15}$$

This estimator is biased, but the bias decreases with increasing sample size. Pugh (1963) gives the minimum variance unbiased estimator, which is not used in practice.

For 34 kV, the estimate of reliability $R(2.0)$ at $y = 2.0$ minutes, is $R^*(2.0) = \exp(-2.0/14.3) = 0.869$.

Limits. Two-sided $100\gamma\%$ confidence limits for $R(y)$ are

$$\underline{R}(y) = \exp(-y/\underline{\theta}) = \exp\big\{-y\chi^2\big[(1+\gamma)/2;2n\big]/(2n\overline{Y})\big\},$$

$$\tilde{R}(y) = \exp(-y/\tilde{\theta}) = \exp\big\{-y\chi^2\big[(1-\gamma)/2;2n\big]/(2n\overline{Y})\big\}, \tag{4.16}$$

where $\chi^2(\delta;2n)$ is the 100δth chi-square percentile with $2n$ degrees of freedom. One-sided limits for $R(y)$ employ the one-sided limits (4.4) and (4.5) for θ. In practice, one is usually interested in just a lower limit.

For 34 kV, two-sided 95% confidence limits for $R(2.0)$ are $\underline{R}(2.0) = \exp(-2.0/9.6) = 0.812$ and $\tilde{R}(2.0) = \exp(-2.0/23.7) = 0.919$. The lines for percentile confidence limits in Figure 4.1 are also confidence limits for the fraction failing or for the reliability.

Choice of Sample Size n

One can determine a necessary sample size to estimate a function of θ with a desired precision. To do this, convert the statement of the precision of the estimate of the function to a statement of the precision of the estimate of θ. Then use the sample size formula (4.9) for estimating θ. For example, suppose y_p is to be estimated within a factor f with $100\gamma\%$ probability, that is, within $\theta[-\ln(1-P)]/f$ to $\theta[-\ln(1-P)]f$. This is equivalent to estimating θ within a factor of f with $100\gamma\%$ probability.

PREDICTION

Suppose that the m observations of a future sample are $X_1,\ldots,X_m$. The following presents predictors and prediction limits for the sample mean $\bar{X}=(X_1+\cdots+X_m)/m$ and any ordered observation $X_{(1)}\le\cdots\le X_{(m)}$. Suppose that the previous sample has n observations $Y_1,\ldots,Y_n$ and a mean $\bar{Y}$ and is statistically **independent** of the future sample. Hahn and Nelson (1973) survey prediction methods for the exponential distribution, and they give references that justify the following results.

Future Mean, Total, or Observation

Predictor.   The predictor for the future mean $\bar{X}$ is

$$\hat{\bar{X}}=\bar{Y}, \tag{4.17}$$

the previous average. This predictor is unbiased, and the variance of the prediction error $\hat{\bar{X}}-\bar{X}$ is

$$\mathrm{Var}(\hat{\bar{X}}-\bar{X})=\theta^2(m+n)/(mn), \tag{4.18}$$

where θ is the true exponential mean. If m and n are large (say, both exceed 15), then the distribution of $(\hat{\bar{X}}-\bar{X})$ is approximately normal with mean 0 and variance (4.18). The predictor for the total $X=m\bar{X}$ of the future sample is

$$\hat{X}=m\hat{\bar{X}}=m\bar{Y}.$$

This is an unbiased predictor for the total, and the variance of the prediction error $(\hat{X}-X)$ is

$$\mathrm{Var}(\hat{X}-X)=\theta^2(m+n)(m/n).$$

The prediction of the total test time for $m=77$ more breakdowns at 34 kV is $\hat{X}=77(14.3)=1101$ minutes.

Limits.   Two-sided $100\gamma\%$ prediction limits for $\overline{X}$ from m observations are

$$\underset{\sim}{\overline{X}}=\overline{Y}/F[(1+\gamma)/2;2n,2m], \qquad \tilde{\overline{X}}=\overline{Y}\cdot F[(1+\gamma)/2;2m,2n], \quad (4.19)$$

where $F(\delta; a, b)$ is the 100δ th F percentile with a degrees of freedom in the numerator and b in the denominator. F percentiles are tabulated in Appendix A5. One-sided $100\gamma\%$ prediction limits for $\overline{X}$ are given by (4.19), where $(1+\gamma)/2$ is replaced by γ. Such prediction limits for the total X of a future sample of m observations are $\tilde{X}=m\tilde{\overline{X}}$ and $\underset{\sim}{X}=m\underset{\sim}{\overline{X}}$.

The 90% prediction limits for the total test time of another $m=77$ breakdowns at 34 kV are $\underset{\sim}{X}=77(14.3)/F[(1+0.90)/2;2(19),2(77)]=744$, and $\tilde{X}=77(14.3)F[(1+0.90)/2;2(77),2(19)]=1750$ minutes.

Smallest Observation of a Future Sample

Predictor.   The predictor for the smallest $X_{(1)}$ of m future observations is

$$\hat{X}_{(1)}=\overline{Y}/m. \qquad (4.20)$$

$X_{(1)}$ would be the life of a series system of m independent components from the same exponential distribution. This predictor is unbiased, and

$$\text{Var}\left(\hat{X}_{(1)}-X_{(1)}\right)=\theta^2(n+1)/(m^2 n).$$

The prediction of $X_{(1)}$ for $m=77$ future breakdowns at 34 kV is $\hat{X}_{(1)}=14.3/77=0.19$ minute.

Limits.   Two-sided $100\gamma\%$ prediction limits for $X_{(1)}$ of m future observations are

$$\underset{\sim}{X}_{(1)}=(\overline{Y}/m)/F[(1+\gamma)/2;2n,2],$$

$$\tilde{X}_{(1)}=(\overline{Y}/m)\cdot F[(1+\gamma)/2;2,2n], \qquad (4.21)$$

where $F(\delta; a, b)$ is the 100δ th F percentile with a degrees of freedom in the numerator and b in the denominator. One-sided $100\gamma\%$ prediction limits for $X_{(1)}$ are given by (4.21), where γ replaces $(1+\gamma)/2$. A lower limit $\underset{\sim}{X}_{(1)}$ for the first failure in a fleet is called a "safe warranty life" or "assurance limit" for the fleet. Also, a one-sided lower limit $X_{(1)}$ can be used as a lower limit for the life of a single series system of m identical components. A producer of such systems can "guarantee" a customer that his one system will survive at least a time $\underset{\sim}{X}_{(1)}$ with $100\gamma\%$ probability.

For 34 kV, the 90% prediction limits for the smallest of $m=77$ future observations are $X_{(1)}=(14.3/77)/F[(1+0.90)/2;2\cdot19,2]=0.0095$ minute and $\tilde{X}_{(1)}=(14.3/77)F[(1+0.90)/2;2,2\cdot19]=0.60$ minute.

jth Observation $X_{(j)}$ of a Future Sample

Lawless (1972a) describes a predictor and prediction limits for the jth observation $X_{(j)}$ of a future sample from an exponential distribution. He tabulates factors for the prediction limits.

jth Observation $Y_{(j)}$ of the Same Sample

Lawless (1971) describes one-sample predictors and prediction limits for the jth observation $Y_{(j)}$ of the same sample, using the first r observations ($r<j$).

Other Exponential Methods

Chapter 3 describes graphical methods for analysis of complete exponential data. Chapter 7 (8) gives methods for singly (multiply) censored data. Chapters 10, 11, and 12 present methods for comparing a number of exponential populations for equality of their means.

5. NORMAL AND LOGNORMAL DATA

Introduction

The normal and lognormal distributions are widely used models for the life, strength, and other properties of products. These distributions have two parameters—the (log) mean μ and the (log) standard deviation σ—and are described in Chapter 2. A **complete** sample of normal data consists of n observations $Y_1,\ldots,Y_n$; for lognormal data, $Y_1,\ldots,Y_n$ are the base 10 logs of the observations.

The following methods provide estimates and confidence intervals for μ, σ, percentiles, and reliabilities. These estimates and limits can be displayed in a (log) normal probability plot similar to Figure 6.1 of Chapter 3. Also, this section provides predictions and prediction intervals for a future observation, sample mean (or total), and smallest or largest sample observation. It is important to use graphical methods of Chapter 3 to check how well the (log) normal distribution fits the data before using the analytic methods below.

ESTIMATES AND CONFIDENCE LIMITS

The following standard estimates and confidence limits appear and are derived in most statistics books.

The (Log) Standard Deviation σ (Variance σ^2)

Variance estimate. The estimate for the variance σ^2 of (log) life is the sample variance

$$S^2 = \left[(Y_1 - \overline{Y})^2 + \cdots + (Y_n - \overline{Y})^2 \right] / (n-1), \qquad (5.1)$$

where $\overline{Y} = (Y_1 + \cdots + Y_n)/n$ is the sample (log) mean. By computer,

$$S^2 = \left((Y_1^2 + \cdots + Y_n^2) - \frac{1}{n}(Y_1 + \cdots + Y_n)^2 \right) / (n-1) \qquad (5.2)$$

is convenient, but it causes greater numerical round-off error than (5.1), since it may involve a small difference of two large numbers. S^2 is an unbiased estimator for σ^2, and

$$\mathrm{Var}(S^2) = 2\sigma^4 / (n-1). \qquad (5.3)$$

The sampling distribution of $(n-1)S^2/\sigma^2$ is chi square with $n-1$ degrees of freedom. If n is large (say, greater than 30), then the distribution of S^2 is approximately normal with a mean σ^2 and a variance (5.3).

Estimate of σ. The estimate of the (log) standard deviation is the sample standard deviation

$$S = (S^2)^{1/2}. \qquad (5.4)$$

This is a biased estimator for σ, but the bias decreases with increasing sample size n. Its variance is approximately

$$\mathrm{Var}(S) \simeq \sigma^2 / [2(n-1)]. \qquad (5.5)$$

The sampling distribution of $(n-1)^{1/2}S/\sigma$ is a chi distribution, with $n-1$ degrees of freedom, that is, the distribution of the square root of a chi-square variable. If n is large (say, greater than 30), then the distribution of S is approximately normal with a mean of σ and a variance (5.5). Johnson and Kotz (1970) give the minimum variance unbiased estimator for σ; it has the form $C_n S$, where C_n depends on n.

Class H insulation example. Table 5.1 shows accelerated test data on log time to failure of a Class H electrical insulation for motors. Life of such insulation is assumed to have a lognormal distribution at each temperature. There is a sample of 10 specimens for each of four test temperatures. The

Table 5.1. Log Life of Class-H Specimens

	190°C	220°C	240°C	260°C
	3.8590	3.2465	3.0700	2.7782
	3.8590	3.3867	3.0700	2.8716
	3.8590	3.3867	3.1821	2.8716
	3.9268	3.3867	3.1956	2.8716
	3.9622	3.3867	3.2087	2.9600
	3.9622	3.3867	3.2214	3.0892
	3.9622	3.4925	3.2214	3.1206
	3.9622	3.4925	3.2338	3.1655
	4.0216	3.4925	3.2458	3.2063
	4.0216	3.4925	3.2907	3.2778
Total:	39.3958	34.1500	31.9395	30.1755

$\bar{Y}_1 = 3.93958$, $\bar{Y}_2 = 3.41500$, $\bar{Y}_3 = 3.19395$, $\bar{Y}_4 = 3.01755$.

Antilog: 8,701   2,600   1,563   1,041

$S_1 = \{[(3.8590-3.93958)^2+\ldots+(4.0216-3.93958)^2]/(10-1)\}^{1/2} = 0.0624891$.
$S_2 = \{[(3.2465-3.41500)^2+\ldots+(3.4925-3.41500)^2]/(10-1)\}^{1/2} = 0.0791775$.
$S_3 = \{[(3.0700-3.19395)^2+\ldots+(3.2907-3.19395)^2]/(10-1)\}^{1/2} = 0.0716720$.
$S_4 = \{[(2.7782-3.01755)^2+\ldots+(3.2778-3.01755)^2]/(10-1)\}^{1/2} = 0.170482$.

specimens were periodically inspected for failure, and a failure time is the midpoint of the interval where the failure occurred. This slight coarseness of the data can be neglected. Figure 6.1 of Chapter 3 displays the data in a lognormal plot. The plotted data tend to follow straight lines, indicating that the lognormal distribution is a reasonable model.

The sample variance of log life at 190°C is $S^2=[(3.8590^2+\cdots+4.0216^2)-\frac{1}{10}(39.3958)^2]/(10-1)=0.00390489$. The sample standard deviation is $S=(0.00390489)^{1/2}=0.0624891$. This small value indicates that the lognormal distribution is close to normal and that the failure rate increases with age.

Confidence limits for σ.   Two-sided $100\gamma\%$ confidence limits for the (log) standard deviation σ are

$$\underline{\sigma}=S\{(n-1)/\chi^2[(1+\gamma)/2; n-1]\}^{1/2},$$
$$\tilde{\sigma}=S\{(n-1)/\chi^2[(1-\gamma)/2; n-1]\}^{1/2}, \tag{5.6}$$

where $\chi^2(\delta; n-1)$ is the 100δth percentile of the chi-square distribution with $n-1$ degrees of freedom and is tabulated in Appendix A3. One-sided lower and upper $100\gamma\%$ confidence limits for σ are

$$\underset{\sim}{\sigma}=S\left[(n-1)/\chi^2(\gamma; n-1)\right]^{1/2}, \qquad \tilde{\sigma}=S\left[(n-1)/\chi^2(1-\gamma; n-1)\right]^{1/2}$$

$$(5.7)$$

If the parent distribution is not (log) normal, (5.6) and (5.7) are poor approximations, even for large n.

Insulation example. For 190°C, two-sided 90% confidence limits for σ are

$$\underset{\sim}{\sigma}=0.0624891\left\{(10-1)/\chi^2[(1+0.90)/2; 10-1]\right\}^{1/2}$$

$$=0.0624891(9/16.92)^{1/2}=0.0456,$$

$$\tilde{\sigma}=0.0624891\left\{(10-1)/\chi^2[(1-0.90)/2; 10-1]\right\}^{1/2}$$

$$=0.0624891(9/3.325)^{1/2}=0.1028.$$

Sample size to estimate σ. The estimate S is between σ/f and σf, with about $100\gamma\%$ probability if the sample size is

$$n \cong 1+0.5\left[K_\gamma/\ln(f)\right]^2, \qquad (5.8)$$

where K_γ is the $[100(1+\gamma)2]$th standard normal percentile. The accuracy of (5.8) increases with the sample size; (5.8) usually satisfies practical purposes if $n > 15$. Mace (1964) gives a sample size formula in terms of the width of the confidence interval for σ.

Insulation example. For 190°C, suppose that σ is to be estimated within a factor $f = 1.20$ [roughly within $100(1.20-1)=20\%$] with 90% probability. The estimate of the needed sample size is

$$n \cong 1+0.5\left[1.645/\ln(1.20)\right]^2 \cong 42.$$

$42 > 15$ and the approximation is satisfactory.

The (Log) Mean μ

Estimate for μ. The estimate for the (log) mean μ is the sample mean

$$\bar{Y}=(Y_1+ \cdots + Y_n)/n. \qquad (5.9)$$

$\overline{Y}$ is an unbiased estimate for μ. Its variance is

$$\text{Var}(\overline{Y})=\sigma^2/n. \tag{5.10}$$

The sampling distribution of $\overline{Y}$ is normal, with a mean of μ and a variance of σ^2/n.

Insulation example. For the 190°C data in Table 5.1, the estimate of μ, the mean log life, is $\overline{Y}=(3.8590+\cdots+4.0216)/10=3.93958$. The estimate of lognormal median life is antilog $(3.93958)=8701$ hours.

Confidence limits for μ. Two-sided $100\gamma\%$ confidence limits for μ are

$$\underset{\sim}{\mu}=\overline{Y}-t[(1+\gamma)/2:n-1](S/n^{1/2}),$$

$$\tilde{\mu}=\overline{Y}+t[(1+\gamma)/2;n-1](S/n^{1/2}), \tag{5.11}$$

where $t[(1+\gamma)/2;n-1]$ is the $[100(1+\gamma)/2]$th percentile of the t-distribution with $n-1$ degrees of freedom and is tabulated in Appendix A4. One-sided lower and upper $100\gamma\%$ confidence limits for μ are

$$\underset{\sim}{\mu}=\overline{Y}-t(\gamma;n-1)(S/n^{1/2}), \qquad \tilde{\mu}=\overline{Y}+t(\gamma;n-1)(S/n^{1/2}). \tag{5.12}$$

These are approximate intervals for the mean of a distribution that is not normal. The "closer" the distribution is to normal and the larger n, the closer the confidence is to $100\gamma\%$. This comes from the central limit theorem for sample averages.

Insulation example. For 190°C, two-sided 90% confidence limits for μ are

$$\underset{\sim}{\mu} = 3.93958 - t[(1 + 0.90)/2; 10 - 1] (0.0624891/10^{1/2}) = 3.90336,$$

$$\tilde{\mu} = 3.93958 + 1.833 (0.0624891 /10^{1/2}) = 3.97580.$$

The antilogs of these limits are 90% confidence limits for the lognormal median, namely, antilog(3.90336) = 8005 hours and antilog(3.97580) = 9458 hours.

Sample size to estimate μ. The estimate $\overline{Y}$ of a normal mean will be within $\pm w$ of the true value μ with $100\gamma\%$ probability if the sample size is

$$n=\sigma^2(K_\gamma/w)^2, \tag{5.13}$$

where K_γ is the $[100(1+\gamma)/2]$th standard normal percentile. To use (5.13), one must approximate the unknown σ^2. Mace (1964) gives a sample size formula in terms of a desired expected width of the confidence interval for μ. If σ^2 is estimated from previous data, then one can use an upper confidence limit for σ^2 to get an upper confidence limit for the desired sample size.

Insulation example. For 190°C, suppose that μ is to be estimated within ± 0.02 with 90% probability. The estimate of the needed sample size is

$$n \simeq 0.003905(1.645/0.02)^2 = 26.$$

Percentile y_P

Percentile estimate. The estimate of the $100P$th normal percentile $y_P = \mu + z_P\sigma$ is

$$Y_P = \bar{Y} + z_P S, \tag{5.14}$$

where z_P is the $100P$th standard normal percentile. This commonly used estimator is biased, since S is a biased estimator for σ. The bias decreases with increasing sample size. The estimate of the $100P$th percentile $t_P = \text{antilog}(\mu + z_P\sigma)$ of the corresponding lognormal distribution is

$$T_P = \text{antilog}(Y_P). \tag{5.15}$$

For example, the estimate of the lognormal median is $T_{50} = \text{antilog}(\bar{Y})$, since $z_{50} = 0$.

Insulation example. For 190°C, the estimate of the fifth percentile is $Y_{.05} = 3.93958 + (-1.645)0.0624891 = 3.83679$. The estimate for the fifth percentile of the lognormal distribution is $\text{antilog}(3.83679) = 6867$ hours. The estimate of the lognormal median is $\text{antilog}(3.93958) = 8701$ hours.

Confidence limits. A lower one-sided $100\gamma\%$ confidence limit for y_P, the $100P$th percentile ($P < 0.5$), is

$$\underset{\sim}{y_P} = \bar{Y} - K(n, \gamma, P)S; \tag{5.16}$$

the factors $K(n, \gamma, P)$ are tabulated by Natrella (1963), Lieberman (1958), Owen (1962), and Odeh and Owen (1980). Approximate factors are given by

$$a = 1 - \left[(z_\gamma^2/2)/(n-1)\right], \qquad b = z_P^2 - (z_\gamma^2/n),$$

$$K(n, \gamma, P) \simeq \left[-z_P + (z_P^2 - ab)^{1/2}\right] / a; \tag{5.17}$$

z_P is the $100P$th standard normal percentile.

The approximate limit is more accurate the larger n and the closer P and γ to 0.5. y_P is also called a "lower tolerance limit for $[100(1-P)]\%$ of the population." It may be regarded as a guaranteed life that at least $[100(1-P)]\%$ of the population will survive with $100\gamma\%$ confidence. This interval is correct only if the parent distribution is (log) normal.

Insulation example. For 190°C, the calculations for a lower 90% confidence limit for $y_{.05}$ are

$$a = 1 - \left[(1.282^2/2)/(10-1)\right] = 0.9087,$$

$$b = (-1.645)^2 - (1.282^2/10) = 2.542,$$

$$K(10, 0.90, 0.05) \cong \left[1.645 + (1.645^2 - 0.9087 \times 2.542)^{1/2}\right]/0.9087 = 2.503.$$

The exact value of this factor is $K(10, 0.90, 0.05) = 2.568$. Then

$$y_{.05} \cong 3.93958 - (2.503)0.0624891 = 3.78317.$$

The lower 90% confidence limit for the fifth lognormal percentile is $t_{.05} =$ antilog$(3.78317) = 6070$ hours.

Reliability $R(y)$

Estimate. The estimate of the fraction failing $F(y) = \Phi[(y-\mu)/\sigma]$ by (log) age y is

$$F^*(y) = \Phi(Z), \tag{5.18}$$

where $\Phi(\ )$ is the standard normal cumulative distribution,

$$Z = (y - \bar{Y})/S, \tag{5.19}$$

and $\bar{Y}$ and S are the sample mean and standard deviation. $F^*(y)$ is a biased estimator for $F(y)$. The estimate of reliability $R(y)$ at (log) age y is

$$R^*(y) = 1 - \Phi(Z). \tag{5.20}$$

Kirkpatrick (1970) tabulates unbiased estimators for $F(y)$ and $R(y)$ in terms of Z. His estimates differ little from (5.18).

Insulation example. For 190°C, the estimate of the reliability at 10,000 hours [$\log(10,000) = 4.00000$] is obtained as

$$Z = (4.00000 - 3.93958)/0.0624891 = 0.9969,$$

$$F^*(10,000) = \Phi(0.9969) = 0.841, \qquad R^*(10,000) = 1 - 0.841 = 0.159.$$

Confidence limits. Two-sided approximate $100\gamma\%$ confidence limits for the fraction failing $F(y)$ by (log) age y are

$$\underline{F}(y)=\Phi(\underline{z}) \quad \text{and} \quad \tilde{F}(y)=\Phi(\tilde{z}), \tag{5.21}$$

where $Z=(y-\bar{Y})/S$,

$$\underline{z} \cong Z-(K_\gamma/\sqrt{n})\{1+[Z^2(n/2)/(n-1)]\}^{1/2},$$

$$\tilde{z} \cong Z+(K_\gamma/\sqrt{n})\{1+[Z^2(n/2)/(n-1)]\}^{1/2}. \tag{5.22}$$

The approximate limits are more accurate the larger n is, and n should be at least 20. Kirkpatrick (1970) tabulates exact limits and unbiased estimates for the proportion in one or two tails of a normal distribution. Exact limits involve the noncentral t-distribution, which is tabulated by Resnikoff and Liebermann (1957) and Locks, Alexander, and Byars (1963). The limits for reliability are

$$\underline{R}(y)=1-\tilde{F}(y) \quad \text{and} \quad \tilde{R}(y)=1-\underline{F}(y). \tag{5.23}$$

These limits are correct only if the parent distribution is (log) normal.

Insulation example. For 190°C, the approximate 90% confidence limits for the fraction failing and reliability at 10,000 hours $[\log(10,000)=4.0000]$ are obtained as

$$Z=(4.0000-3.93958)/0.0624891=0.9969,$$

$$\underline{z}=0.9969-(1.645/\sqrt{10})\{1+[0.9969^2(10/2)/(10-1)]\}^{1/2}=0.3488,$$

$$\tilde{z}=0.9969+(1.645/\sqrt{10})\{1+[0.9969^2\cdot5/9]\}^{1/2}=1.6450,$$

$$\underline{F}(10,000)=\Phi(0.3488)=0.636, \qquad \tilde{F}(10,000)=\Phi(1.6450)=0.950,$$

$$\underline{R}(10,000)=1-0.950=0.050, \qquad \tilde{R}(10,000)=1-0.636=0.364.$$

The exact limits from Kirkpatrick (1970) are

$$\underline{R}(10,000)=0.052 \quad \text{and} \quad \tilde{R}(10,000)=0.369.$$

PREDICTION

Suppose the m (log) observations of a future sample are $X_1, \ldots, X_m$. The following presents predictors and prediction limits for the sample total $X = X_1 + \cdots + X_m$ and mean $\bar{X} = X/m$ and for the smallest observation $X_{(1)}$ and largest observation $X_{(m)}$. Suppose that the previous sample of n observations has an average $\bar{Y}$ and a standard deviation S. Hahn (1970) and Hahn and Nelson (1973) survey prediction methods for the normal and lognormal distributions, and their references give derivations of the following results.

Future Observation, Mean, or Total

Predictor. For the future mean $\bar{X}$ of m (log) observations or a single future observation ($m=1$) from a normal distribution, the predictor is

$$\hat{\bar{X}} = \bar{Y}. \tag{5.24}$$

This is an unbiased predictor; that is, $E(\hat{\bar{X}} - \bar{X}) = 0$. The variance of the prediction error $(\hat{\bar{X}} - \bar{X})$ is

$$\text{Var}(\hat{\bar{X}} - \bar{X}) = \sigma^2 [(1/m) + (1/n)]. \tag{5.25}$$

The distribution of the prediction error $(\hat{\bar{X}} - \bar{X})$ is normal, with a mean of zero and a variance (5.25). The prediction for the total is $\hat{X} = m\bar{Y}$, and $\text{Var}(\hat{X} - X) = m\sigma^2 [1 + (m/n)]$.

Insulation example. For another specimen at 190°C, the prediction of its log life is $\hat{\bar{X}} = 3.93958$, the average log life of the previous sample. The prediction of its life is antilog $(3.93958) = 8701$ hours.

Limits. Two-sided $100\gamma\%$ prediction limits for the future mean $\bar{X}$ are

$$\underset{\sim}{\bar{X}} = \bar{Y} - t[(1+\gamma)/2; n-1] S[(1/m) + (1/n)]^{1/2},$$
$$\tilde{\bar{X}} = \bar{Y} + t[(1+\gamma)/2; n-1] S[(1/m) + (1/n)]^{1/2}, \tag{5.26}$$

where $t[(1+\gamma)/2; n-1]$ is the $[100(1+\gamma)/2]$th percentile of the t-distribution with $n-1$ degrees of freedom and is tabulated in Appendix A4. A one-sided $100\gamma\%$ prediction limit is obtained from (5.26) by replacing $(1+\gamma)/2$ by γ. The limits for the total are $\underset{\sim}{X} = m\underset{\sim}{\bar{X}}$ and $\tilde{X} = m\tilde{\bar{X}}$.

Insulation example. For another specimen at 190°C, 90% prediction limits for its log life are

$$\underset{\sim}{X}=3.93958-t[(1+0.90)/2; 10 \div 1]0.0624891[(1/1)+(1/10)]^{1/2}$$

$$=3.79133,$$

$$\tilde{X}=3.93958+(2.262)0.0624891(1.1)^{1/2}=4.08783.$$

90% prediction limits for the specimen life are antilog(3.79133)=6185 hours and antilog(4.08783)=12,241 hours. That is, such limits enclose the specimen life with 90% probability.

Smallest or Largest Future Observation

Predictors. For the smallest $X_{(1)}$ and largest $X_{(m)}$ future observation in a normal sample of size m, the predictors are

$$\hat{X}_{(1)}=\overline{Y}-z[(m-0.5)/m]S, \qquad (5.27)$$

$$\hat{X}_{(m)}=\overline{Y}+z[(m-0.5)/m]S, \qquad (5.28)$$

where $z(\delta)$ is the 100δth standard normal percentile. These predictors are biased. $X_{(1)}$ would be the life of a series system of m such components. $X_{(m)}$ would be the time to complete a test of m units that start running simultaneously. Also, $X_{(m)}$ would be the life of a parallel system of m such components.

Insulation example. A sample of $m=10$ new specimens is to be tested at 190°C. The prediction of the log life $X_{(10)}$ of the longest running specimen is

$$\hat{X}_{(10)}=3.93958+z[(10-0.5)/10]0.0624891=4.04237.$$

Then antilog(4.04237)=11,024 hours is a prediction of the time required to run all 10 specimens to failure when they are put on test at the same time.

Limits. One-sided approximate $100\gamma\%$ prediction limits for the smallest $X_{(1)}$ and largest $X_{(m)}$ future observations in a normal sample of size m are

$$\underset{\sim}{X}_{(1)} \cong \overline{Y}-t[(m-1+\gamma)/m; n-1]S[1+(1/n)]^{1/2},$$
$$\qquad (5.29)$$
$$\tilde{X}_{(m)} \cong \overline{Y}+t[(m-1+\gamma)/m; n-1]S[1+(1/n)]^{1/2},$$

where $t(\delta; n-1)$ is the 100δth percentile of the t-distribution with $n-1$ degrees of freedom. These limits are conservative, since the corresponding probability is at least γ.

Hahn (1970) gives tables for exact one-sided prediction limits. Hahn and Nelson (1973) reference tables for two-sided prediction limits.

For a nonnormal parent distribution, such intervals may be crude, particularly for large m and γ. $X_{(1)}$ can be used as a "guaranteed" or warranty life for a fleet of m units; that is, all m units will survive that age with probability γ. Also, $X_{(1)}$ is a lower limit for the life of a single series system of m units.

Insulation example. $m = 10$ new insulation specimens are tested together at 190°C. The upper 90% prediction limit for the log time to complete the test is

$$\tilde{X}_{(10)} \cong 3.93958 + t\big[(10-1+0.90)/10; 10-1\big]\cdot 0.0624891\big(1+(1/10)\big)^{1/2}$$

$$= 4.15258.$$

The corresponding actual time is antilog(4.15258) = 14,210 hours.

OTHER METHODS FOR NORMAL AND LOGNORMAL DATA

Most statistics books present a variety of methods for analysis of complete normal data. Aitchison and Brown (1957) present a variety of data analysis methods for the lognormal distribution. Hahn (1970) surveys statistical intervals for a normal distribution; he provides brief tables of the necessary factors. Ellison (1964) gives a detailed encyclopedic presentation of the mathematical theory of statistical methods for the normal distribution. Johnson and Kotz (1970) survey statistical methods for the normal distribution. Chapter 7 presents one-sample prediction for later observations in the same sample. Chapters 10, 11, and 12 present methods for comparing a number of samples.

6.   WEIBULL AND EXTREME VALUE DATA

The Weibull and the related smallest extreme value distributions are widely used as models for the life, strength, and other properties of products. These distributions and their relationship are described in detail in Chapter 2.

This section provides estimates and confidence limits for distribution parameters, percentiles, and reliabilities. Also, it provides predictions and prediction limits for a future observation or smallest observation in a future sample. Unlike the normal and lognormal distributions, the Weibull and extreme value distributions have no obviously "best" methods for data analysis. This section presents simple but crude methods. Chapters 7 and 8 present more accurate but more laborious methods. Chapter 3 provides graphical checks for how well the Weibull (or extreme value) distribution fits data. Check this before using the analytic methods below. Many of the proposed methods are surveyed by Mann (1968).

Complete data from an extreme value distribution consist of n observations $Y_1,\ldots, Y_n$; these are the (base e) logs of Weibull data. The same methods apply to extreme value and Weibull data, but one works with the log values of Weibull data. The Weibull shape parameter β in terms of the corresponding extreme value scale parameter δ is $\beta = 1/\delta$, and the Weibull scale parameter α in terms of the corresponding extreme value location parameter λ is $\alpha = \exp(\lambda)$.

ESTIMATES AND CONFIDENCE LIMITS

Distribution parameters

Extreme value scale parameter δ.   A simple estimate of δ is

$$D = 0.7797S, \tag{6.1}$$

where $0.7797\ldots = 6^{1/2}/\pi$, S is the sample standard deviation

$$S = \left\{ \left[(Y_1 - \bar{Y})^2 + \cdots + (Y_n - \bar{Y})^2 \right] / (n-1) \right\}^{1/2}, \tag{6.2}$$

and

$$\bar{Y} = (Y_1 + \cdots + Y_n)/n \tag{6.3}$$

is the sample average.

Weibull shape parameter β.   An estimate of the corresponding Weibull shape parameter β is

$$B = 1/D = 1.283/S. \tag{6.4}$$

D and B are biased estimators, and their sampling distributions have not been derived. Lieblein (1954) and Thoman, Bain, and Antle (1969) give Monte Carlo approximations of the distributions. Menon (1963) shows that the large-sample distribution of D is approximately normal, with a mean of δ and a variance of $1.100\delta^2/n$. More accurate and more laborious estimators for δ and β appear in Chapters 7 and 8.

Insulating fluid example.   Table 6.1 shows ln times to breakdown of an insulating fluid in an accelerated test. Suppose that time to breakdown has a Weibull distribution at each test voltage. A purpose of the test is to assess

Table 6.1.   Ln Times and Summary Statistics for Insulating Fluid

	26 kV	28 kV	30 kV	32 kV	34 kV	36 kV	38 kV
	1.7561	4.2319	2.0464	-1.3094	-1.6608	-1.0499	-2.4080
	7.3648	4.6848	2.8361	-0.9163	-0.2485	-0.5277	-0.9417
	7.7509	4.7031	3.0184	-0.3711	-0.0409	-0.0409	-0.7551
		6.0546	2.0454	-0.2358	0.2700	-0.0101	-0.3148
		6.9731	3.1206	1.0116	1.0224	0.5247	-0.3012
			3.7704	1.3635	1.1505	0.6780	0.1222
			3.8565	2.2905	1.4231	0.7275	0.3364
			4.9349	2.6354	1.5411	0.9477	0.8671
			4.9706	2.7682	1.5789	0.9969	
			5.1698	3.3250	1.8718	1.0647	
			5.2724	3.9748	1.9947	1.3001	
				4.4170	2.0806	1.3837	
				4.4918	2.1126	1.6770	
				4.6109	2.4898	2.6224	
				5.3711	3.4578	3.2386	
					3.4818		
					3.5237		
					3.6030		
					4.2889		
Total:	16.8718	26.6475	42.0415	33.4272	33.9405	13.5327	-3.3951
Average $\bar{Y}_j$:	5.62393	5.32950	3.82195	2.22848	1.78634	0.90218	-0.424388
Antilog:	276.977	206.335	45.6934	9.28574	5.96758	2.46497	0.65417
No. in sample:	$n_1=3$	$n_2=5$	$n_3=11$	$n_4=15$	$n_5=19$	$n_6=15$	$n_7=8$

$$S_1 = \{[(1.7561-5.62393)^2+\ldots+(7.7509-5.62393)^2]/(3-1)\}^{1/2} = 3.35520$$
$$S_2 = \{[(4.2319-5.32950)^2+\ldots+(6.9731-5.32950)^2]/(5-1)\}^{1/2} = 1.14455$$
$$S_3 = \{[(2.046+-3.82195)^2+\ldots+(5.2724-3.82195)^2]/(11-1)\}^{1/2} = 1.11119$$
$$S_4 = \{[(-1.3094-2.22848)^2+\ldots+(5.3711-2.22848)^2]/(15-1)\}^{1/2} = 2.19809$$
$$S_5 = \{[(-1.6608-1.78634)^2+\ldots+(4.2889-1.78634)^2]/(19-1)\}^{1/2} = 1.52521$$
$$S_6 = \{[(-1.0499-0.90218)^2+\ldots+(3.2386-0.90218)^2]/(15-1)\}^{1/2} = 1.10989$$
$$S_7 = \{[(-2.4080+0.424388)^2+\ldots+(0.8671+0.424388)^2]/(8-1)\}^{1/2} = 0.991707$$

whether the distribution is exponential; that is, $\beta=1$. Then the fluid has a constant failure rate, consistent with engineering opinion that such fluids do not age. The table shows that the mean of the 19 log observations at 34 kV is $\overline{Y}=[(-1.6608)+(-0.2485)+ \cdots +4.2889]/19=1.78634$, and their standard deviation is $S=\{[(-1.6608-1.78634)^2 + \cdots +(4.2889-1.78634)^2]/(19-1)\}^{1/2}=1.52521$. The estimate of the corresponding extreme value scale parameter is $D=0.7797(1.52521)=1.1892$, and the estimate of the Weibull shape parameter is $B=1/1.1892=0.8409$, slightly less than unity.

Extreme value location parameter λ.   A simple estimate for $\lambda=EY+0.5772\delta$ is

$$L=\overline{Y}+0.5772D,\tag{6.5}$$

where $0.5772\dots$ is Euler's constant.

Weibull scale parameter α.   An estimate of $\alpha=\exp(\lambda)$ is

$$A=\exp(L).\tag{6.6}$$

L and A are biased, and their sampling distributions have not been derived. Lieblein (1954) gives Monte Carlo results on the distribution of L.

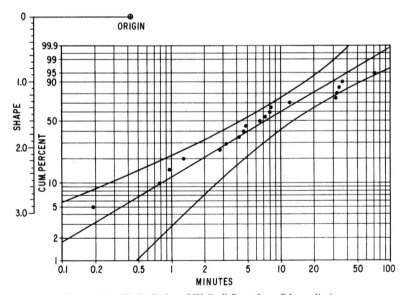

Figure 6.1.   Weibull plot of Weibull fit and confidence limits.

Menon (1963) shows that the large-sample distribution of L is approximately normal, with a mean of λ and a variance of $1.168\delta^2/n$. More accurate (smaller variance) and more laborious estimators for λ and α appear in Chapters 7 and 8.

Insulating fluid example.   For the extreme value location parameter at 34 kV, $L = 1.78634 + (0.5772)1.1892 = 2.4727$. The corresponding estimate of the Weibull scale parameter is $A = \exp(2.4727) = 11.85$ minutes.

Figure 6.1 shows a Weibull probability plot of the 34-kV data. The fitted distribution with the parameter estimates above is the straight line. The confidence limits for percentiles (and reliabilities) are curves and are narrowest near the center of the data. Compare this with the exponential fit in Figure 4.1, which has straight lines for confidence limits.

Confidence Limits for the Distribution Parameters

Extreme value scale parameter δ.   Two-sided approximate $100\gamma\%$ confidence limits for δ are

$$\underset{\sim}{\delta} \cong D/\exp\left(K_\gamma 1.049/n^{1/2}\right), \qquad \tilde{\delta} \cong D \cdot \exp\left(K_\gamma 1.049/n^{1/2}\right), \qquad (6.7)$$

where K_γ is the $[100(1+\gamma)/2]$th standard normal percentile. One-sided approximate $100\gamma\%$ confidence limits are obtained from (6.7), where K_γ is replaced by z_γ, the 100γth standard normal percentile. Here $\ln(D)$ is treated as approximately normally distributed for large n. Here n must be fairly large for a good approximation.

Weibull shape parameter β.   For the related Weibull shape parameter, corresponding limits are

$$\underset{\sim}{\beta} = 1/\tilde{\delta}, \qquad \tilde{\beta} = 1/\underset{\sim}{\delta}.$$

Insulating fluid example.   For 34 kV, approximate 90% confidence limits for δ are

$$\underset{\sim}{\delta} \cong 1.1892/\exp\left[1.645(1.049)/19^{1/2}\right] = 0.8004,$$

$$\tilde{\delta} \cong 1.1892 \cdot \exp(0.3959) = 1.767.$$

Corresponding limits for the Weibull shape parameter are $\underset{\sim}{\beta} = 1/1.767 \cong 0.566$ and $\tilde{\beta} = 1/0.8004 = 1.249$. This interval encloses 1, the shape parameter value for an exponential distribution. Thus, the 34-kV data are consistent with an exponential distribution and a constant failure rate. The great width of the interval shows that one cannot claim that the true shape parameter value is close to unity.

Extreme value location parameter λ. Two-sided approximate $100\gamma\%$ confidence limits for λ are

$$\underset{\sim}{\lambda} \cong L - K_\gamma 1.081 (D/n^{1/2}), \qquad \tilde{\lambda} \cong L + K_\gamma 1.081 (D/n^{1/2}), \qquad (6.8)$$

where K_γ is the $[100(1+\gamma)/2]$th standard normal percentile. One-sided approximate $100\gamma\%$ confidence limits are obtained from (6.8), where K_γ is replaced by z_γ, the 100γth standard normal percentile.

Weibull scale parameter α. For the related Weibull scale parameter $\alpha = \exp(\lambda)$, corresponding limits are

$$\underset{\sim}{\alpha} = \exp(\underset{\sim}{\lambda}), \qquad \tilde{\alpha} = \exp(\tilde{\lambda}).$$

Insulating fluid example. For 34 kV, approximate 90% confidence limits for λ are

$$\underset{\sim}{\lambda} \cong 2.4727 - (1.645)1.081(1.1892/19^{1/2}) = 1.9876,$$

$$\tilde{\lambda} \cong 2.4727 + (1.645)1.081(1.1892/19^{1/2}) = 2.9578.$$

Corresponding limits for the Weibull scale parameter are $\underset{\sim}{\alpha} = \exp(1.9876) = 7.3$ minutes and $\tilde{\alpha} = \exp(2.9578) = 19.3$ minutes. Each of these limits is a one-sided approximate 95% confidence limit.

The approximate limits (6.7) and (6.8) are often good enough in practice, but are crude unless n is quite large, say, greater than 100. More accurate but more laborious approximations appear in Chapters 7 and 8. Lawless (1972b, 1978) gives a method for exact conditional fiducial limits for the distribution parameters; the laborious calculation of his limits requires a special computer program.

Sample Size to Estimate a Weibull Shape Parameter

The estimate B of a Weibull shape parameter β is between $f\beta$ and β/f, with about $100\gamma\%$ probability, if the sample size is

$$n \simeq 1.100 \left[K_\gamma / \ln(f) \right]^2, \qquad (6.9)$$

where K_γ is the $[100(1+\gamma)/2]$th standard normal percentile. The accuracy of this formula increases with the sample size. It is based on the approximate normal distribution of $\ln(B)$ for large n.

For 34 kV, suppose the β is to be estimated within a factor of 1.25 [roughly within $100(1.25-1)=25\%$] with 90% probability. The estimate of the needed sample size is

$$n \cong 1.100[1.645/\ln(1.25)]^2 \cong 60.$$

Percentiles

Percentile estimate.   The estimate of the $100P$th extreme value percentile $y_P = \lambda + u_P \delta$ is

$$Y_P = L + u_P D = \overline{Y} + (0.5772 + u_P)0.7797S, \qquad (6.10)$$

where $u_P = \ln[-\ln(1-P)]$ is the $100P$th percentile of the standard extreme value distribution. The estimate of the $100P$th percentile of the corresponding Weibull distribution is

$$T_P = \exp(Y_P). \qquad (6.11)$$

Insulating fluid example.   For 34 kV, the estimate of the 10th extreme value percentile is $Y_{.10} = 2.4727 + (-2.2504)1.1892 = -0.2034$, where $u_{.10} = \ln[-\ln(1-0.10)] = -2.2504$. For the Weibull distribution, $T_{.10} = \exp(-0.2034) = 0.82$ minute.

Percentile limits.   Two-sided approximate $100\gamma\%$ confidence limits for the extreme value percentile y_P are

$$\underline{y}_P \cong Y_P - K_\gamma D\left[(1.1680 + u_P^2 1.1000 - u_P 0.1913)/n\right]^{1/2},$$
$$\qquad (6.12)$$
$$\tilde{y}_P \cong Y_P + K_\gamma D\left[(1.1680 + u_P^2 1.1000 - u_P 0.1913)/n\right]^{1/2},$$

where K_γ is the $[100(1+\gamma)/2]$th standard normal percentile, and $u_P = \ln[-\ln(1-P)]$. One-sided approximate $100\gamma\%$ confidence limits are obtained from (6.12), where K_γ is replaced by z_γ, the 100γth standard normal percentile. Based on the normal approximation for Y_P, (6.12) is rough unless n is large, say, greater than 100. More accurate but more laborious confidence limits are given in Chapters 7 and 8. Also, Lawless (1974, 1978) gives a method for exact conditional confidence limits for percentiles; the laborious calculation requires a special computer program. Corresponding limits for the Weibull $100P$th percentile $t_P = \exp(y_P)$ are

$$\underline{t}_P = \exp(\underline{y}_P) \quad \text{and} \quad \tilde{t}_P = \exp(\tilde{y}_P). \tag{6.13}$$

Insulating fluid example. For 34 kV, such 90% confidence limits for the extreme value $y_{.10}$ are

$$\underline{y}_{.10} = -0.2034 - 1.645(1.1892)\left\{\left[1.1680 + (-2.2504)^2 1.1000\right.\right.$$

$$\left.\left. - (-2.2504)0.1913\right]/19\right\}^{1/2} = -1.4050,$$

$$\tilde{y}_{.10} = -0.2034 + 1.2016 = 0.9982.$$

Corresponding limits for the Weibull 10th percentile are $\underline{t}_{.10} = \exp(-1.4050) = 0.25$ minute and $\tilde{t}_{.10} = \exp(0.9982) = 2.7$ minutes. Each of these limits is a one-sided 95% confidence limit.

Reliability

Estimate. For an extreme value distribution, the estimate of reliability $R(y) = \exp\{-\exp[(y-\lambda)/\delta]\}$ at age y is

$$R^*(y) = \exp\{-\exp[(y-L)/D]\}. \tag{6.14}$$

The same estimate applies to the Weibull distribution, but y is then the natural log of age. Equivalently, the estimate of the Weibull reliability at age t is

$$R^*(t) = \exp\left[-(t/A)^B\right]. \tag{6.15}$$

Insulating fluid example. For 34 kV, the estimate of the fluid reliability at 2.0 minutes is

$$R^*(2.0) = \exp\left[-(2.0/11.85)^{0.8409}\right] = 0.799.$$

Limits. Two-sided approximate $100\gamma\%$ confidence limits for the extreme value reliability $R(y)$ at age y are

$$\underline{R}(y) = \exp\left[-\exp(\tilde{u})\right] \quad \text{and} \quad \tilde{R}(y) = \exp\left[-\exp(\underline{u})\right], \tag{6.16}$$

where $U = (y - L)/D$,

$$\underline{u} = U - K_\gamma \left[(1.1680 + U^2 1.1000 - U 0.1913)/n \right]^{1/2},$$

$$\tilde{u} = U + K_\gamma \left[(1.1680 + U^2 1.1000 - U 0.1913)/n \right]^{1/2},$$

and K_γ is the $[100(1+\gamma)/2]$th standard normal percentile. One-sided approximate $100\gamma\%$ confidence limits are obtained from (6.16), where K_γ is replaced by z_γ, the 100γth standard normal percentile. The approximation is crude unless n is large, say, greater than 100. More accurate but more laborious limits appear in Chapters 7 and 8. Also, Lawless (1974) gives a method for exact conditional confidence limits for reliability; the laborious calculation requires a special computer program. Corresponding limits for the Weibull reliability at age t are the same, but $y = \ln(t)$.

For 34 kV, two-sided approximate 90% confidence limits for the fluid reliability at 2.0 minutes are obtained as

$$U = \left[\ln(2.0) - 2.4727 \right]/1.1892 = -1.4964,$$

$$\underline{u} = -1.4964 - 1.645 \left\{ \left[1.1680 + (-1.4964)^2 1.1000 \right. \right.$$

$$\left. \left. - (-1.4964)0.1913 \right]/19 \right\}^{1/2} = -2.2433,$$

$$\tilde{u} = -1.4964 + 0.7469 = -0.7495,$$

$$\underline{R}(2.0) = \exp\left[-\exp(-0.7495) \right] = 0.623,$$

$$\tilde{R}(2.0) = \exp\left[-\exp(-2.2433) \right] = 0.899.$$

PREDICTION

Suppose that a future sample from an extreme value distribution has m observations $X_{(1)}, \ldots, X_{(m)}$, ordered from smallest to largest. Suppose the previous sample of n observations yields the estimates D for δ and L for λ. Chapters 7 and 8 give further prediction methods.

Predictor. For the smallest future observation $X_{(1)}$, the predictor is

$$\hat{X}_{(1)} = L - D \ln(m). \tag{6.17}$$

This predictor estimates the mode (most likely value) of the distribution of

$X_{(1)}$. It is a biased predictor. This predicts the life of a series system of m components from the same extreme value life distribution (see Chapter 5).

Limits. Lawless (1973) gives a method for conditional prediction limits for $X_{(1)}$; the laborious calculation requires a special computer program. Gumbel (1958, pp. 234–235) presents material on prediction of $X_{(1)}$.

KNOWN SHAPE PARAMETER

At times one may wish to assume that the shape parameter value β is known and the scale parameter α is unknown. The β value may come from previous or related data or may be a widely accepted value for the product or even an engineering guess. The transformed data values $U_i = Y_i^\beta$ come from an exponential distribution with a mean $\theta = \alpha^\beta$. Then methods for exponential data can be used to obtain estimates, confidence limits, predictions, etc. Of course, the accuracy of this approach depends on the accuracy of the β value. Different β values may be tried to assess their effects on the results.

OTHER METHODS

The methods above are the simplest ones for Weibull and extreme value data. Exact, more efficient (and more laborious) methods appear in Chapters 7 and 8. Mann (1968) surveys older methods for analysis of Weibull (and extreme value) data. Chapters 10, 11, and 12 present methods for comparing a number of Weibull or extreme value populations.

7. DISTRIBUTION-FREE METHODS

Sometimes one does not wish to use a parametric distribution as a model for a population. This section describes analytic methods that apply to complete data from any distribution. These so-called **distribution-free** or **nonparametric** methods are usually not as efficient as previous parametric ones designed for particular distributions; that is, most nonparametric confidence and prediction intervals are much wider than the parametric ones. However, parametric methods may be inaccurate if the true population distribution differs from the assumed parametric distribution. Even if a parametric distribution is correct, it is often easier to use a nonparametric method. Nonparametric methods for life data have primarily been developed and used for biomedical applications; see Gross and Clark (1975). Parametric methods are usually used in engineering applications, as they (if correct) yield more precise information than nonparametric methods and permit extrapolation outside the range of data.

A complete sample consists of the n observations $Y_1, \ldots, Y_n$, which may be arranged as the ordered values $Y_{(1)} \leq Y_{(2)} \leq \cdots \leq Y_{(n)}$.

The methods below provide estimates and confidence intervals for the distribution mean, standard deviation, percentiles, and reliabilities. Also, they provide predictions and prediction limits for a future sample observation, average (or total), and smallest or largest observation. These methods are exact for continuous distributions, but they can also be used for discrete distributions. Then the true probability for an interval is the same as or higher than that given here for a continuous distribution. Justifications for these methods appear in books on nonparametric methods, for example, in Gibbons (1976), Hollander and Wolfe (1973), and Lehmann (1975).

Mean

Estimate. The estimate for the mean μ of a distribution is the sample average

$$\overline{Y} = (Y_1 + \cdots + Y_n)/n. \tag{7.1}$$

This is unbiased, and

$$\text{Var}(\overline{Y}) = \sigma^2/n, \tag{7.2}$$

where σ^2 is the distribution variance, assumed to be finite. For large sample size n, the sampling distribution of $\overline{Y}$ is approximately normal with a mean of μ and a variance of σ^2/n.

Warranty cost example. The manufacturer of a type of large motor must pay warranty costs. Management wanted the average warranty cost per motor and a prediction of the total warranty cost for the coming year. Management also wanted to know the uncertainty in this information. Data on $n = 184$ motors had an average warranty cost per motor per year of $\overline{Y} = \$751$ (desired information) and a standard deviation of $S = \$944$. The standard deviation is larger than the average; the distribution is very skewed, since most motors have zero warranty cost and a few have a very high cost.

Confidence limits. Two-sided approximate $100\gamma\%$ confidence limits for a distribution mean μ are

$$\underset{\sim}{\mu} \cong \overline{Y} - t[(1+\gamma)/2; n-1](S/n^{1/2}),$$
$$\tilde{\mu} \cong \overline{Y} + t[(1+\gamma)/2; n-1](S/n^{1/2}), \tag{7.3}$$

where $t[(1+\gamma)/2; n-1]$ is the $[100(1+\gamma)/2]$th percentile of the t-

distribution with $n-1$ degrees of freedom. One-sided approximate $100\gamma\%$ confidence limits are obtained from (7.3), where $(1+\gamma)/2$ is replaced by γ. These limits are satisfactory if the sample size n is large enough for the sampling distribution of $\bar{Y}$ to be approximately normal. If the population is close to normal, then the approximation is often satisfactory for small n.

Warranty cost example.   A two-sided approximate 90% confidence interval for the true mean yearly warranty cost of such motors is

$$\underaccent{\tilde}{\mu}=751-t[(1+0.90)/2;184-1](944/184^{1/2})=\$637,$$

$$\tilde{\mu}=751+(1.645)(944/184^{1/2})=\$866.$$

Each of these limits is a one-sided approximate 95% confidence limit. These limits do not take into account yearly inflation of costs.

Variance and Standard Deviation

Variance.   The estimate for the population variance σ^2 is the sample variance

$$S^2=\left[(Y_1-\bar{Y})^2+\cdots+(Y_n-\bar{Y})^2\right]/(n-1), \qquad (7.4)$$

where $\bar{Y}$ is the sample mean. Equivalently,

$$S^2=\left[(Y_1^2+\cdots+Y_n^2)-n\bar{Y}^2\right]/(n-1). \qquad (7.5)$$

This is an unbiased estimator for σ^2 for any distribution. The distribution of $(n-1)S^2/\sigma^2$ does **not** approach a chi-square distribution for large n, but it does approach a normal distribution if the fourth moment of the population distribution is finite.

Standard deviation.   The estimate for σ is the sample standard deviation

$$S=(S^2)^{1/2}. \qquad (7.6)$$

This is a biased estimator for σ. However, the bias decreases as sample size increases. The warranty cost example above employs this estimate.

There are no standard nonparametric confidence limits for the variance and standard deviation of any continuous distribution. One can use some other measure of distribution spread, such as the interquartile range ($y_{.75}-y_{.25}$) or interdecile range ($y_{.90}-y_{.10}$); there are nonparametric confidence limits for them. However, they are not used for life data.

Reliability

Estimate.   A distribution-free estimate of the reliability $R(y)$ at age y is the sample fraction that survive an age y. That is, if X of the n times to failure are beyond age y, then the estimate of the reliability at age y is

$$R^*(y) = X/n.$$

This is unbiased, and

$$\mathrm{Var}[R^*(y)] = R(y)[1 - R(y)]/n. \qquad (7.7)$$

The sampling distribution of X is binomial, with sample size n and probability $R(y)$ of "success." Estimation of such a binomial proportion is described in Section 3. The entire sample reliability function is obtained by estimating $R(y)$ for all y values. The sample reliability function $R^*(y)$ is a decreasing staircase function that decreases by $1/n$ at each data value and is constant between data values. So it needs to be calculated only at each data value. The sample cumulative distribution function is the increasing staircase function

$$F^*(y) = 1 - R^*(y) = (n - X)/n. \qquad (7.8)$$

Such a function appears in Figure 7.1.

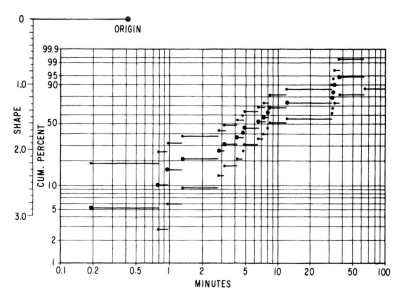

Figure 7.1.   Weibull plot of nonparametric estimate and confidence limits.

Confidence limits. Binomial limits apply to reliability $R(y)$ at age y (Section 3). Usually one wants a one-sided lower limit $\underline{R}(y)$ for reliability. A lower (and upper) limit $\underline{R}(y)$ for the reliability function at all y values is also a decreasing staircase function that decreases at each data value and is constant between data values. Thus it needs to be calculated only at each of the n data values.

Insulating fluid example. Figure 7.1 depicts the nonparametric estimate of the cumulative distribution function at 34 kV. It also shows 90% confidence limits for the fraction failing as a function of age. Compare this plot with Figures 4.1 and 6.1. The nonparametric limits are wider.

Percentile

Estimate. Suppose the ordered observations are $Y_{(1)} \leq Y_{(2)} \leq \cdots \leq Y_{(n)}$. Also, suppose one wishes to estimate the $100P$th percentile of a continuous distribution where $i \leq (n+1)P \leq i+1$ for some $i=1,\ldots,n-1$. Then an estimate is

$$Y_P = \left[i+1-(n+1)P\right]Y_{(i)} + \left[(n+1)P-i\right]Y_{(i+1)}. \tag{7.9}$$

If $P=0.50$, the estimate of the population median is the sample median $Y_{((n+1)/2)}$ when n is odd, and it is the average of the two middle observations when n is even.

Insulating fluid example. For 34 kV, a nonparametric estimate of the 63.21st percentile is obtained as follows. First, $(n+1)P=(19+1)0.6321=12.642$. Thus the 12th and 13th ordered observations (8.01 and 8.27) are used. Then $Y_{.6321}=(12+1-12.642)8.01+(12.642-12)8.27=8.18$ minutes. In Section 6, the parametric estimate of the Weibull scale parameter (63.21st percentile) is 11.85 minutes.

Two-sided confidence limits. For the $100P$th percentile of a continuous distribution, use the rth and sth order statistics $Y_{(r)} \leq Y_{(s)}$ as the limits. The confidence level is the binomial probability

$$\gamma = \sum_{x=r}^{s-1} \frac{n!}{x!(n-x)!} P^x (1-P)^{n-x}. \tag{7.10}$$

The basis for (7.10) is that the interval encloses the $100P$th percentile if anywhere from r to $s-1$ observations fall below the percentile, and each observation is a binomial trial that falls below the percentile with probability P. Using a binomial table, one chooses r and s to make γ close to the desired confidence level.

For large n, approximate r and s are

$$r \cong nP - 0.5 - K_\gamma \left[nP(1-P) \right]^{1/2}, \qquad s \cong nP - 0.5 + K_\gamma \left[nP(1-P) \right]^{1/2},$$

$$(7.11)$$

where K_γ is the $[100(1+\gamma)/2]$th standard normal percentile. These values come from a normal approximation to (7.10).

Insulating fluid example. For 34 kV, the ranks of order statistics for approximate 90% confidence limits for the 63.21st percentile are

$$r \approx 19(0.6321) - 0.5 - 1.645 \left[19 \times 0.6321(1-0.6321) \right]^{1/2} \approx 8.05,$$

$$s \approx 19(0.6321) - 0.5 + 3.46 \approx 14.97.$$

The corresponding 8th and 15th order statistics are 4.67 and 31.75 minutes; they are the confidence limits for $y_{.6321}$. The exact binomial probability (7.10) for these order statistics is $\gamma = 0.866$ instead of 0.900. In Section 6, the parametric approximate 90% confidence limits for the Weibull scale parameter (63.2nd percentile) are 7.3 and 19.3, a narrower interval than 4.67 to 31.75.

One-sided lower confidence limit. For the $100P$th percentile of a continuous distribution, use the rth order statistic $Y_{(r)}$. The confidence level is the binomial probability.

$$\gamma = \sum_{x=r}^{n} \frac{n!}{x!(n-x)!} P^x (1-P)^{n-x}. \qquad (7.12)$$

For example, if $Y_{(1)}$ is used as a lower limit for the $100P$th percentile, the confidence is

$$\gamma = 1 - P^n. \qquad (7.13)$$

For large n, an approximate r is

$$r \cong nP - 0.5 - z_\gamma \left[nP(1-P) \right]^{1/2},$$

where z_γ is the 100γth standard normal percentile.

PREDICTION

Suppose a future sample has m observations $X_1, \ldots, X_m$. There are predictors and prediction limits for the sample total $X = X_1 + \cdots + X_m$ and mean $\bar{X} = X/m$ and for any ordered observation $X_{(1)} \leqslant X_{(2)} \leqslant \cdots \leqslant X_{(m)}$. Suppose the past sample of n observations has an average $\bar{Y}$ and a standard deviation S and the ordered observations are $Y_{(1)} \leqslant Y_{(2)} \leqslant \cdots \leqslant Y_{(n)}$. The references of Hahn and Nelson (1973) justify the following results.

Prediction and Limits for the Mean or Total

Prediction. The prediction of the future mean $\bar{X}$ is

$$\hat{\bar{X}} = \bar{Y}, \tag{7.14}$$

the mean of the previous sample. This is unbiased, and

$$\mathrm{Var}(\hat{\bar{X}} - \bar{X}) = \sigma^2[(1/m) + (1/n)], \tag{7.15}$$

where σ^2 is the distribution variance. For m and n large, the distribution of the prediction error $(\hat{\bar{X}} - \bar{X})$ is approximately normal with a mean of 0 and variance (7.15).

The prediction of the total X of m future observations is

$$\hat{X} = m\bar{Y}. \tag{7.16}$$

This is unbiased, and

$$\mathrm{Var}(\hat{X} - X) = m^2\sigma^2[(1/m) + (1/n)]. \tag{7.17}$$

For m and n large, the distribution of $(\hat{X} - X)$ is approximately normal, with a mean of 0 and variance (7.17)

Motor warranty example. For the coming year, there would be $m = 70$ new motors on warranty. The prediction of the total warranty cost of those motors is $\hat{X} = 70 \times \$751 = \$52,570$.

Prediction limits. Two-sided approximate $100\gamma\%$ prediction limits for a future mean $\bar{X}$ are

$$\underset{\sim}{\bar{X}} \cong \bar{Y} - t[(1+\gamma)/2; n-1] S[(1/m) + (1/n)]^{1/2},$$
$$\tilde{\bar{X}} \cong \bar{Y} + t[(1+\gamma)/2; n-1] S[(1/m) + (1/n)]^{1/2}, \tag{7.18}$$

where $t[(1+\gamma)/2; n-1]$ is the $[100(1+\gamma)/2]$th percentile of the t-distribution with $n-1$ degrees of freedom and S is the past sample standard deviation. One-sided approximate $100\gamma\%$ prediction limits are obtained from (7.18), where $(1+\gamma)/2$ is replaced by γ. This approximation is more accurate the larger m and n are; so the sampling distributions of $\overline{Y}$ and $\overline{X}$ are close to normal. Two-sided approximate $100\gamma\%$ prediction limits for a total X of m future observations are

$$\underset{\sim}{X} \cong m\overline{X} \quad \text{and} \quad \tilde{X} \cong m\tilde{X}. \tag{7.19}$$

Warranty cost example.   For the total warranty cost of the motors in the coming year, 90% prediction limits are

$$\underset{\sim}{X} \cong 70\left\{751 - t\left[(1+0.90)/2; 184 - 1\right]944(1/70 + 1/184)^{1/2}\right\} = \$37{,}310,$$

$$\tilde{X} \cong 70\left[751 + 1.645(944)(1/70 + 1/184)^{1/2}\right] = \$67{,}830.$$

Each of these is a one-sided approximate 95% prediction limit. The prediction limits divided by the predicted cost are $37{,}310/52{,}570 = 0.71$ and $67{,}830/52{,}570 = 1.29$. So the uncertainty in the warranty costs is $\pm 29\%$ and is large. Management was able to plan the warranty budget accordingly.

Prediction and Limits for Ordered Observations

Prediction.   To predict the rth largest future observation $X_{(r)}$, suppose $i \leqslant (n+1)r/(m+1) \leqslant i+1$ for some $i = 1, \ldots, n-1$. Then a predictor for $X_{(r)}$ is

$$\hat{X}_{(r)} = \left\{i + 1 - (n+1)\left[r/(m+1)\right]\right\} Y_{(i)} + \left\{\left[(n+1)r/(m+1)\right] - i\right\} Y_{(i+1)}. \tag{7.20}$$

That is, the predictor is the $(100r/m)$th percentile of the past sample, perhaps, interpolating between two past observations.

Insulating fluid example.   For 34 kV, suppose one wants a prediction of the smallest observation ($r = 1$) in a future sample of $m = 10$ observations. First calculate $(n+1)r/(m+1) = (19+1)1/(10+1) = 1.818$. This indicates that the first and second order statistics (0.19 and 0.78 minute) are used. Then $\hat{X}_{(1)} = (1 + 1 - 1.818)0.19 + (1.818 - 1)0.78 = 0.67$ minute.

One-sided lower prediction limit.   For the smallest future observation $X_{(1)}$, use the smallest observation $Y_{(1)}$ of the past sample. The probability of $Y_{(1)}$ being below $X_{(1)}$ is

$$\gamma = n/(m+n). \tag{7.21}$$

Similarly, $Y_{(n)}$ is an upper prediction limit for $X_{(m)}$, and the probability of it being above $X_{(m)}$ is γ.

Insulating fluid example.   For 34 kV, the probability of the smallest, 0.19 minute, of $n=19$ observations being below the smallest of $m=10$ observations is $\gamma=19/(10+19)=0.66$, not a high probability.

More general prediction limits are given by Lieberman and Owen (1961). Their hypergeometric tables can be used to obtain the distribution of the number of exceedances, that is, to obtain the probability of the rth largest among m future observations exceeding the sth largest among n previous observations.

OTHER DISTRIBUTION-FREE METHODS

Chapter 3 describes graphical methods that can be used for nonparametric analysis of data. Most statistics books present some nonparametric methods. Books on nonparametric methods include those by Gibbons (1976), Hollander and Wolfe (1973), and Lehmann (1975).

For some work it may be possible to make a stronger but nonparametric assumption that the distribution has (1) an increasing (or else decreasing) failure rate or (2) an increasing (or else decreasing) failure rate **on the average**. Methods for such distributions have been developed by Barlow and Proschan (1965). Hanson and Koopmans (1964) give simple (conservative) nonparametric lower confidence limits for any percentile when $-\ln[F(y)]$, minus log cdf, is a convex function.

PROBLEMS

6.1.   Battery failures.   In five years, 500,000 batteries were put into service at a uniform rate. In that period, 5000 of them failed and were promptly replaced.

(a)   How many battery-years of exposure were accumulated by the end of the five-year period?

(b)   Assuming Poisson data, estimate the failure rate of such batteries.

(c)   Give an upper 95% confidence limit for the true failure rate.

(d)   Is the Poisson distribution appropriate? State your reasons.

(e)   100,000 more units will be manufactured and go into service over the coming year. Give a prediction of the number of failures in the entire population in service in the coming year.

(f)   Give the formula for the variance of the prediction error for (e) in terms of the failure rate λ, previous exposure t, and future exposure s.

(g)   Calculate approximate 95% prediction limits for (e).

6.2. Power line outages. Seven power lines of different lengths had outages as shown below. The exposure of a line equals its length times the years observed. Assume that the number of outages has a Poisson distribution.

(a) Estimate the outage rates (failures per year per mile) for lines 1 and 5.

(b) Calculate exact two-sided 95% confidence limits for the outage rate of line 1.

(c) Calculate approximate two-sided confidence limits for the outage rate of line 5.

(d) Calculate a prediction and exact 90% prediction limits for the number of outages on line 1 in the coming year.

(e) Calculate a prediction and approximate 90% prediction limits for the number of outages on line 5 in the coming year.

(f) Criticize all assumptions in (a) through (e).

Line k	Length L_k Miles		Years N_k		Exposure $t_k = N_k L_k$	Outages Y_k
1	10	×	5	=	50	10
2	13		1		13	13
3	17		5		85	17
4	24		9		216	102
5	42		6		252	124
6	61		2		122	53
7	69		2		138	44
					$t = 876$	$363 = Y$

6.3. Insulating fluid (exponential). For the 38-kV data in Table 1.1 of Chapter 3, do the following, assuming that time to breakdown has an exponential distribution.

(a) Estimate the mean time to failure.

(b) Calculate two-sided exact 95% confidence limits for the mean.

(c) Use the normal approximation to calculate such limits.

(d) Calculate the corresponding estimate and limits for the failure rate.

(e) Estimate the 10th percentile, and calculate two-sided exact 95% confidence limits for it.

(f)   Calculate a prediction for a single future observation.

(g)   Calculate an upper 90% prediction limit for a single future observation.

(h)   Calculate the sample size needed so that the estimate is within a factor $f = 1.5$ of the true mean.

(i)   Repeat (a) through (h) for data from other voltages in Table 1.1.

(j)   Plot the data on Weibull probability paper, plot the fitted exponential distribution, and assess the assumption of an exponential distribution.

6.4.   Use the life data on insulation system 3 in Problem 3.1. Assume life has a lognormal distribution and use base 10 logarithms.

(a)   For each temperature, estimate μ and σ.

(b)   For each temperature, calculate two-sided 95% confidence limits for μ and σ.

(c)   Estimate the 1% point for each temperature.

(d)   How large a sample is needed to estimate σ at 225°C within a factor of 2 with 90% confidence? Is the approximation satisfactory?

(e)   Calculate a prediction and 90% prediction limits for a single unit at 225°C.

6.5.   Insulating fluid (Weibull).   For the 38-kV data in Table 4.1, do the following, assuming that time to breakdown has a Weibull distribution.

(a)   Estimate the Weibull parameters.

(b)   Calculate two-sided approximate 90% confidence limits for the shape parameter.

(c)   Estimate the 10th percentile.

(d)   Calculate two-sided approximate 90% confidence limits for the 10th percentile.

(e)   Plot the data and the estimate of the distribution on Weibull paper.

(f)   Calculate two-sided approximate 90% confidence limits for other percentiles and draw smooth curves for the limits on the Weibull plot.

6.6.   Oil breakdown voltage.   The following summary statistics were calculated from the base e logs of data like that of Problem 5.3, but all eight sample sizes are 25 here.

	1000 V/second		100 V/second	
	1 in.	3 in.	1 in.	3 in.
Average	4.0296	2.4563	3.8996	3.7942
Standard Deviation	0.1044	0.0900	0.1057	0.0745

	10 V/second		1 V/second	
	1 in.	3 in.	1 in.	3 in.
	3.7268	3.6046	3.5974	3.4873
	0.0946	0.1168	0.1019	0.2279

(a) Choose a test condition and estimate the Weibull parameters.

(b) Calculate two-sided approximate 90% confidence limits for each parameter at your test condition.

(c) Estimate all shape parameters.

(d) Calculate two-sided approximate 90% confidence limits for all shape parameters.

(e) Plot all shape parameter estimates and confidence limits on semi-log paper. Are the estimates consistent with theory, which says that all eight distributions have the same shape parameter value?

6.7. Insulating fluid (nonparametric). For the 38-kV data in Table 4.1, do the following using nonparametric methods.

(a) Estimate the mean time to failure.

(b) Calculate two-sided approximate 95% confidence limits for the mean.

(c) Estimate the distribution median.

(d) If the smallest and largest observations are used as confidence limits for the distribution median, what is the confidence level?

(e) Plot the nonparametric estimate of the cumulative distribution function on Weibull paper.

(f) Plot nonparametric two-sided 95% confidence limits for the cumulative distribution function.

(g) Calculate a prediction for a single future observation.

(h) If the smallest and largest observations are used as prediction limits for a single future observation, what is the probability of them enclosing the observation?

(i*) Repeat (a) through (h) for data from other voltages in Table 4.1.

*Asterisk denotes laborious or difficult.

7

Linear Methods for Singly Censored Data

This chapter presents linear methods, which employ linear combinations of the ordered observations. Such numerical analyses apply to singly censored and complete life data. **Singly censored data** consist of the earliest r failure times in a sample of n units. Such data come from life tests where units are put on test together and the data analyzed before all units fail. Then the unfailed units all have the same running time, called the **censoring time**. Of course, a complete sample (all units run to failure) yields more precise estimates than a censored sample of the same size. However, the reduced precision from a censored sample is often compensated for by the time saved from analyzing the data before all units fail. Moreover, the reduction in precision is usually small for estimates of low percentiles and high reliabilities—usually of greatest interest. Also, one may have a complete sample and wish to use only the early observations to estimate the lower tail (percentiles and reliabilities), especially if the theoretical distribution does not adequately describe the upper part of the true distribution. Then the estimates and confidence limits for the lower tail, although approximate, are not badly biased by data from the upper tail.

Chapter 3 presents simple graphical methods for analyzing such data; a combination of graphical and numerical methods is often informative. Chapter 11 presents methods for combining or comparing estimates from a number of such samples. Chapter 8 presents maximum likelihood analyses of multiply censored data, that is, data with differing running times on some units; those methods also apply to singly censored data. However, the linear methods presented in this chapter are easier to use (usually hand calculations) than are maximum likelihood methods, which usually require special computer programs. In general, linear and maximum likelihood methods involve an assumed parametric distribution. However, the nonparametric methods of Chapter 6 apply to singly censored data, provided that the needed order statistics are observed. Linear methods also apply to complete data ($r=n$) and are alternatives to the methods for complete data in Chapter 6. To use linear methods, one should be acquainted with the basic concepts of estimation and prediction in Chapter 6. This chapter is more difficult than earlier ones.

This chapter first explains basic concepts for linear estimation from singly censored data. It then presents methods for obtaining estimates, confidence intervals, predictions, and prediction limits for the exponential (Section 2), normal and lognormal (Section 3), and Weibull and extreme value (Section 4) distributions. Section 5 presents theory for order statistics and linear estimates and is advanced.

1. BASIC CONCEPTS

This section presents basic concepts for singly censored data, order statistics, and linear methods for data analysis. Later sections apply these concepts to various distributions. Readers interested only in applying the methods can skip this technical section and proceed directly to sections on particular distributions.

Singly Censored Data

A sample is **singly censored on the right** if only its r smallest failure times are observed, and the unobserved times are known only to be above their common running time, called the "censoring time." Such data arise when units start on test together and the data are analyzed before all units fail. Such a test can save time. Epstein and Sobel (1953) investigate the expected time savings of a censored test over one running all units to failure for an exponential distribution.

Such data have **Type I** censoring (or **time** censoring) if the censoring time is prespecified. For example, time constraints may require that a test be stopped at 1000 hours. Such censoring is common in practice. Then the censoring time is fixed, and the number of failures is random. Data singly censored on the right have **Type II** censoring (or **failure** censoring) if the test is stopped at the time of the rth failure. Then the number of failures is fixed, and the length of the test is random. Such censoring is common in the literature, since it is mathematically simpler than Type I censoring. The methods of this chapter are exact for Type II data. Many people analyze Type I data as if it were Type II data; this is often satisfactory in practice and is done in this chapter. Chapter 8 gives maximum likelihood methods for Type I data.

Order Statistics

When n units start a test together, the shortest sample life is observed first, the second shortest is observed second, etc. The smallest sample value is the first order statistic, the second smallest value is the second order statistic, etc., and the largest value is the nth order statistic. These are denoted by $Y_{(1)}, Y_{(2)}, \ldots, Y_{(n)}$. For example, if n is odd, the sample median $Y_{((n+1)/2)}$ is such an order statistic. If a sample is singly right censored, only the first r order statistics $Y_{(1)}, Y_{(2)}, \ldots, Y_{(r)}$ are observed.

Suppose that random samples of size n are taken from the same distribution. Then the ith order statistic has different values in the different samples. That is, it has a sampling distribution, which depends on the population distribution, i, and n. Order statistics can be used to estimate the parameters and other characteristics of the population. This chapter explains how to do this. The sampling distributions and other properties of order statistics appear in Section 5.

Linear Methods

A linear combination of K order statistics of a sample with Type II censoring is called a **systematic statistic**. It has the form

$$a_{i_1} Y_{(i_1)} + a_{i_2} Y_{(i_2)} + \cdots + a_{i_K} Y_{(i_K)}; \qquad (1.1)$$

this contains any selected order statistics and known coefficients a_{i_K}. For example, if the sample is singly right censored, a linear combination of the r smallest order statistics has the form

$$a_1 Y_{(1)} + a_2 Y_{(2)} + \cdots + a_r Y_{(r)}. \qquad (1.2)$$

The coefficients are chosen so the linear combination is a good estimator for some parameter of the parent distribution. The coefficients depend on the parent distribution (exponential, normal, etc.), the quantity being estimated (mean, standard deviation, percentile, etc.), the size of the sample, the order statistics being used, and the properties of the estimator (unbiased, invariant, etc.).

A linear estimator θ^* is called a **best linear unbiased estimator** (BLUE) for a parameter θ if it is **unbiased** (its mean $E\theta^* = \theta$) and has minimum variance among linear unbiased estimators. A linear estimator θ^{**} is called a **best linear invariant estimator** (BLIE) for a parameter θ if it has a minimum mean squared error, $E(\theta^{**} - \theta)^2$, among linear estimators. There are tables of coefficients for both types of estimators, and one type may be obtained from the other (Section 5.5). The two estimators differ little for practical purposes, compared to the scatter in the data. The choice of either estimator is mostly a matter of taste. The BLUEs are presented below. The BLIEs are used the same way, only the coefficients differ. BLIEs are briefly described in Section 5.5 and used in Section 4, as tables for confidence limits for the Weibull distribution are in terms of BLIEs. Mann, Schafer, and Singpurwalla (1974) and Bain (1978) present these and other linear estimators.

The linear methods below are exact only for data with Type II (failure) censoring. However, they are often used for data with Type I (time) censoring. Then they usually provide a satisfactory approximation for practical purposes.

BLUEs have good properties compared to other estimators, such as maximum likelihood ones. For small samples, their variances and mean squared errors are comparable to those of other estimators. For large samples, their asymptotic variances and mean squared errors usually equal the theoretical minimum variance (Cramer–Rao lower bound) for any estimators.

Linear estimators can readily be derived for two-parameter distributions that have a **scale parameter** σ and a **location parameter** μ. That is, the cumulative distribution function $F(y; \mu, \sigma)$ can be written as $F(y; \mu, \sigma) = G[(y - \mu)/\sigma]$, where $G(\ )$ is a function that does not depend on μ or σ. The normal and smallest extreme value distributions are such distributions. Linear estimators are used with the logs of lognormal and Weibull data, since the log data come from a normal and extreme value distribution, respectively. Then the true variances and covariance of the BLUEs μ^* and σ^* for the location and scale parameters have the form

$$\text{Var}(\mu^*) = A\,\sigma^2, \qquad \text{Var}(\sigma^*) = B\,\sigma^2, \qquad \text{Cov}(\mu^*, \sigma^*) = C\,\sigma^2,$$

where A, B, and C depend on r and n and the distribution, but not on μ and σ. This is sometimes indicated with subscripts on the factors $A_{n,r}$, $B_{n,r}$, and $C_{n,r}$. Such factors for pooled estimates (Section 5.5) lack subscripts. Linear estimators can readily be derived for one-parameter distributions that have a scale parameter or else a location parameter. The exponential distribution is such a distribution. This chapter applies only to distributions whose parameters are only a location and/or a scale parameter.

Other Distributions

This chapter presents linear methods only for the exponential, normal, lognormal, Weibull, and extreme value distributions. Such methods for other distributions are surveyed by Sarhan and Greenberg (1962) and David (1970). Current literature (Chapter 13) gives further information. Gupta and others (1967) and Engelhardt (1975) give linear methods for the logistic distribution. Nonparametric methods of Chapter 6 apply to singly censored data when the method employs the early order statistics; for example, see Barlow and Proschan (1965), Hanson and Koopmans (1964) and Lawless (1982, Chap. 8).

2. EXPONENTIAL DATA

The exponential distribution is a model for the life of products with a true constant failure rate λ and a mean time to failure $\theta = 1/\lambda$. This distribution is described in detail in Chapter 2. A singly censored sample of n units consists of the r smallest ordered observations $Y_{(1)}, \ldots, Y_{(r)}$.

The following methods provide best linear unbiased estimates (BLUEs) and confidence intervals for θ, λ, percentiles, and reliabilities. Also, they provide predictions and prediction intervals for a future observation, sample average (or total), and smallest observation.

ESTIMATES AND CONFIDENCE INTERVALS

The Mean

Estimate. For failure censoring, the best linear unbiased estimator (BLUE) for the exponential mean θ is

$$\theta^* = \left[Y_{(1)} + \cdots + Y_{(r)} + (n-r)Y_{(r)} \right]/r. \qquad (2.1)$$

This is the total running time divided by the number of failures. For time censoring, θ^* is biased. θ^* is also the maximum likelihood estimator for θ; this is discussed further in Chapter 8. Its variance for failure censoring is

$$\text{Var}(\theta^*) = \theta^2/r. \qquad (2.2)$$

For failure censoring, the sampling distribution of $2r\theta^*/\theta$ is chi square with $2r$ degrees of freedom. If r is large (say, greater than 15), then the distribution of θ^* is approximately normal. Section 5.5 shows how to pool such estimates.

Table 2.1.   Seconds to Insulating Fluid Breakdown

25 kV	30 kV	35 kV	40 kV	45 kV
521	50	30	1	1−
2520	134	33	1	1−
4060	187	41	2	1−
12600	882	87	3	2
40300	1450	93	12	2
50600+	1470	98	25	3
52900+	2290	116	46	9
67300+	2930	258	56	13
84000	4180	461	68	47
85500+	15800	1180	109	50
85700+	29200+	1350	323	55
86400+	86100+	1500	417	71

```
+ running time without breakdown.
- breakdown occurred earlier.
```

Insulating fluid example.   Table 2.1 shows data on time to breakdown of an insulating fluid at five voltages. Such a distribution of time to breakdown is usually assumed to be exponential in engineering theory. For 30 kV, the data are singly time censored on the right. The estimate of the mean time to failure, based on the 10 failure times, is $\theta^* = [50 + \cdots + 15,800 + (12 - 10)15,800]/10 = 6097.3$ seconds (two- or three-figure accuracy would do). $\mathrm{Var}(\theta^*) \cong \theta^2/10$ approximately, since the data are time censored.

Exact confidence limits.   Two-sided $100\gamma\%$ confidence limits for the exponential mean θ are

$$\underset{\sim}{\theta} = 2r\theta^*/\chi^2[(1+\gamma)/2; 2r], \qquad \tilde{\theta} = 2r\theta^*/\chi^2[(1-\gamma)/2; 2r], \quad (2.3)$$

where $\chi^2(\delta; 2r)$ is the 100δth chi-square percentile for $2r$ degrees of freedom. (2.3) yields a one-sided $100\gamma\%$ confidence limit when γ replaces $(1+\gamma)/2$ or $1-\gamma$ replaces $(1-\gamma)/2$. The limits are exact (approximate) for failure (time) censoring.

Insulating fluid example.   At 30 kV, the one-sided lower 90% confidence limit for θ is $\underset{\sim}{\theta} \cong 2(10)6097.3/\chi^2[0.90; 2(10)] = 2(10)6097.3/28.412 = 4092.1$ seconds. This limit is approximate, since the sample is time censored.

Samples with no failures. For some samples, there are no failures when the data must be analyzed (time censored). Then (2.1) does not yield an estimate, and (2.3) does not yield confidence limits. A commonly used one-sided lower $100\gamma\%$ confidence limit for θ is

$$\underline{\theta} = 2(Y_1 + \cdots + Y_n)/\chi^2(\gamma; 2) = -(Y_1 + \cdots + Y_n)/\ln(1-\gamma),$$

where $Y_1, \ldots, Y_n$ are the sample running times. For an estimate, some use a 50% confidence limit; this method of estimation has no theoretical basis and can be seriously misleading. The confidence limit has a difficulty. As stated, it applies only when there are no failures. The limit must also be defined when there is one failure or two or three, etc.

Approximate confidence limits. Suppose r is large (say, greater than 15). Then two-sided approximate $100\gamma\%$ confidence limits for θ are

$$\underline{\theta} \cong \theta^*/\exp\left(K_\gamma/r^{1/2}\right), \qquad \tilde{\theta} \cong \theta^* \cdot \exp\left(K_\gamma/r^{1/2}\right), \tag{2.4}$$

where K_γ is the $[100(1+\gamma)/2]$th standard normal percentile. (2.4) yields a one-sided $100\gamma\%$ confidence limit when K_γ is replaced by z_γ, the 100γth standard normal percentile.

Sample size. The estimate θ^* will be within a factor f of θ with approximate probability $100\gamma\%$ if the observed number r (assumed large, say, over 15) of failures is

$$r \cong \left[K_\gamma/\ln(f)\right]^2. \tag{2.5}$$

where K_γ is the $[100(1+\gamma)/2]$th standard normal percentile. That is, θ^* will be between θ/f and θf with $100\gamma\%$ probability. This result for the exponential mean does not depend on the sample size n. However, the larger n is, the shorter the waiting time to the rth failure. Epstein and Sobel (1953) compare the test times of complete and censored tests in detail.

Insulating fluid example. Suppose the mean time to breakdown of the insulating fluid at 30 kV is to be estimated within 20% with 90% probability. Then $f \cong 1.20$, and the needed number of failures is $r \cong [1.645/\ln(1.20)]^2 \cong 81$.

The Failure Rate

Estimate. The usual estimate of the exponential failure rate λ is

$$\lambda^* = 1/\theta^* = r/\left[Y_{(1)} + \cdots + Y_{(r)} + (n-r)Y_{(r)}\right]. \tag{2.6}$$

This is the number of failures r divided by the total running time, that is, the observed failure rate. λ^* is not a linear function of the order statistics; it is a transformed linear estimator. λ^* is biased. The bias decreases with increasing r. For large r, the cumulative distribution of λ^* is close to a normal one with mean λ and variance λ^2/r. Equation (2.6) is often used incorrectly to estimate a failure rate for life distributions that are not exponential.

Insulating fluid example. At 30 kV, $\lambda^* = 1/6097.3 = 0.16 \times 10^{-3}$ failures per second.

Confidence limits. Two-sided $100\gamma\%$ confidence limits for the failure rate λ are

$$\underset{\sim}{\lambda} = 1/\tilde{\theta} = \lambda^* \chi^2[(1-\gamma)/2; 2r]/(2r),$$

$$\tilde{\lambda} = 1/\underset{\sim}{\theta} = \lambda^* \chi^2[(1+\gamma)/2; 2r]/(2r), \tag{2.7}$$

where $\chi^2(\delta; 2r)$ is the 100δ th chi-square percentile for $2r$ degrees of freedom. One-sided limits for λ are the reciprocals of the one-sided limits for θ. These limits are exact (approximate) for failure (time) censoring.

Insulating fluid example. At 30 kV, the one-sided upper 90% confidence limit for λ is $\tilde{\lambda} = 1/4092.1 = 0.24 \times 10^{-3}$ failures per second. This limit is approximate, since the sample is time censored.

Percentile

Estimate. The BLUE for the $100P$ th percentile $y_P = -\theta \ln(1-P)$ of an exponential distribution is

$$y_P^* = -\theta^* \ln(1-P). \tag{2.8}$$

Its variance is

$$\text{Var}(y_P^*) = [\theta \ln(1-P)]^2/r. \tag{2.9}$$

Insulating fluid example. At 30 kV, the estimate of the 10th percentile is $y_{.10}^* = -6097.3 \ln(1-0.10) = 642$ seconds. $\text{Var}(y_{.10}^*) = [\theta \ln(1-0.10)]^2/10 = 0.0011\theta^2$.

Confidence limits. Two-sided $100\gamma\%$ confidence limits for the $100P$ th exponential percentile are

$$\underset{\sim}{y_P} = -\underset{\sim}{\theta} \ln(1-P) = y_P^*\{2r/\chi^2[(1+\gamma)/2; 2r]\},$$

$$\tilde{y}_P = -\tilde{\theta} \ln(1-P) = y_P^*\{2r/\chi^2[(1-\gamma)/2; 2r]\}, \tag{2.10}$$

where $\chi^2(\delta;2r)$ is the 100δth chi-square percentile with $2r$ degrees of freedom. One-sided limits for y_P employ the one-sided limits for θ. These limits are exact (approximate) for failure (time) censoring.

Insulating fluid example. At 30 kV, the one-sided 90% confidence limit for the 10th percentile is $y_{.10} = -4092.1 \ln(1-0.10) = 431$ seconds. This limit is approximate, since the sample is time censored.

Reliability

Estimate. The transformed linear estimate of the reliability $R(y) = \exp(-y/\theta)$ for age y is

$$R^*(y) = \exp(-y/\theta^*) = \exp(-\lambda^* y). \qquad (2.11)$$

This estimator is biased, but the bias decreases with increasing sample size.

Insulating fluid example. At 30 kV, the estimate of the reliability at $y = 60$ seconds is $R^*(60) = \exp(-60/6097.3) = 0.990$.

Confidence limits. Two-sided $100\gamma\%$ confidence limits for reliability at age y are

$$\underset{\sim}{R}(y) = \exp(-y/\underset{\sim}{\theta}) = \exp(-\tilde{\lambda}y) = \exp\{-y\chi^2[(1+\gamma)/2;2r]/(2r\theta^*)\},$$

$$\tilde{R}(y) = \exp(-y/\tilde{\theta}) = \exp(-\underset{\sim}{\lambda}y) = \exp\{-y\chi^2[(1-\gamma)/2;2r]/(2r\theta^*)\},$$

$$(2.12)$$

where $\chi^2(\delta;2r)$ is the 100δth chi-square percentile for $2r$ degrees of freedom. One-sided limits for $R(y)$ employ the one-sided limits for θ.

Insulating fluid example. At 30 kV, the one-sided lower 90% confidence limit for reliability at 60 seconds is $\underset{\sim}{R}(60) = \exp(-60/4092.1) = 0.985$. This limit is approximate, since the sample is time censored.

Estimates of the Mean from Selected Order Statistics

Previous linear methods for a singly censored sample from an exponential distribution employ all r observed order statistics. Linear estimates for θ based on selected order statistics are sometimes useful. These estimates and confidence limits for θ can be used to obtain estimates and limits for the failure rate, percentiles, reliabilities, and other functions of θ as described above. These estimates are easy to calculate, particularly for large samples outside the tables for BLUEs.

Estimates from a single order statistic.   The BLUE for θ based on the jth order statistics $Y_{(j)}$ is

$$\theta^*_{(j)} = Y_{(j)}/E_j, \tag{2.13}$$

where $E_j = [n^{-1} + (n-1)^{-1} + \cdots + (n-j+1)^{-1}]$. Its variance is

$$\mathrm{Var}(\theta^*_{(j)}) = \theta^2 [n^{-2} + (n-1)^{-2} + \cdots + (n-j+1)^{-2}]/E_j^2. \tag{2.14}$$

These results are derived in Sarhan and Greenberg (1962). For j and $(n-j)$ large,

$$\theta^*_{(j)} \simeq Y_{(j)}/\ln[(n+\tfrac{1}{2})/(n-j+\tfrac{1}{2})],$$

$$\mathrm{Var}(\theta^*_{(j)}) \simeq [\theta^2/(n+\tfrac{1}{2})][j/(n-j+\tfrac{1}{2})]/\{\ln[(n+\tfrac{1}{2})/(n-j+\tfrac{1}{2})]\}^2.$$

The distribution of $\theta^*_{(j)}$ is approximately normal. $\mathrm{Var}(\theta^*_{(j)})$ is a minimum (for large samples) if $j = \langle 0.80n \rangle$. The notation $\langle x \rangle$ indicates that x is rounded up to the nearest integer. That is, the 80th sample percentile (more precisely, 79.68%) provides the best large-sample linear estimator based on a single order statistic. Then

$$\theta^*_{(.80n)} = 0.6275 Y_{(\langle .80n \rangle)} \tag{2.15}$$

and

$$\mathrm{Var}(\theta^*_{(.80n)}) \cong 1.544\theta^2/n. \tag{2.16}$$

For comparison, the variance of $\overline{Y}$, the BLUE based on the entire sample, is $\mathrm{Var}(\overline{Y}) = \theta^2/n$. If less than 80% of the sample is observed, the estimate should use the largest order statistic.

The estimate based on one order statistic is useful in graphical displays of data. An order statistic can then be marked and used as an estimate of θ. This helps one graphically compare a number of samples, as described in Chapter 10.

Insulating fluid example.   At 30 kV, $n = 12$ and the estimate is based on the $j = \langle 0.80(12) \rangle = 10$th order statistic. $Y_{(10)} = 15,800$ and $\theta^*_{(10)} \cong 0.6275$ $(15,800) = 9914.5$ seconds. Then $\mathrm{Var}(\theta^*_{(10)}) \cong \theta^2 1.544/12 \cong 0.1287\theta^2$.

Confidence limits.   Exact confidence intervals for θ from $\theta^*_{(j)}$ can be derived from the sampling distribution of $Y_{(j)}$. Such sampling distributions are presented in Sarhan and Greenberg (1962). Two-sided approximate $100\gamma\%$ confidence limits for θ from the normal approximation for $\theta^*_{(j)}$ are

$$\underline{\theta}_{(j)} \cong \theta^*_{(j)}/\exp(K_\gamma C_j), \qquad \tilde{\theta}_{(j)} \cong \theta^*_{(j)} \cdot \exp(K_\gamma C_j), \qquad (2.17)$$

where $C_j = [\mathrm{Var}(\theta^*_{(j)})/\theta^2]^{1/2}$ from (2.14) and K_γ is the $[100(1+\gamma)/2]$th standard normal percentile. (2.17) yields one-sided limits when K_γ is replaced by z_γ, the 100γth standard normal percentile.

Insulating fluid example. At 30 kV, the one-sided lower approximate 90% confidence limit for θ is $\underline{\theta}_{(10)} = 9914.5/\exp[1.282(0.1287)] = 8406.5$ seconds.

Estimates based on the best K order statistics. For large n, Sarhan and Greenberg (1962, Sec. 11D) give the asymptotically BLUEs θ^*_K for θ that are based on the best $K = 1, \ldots, 10$ order statistics that minimize the asymptotic variance of θ^*_K. The estimator for $K = 1$ was given above. For $K = 2$, the asymptotically BLUE is

$$\theta^*_2 = 0.5232 Y_{(\langle .6386n \rangle)} + 0.1790 Y_{(\langle .9266n \rangle)},$$

and

$$\mathrm{Var}(\theta^*_2) \cong 1.219\theta^2/n. \qquad (2.18)$$

These estimators generally employ high sample percentiles. So they are usually useful only for complete samples. For large n, these estimators are approximately normally distributed.

PREDICTION

Suppose the earliest r of n failure times from a past sample from an exponential distribution are $Y_{(1)}, \ldots, Y_{(r)}$. The following presents predictors and prediction limits for the later order statistics $Y_{(r+1)}, \ldots, Y_{(n)}$ of the same sample and, for an independent future sample of size m, the average (or total) and the order statistics $X_{(1)}, \ldots, X_{(m)}$. θ^* denotes the BLUE (2.1) for the exponential mean θ. The methods are exact for failure censoring.

jth Observation of the Same Sample

Predictor. The predictor $\hat{Y}_{(j)}$ for how long a test will run until the jth observation of the same sample is

$$\hat{Y}_{(j)} = Y_{(r)} + \theta^* \left[(n-r)^{-1} + (n-r-1)^{-1} + \cdots + (n-j+1)^{-1} \right]. \qquad (2.19)$$

This predictor is unbiased, and the prediction error variance is

$$\text{Var}\left[\hat{Y}_{(j)} - Y_{(j)}\right] = \theta^2 \big\{ (1/r)\big[(n-r)^{-1} + (n-r-1)^{-1}$$

$$+ \cdots + (n-j+1)^{-1}\big]^2 + \big[(n-r)^{-2} + (n-r-1)^{-2}$$

$$+ \cdots + (n-j+1)^{-2}\big]\big\}. \tag{2.20}$$

$\hat{Y}_{(n)} - Y_{(r)}$ is usually of greatest interest, as it is the remaining time required for all sample units to fail.

Prediction limits for the $(r+1)$th observation. Two-sided $100\gamma\%$ prediction limits for the $(r+1)$th observation of the same sample are

$$\underset{\sim}{Y}_{(r+1)} = Y_r + \big\{ \big[\theta^*/(n-r)\big] / F\big[(1+\gamma)/2; 2r, 2\big]\big\},$$

$$\tilde{Y}_{(r+1)} = Y_r + \big\{ \big[\theta^*/(n-r)\big] F\big[(1+\gamma)/2; 2, 2r\big]\big\}, \tag{2.21}$$

where $F(\delta; a, b)$ is the 100δth F percentile with a degrees of freedom in the numerator and b in the denominator. Note that the degrees of freedom in the two limits differ. (2.21) gives a one-sided $100\gamma\%$ prediction limit for $Y_{(r+1)}$ when γ replaces $(1+\gamma)/2$.

Insulating fluid example. At 30 kV, the one-sided 90% upper prediction limit for the waiting time to the next failure is $\tilde{Y}_{(11)} = 15{,}800 + \{[6097.3/(12 - 10)]\ F[0.90; 2, 2(10)]\} = 15{,}800 + [(6097.3/2)(2.589)] = 23693$ seconds.

Prediction limits for the jth observation. Lawless (1971) provides exact prediction limits for the jth observation of the same sample. The limits for the $(r+1)$th observation are a special case of his results. An upper prediction limit for the nth observation is related to an outlier test. If the nth observation is above a $100\gamma\%$ limit, it is a statistically significant outlier at the $[100(1-\gamma)]\%$ level.

An Observation, Mean, or Total of a Future Sample

Predictor. The predictor for the mean $\bar{X}$ of an independent future sample of size m is

$$\hat{\bar{X}} = \theta^* = \big[Y_{(1)} + \cdots + Y_{(r)} + (n-r)Y_{(r)}\big]/r, \tag{2.22}$$

where θ^* is the BLUE for the exponential mean θ. If $m=1$, this is the predictor for a single future observation. This predictor is unbiased, and

$$\text{Var}(\hat{\bar{X}}-\bar{X})=\theta^2[(1/r)+(1/m)]. \tag{2.23}$$

If r and m are large (say, both greater than 15), then the distribution of the prediction error $(\hat{\bar{X}}-\bar{X})$ is approximately normal.

The prediction of the total $X=X_1+\cdots+X_m$ of m future observations is $\hat{X}=m\bar{X}=m\theta^*$ and $\text{Var}(\hat{X}-X)=m^2\theta^2[(1/r)+(1/m)]$.

Insulating fluid example. At 30 kV, the prediction of the time to run $m=1$ more breakdown is $\hat{X}=6097.3$ seconds, the estimate of the mean based on the past sample.

Prediction limits. Two-sided $100\gamma\%$ prediction limits for the average $\bar{X}$ of a future sample of m observations are

$$\underset{\sim}{\bar{X}}=\theta^*/F[(1+\gamma)/2;2r,2m], \qquad \tilde{\bar{X}}=\theta^*\cdot F[(1+\gamma)/2;2m,2r], \tag{2.24}$$

where $F(\delta; a, b)$ is the 100δth F percentile with a degrees of freedom above and b below. The degrees of freedom in the two limits differ. (2.24) gives a one-sided prediction limit for $\bar{X}$ when γ replaces $(1+\gamma)/2$. Such prediction limits for the total $X=m\bar{X}$ of a future sample of m observations are

$$\underset{\sim}{X}=m\underset{\sim}{\bar{X}} \quad \text{and} \quad \tilde{X}=m\tilde{\bar{X}}. \tag{2.25}$$

The formulas apply to a single observation; then $m=1$.

Insulating fluid example. At 30 kV, two-sided 90% prediction limits for a single future observation are $\underset{\sim}{X}=6097.3/F[(1+0.90)/2; 2(10), 2(1)]=6097.3/19.446=314$ and $\tilde{X}=6097.3\cdot F[(1+0.90)/2; 2(1), 2(10)]=6097.3\cdot3.4928=21,297$ seconds.

jth Observation of a Future Sample

Predictor. The predictor of the jth ordered observation $X_{(j)}$ of an independent future sample is

$$\hat{X}_{(j)}=\theta^*[m^{-1}+(m-1)^{-1}+\cdots+(m-j+1)^{-1}]. \tag{2.26}$$

This linear predictor is unbiased, and its prediction error variance is minimum (but can be large) and equals

$$\text{Var}\left[\hat{X}_{(j)} - X_{(j)}\right] = \theta^2 \left\{ (1/r)\left[m^{-1} + (m-1)^{-1} + \cdots + (m-j+1)^{-1}\right]^2 \right.$$

$$\left. + \left[m^{-2} + (m-1)^{-2} + \cdots + (m-j+1)^{-2}\right]\right\}. \quad (2.27)$$

Insulating fluid example. At 30 kV, the prediction of the first failure time in a future sample of size $m = 12$ is $\hat{X}_{(1)} = 6097.3(12^{-1}) = 508$ seconds.

Prediction limits for the smallest observation. Two-sided $100\gamma\%$ prediction limits for the smallest observation $X_{(1)}$ in an independent future sample of m observations are

$$\underset{\sim}{X}_{(1)} = (\theta^*/m)/F[(1+\gamma)/2; 2r, 2], \qquad \tilde{X}_{(1)} = (\theta^*/m) \cdot F[(1+\gamma)/2; 2, 2r],$$

$$(2.28)$$

where $F(\delta; a, b)$ is the 100δth F percentile with a degrees of freedom above and b below. (2.28) gives a one-sided $100\gamma\%$ prediction limit when γ replaces $(1+\gamma)/2$. A lower limit $\underset{\sim}{X}_{(1)}$ for a fleet of m units is called a **safe warranty life** or an **assurance limit** for the fleet. Also, $X_{(1)}$ is a lower limit for the life of a series system of m identical components. That is, a producer of such systems can "guarantee" that a customer's one system will survive at least a time $\underset{\sim}{X}_{(1)}$ with $100\gamma\%$ probability.

Insulating fluid example. At 30 kV, the one-sided lower 90% prediction limit for the smallest of $m = 12$ future observations is $\underset{\sim}{X}_{(1)} = (6097.3/12)$ $/F[(1+0.90)/2; 2(10), 2] = (6097.3/12)/19.446 = 26.1$ seconds.

Prediction limits for the jth observation. Lawless (1971) presents tables for prediction limits for $X_{(j)}$ of an independent future sample from an exponential distribution. These limits are related to those for a future ordered observation of the same sample.

OTHER EXPONENTIAL METHODS

Chapter 3 describes probability plotting methods for graphical analysis of singly censored exponential data. Chapter 11 presents linear methods for comparing a number of exponential populations for equality of their means with such data. Chapters 8 and 12 presents maximum likelihood methods.

3. NORMAL AND LOGNORMAL DATA

The normal and lognormal distributions serve as models for the life, strength, and other properties of many products and materials. These distributions have two parameters—the (log) mean μ and the (log) standard deviation σ—and are described in Chapter 2. A singly censored sample of n units consists of the r smallest ordered (log) observations $Y_{(1)}, \ldots, Y_{(r)}$.

This section presents best linear unbiased estimates (BLUEs) and confidence intervals for the population μ, σ, percentiles, and reliabilities. The methods are exact (approximate) for failure (time) censoring. Also, this section presents predictors and prediction limits for future ordered observations from the same or another sample. The methods apply to both normal and lognormal data, but one works with the (base 10) log values of lognormal data. Chapter 3 provides graphical methods for checking how well the (log) normal distribution fits such singly censored data. This should be checked before one uses the analytic methods below.

ESTIMATES

Estimate of μ

Estimate.   The BLUE for μ is

$$\mu^* = a(1; n, r) Y_{(1)} + \cdots + a(r; n, r) Y_{(r)}; \qquad (3.1)$$

the coefficients $a(i; n, r)$ are in Appendix A11 for $n = 2(1)12$ and $r = 2(1)n$. The variance of μ^* is

$$\mathrm{Var}(\mu^*) = \sigma^2 A_{n,r}; \qquad (3.2)$$

the variance factors $A_{n,r}$ are in Appendix A11. Sarhan and Greenberg (1962) give larger tables for $n = 2(1)20$ and $r = 2(1)n$. For r large, μ^* is approximately normally distributed, with mean μ and variance (3.2). Section 5.5 shows how to pool such estimates of the same μ.

Class-B insulation example.   Table 3.1 shows singly censored data from an accelerated life test of a Class-B electrical insulation for motors. Such insulation is assumed to have a lognormal life distributions at each temperature. The table gives the base 10 logs of the hours to failure. For 170°C, 7 of 10 failed. The estimate of the log mean is $\mu^* = (0.0244)3.2465 + \cdots + (0.5045)3.7157 = 3.6381$. Here $\mathrm{Var}(\mu^*) = 0.1167\,\sigma^2$. The estimate of the lognormal median is antilog $(3.6381) = 4346$ hours.

Table 3.1.   Class-B Insulation Log Failure Times

150°C	170°C	190°C	220°C
3.9066+	3.2465	2.6107	2.6107
3.9066+	3.4428	2.6107	2.6107
3.9066+	3.5371	3.1284	2.7024
3.9066+	3.5492	3.1284	2.7024
3.9066+	3.5775	3.1584	2.7024
3.9066+	3.6866	3.2253+	2.7226+
3.9066+	3.7157	3.2253+	2.7226+
3.9066+	3.7362+	3.2253+	2.7226+
3.9066+	3.7362+	3.2253+	2.7226+
3.9066+	3.7362+	3.2253+	2.7226+

+ denotes running time without failure.

Estimate of σ

Estimate. The BLUE for σ is

$$\sigma^* = b(1; n, r)Y_{(1)} + \cdots + b(r; n, r)Y_{(r)}; \qquad (3.3)$$

the coefficients $b(i; n, r)$ are in Appendix A11 for $n = 2(1)12$ and $r = 2(1)n$. The variance of σ^* is

$$\text{Var}(\sigma^*) = \sigma^2 B_{n,r}; \qquad (3.4)$$

the variance factors $B_{n,r}$ are in Appendix A11. Sarhan and Greenberg (1962) give larger tables for $n = 2(1)20$ and $r = 2(1)n$. For large r, σ^* is approximately normally distributed, with mean σ and variance (3.4). Section 5.5 shows how to pool such estimates of the same σ.

Class-B insulation example. For 170°C, the estimate of the log standard deviation is $\sigma^* = (-0.3252)3.2465 + \cdots + (0.6107)3.7157 = 0.2265$. Here $\text{Var}(\sigma^*) = 0.0989 \, \sigma^2$.

Estimate of a Percentile

Estimate. The BLUE for the $100P$th normal percentile $y_P = \mu + z_P \sigma$ is

$$y_P^* = \mu^* + z_P \sigma^*, \qquad (3.5)$$

where z_P is the 100Pth standard normal percentile. The variance of y_P^* is

$$\text{Var}(y_P^*) = \sigma^2 D_{n,r}, \tag{3.6}$$

where

$$D_{n,r} = \left(A_{n,r} + z_P^2 B_{n,r} + 2z_P C_{n,r}\right), \tag{3.7}$$

where $C_{n,r}$ is defined by $\text{Cov}(\mu^*, \sigma^*) = C_{n,r}\sigma^2$. The $C_{n,r}$ are in Appendix A11 for $n=2(1)12$ and $r=2(1)n$. Sarhan and Greenberg (1962) give a larger table of $C_{n,r}$ for $n=2(1)20$ and $r=2(1)n$. For large r, y_P^* is approximately normally distributed, with mean y_P and variance (3.6). The transformed linear estimate for the 100Pth lognormal percentile $t_P = \text{antilog}(y_P)$ is

$$t_P^* = \text{antilog}(y_P^*). \tag{3.8}$$

This estimator is biased. For r large, t_P^* is approximately normally distributed, with approximate mean t_P and approximate variance

$$\text{Var}(t_P^*) \simeq \left[t_P \ln(10)\right]^2 \text{Var}(y_P^*)$$

$$= \left[t_P \ln(10)\right]^2 \left(A_{n,r} + z_P^2 B_{n,r} + 2z_P C_{n,r}\right)\sigma^2. \tag{3.9}$$

Class-B insulation example. For 170°C, the estimate of the 10th percentile of the log data is $y_{10}^* = 3.6381 + (-1.282)0.2265 = 3.3477$. The estimate of the lognormal 10th percentile is antilog(3.3477) = 2227 hours. Here $\text{Var}(y_{10}^*) = \sigma^2[0.1167 + (-1.282)^2 0.0989 + 2(-1.282)0.0260] = 0.2126\ \sigma^2$.

Estimate of Reliability

Estimate. The transformed linear estimate of the fraction $F(y) = \Phi[(y-\mu)/\sigma]$ failing by (log) age y is

$$F^*(y) = \Phi(z^*), \tag{3.10}$$

where $\Phi(\ )$ is the standard normal cumulative distribution function,

$$z^* = (y - \mu^*)/\sigma^* \tag{3.11}$$

estimates the standardized normal deviate $z = (y-\mu)/\sigma$, and μ^* and σ^* are the BLUEs. This estimator is biased. For r large, the cumulative distribution of z^* is close to a normal one, with mean $z = (y-\mu)/\sigma$ and approximate variance

$$\text{Var}(z^*) \simeq A_{n,r} + z^2 B_{n,r} + 2zC_{n,r}, \tag{3.12}$$

where $\text{Var}(\mu^*) = A_{n,r} \sigma^2$, $\text{Var}(\sigma^*) = B_{n,r} \sigma^2$, and $\text{Cov}(\mu^*, \sigma^*) = C_{n,r} \sigma^2$, which are tabulated in Appendix A11. The corresponding (biased) estimator for reliability at (log) age y is

$$R^*(y) = 1 - \Phi(z^*) = \Phi(-z^*). \tag{3.13}$$

Class-B insulation example. At 170°C, the estimate of the reliability for 3000 hours $[y = \log(3000) = 3.4771]$ is obtained from $z^* = (3.4771 - 3.6381)/0.2265 = -0.7108$ and $R^*(3000) = \Phi(0.7108) = 0.761$.

Estimates Based on Selected Order Statistics

Previous linear methods for a singly censored sample from a (log) normal distribution employ all r observed order statistics in a sample of size n. Linear estimates of μ and σ based on selected order statistics are sometimes useful, particularly for sample sizes outside the tables. Selected order statistics provide estimates and confidence limits for percentiles, reliabilities, and other functions of μ and σ as described above. Such estimates are convenient, particularly for large samples and data plots.

Estimates for μ. The BLUE for μ based on one order statistic is the sample median; that is,

$$\mu_1^* = \begin{cases} Y_{((n+1)/2)} & \text{for } n \text{ odd,} \\ \left[0.5Y_{(0.5n)} + 0.5Y_{(0.5n+1)}\right] & \text{for } n \text{ even.} \end{cases} \tag{3.14}$$

The median is easy to mark in a plot of a sample for easy comparison with other sample medians. Its variance is

$$\text{Var}(\mu_1^*) = D_n \sigma^2, \tag{3.15}$$

where D_n is tabled by Sarhan and Greenberg (1962). For n large (say, $n \geqslant 20$), μ_1^* is approximately normally distributed and

$$\text{Var}(\mu_1^*) \simeq (\pi/2)\sigma^2/n = 1.571 \, \sigma^2/n. \tag{3.16}$$

For comparison, the variance of the BLUE $\bar{Y}$ based on a complete sample of size n is $\text{Var}(\bar{Y}) = \sigma^2/n$.

Class-B insulation example. For 170°C, the sample median estimates μ: $\mu_1^* = (3.5775 + 3.6866)/2 = 3.6321$. The approximate variance is $\text{Var}(\mu_1^*) \cong \sigma^2 1.571/10 = \sigma^2 0.1571$. The estimate based on the seven observed values has $\text{Var}(\mu^*) = 0.1167\,\sigma^2$. The estimate of the lognormal median is antilog $(3.6321) = 4286$ hours. The estimate based on the seven observed values is 4346 hours.

For large n, the asymptotically BLUE for μ based on two order statistics is

$$\mu_2^* = 0.5 Y_{(\langle 0.270n \rangle)} + 0.5 Y_{(\langle 0.730n \rangle)}, \tag{3.17}$$

where $\langle x \rangle$ denotes x rounded up to the nearest integer. For n large, this estimator is approximately normally distributed with a variance

$$\text{Var}(\mu_2^*) \simeq 1.235\,\sigma^2/n. \tag{3.18}$$

Estimates for σ. The BLUE for σ based on the jth and kth order statistics ($j < k$) is

$$\sigma_2^* = (Y_{(k)} - Y_{(j)})/(EZ_{(k)} - EZ_{(j)}), \tag{3.19}$$

where $EZ_{(j)}$ and $EZ_{(k)}$ are the expectations of the standard normal order statistics. The expectations are tabulated by Sarhan and Greenberg (1962) and Owen (1962). The variance of this estimator is

$$\text{Var}(\sigma_2^*) = \sigma^2 \big[\text{Var}(Z_{(k)}) + \text{Var}(Z_{(j)}) $$
$$ - 2\text{Cov}(Z_{(j)}, Z_{(k)}) \big] / (EZ_{(k)} - EZ_{(j)})^2; \tag{3.20}$$

this is in terms of the variances and covariances of the standard normal order statistics, which are tabulated by Sarhan and Greenberg (1962) and Owen (1962).

When samples of the same size are plotted, their standard deviations can easily be graphically compared. Mark the selected jth and kth observations of each sample, and visually compare the differences $Y_{(k)} - Y_{(j)}$.

For n large, the estimator is approximately normally distributed and approximated by

$$\sigma_2^* \simeq (Y_{(k)} - Y_{(j)})/(z_{k/(n+1)} - z_{j/(n+1)}), \tag{3.21}$$

where z_δ is the 100δth standard normal percentile. The approximation is valid for large j and $n-k$. The approximate large-sample variance is

$$\text{Var}(\sigma_2^*) \simeq (\sigma^2/n)\left\{(k/n)[1-(k/n)][\phi(z_{k/n})]^{-2}\right.$$

$$+ (j/n)[1-(j/n)][\phi(z_{j/n})]^{-2} - 2(j/n)[1-(k/n)]$$

$$\left. \times [\phi(z_{k/n})\phi(z_{j/n})]^{-1}\right\}/(z_{k/(n+1)}-z_{j/(n+1)})^2, \qquad (3.22)$$

where $\phi(\ )$ is the standard normal probability density.

This variance is minimized for $j = \langle 0.0694n \rangle$ and $k = \langle 0.9306n \rangle$. Then the asymptotically BLUE for σ based on two order statistics is

$$\sigma_2^* \cong 0.338 Y_{(\langle .9306n \rangle)} - 0.338 Y_{(\langle .0694n \rangle)}. \qquad (3.23)$$

Its approximate large-sample variance is

$$\text{Var}(\sigma_2^*) \simeq 0.767 \, \sigma^2/n. \qquad (3.24)$$

This estimator works only for nearly complete samples. This estimator suggests that the largest available order statistic be used for $Y_{(k)}$ when the sample is censored below the 93rd percentile. For comparison, the "best" estimator of σ is S, the standard deviation of a complete sample of size n. Its large-sample variance is

$$\text{Var}(S) \cong 0.500 \, \sigma^2/n. \qquad (3.25)$$

Class-B insulation example.   For 170°C, the approximate estimate based on the first and seventh order statistics (smallest and largest available) is $\sigma_2^* \cong (3.7157 - 3.2465)/[0.3758 - (-1.5388)] = 0.2453$. Its variance is $\text{Var}(\sigma_2^*) = \sigma^2[0.1579 + 0.3433 - 2(0.0489)]/[0.3758 - (-1.5388)]^2 = 0.1103 \, \sigma^2$. For the BLUE based on all seven order statistics, $\text{Var}(\sigma^*) = 0.0989 \, \sigma^2$, about 12% smaller.

Sarhan and Greenberg (1962, Sec. 10E) give the asymptotically BLUEs for μ and σ and their variances for the two, three,..., and ten best order statistics.

CONFIDENCE LIMITS

The following paragraphs present exact and approximate confidence limits, based on BLUEs, for (log) normal distribution parameters, percentiles, and reliabilities. The exact intervals were developed by Nelson and Schmee

(1979). The approximate limits will do for many practical problems outside the ranges of the tables of Nelson and Schmee (1979). Also, the approximate limits can be used with pooled estimates described in Section 5.5.

Confidence Limits for σ

Exact limits. Two-sided exact $100\gamma\%$ confidence limits for σ are

$$\underline{\sigma}=\sigma^*/w^*\big[(1+\gamma)/2; n, r\big], \qquad \tilde{\sigma}=\sigma^*/w^*\big[(1-\gamma)/2; n, r\big], \quad (3.26)$$

where $w^*(\delta; n, r)$ is the 100δth percentile of the distribution of $w^*=\sigma^*/\sigma$. These percentiles appear in Table 3.2 for $n=2(1)10$, $r=2(1)n$, and $1-\delta$ and $\delta=.005,.01,.025,.05,.1,.5$. (3.26) yields a one-sided $100\gamma\%$ confidence limit when γ replaces $(1+\gamma)/2$ or $1-\gamma$ replaces $(1-\gamma)/2$. These limits are exact (approximate) for failure (time) censored samples.

Class-B insulation example. Two-sided 90% confidence limits for σ at 170°C are $\underline{\sigma}=0.2265/w^*(0.95; 10, 7)=0.2265/1.559=0.1453$ and $\tilde{\sigma}=0.2265/0.521=0.4347$. The sample is time censored; so the limits are approximate.

Approximate limits. Suppose that σ^* is any linear unbiased estimator for σ and $\text{Var}(\sigma^*)=B\sigma^2$. Then two-sided approximate $100\gamma\%$ confidence limits for σ are

$$\underline{\sigma}\cong\sigma^*/\big[1-(B/9)+K_\gamma(B/9)^{1/2}\big]^3, \qquad \tilde{\sigma}\cong\sigma^*/\big[1-(B/9)-K_\gamma(B/9)^{1/2}\big]^3,$$

$$(3.27)$$

where K_γ is the $[100(1+\gamma)/2]$th standard normal percentile. Mann, Schafer, and Singpurwalla (1974) give this (Wilson–Hilferty chi-square) approximation.

Class-B insulation example. For 170°C, $\sigma^*=0.2265$, and $\text{Var}(\sigma^*)=0.0989\,\sigma^2$. Two-sided approximate 90% confidence limits for σ are $\underline{\sigma}=0.2265/[1-(0.0989/9)+1.645(0.0989/9)^{1/2}]^3=0.145$ and $\tilde{\sigma}=0.2265/[1-(0.0989/9)-1.645(0.0989/9)^{1/2}]^3=0.416$. The pooled BLUE (Section 5.5) of σ for all test temperatures is $\sigma^*=0.2336$, and $\text{Var}(\sigma^*)=0.04442\,\sigma^2$. Corresponding two-sided 90% confidence limits for σ are $\underline{\sigma}=0.2336/[1-(0.04442/9)+1.645(0.04442/9)^{1/2}]^3 = 0.171$ and $\tilde{\sigma}=0.2336/[1-(0.04442/9)-1.645(0.04442/9)^{1/2}]^3=0.343$.

Table 3.2. Percentiles $w^*(\delta; n, r)$ for Limits for σ

n	r	δ:	.005	.01	.025	.05	.10	.50	.90	.95	.975	.99	.995
2	2		.0065	.012	.036	.072	.156	.852	2.057	2.450	2.803	3.233	3.488
3	2		.0075	.015	.035	.069	.136	.788	2.105	2.544	2.938	3.388	3.697
3	3		.067	.106	.174	.246	.354	.916	1.697	1.926	2.150	2.412	2.571
4	2		.0069	.013	.031	.067	.133	.794	2.180	2.669	3.119	3.622	3.961
4	3		.069	.099	.161	.227	.340	.934	1.763	2.025	2.260	2.536	2.740
4	4		.158	.210	.289	.373	.481	.976	1.568	1.767	1.936	2.128	2.283
5	2		.0063	.013	.032	.065	.140	.803	2.212	2.682	3.140	3.658	3.909
5	3		.079	.102	.164	.237	.340	.933	1.799	2.060	2.305	2.605	2.844
5	4		.149	.201	.283	.368	.471	.976	1.617	1.810	1.993	2.205	2.355
5	5		.234	.282	.366	.451	.553	.983	1.508	1.671	1.808	1.984	2.094
6	2		.0050	.012	.030	.062	.128	.784	2.168	2.681	3.153	3.722	4.229
6	3		.067	.093	.152	.216	.321	.907	1.769	2.071	2.301	2.589	2.815
6	4		.153	.181	.259	.343	.444	.943	1.601	1.813	2.003	2.213	2.386
6	5		.223	.265	.354	.433	.527	.961	1.510	1.671	1.810	1.992	2.106
6	6		.291	.347	.427	.504	.594	.972	1.422	1.571	1.689	1.842	1.930
7	2		.0068	.011	.032	.061	.126	.778	2.213	2.696	3.167	3.728	4.188
7	3		.061	.087	.144	.216	.320	.914	1.803	2.074	2.321	2.643	2.846
7	4		.138	.183	.267	.336	.444	.948	1.628	1.841	2.024	2.262	2.446
7	5		.218	.263	.340	.418	.517	.966	1.526	1.714	1.857	2.052	2.165
7	6		.298	.343	.415	.489	.582	.976	1.461	1.617	1.744	1.885	1.990
7	7		.349	.401	.470	.537	.628	.986	1.406	1.536	1.637	1.741	1.839
8	2		.0056	.011	.030	.067	.130	.779	2.219	2.697	3.179	3.791	4.189
8	3		.067	.097	.156	.231	.326	.917	1.815	2.121	2.385	2.677	2.965
8	4		.138	.196	.271	.345	.449	.958	1.641	1.859	2.053	2.290	2.440
8	5		.221	.267	.351	.428	.529	.968	1.528	1.700	1.852	2.043	2.174
8	6		.286	.343	.420	.493	.584	.979	1.462	1.609	1.724	1.872	1.958
8	7		.335	.385	.470	.534	.623	.989	1.409	1.534	1.644	1.764	1.858
8	8		.395	.445	.511	.578	.660	.989	1.357	1.472	1.574	1.687	1.757
9	2		.0061	.014	.032	.064	.131	.757	2.176	2.731	3.257	3.801	4.134
9	3		.061	.089	.141	.213	.318	.902	1.803	2.096	2.365	2.669	2.852
9	4		.137	.184	.259	.338	.433	.944	1.628	1.853	2.035	2.273	2.486
9	5		.213	.260	.338	.412	.514	.963	1.529	1.706	1.872	2.046	2.181
9	6		.273	.319	.404	.473	.566	.971	1.466	1.617	1.742	1.897	2.006
9	7		.324	.375	.452	.520	.609	.974	1.411	1.539	1.647	1.776	1.892
9	8		.382	.421	.498	.563	.641	.978	1.370	1.482	1.573	1.693	1.801
9	9		.425	.469	.535	.598	.671	.981	1.337	1.434	1.535	1.650	1.733
10	2		.0068	.013	.032	.060	.124	.792	2.191	2.728	3.208	3.743	4.285
10	3		.066	.092	.147	.205	.311	.905	1.813	2.108	2.388	2.741	3.024
10	4		.147	.184	.253	.328	.429	.940	1.639	1.872	2.066	2.323	2.498
10	5		.214	.257	.333	.408	.509	.966	1.542	1.717	1.878	2.074	2.230
10	6		.282	.318	.402	.478	.571	.977	1.475	1.616	1.767	1.943	2.041
10	7		.328	.367	.452	.521	.610	.981	1.420	1.559	1.672	1.816	1.908
10	8		.375	.427	.494	.564	.644	.983	1.377	1.494	1.606	1.724	1.791
10	9		.426	.473	.535	.598	.674	.987	1.342	1.457	1.559	1.659	1.735
10	10		.450	.499	.566	.624	.699	.988	1.312	1.415	1.508	1.607	1.663
n	r	δ:	.005	.01	.025	.05	.10	.50	.90	.95	.975	.99	.995

From Nelson and Schmee (1979) with permission of the American Statistical Assoc.

Confidence Limits for μ

Exact limits. Two-sided exact $100\gamma\%$ confidence limits for μ are

$$\underset{\sim}{\mu}=\mu^*-t^*[(1+\gamma)/2;n,r]\sigma^*,$$

$$\tilde{\mu}=\mu^*-t^*[(1-\gamma)/2;n,r]\sigma^*,$$

(3.28)

where $t^*(\delta;n,r)$ is the 100δth percentile of the distribution of the "t-like" statistic $t^*=(\mu^*-\mu)/\sigma^*$. These percentiles appear in Table 3.3 for the same n,r, and δ as in Table 3.2. A t^*-distribution for $r<n$ is not symmetric; that is, $t^*(\delta;n,r)\neq-t^*(1-\delta;n,r)$. (3.28) yields a one-sided $100\gamma\%$ confidence limit when γ replaces $(1+\gamma)/2$ or $1-\gamma$ replaces $(1-\gamma)/2$.

These limits are exact (approximate) for failure (time) censored samples. If limits (3.28) are calculated from the logs of lognormal data, the antilogs of the limits are confidence limits for the lognormal median.

Class-B insulation example. Two-sided 90% confidence limits for μ at 170°C are $\underset{\sim}{\mu}=3.6381-t^*(0.95;10,7)0.2265=3.6381-(0.565)0.2265=3.5101$ and $\tilde{\mu}=3.6381-t^*(0.05;10,7)0.2265=3.6381-(-0.739)0.2265=3.8055$. The limits for lognormal median life are antilog(3.5101)=3237 and antilog (3.8055)=6390 hours. The sample is time censored; so the limits are approximate.

Approximate limits. Two-sided approximate $100\gamma\%$ confidence limits for μ are

$$\underset{\sim}{\mu}\cong\mu^*-K_\gamma A^{1/2}\sigma^*, \qquad \tilde{\mu}\cong\mu^*+K_\gamma A^{1/2}\sigma^*,$$

(3.29)

where σ^* estimates σ, K_γ is the $[100(1+\gamma)/2]$th standard normal percentile, and $\mathrm{Var}(\mu^*)=A\sigma^2$. Although sometimes much narrower than the corresponding exact interval, the approximate one may be wide enough to warn when an estimate is too uncertain. The limits are more exact the larger r is. (3.29) yields a one-sided confidence limit when K_γ is replaced by z_γ, the standard normal 100γth percentile.

Class-B insulation example. For 170°C, $\mu^*=3.6381$, $\sigma^*=0.2265$, $\mathrm{Var}(\mu^*)=0.1167\,\sigma^2$, and $\mathrm{Var}(\sigma^*)=0.0989\,\sigma^2$. Two-sided approximate 90% confidence limits for μ are $\underset{\sim}{\mu}=3.6381-1.645(0.1167)^{1/2}0.2265=3.5108$ and $\tilde{\mu}=3.6381+0.1273=3.7654$. The limits for the lognormal median life are antilog(3.5108)=3242 and antilog(3.7654)=5826 hours. A pooled estimate (Section 5.5) of σ could be used, assuming that σ has the same value at all test temperatures.

Table 3.3. Percentiles $t^*(\delta; n, r)$ for Limits for μ

n	r	δ:	.005	.01	.025	.05	.10	.50	.90	.95	.975	.99	.995
2	2		-40.68	-20.52	-7.864	-3.486	-1.752	.007	1.749	3.523	7.107	17.92	35.65
3	2		-59.52	-30.09	-12.15	-5.996	-2.952	-.026	1.037	1.990	3.499	7.613	14.53
3	3		-4.965	-3.406	-2.328	-1.543	-.993	-.009	.967	1.489	2.127	3.497	5.279
4	2		-94.47	-46.01	-17.38	-8.244	-3.946	-.049	.740	1.156	1.980	4.339	8.478
4	3		-6.868	-5.079	-3.091	-1.876	-1.120	-.012	.742	1.098	1.579	2.615	3.851
4	4		-2.650	-2.023	-1.412	-1.061	-.750	-.006	.739	1.039	1.421	2.016	2.561
5	2		-99.43	-53.21	-21.28	-9.928	-4.354	-.066	.669	.879	1.226	2.326	3.966
5	3		-7.629	-5.583	-3.289	-2.164	-1.284	.002	.626	.874	1.194	1.739	2.318
5	4		-3.237	-2.343	-1.578	-1.119	-.752	.006	.625	.865	1.164	1.614	1.953
5	5		-1.892	-1.613	-1.168	-.863	-.630	.006	.620	.866	1.134	1.589	1.915
6	2		-164.75	-68.57	-25.60	-11.90	-5.394	-.106	.664	.821	1.081	1.744	2.519
6	3		-10.27	-6.801	-4.077	-2.541	-1.535	-.008	.592	.801	1.076	1.570	2.128
6	4		-3.642	-2.847	-1.933	-1.361	-.865	.002	.578	.802	1.059	1.555	1.961
6	5		-2.174	-1.774	-1.228	-.972	-.660	-.004	.577	.796	1.045	1.442	1.778
6	6		-1.682	-1.360	-1.034	-.786	-.579	-.003	.579	.788	1.009	1.336	1.592
7	2		-136.32	-76.22	-28.50	-13.44	-6.063	-.162	.656	.777	.909	1.145	1.462
7	3		-11.78	-7.840	-4.645	-2.845	-1.672	-.024	.546	.705	.891	1.145	1.410
7	4		-4.237	-3.124	-2.053	-1.440	-.953	-.002	.517	.697	.887	1.149	1.372
7	5		-2.308	-1.849	-1.343	-1.014	-.695	-.003	.512	.689	.889	1.123	1.412
7	6		-1.614	-1.357	-1.018	-.797	-.576	-.004	.509	.690	.890	1.100	1.324
7	7		-1.316	-1.105	-.872	-.696	-.512	-.005	.508	.688	.875	1.097	1.259
8	2		-193.49	-88.26	-30.88	-14.60	-6.495	-.169	.678	.798	.892	1.032	1.232
8	3		-11.22	-7.710	-4.528	-2.986	-1.758	-.021	.541	.671	.795	1.007	1.236
8	4		-4.475	-3.338	-2.123	-1.460	-.939	-.000	.500	.643	.794	1.006	1.196
8	5		-2.509	-1.966	-1.380	-1.013	-.684	.001	.487	.637	.795	1.005	1.206
8	6		-1.661	-1.381	-1.032	-.798	-.558	.003	.480	.635	.793	1.008	1.199
8	7		-1.462	-1.206	-.891	-.689	-.502	.002	.481	.635	.794	1.010	1.186
8	8		-1.212	-1.021	-.806	-.644	-.475	.000	.481	.635	.796	1.005	1.146
9	2		-166.51	-79.57	-31.64	-15.40	-6.837	-.226	.709	.822	.907	1.028	1.138
9	3		-13.03	-9.483	-5.288	-3.139	-1.873	-.034	.557	.686	.805	.983	1.154
9	4		-5.161	-3.667	-2.279	-1.635	-1.056	.008	.502	.639	.781	.989	1.151
9	5		-2.896	-2.159	-1.494	-1.094	-.738	.005	.479	.627	.774	.990	1.139
9	6		-1.868	-1.518	-1.149	-.871	-.598	.008	.470	.622	.772	.980	1.139
9	7		-1.462	-1.231	-.924	-.725	-.522	.006	.465	.621	.772	.973	1.139
9	8		-1.247	-1.019	-.827	-.660	-.480	.007	.462	.621	.771	.981	1.116
9	9		-1.044	-.974	-.746	-.612	-.462	.007	.461	.619	.769	.985	1.107
10	2		-176.92	-95.69	-34.88	-17.22	-7.497	-.243	.725	.835	.915	1.037	1.104
10	3		-14.20	-9.493	-5.638	-3.596	-2.077	-.044	.552	.662	.773	.937	1.059
10	4		-5.044	-3.555	-2.568	-1.776	-1.171	-.016	.481	.607	.724	.894	1.041
10	5		-2.852	-2.228	-1.652	-1.189	-.807	.000	.445	.591	.706	.884	1.040
10	6		-1.932	-1.590	-1.173	-.889	-.615	.000	.429	.578	.704	.884	1.041
10	7		-1.524	-1.245	-.954	-.739	-.534	.001	.429	.565	.702	.887	1.035
10	8		-1.231	-1.069	-.817	-.652	-.476	-.003	.424	.566	.701	.888	1.024
10	9		-1.075	-.958	-.738	-.597	-.440	-.003	.424	.567	.701	.896	1.027
10	10		-.977	-.848	-.704	-.566	-.428	.001	.423	.566	.700	.884	1.010
n	r	δ:	.005	.01	.025	.-5	.10	.50	.90	.95	.975	.99	.995

From Nelson and Schmee (1979) with permission of the American Statistical Assoc.

Confidence Limits for Percentiles

Exact limits. Two-sided exact $100\gamma\%$ confidence limits for the $100P$th normal percentile $y_P = \mu + z_P\,\sigma$ are

$$\underset{\sim}{y}_P = \mu^* + t^*[(1-\gamma)/2; P, n, r]\sigma^*, \qquad \tilde{y}_P = \mu^* + t^*[(1+\gamma)/2; P, n, r]\sigma^*,$$

$$(3.30)$$

where $t^*(\delta; P, n, r)$ is the 100δth percentile of the distribution of $t^* = (y_P - \mu^*)/\sigma^*$. These percentiles appear in Table 3.4 for $P = 0.10$, $n = 2(1)10$, $r = 2(1)n$, and δ and $1 - \delta = .005, .01, .025, .05, .10, .50$. Further tables for $P = 10^{-6}, .001, .01, .05$ are in Nelson and Schmee (1976, 1979). The confidence limits for μ are confidence limits for $y_{.50}$. (3.30) yields a one-sided $100\gamma\%$ confidence limit if γ replaces $(1+\gamma)/2$ or $1-\gamma$ replaces $(1-\gamma)/2$. (3.30) is exact (approximate) for failure (time) censored samples. If limits (3.30) are calculated from the logs of lognormal data, the antilogs of the limits are confidence limits for the lognormal percentile.

Class-B insulation example. For 170°C, two-sided 90% confidence limits for $y_{.10}$ are $\underset{\sim}{y}_{.10} = 3.6381 + t^*(0.05; 0.10, 10, 7)0.2265 = 3.6381 + (-2.497)$ $0.2265 = 3.0725$ and $\tilde{y}_{.10} = 3.6381 + t^*(0.95; 0.10, 10, 7)0.2265 = 3.6381 + (-0.689)0.2265 = 3.4820$. The limits for the lognormal percentile are antilog $(3.0725) = 1182$ and antilog$(3.4820) = 3034$ hours.

Approximate limits for low percentiles. Suppose that μ^* and σ^* are the BLUEs for μ and σ. Also, $\mathrm{Var}(\mu^*) = A\sigma^2$, $\mathrm{Var}(\sigma^*) = B\sigma^2$, and $\mathrm{Cov}(\mu^*, \sigma^*) = C\sigma^2$. Then a one-sided lower approximate $100\gamma\%$ confidence limit for the $100P$th percentile $y_P = \mu + z_P\sigma$ is

$$\underset{\sim}{y}_P \cong y_P^* + \sigma^*[F(\gamma; a, b) - 1][(C/B) + z_P], \qquad (3.31)$$

where z_P is the $100P$th standard normal percentile and $F(\gamma; a, b)$ is the 100γth F percentile with $a = 2[z_P + (C/B)]^2/[A - (C^2/B)]$ degrees of freedom above and $b = 2/B$ below. Mann, Schafer, and Singpurwalla (1974) state that this approximation is satisfactory for r large (say, greater than 10) and $P < 0.10$. P must be small to insure that $[(C/B) + z_P]$ is negative. The lower limit for the $100P$th lognormal percentile is antilog(y_P).

For percentiles that are not low, approximate limits like (3.29) can be used. Then y_P^* replaces μ^* and D of (3.7) replaces A in (3.29).

Class-B insulation example. For 170°C, $\mu^* = 3.6381$, $\sigma^* = 0.2265$, $A = 0.1167$, $B = 0.0989$, $C = 0.0260$, and $y_{.10}^* = 3.3477$. A lower approximate 95%

Table 3.4.   Percentiles $t^*(\delta; 10, n, r)$ for Limits for $y_{.10}$

n	r	δ: .005	.01	.025	.05	.10	.50	.90	.95	.975	.99	.995
2	2	-183.9	-88.09	-36.30	-16.40	-7.869	-1.405	-.327	-.118	.0830	.464	.888
3	2	-117.6	-59.41	-26.14	-13.45	-6.786	-1.380	-.475	-.256	-.0715	.898	2.072
3	2	-17.60	-12.28	-7.761	-5.497	-3.691	-1.326	-.473	-.305	-.154	.0471	.172
4	2	-113.5	-56.59	-21.50	-11.55	-5.934	-1.348	-.537	-.280	.236	1.864	4.305
4	3	-15.58	-11.52	-7.112	-5.071	-3.503	-1.313	-.558	-.399	-.252	-.0573	.107
4	4	-9.617	-7.008	-4.985	-3.792	-2.921	-1.290	-.559	-.409	-.265	-.124	-.0141
5	2	-112.9	-48.61	-18.13	-9.490	-4.988	-1.322	-.567	-.182	.519	2.436	5.585
5	3	-13.33	-9.911	-6.294	-4.598	-3.208	-1.306	-.622	-.463	-.285	-.00405	.271
5	4	-7.956	-6.415	-4.652	-3.636	-2.778	-1.292	-.624	-.475	-.335	-.173	-.0587
5	5	-5.565	-4.871	-3.892	-3.184	-2.585	-1.287	-.622	-.472	-.344	-.202	-.109
6	2	-76.15	-37.64	-17.14	-8.573	-4.669	-1.313	-.570	-.0928	.779	3.427	6.336
6	3	-14.20	-10.33	-6.289	-4.393	-3.107	-1.303	-.667	-.503	-.323	-.0332	.375
6	4	-7.414	-5.870	-4.386	-3.487	-2.721	-1.302	-.673	-.546	-.427	-.286	-.197
6	5	-5.450	-4.681	-3.746	-3.083	-2.498	-1.297	-.673	-.547	-.440	-.309	-.219
6	6	-4.663	-4.102	-3.336	-2.826	-2.371	-1.300	-.678	-.546	-.438	-.322	-.246
7	2	-68.12	-32.41	-15.19	-8.208	-4.340	-1.303	-.561	-.0160	.989	4.067	9.640
7	3	-12.82	-9.162	-5.602	-4.099	-3.005	-1.301	-.710	-.529	-.330	-.0654	.292
7	4	-7.571	-5.812	-4.171	-3.346	-2.622	-1.298	-.722	-.587	-.464	-.289	-.120
7	5	-5.523	-4.604	-3.705	-3.034	-2.417	-1.295	-.723	-.595	-.495	-.361	-.243
7	6	-4.636	-4.112	-3.351	-2.791	-2.320	-1.288	-.722	-.597	-.501	-.381	-.294
7	7	-4.282	-3.693	-3.041	-2.641	-2.228	-1.285	-.725	-.595	-.501	-.376	-.287
8	2	-63.58	-29.95	-13.34	-6.948	-3.728	-1.293	-.472	.226	1.645	6.157	14.50
8	3	-10.54	-7.628	-4.967	-3.637	-2.756	-1.297	-.732	-.549	-.309	.130	.540
8	4	-6.590	-5.172	-3.937	-3.119	-2.484	-1.286	-.752	-.609	-.468	-.266	-.110
8	5	-5.363	-4.296	-3.443	-2.885	-2.361	-1.287	-.757	-.629	-.507	-.358	-.242
8	6	-4.515	-3.855	-3.220	-2.756	-2.282	-1.287	-.758	-.633	-.514	-.386	-.306
8	7	-4.086	-3.583	-3.019	-2.602	-2.205	-1.284	-.755	-.632	-.514	-.399	-.336
8	8	-3.781	-3.413	-2.851	-2.485	-2.162	-1.290	-.757	-.630	-.514	-.404	-.338
9	2	-56.28	-30.29	-11.44	-6.353	-3.550	-1.314	-.465	.353	2.089	6.581	11.57
9	3	-11.84	-8.048	-5.062	-3.693	-2.704	-1.312	-.739	-.551	-.304	.209	.744
9	4	-6.562	-5.165	-3.775	-3.043	-2.457	-1.310	-.772	-.640	-.505	-.300	-.167
9	5	-5.034	-4.316	-3.446	-2.851	-2.318	-1.307	-.777	-.660	-.548	-.427	-.326
9	6	-4.437	-3.901	-3.210	-2.714	-2.232	-1.305	-.779	-.663	-.566	-.444	-.365
9	7	-4.103	-3.651	-3.080	-2.605	-2.177	-1.303	-.780	-.665	-.570	-.457	-.373
9	8	-3.813	-3.408	-2.878	-2.490	-2.135	-1.300	-.776	-.665	-.572	-.459	-.381
9	9	-3.624	-3.228	-2.752	-2.413	-2.087	-1.298	-.780	-.666	-.571	-.458	-.396
10	2	-60.81	-27.39	-10.07	-5.683	-3.326	-1.299	-.340	.770	2.744	7.973	15.67
10	3	-9.675	-6.990	-4.618	-3.420	-2.572	-1.301	-.758	-.532	-.255	.306	1.018
10	4	-6.388	-5.028	-3.792	-2.993	-2.353	-1.297	-.798	-.657	-.528	-.332	-.134
10	5	-5.258	-4.324	-3.357	-2.749	-2.257	-1.297	-.801	-.684	-.581	-.426	-.315
10	6	-4.559	-3.701	-3.029	-2.569	-2.182	-1.294	-.802	-.689	-.590	-.488	-.405
10	7	-3.838	-3.430	-2.853	-2.497	-2.121	-1.293	-.803	-.689	-.593	-.497	-.428
10	8	-3.706	-3.173	-2.752	-2.401	-2.091	-1.292	-.803	-.688	-.598	-.502	-.431
10	9	-3.458	-3.118	-2.636	-2.353	-2.046	-1.288	-.804	-.689	-.598	-.502	-.438
10	10	-3.297	-2.996	-2.561	-2.284	-2.020	-1.290	-.804	-.693	-.598	-.501	-.437

From Nelson and Schmee (1979) with permission of the American Statistical Assoc.

confidence limit for $y_{.10}$ is calculated as follows:

$$a = 2\left[(-1.282) + (0.0260/0.0989)\right]^2 / \left[0.1167 - (0.0260^2/0.0989)\right] = 18.9,$$

$$b = 2/0.0989 = 20.2, \qquad F(0.95; 18.9, 20.2) = 2.136,$$

$$y_{.10} \cong 3.3477 + 0.2265(2.136 - 1)\left[(0.0260/0.0989) + (-1.282)\right] = 3.0855.$$

The lower limit for the 10th lognormal percentile is antilog(3.0855) = 1216 hours.

Confidence Limits for Reliability

Exact limits. A one-sided lower 95% confidence limit $R(y)$ for reliability $R(y)=1-\Phi[(y-\mu)/\sigma]$ by (log) age y is obtained as follows. First calculate the estimate $z^*=(y-\mu^*)/\sigma^*$ of the standard deviate. Choose the chart for the r and desired confidence. Nelson and Schmee (1977a, 1979) give charts for lower 99, 95, 90, and 50% confidence limits for $n=2(1)10$ and $r=2(1)n$. Figure 3.1 shows the chart for $r=7$ and 95% confidence. Enter the chart on the axis at z^*, go to the curve for n, and then go to the probability axis to read the lower limit $R(y)$. The corresponding one-sided upper limit for the fraction failing $F(y)=1-R(y)$ is $\tilde{F}(y)=1-R(y)$. These limits are exact (approximate) for failure (time) censored samples.

Class-B insulation example. The lower 95% confidence limit for the reliability at 3000 hours is desired. Then $y=\log(3000)=3.4771$ and $z^*=(3.4771-3.6381)/0.2265=-0.7108$. From the chart for 95% confidence, we find the lower limit $R(3.4771)=0.56$.

Approximate limits. Two-sided approximate $100\gamma\%$ confidence limits for reliability $R(y)$ at (log) age y are calculated as follows. Estimate the standardized deviate $z=(y-\mu)/\sigma$ with

$$z^*=(y-\mu^*)/\sigma^*, \qquad (3.32)$$

where $\text{Var}(\mu^*)=A\sigma^2, \text{Var}(\sigma^*)=B\sigma^2$, and $\text{Cov}(\mu^*,\sigma^*)=C\sigma^2$. Calculate

$$D^*=A+z^{*2}B+2z^*C, \qquad (3.33)$$

$$z=z^*-K_\gamma D^{*1/2}, \qquad \tilde{z}=z^*+K_\gamma D^{*1/2}, \qquad (3.34)$$

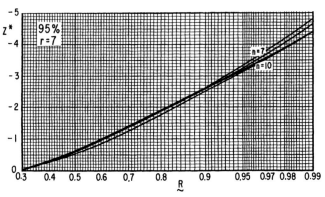

Figure 3.1. Chart for the lower confidence limit for reliability. From Nelson and Schmee (1979) with permission of the American Statistical Assoc.

where K_γ is the standard normal $[100(1+\gamma)/2]$th percentile. The two-sided limits for $R(y)$ are

$$\underset{\sim}{R}(y) \cong 1 - \Phi(\tilde{z}), \qquad \tilde{R}(y) \cong 1 - \Phi(\underset{\sim}{z}). \tag{3.35}$$

These limits are more accurate the larger r is. The interval tends to be short. For a one-sided $100\gamma\%$ confidence limit, replace K_γ in (3.34) by z_γ, the standard normal 100γth percentile.

Class-B insulation example. For 170°C, two-sided 90% confidence limits for insulation reliability at 3000 hours are calculated as follows:

$$y = \log(3000) = 3.4771, \qquad z^* = (3.4771 - 3.6381)/0.2265 = -0.7108,$$

$$D^* = 0.1167 + (-0.7108)^2 0.0989 + 2(-0.7108)0.0260 = 0.1297,$$

$$\underset{\sim}{z} = -0.7108 - 1.645(0.1297)^{1/2} = -1.3032,$$

$$\tilde{z} = -0.7108 + 0.5924 = -0.1184,$$

$$\underset{\sim}{R}(3.4771) = 1 - \Phi(-0.1184) = 0.55, \quad \tilde{R}(3.4771) = 1 - \Phi(-1.3032) = 0.90.$$

PREDICTION

*j*th Observation of the Same Sample

Predictor. A simple linear unbiased predictor for the jth (log) order statistic $Y_{(j)}$ of the same sample is

$$\hat{Y}_{(j)} = Y_{(r)} + \left(E\dot{Z}_{(j)} - EZ_{(r)} \right) \sigma^*, \tag{3.36}$$

where $EZ_{(j)}$ and $EZ_{(r)}$ are the expectations of the jth and rth standard normal order statistics, and σ^* is the BLUE for σ. This predictor is unbiased, and prediction error variance is comparable to that of the best linear unbiased predictor given by Schmee and Nelson (1979). Owen (1962) and Sarhan and Greenberg (1962) tabulate the expectations, and David (1970) lists other tables.

Class-B insulation example. For 170°C, the prediction of the log of the 10th order statistic is $\hat{Y}_{(10)} = 3.7157 + (1.5388 - 0.3758)0.2265 = 3.9791$. Converted to hours, the prediction is antilog(3.9791) = 9530 hours, the time when the test will end.

Prediction limits. Two-sided approximate $100\gamma\%$ prediction limits for $Y_{(j)}$ are

$$Y_{(j)} \approx Y_{(r)} + \{1/F[(1+\gamma)/2; b, a]\}(EZ_{(j)} - EZ_{(r)})\sigma^*,$$
$$\tilde{Y}_{(j)} \approx Y_{(r)} + F[(1+\gamma)/2; a, b](EZ_{(j)} - EZ_{(r)})\sigma^*, \qquad (3.37)$$

where $F(\delta; a, b)$ is the 100γth F percentile with $a = 2(EZ_{(j)} - EZ_{(r)})^2$ $/[\text{Var}(Z_{(j)}) + \text{Var}(Z_{(r)}) - 2\,\text{Cov}(Z_{(j)}, Z_{(r)})]$ degrees of freedom above and $b = 2/B$ below, where $\text{Var}(\sigma^*) = B\sigma^2$. Prediction limits for the jth lognormal order statistic are $\text{antilog}(Y_{(j)})$ and $\text{antilog}(\tilde{Y}_{(j)})$. Owen (1962) and Sarhan and Greenberg (1962) tabulate the variances and covariances, and David (1970) lists other tables.

Exact prediction limits for $Y_{(n)}$ from a singly censored sample from a (log) normal distribution are given by Nelson and Schmee (1977b). $Y_{(n)}$ is usually of greatest interest, as it is the time required for all sample units to fail. The approximate limits (3.37) are based on Mann, Schafer, and Singpurwalla (1974, Section 5.2.4).

An upper prediction limit for the nth observation is related to an outlier test. If the nth observation is above a $100\gamma\%$ limit, it is a statistically significant outlier at the $[100(1-\gamma)]\%$ level.

Class-B insulation example. For $170°C$, calculations of two-sided approximate 90% prediction limits for the log of the 10th order statistic are $a = 2(1.5388 - 0.3758)^2/[0.3443 + 0.1579 - 2(0.0882)] = 3.99$ and $b = 2/0.0989 = 20.22$, $F(0.95; 20.22, 3.99) = 5.830$ and $F(0.95; 3.99, 20.22) = 2.863$, $Y_{(10)} \approx 3.7157 + (1/5.830)(1.5388 - 0.3758)0.2265 = 3.7609$, and $\tilde{Y}_{(10)} \approx 3.7157 + 2.863(1.5388 - 0.3758)0.2265 = 4.4699$. Converted to hours, the prediction limits are $\text{antilog}(3.7609) = 5766$ hours and $\text{antilog}(4.4699) = 29,500$ hours.

First Observation of a Future Sample

Predictor. For the first (log) order statistic $X_{(1)}$ of an independent future sample of size m, the best linear unbiased predictor is

$$\hat{X}_{(1)} = \mu^* + EZ_{(1)}\sigma^*, \qquad (3.38)$$

where μ^* and σ^* are the BLUEs for μ and σ, and $EZ_{(1)}$ is the expectation of the first standard normal order statistic of a sample of size m. This predictor is unbiased, and the variance of its prediction error ($\hat{X}_{(1)} - X_{(1)}$) is

$$\text{Var}(\hat{X}_{(1)} - X_{(1)}) = \sigma^2[A_{n,r} + B_{n,r}(EZ_{(1)})^2 + 2C_{n,r}EZ_{(1)} + \text{Var}(Z_{(1)})].$$

$$(3.39)$$

where $\mathrm{Var}(\mu^*) = A_{n,r}\sigma^2$, $\mathrm{Var}(\sigma^*) = B_{n,r}\sigma^2$, $\mathrm{Cov}(\mu^*, \sigma^*) = C_{n,r}\sigma^2$, and $\mathrm{Var}(Z_{(1)})$ is the variance of the first standard normal order statistic of a sample of size m. The prediction of the first lognormal order statistic is antilog($\hat{X}_{(1)}$).

Class-B insulation example. For 170°C, the prediction of the first log time to failure of a second sample of $m=10$ motorettes is $\hat{X}_{(1)}=3.6381 + (-1.5388)0.2265 = 3.2896$. The prediction is then antilog(3.2896) = 1948 hours.

Prediction limits. A one-sided lower approximate $100\gamma\%$ prediction limit for $X_{(1)}$ is

$$\underset{\sim}{X}_{(1)} = \hat{X}_{(1)} - \sigma^*\left[F(\gamma; a, b) - 1\right]\left[-EZ_{(1)} - (C_{n,r}/B_{n,r})\right], \qquad (3.40)$$

where $F(\gamma; a, b)$ is the 100γth F percentile with $a = 2[EZ_{(1)} + (C_{n,r}/B_{n,r})]^2/[A_{n,r} - (C^2_{n,r}/B_{n,r}) + \mathrm{Var}(Z_{(1)})]$ degrees of freedom above and $b = 2/B_{n,r}$ below. A lower prediction limit for the first lognormal order statistic is antilog($\underset{\sim}{X}_{(1)}$). Mann, Schafer, and Singpurwalla (1974) present this approximate limit, which they call a $100\gamma\%$ *warranty period* for a future lot of size m.

Class-B insulation example. For 170°C, the calculation of an approximate 90% prediction limit is $a = 2[-1.5388 + (0.0260/0.0989)]^2/[0.1167 - (0.0260^2/0.0989) + 0.3443] = 7.17$, $b = 2/0.0989 = 20.2$, $F(0.90; 7.17, 20.2) = 2.5648$, and $\underset{\sim}{X}_{(1)} = 3.2896 - 0.2265(2.5648 - 1)[-(-1.5388) - (0.0260/0.0989)] = 2.8374$. Converted to hours, the prediction limit is antilog(2.8374) = 688 hours.

4. WEIBULL AND EXTREME VALUE DATA

The Weibull and extreme value distributions serve as models for the life, strength, and other properties of many products and materials. The Weibull distribution parameters are the shape parameter β and the scale parameter α. The corresponding parameters of the extreme value distribution are the location parameter λ and the scale parameter δ. A singly censored sample of n units consists of the r smallest ordered (ln) observations $Y_{(1)}, \ldots, Y_{(r)}$.

The methods below provide best linear unbiased estimates (BLUEs) and confidence intervals for the distribution parameters, percentiles, and reliabilities. The methods are exact (approximate) for failure (time) censoring. The methods apply to both the Weibull and extreme value distributions, but one works with the base e log values of Weibull data. Chapter 3 provides graphical methods for checking how well a Weibull or extreme value distribution fits such singly censored data.

ESTIMATION

Extreme Value Location λ and Weibull Scale α Parameters

Estimate λ.   The BLUE for the extreme value location parameter λ is

$$\lambda^* = a(1; n, r) Y_{(1)} + \cdots + a(r; n, r) Y_{(r)}; \tag{4.1}$$

the coefficients $a(i; n, r)$ are in Appendix A12 for $n = 2(1)12$ and $r = 2(1)n$. The variance of λ^* is

$$\mathrm{Var}(\lambda^*) = \delta^2 A_{n, r}; \tag{4.2}$$

the variance factors $A_{n, r}$ are in Appendix A12. White (1964) gives a larger table of the coefficients and variance factors for $n = 2(1)20$ and $r = 2(1)n$. Lieblein and Zelen (1956) gave the first such table with $n = 2(1)6$ and $r = 2(1)n$. Section 5.5 shows how to pool such estimates of the same λ. Mann, Schafer, and Singpurwalla (1974) give a table of the coefficients and factors for mean squared error for the best linear invariant estimators for λ. For r large, λ^* is approximately normally distributed, with mean λ and variance (4.2).

Estimate α.   The transformed linear estimator for the Weibull scale parameter α is

$$\alpha^* = \exp(\lambda^*). \tag{4.3}$$

This estimator is biased. For r large, the distribution of α^* is close to a normal one, with mean α and variance

$$\mathrm{Var}(\alpha^*) \simeq A_{n, r}(\alpha/\beta)^2. \tag{4.4}$$

Insulating fluid example.   Table 4.1 shows the base e logs of times to breakdown of an insulting fluid at five voltages. The data are to be analyzed with the Weibull distribution. For 30 kV, the estimate of the extreme value location parameter for the log data is $\lambda^* = 0.00660(3.912) + \cdots + 0.50224(9.668) = 8.583$, and $\mathrm{Var}(\lambda^*) = 0.10574\delta^2$. The estimate of the Weibull scale parameter is $\alpha^* = \exp(8.583) = 5340$ seconds.

Table 4.1.   Ln Seconds to Insulating Fluid Breakdown

25 kV	30 kV	35 kV	40 kV	45 kV
6.256	3.912	3.401	0.000	0.000−
7.832	4.898	3.497	0.000	0.000−
8.309	5.231	3.715	0.693	0.000−
9.441	6.782	4.466	1.099	0.693
10.604	7.279	4.533	2.485	0.693
10.832+	7.293	4.585	3.219	1.099
10.876+	7.736	4.754	3.829	2.197
11.117+	7.983	5.553	4.025	2.565
11.339	8.338	6.133	4.220	3.850
11.356+	9.668	7.073	4.691	3.912
11.359+	10.282+	7.208	5.778	4.007
11.367+	11.363+	7.313	6.033	4.263

```
+ running time without breakdown.
- breakdown occurred earlier.
```

Extreme Value Scale δ and Weibull Shape β Parameters

Estimate δ.   The BLUE for the extreme value scale parameter δ is

$$\delta^* = b(1; n, r)Y_{(1)} + \cdots + b(r; n, r)Y_{(r)}; \qquad (4.5)$$

the coefficients $b(i; n, r)$ are in Appendix A12 for $n = 2(1)12$ and $r = 2(1)n$. The variance of the estimator is

$$\mathrm{Var}(\delta^*) = B_{n, r}\, \delta^2; \qquad (4.6)$$

the variance factors $B_{n, r}$ are tabulated in Appendix A12. White (1964) gives a larger table of the coefficients and variance factors for $n = 2(1)20$ and $r = 2(1)n$. Section 5.5 shows how to pool such estimates of the same δ. For large r, δ^* is approximately normally distributed, with mean δ and variance (4.6). Mann, Schafer, and Singpurwalla (1974) tabulate the coefficients and mean squared errors for the best linear invariant estimators for δ.

Estimate β. The transformed linear estimator for the Weibull shape parameter β is

$$\beta^* = 1/\delta^*. \qquad (4.7)$$

This estimator is biased. For large r, the cumulative distribution of β^* is close to a normal one, with mean β and variance

$$\text{Var}(\beta^*) \simeq \beta^2 B_{n,r}. \qquad (4.8)$$

Insulating fluid example. For 30 kV, the estimate of the extreme value location parameter for the log data is $\delta^* = -0.08749(3.912) + \cdots + 0.52092(9.668) = 1.984$. Then $\text{Var}(\delta^*) = 0.08175\delta^2$. The estimate of the Weibull shape parameter is $\beta^* = 1/1.984 = 0.504$.

Covariance. The covariance of λ^* and δ^* is

$$\text{Cov}(\lambda^*, \delta^*) = C_{n,r}\delta^2; \qquad (4.9)$$

the covariance factors $C_{n,r}$ are tabulated in Appendix A12. White (1964) gives a larger table for $n = 2(1)20$ and $r = 2(1)n$. For large r, λ^* and δ^* are approximately jointly normally distributed, with the means, variances, and covariance given above. Also, for large r, α^* and β^* are approximately jointly normally distributed, with approximate means and variances above and approximate covariance

$$\text{Cov}(\alpha^*, \beta^*) \simeq -\alpha C_{n,r}. \qquad (4.10)$$

Insulating fluid example. For 30 kV, $\text{Cov}(\lambda^*, \delta^*) = -0.00228\delta^2$ from Appendix A12.

Percentile

Estimate. The BLUE for the $100P$th extreme value percentile $y_P = \lambda + u_P \delta$ is

$$y_P^* = \lambda^* + u_P \delta^*, \qquad (4.11)$$

where $u_P = \ln[-\ln(1-P)]$ is the $100P$th standard extreme value percentile. The variance of y_P^* is

$$\text{Var}(y_P^*) = D_{n,r}\delta^2, \qquad (4.12)$$

where

$$D_{n,r} = \left(A_{n,r} + u_P^2 B_{n,r} + 2u_P C_{n,r} \right). \tag{4.13}$$

For large r, y_P^* is approximately normally distributed, with mean y_P and variance (4.12).

The transformed linear estimator for the Weibull $100P$th percentile is

$$t_P^* = \exp(y_P^*). \tag{4.14}$$

For large r, the distribution of t_P^* is close to a normal one, with mean t_P and variance

$$\mathrm{Var}(t_P^*) \simeq t_P^2 \mathrm{Var}(y_P^*) = (t_P/\beta)^2 \left(A_{n,r} + u_P^2 B_{n,r} + 2u_P C_{n,r} \right). \tag{4.15}$$

Insulating fluid example. For 30 kV, the estimate of the 10th percentile for the log data is $y_{.10}^* = 8.583 + \ln[-\ln(1-0.10)]1.984 = 4.118$. Here $\mathrm{Var}(y_{.10}^*) = \delta^2[0.10574 + (-2.2504)^2 0.08175 + 2(-2.2504)(-0.00228)] = 0.5300\delta^2$. The estimate of the 10th Weibull percentile is $t_{.10}^* = \exp(4.118) = 61$ seconds.

Reliability

Estimate.   The transformed linear estimate of reliability at (ln) age y is

$$R^*(y) = \exp[-\exp(u^*)], \tag{4.16}$$

where

$$u^* = (y - \lambda^*)/\delta^*. \tag{4.17}$$

u^* estimates the standardized extreme value deviate $u = (y-\lambda)/\delta$, and λ^* and δ^* are the BLUEs. This estimator is biased. The corresponding estimate of the fraction failing by (log) age y is

$$F^*(y) = 1 - \exp[-\exp(u^*)]. \tag{4.18}$$

For large r, u^* has a cumulative distribution close to a normal one, with mean $u = (y-\lambda)/\delta$ and approximate variance

$$\mathrm{Var}(u^*) \simeq A_{n,r} + u^2 B_{n,r} + 2u C_{n,r}, \tag{4.19}$$

where $\text{Var}(\lambda^*) = A_{n,r} \delta^2$, $\text{Var}(\delta^*) = B_{n,r} \delta^2$, and $\text{Cov}(\lambda^*, \delta^*) = C_{n,r} \delta^2$, which are tabulated in Appendix A12.

Insulating fluid example. For 30 kV, the estimate of reliability at five seconds is $R^*(5) = \exp\{-\exp[(\ln(5) - 8.583)/1.984]\} = 0.971$.

Estimates from Selected Order Statistics

Linear estimates based on selected order statistics are sometimes useful. They are easy to calculate, they are easy to use with data plots, and they are useful for sample sizes outside of the tables for BLUEs. Such estimates of the extreme value parameters λ and δ (or the Weibull α and β) can be used to obtain estimates and confidence limits for percentiles, reliabilities, and other functions of the parameters as described above. Large-sample theory for such estimates was given by Dubey (1967).

Estimates for λ and α. For large sample size n, the asymptotically BLUE for λ based on two order statistics is

$$\lambda_2^* = 0.44431 Y_{(\langle .3978n \rangle)} + 0.55569 Y_{(\langle .8211n \rangle)}; \tag{4.20}$$

$\langle x \rangle$ is x rounded up to the nearest integer. For large n, λ_2^* is approximately normally distributed, with mean λ and

$$\text{Var}(\lambda_2^*) \cong 1.359 \, \delta^2/n. \tag{4.21}$$

For comparison, the asymptotic variance of the BLUE for λ based on a complete sample of size n is $\text{Var}(\lambda^*) \cong 1.109 \, \delta^2/n$. The estimate of the Weibull scale parameter is $\alpha_2^* = \exp(\lambda_2^*)$.

Insulating fluid example. For 30 kV, the order statistics are $\langle 0.3978 \times 12 \rangle = 7$ and $\langle 0.8211 \times 12 \rangle = 10$. The estimate of λ is $\lambda_2^* = 0.44431(7.736) + 0.55569(9.668) = 8.810$. Its approximate variance is $\text{Var}(\lambda_2^*) \cong 1.359 \, \delta^2/12 = 0.1132 \, \delta^2$. The estimate of the Weibull scale parameter is $\alpha_2^* = \exp(8.810) = 6700$ seconds.

Estimates for δ and β. The BLUE for δ based on the jth and kth order statistics ($j < k$) is

$$\delta_2^* = (Y_{(k)} - Y_{(j)})/(EU_{(k)} - EU_{(j)}), \tag{4.22}$$

where $EU_{(j)}$ and $EU_{(k)}$ are the expectations of the standard extreme value order statistics, which are tabulated by White (1967) for $n = 1(1)50(5)100$.

The variance of δ_2^* is

$$\mathrm{Var}(\delta_2^*) = \delta^2 \big(EU_{(k)} - EU_{(j)} \big)^{-2} \big[V(n; k, k) + V(n; j, j) - 2V(n; j, k) \big],$$

$$(4.23)$$

where $V(n; k, k), V(n; j, j)$, and $V(n; j, k)$ are the variances and covariance of the jth and kth standard order statistics. These variances are tabulated by White (1967) for $n = 1(1)50(5)100$ and by Mann (1968), who gives the second moments for $n = 1(1)25$.

For large sample size n, the asymptotically BLUE for δ based on two order statistics is

$$\delta_2^* = 0.33457 \big(Y_{(\langle .9737n \rangle)} - Y_{(\langle .1673n \rangle)} \big).$$

$$(4.24)$$

For large n, δ_2^* is approximately normally distributed, and

$$\mathrm{Var}(\delta_2^*) \approx 0.9163 \, \delta^2/n.$$

$$(4.25)$$

For comparison, the asymptotic variance of the BLUE for δ based on a complete sample of size n is $\mathrm{Var}(\delta^*) = 0.6079 \, \delta^2/n$. The asymptotic efficiency of the δ_2^* relative to δ^* is $100(0.6079/0.9163) = 66\%$. The estimate of the Weibull shape parameter is $\beta_2^* = 1/\delta_2^*$. δ_2^* is suitable only for nearly complete samples. This estimators suggests that the largest available order statistic be used in place of $Y_{(\langle .9737n \rangle)}$ when the sample is censored below the 97.37th percentile.

Insulating fluid example. The needed order statistics are $\langle 0.9737 \times 12 \rangle = 12$ and $\langle 0.1673 \times 12 \rangle = 3$. At 30 kV, the 12th order statistics was not observed, and δ_2^* cannot be used. For 35 kV, $\delta_2^* = 0.33457(7.313 - 3.715) = 1.204$. Its approximate variance is $\mathrm{Var}(\delta_2^*) \cong 0.9163 \, \delta^2/12 = 0.07638 \, \delta^2$. The estimate of the Weibull shape parameter is $\beta_2^* = 1/1.204 = 0.831$.

CONFIDENCE LIMITS

The following paragraphs present exact and approximate confidence limits, based on BLUEs, for Weibull and extreme value distribution parameters, percentiles, and reliabilities. The exact limits were developed by Mann and Fertig (1973) and are expressed in terms of the best linear invariant estimates (BLIEs) λ^{**} and δ^{**} for the extreme value parameters λ and δ. The BLIEs are described in Section 5.5. The approximate limits will do for many practical problems outside the range of their tables. Also, the approximate limits can be used with pooled estimates (Section 5.5).

Confidence Limits for the Extreme Value Scale δ and Weibull Shape β Parameters

Exact limits. Two-sided exact $100\gamma\%$ confidence limits for δ based on the BLIE δ^{**} are

$$\underline{\delta}=\delta^{**}/w^{**}\big[(1+\gamma)/2;n,r\big], \qquad \tilde{\delta}=\delta^{**}/w^{**}\big[(1-\gamma)/2;n,r\big], \quad (4.26)$$

where $w^{**}(\varepsilon;n,r)$ is the 100εth percentile of the distribution of $w^{**}=\delta^{**}/\delta$. These percentiles appear in Table 4.2 for $n=3(1)12, r=3(1)n$, and $1-\varepsilon$ and $\varepsilon=0.02,0.05,0.10,0.25,0.40,0.50$. Tables are given by Mann and Fertig (1973) for up to $n=16$ and by Mann, Fertig, and Scheuer (1971) up to $n=25$.

Such limits in terms of the BLUE δ^* are

$$\underline{\delta}=\delta^*/\big\{(1+B)w^{**}\big[(1+\gamma)/2;n,r\big]\big\},$$
$$\tilde{\delta}=\delta^*/\big\{(1+B)w^{**}\big[(1-\gamma)/2;n,r\big]\big\}, \qquad (4.27)$$

where $\mathrm{Var}(\delta^*)=B\delta^2$.

Corresponding limits for the Weibull shape parameter β are

$$\underline{\beta}=1/\tilde{\delta}=(1+B)w^{**}\big[(1-\gamma)/2;n,r\big]\beta^*,$$
$$\tilde{\beta}=1/\underline{\delta}=(1+B)w^{**}\big[(1+\gamma)/2;n,r\big]\beta^*. \qquad (4.28)$$

The limits above yield one-sided $100\gamma\%$ confidence limits when γ replaces $(1+\gamma)/2$ or $1-\gamma$ replaces $(1-\gamma)/2$. The limits are exact (approximate) for failure (time) censored data.

Insulating fluid example. For 30 kV, $\delta^*=1.984$ and $B=0.08175$. For a two-sided 90% confidence interval, $\underline{\delta}=1.984/\{(1+0.08175)w^{**}[(1+0.90)/2;12,10]\}=1.984/[(1.08175)1.40]=1.310$ and $\tilde{\delta}=1.984/[(1.08175)0.53]=3.460$. Corresponding limits for the Weibull shape parameter are $\tilde{\beta}=1/1.310=0.763$ and $\underline{\beta}=1/3.460=0.289$. The interval does not enclose 1, a statistically significant indication that the distribution is not exponential.

Approximate limits. Suppose that δ^* is any linear unbiased estimator for the extreme value scale parameter δ, and suppose that its variance is

Table 4.2.   Percentiles of the Distribution of $w^{**} = \delta^{**}/\delta$

n	r	0.02	0.05	0.10	0.25	0.40	0.50	0.60	0.75	0.90	0.95	0.98
3	3	0.11	0.17	0.25	0.42	0.57	0.67	0.78	0.99	1.33	1.56	1.86
4	3	0.10	0.15	0.22	0.39	0.53	0.64	0.75	0.96	1.32	1.56	1.90
	4	0.20	0.28	0.37	0.54	0.68	0.77	0.86	1.05	1.33	1.53	1.77
5	3	0.09	0.14	0.21	0.37	0.51	0.61	0.73	0.94	1.32	1.59	1.93
	4	0.18	0.26	0.34	0.50	0.64	0.74	0.84	1.03	1.35	1.55	1.82
	5	0.28	0.36	0.44	0.60	0.73	0.82	0.91	1.07	1.33	1.50	1.70
6	3	0.09	0.14	0.21	0.36	0.50	0.61	0.72	0.93	1.32	1.59	1.92
	4	0.18	0.25	0.32	0.49	0.62	0.72	0.82	1.01	1.33	1.55	1.84
	5	0.25	0.33	0.42	0.58	0.71	0.79	0.89	1.05	1.33	1.51	1.73
	6	0.33	0.41	0.50	0.65	0.77	0.85	0.93	1.07	1.31	1.46	1.64
7	3	0.08	0.14	0.20	0.35	0.49	0.59	0.71	0.92	1.30	1.56	1.92
	4	0.17	0.24	0.31	0.48	0.62	0.71	0.81	1.01	1.32	1.54	1.82
	5	0.25	0.32	0.40	0.56	0.70	0.78	0.88	1.05	1.33	1.52	1.75
	6	0.32	0.39	0.47	0.63	0.75	0.84	0.92	1.07	1.32	1.48	1.67
	7	0.38	0.46	0.54	0.69	0.80	0.87	0.95	1.08	1.30	1.43	1.60
8	3	0.08	0.13	0.19	0.35	0.49	0.59	0.70	0.92	1.31	1.58	1.95
	4	0.16	0.23	0.31	0.47	0.61	0.70	0.81	1.00	1.33	1.55	1.83
	5	0.23	0.31	0.39	0.55	0.68	0.77	0.87	1.05	1.33	1.52	1.76
	6	0.30	0.38	0.46	0.62	0.74	0.82	0.91	1.06	1.32	1.49	1.69
	7	0.36	0.44	0.52	0.67	0.78	0.86	0.94	1.08	1.30	1.45	1.62
	8	0.42	0.50	0.58	0.71	0.82	0.89	0.96	1.09	1.28	1.41	1.56
9	3	0.08	0.13	0.19	0.34	0.49	0.59	0.70	0.92	1.31	1.58	1.92
	4	0.16	0.23	0.31	0.47	0.60	0.70	0.80	1.00	1.33	1.55	1.84
	5	0.23	0.31	0.39	0.54	0.68	0.77	0.86	1.04	1.33	1.52	1.76
	6	0.30	0.38	0.45	0.60	0.73	0.81	0.90	1.06	1.31	1.48	1.70
	7	0.35	0.43	0.50	0.66	0.77	0.85	0.93	1.07	1.30	1.46	1.65
	8	0.40	0.48	0.55	0.70	0.81	0.88	0.95	1.08	1.28	1.42	1.59
	9	0.45	0.53	0.60	0.74	0.84	0.90	0.97	1.08	1.27	1.39	1.53
10	3	0.08	0.13	0.19	0.34	0.48	0.59	0.71	0.93	1.31	1.59	1.92
	4	0.16	0.23	0.30	0.46	0.60	0.70	0.80	1.00	1.33	1.57	1.86
	5	0.23	0.30	0.38	0.54	0.68	0.77	0.86	1.04	1.33	1.53	1.77
	6	0.29	0.37	0.45	0.60	0.73	0.81	0.90	1.06	1.32	1.49	1.71
	7	0.34	0.42	0.50	0.65	0.77	0.84	0.92	1.07	1.31	1.46	1.66
	8	0.39	0.47	0.54	0.69	0.80	0.87	0.95	1.08	1.29	1.43	1.60
	9	0.43	0.51	0.59	0.73	0.83	0.89	0.96	1.08	1.28	1.40	1.55
	10	0.48	0.55	0.62	0.76	0.85	0.91	0.98	1.09	1.26	1.38	1.51
11	3	0.08	0.13	0.19	0.34	0.48	0.59	0.71	0.92	1.31	1.60	1.97
	4	0.15	0.22	0.30	0.46	0.60	0.70	0.80	1.00	1.34	1.58	1.87
	5	0.22	0.30	0.38	0.54	0.67	0.76	0.86	1.04	1.34	1.54	1.82
	6	0.28	0.36	0.44	0.60	0.73	0.81	0.90	1.07	1.33	1.52	1.73
	7	0.33	0.41	0.49	0.65	0.76	0.84	0.92	1.08	1.32	1.48	1.67
	8	0.38	0.46	0.54	0.68	0.80	0.87	0.95	1.08	1.31	1.45	1.62
	9	0.42	0.50	0.57	0.71	0.82	0.89	0.96	1.09	1.29	1.42	1.58
	10	0.46	0.54	0.61	0.74	0.85	0.91	0.98	1.09	1.27	1.38	1.53
	11	0.50	0.57	0.64	0.77	0.87	0.93	0.99	1.09	1.25	1.36	1.49
12	3	0.08	0.13	0.19	0.34	0.48	0.58	0.70	0.92	1.30	1.56	1.87
	4	0.16	0.22	0.30	0.46	0.60	0.70	0.80	1.00	1.33	1.55	1.82
	5	0.23	0.30	0.38	0.54	0.67	0.76	0.86	1.04	1.33	1.53	1.78
	6	0.29	0.36	0.44	0.60	0.72	0.81	0.90	1.06	1.33	1.49	1.72
	7	0.34	0.41	0.50	0.65	0.76	0.84	0.93	1.08	1.31	1.47	1.66
	8	0.38	0.46	0.54	0.68	0.79	0.87	0.95	1.08	1.30	1.45	1.61
	9	0.42	0.50	0.57	0.71	0.82	0.89	0.96	1.09	1.29	1.43	1.58
	10	0.45	0.53	0.61	0.74	0.84	0.90	0.97	1.09	1.28	1.40	1.55
	11	0.49	0.56	0.64	0.76	0.86	0.92	0.98	1.09	1.27	1.37	1.51
	12	0.53	0.60	0.66	0.78	0.87	0.93	0.99	1.09	1.24	1.35	1.46

From N. R. Mann, R. E. Schafer, and N. D. Singpurwalla, *Methods for Statistical Analysis of Reliability and Life Data*, Wiley, New York, 1974, pp. 222–225.

$\text{Var}(\delta^*) = B\delta^2$. For example, δ^* may be a pooled estimate (Section 5.5). Then two-sided approximate $100\gamma\%$ confidence limits for δ are

$$\underline{\delta} \cong \delta^* / \left[1 - (B/9) + K_\gamma (B/9)^{1/2} \right]^3, \qquad \tilde{\delta} \cong \delta^* / \left[1 - (B/9) - K_\gamma (B/9)^{1/2} \right]^3,$$

$$(4.29)$$

where K_γ is the $[100(1+\gamma)/2]$th standard normal percentile. Mann, Schafer, and Singpurwalla (1974) justify this (Wilson–Hilferty chi-square) approximation. The limits for the Weibull shape parameter β are

$$\underline{\beta} = 1/\tilde{\delta}, \qquad \tilde{\beta} = 1/\underline{\delta}. \qquad (4.30)$$

Insulating fluid example. For 30 kV, $\delta^* = 1.984$, and $\text{Var}(\delta^*) = 0.08175\,\delta^2$. The two-sided approximate 90% confidence limits for δ are $\underline{\delta} \cong 1.984/[1 - (0.08175/9) + 1.645(0.08175/9)^{1/2}]^3 = 1.312$ and $\tilde{\delta} \cong 1.984/[1 - (0.08175/9) - 1.645(0.08175/9)^{1/2}]^3 = 3.418$. The limits for the Weibull shape parameter are $\underline{\beta} \cong 1/3.418 = 0.293$ and $\tilde{\beta} \cong 1/1.312 = 0.762$, not consistent with $\beta = 1$, stated by engineering theory.

Confidence Limits for the Extreme Value Location λ and Weibull Scale α Parameters

Exact limits. Two-sided exact $100\gamma\%$ confidence limits for λ based on the BLIEs λ^{**} and δ^{**} are

$$\underline{\lambda} = \lambda^{**} - t^{**} \left[(1+\gamma)/2; n, r \right] \delta^{**}, \qquad \tilde{\lambda} = \lambda^{**} - t^{**} \left[(1-\gamma)/2; n, r \right] \delta^{**},$$

$$(4.31)$$

where $t^{**}(\varepsilon; n, r)$ is the 100εth percentile of the distribution of $t^{**} = (\lambda^{**} - \lambda)/\delta^{**}$. These percentiles appear in Table 4.3 for $n = 3(1)12$, $r = 3(1)n$, and $1 - \varepsilon$ and $\varepsilon = .02, .05, .10, .25, .40, .50$. Tables are given by Mann and Fertig (1973) up to $n = 16$ and by Mann, Fertig, and Scheuer (1971) up to $n = 25$. Such limits in terms of the BLUEs λ^* and δ^* are

$$\underline{\lambda} = \lambda^* - \left\{ C + t^{**} \left[(1+\gamma)/2; n, r \right] \right\} (1 + B)^{-1} \delta^*,$$

$$(4.32)$$

$$\tilde{\lambda} = \lambda^* - \left\{ C + t^{**} \left[(1-\gamma)/2; n, r \right] \right\} (1 + B)^{-1} \delta^*,$$

where $\text{Var}(\delta^*) = B\delta^2$ and $\text{Cov}(\lambda^*, \delta^*) = C\delta^2$.

Table 4.3. Percentiles of the Distribution of $t^{**} = (\lambda^{**} - \lambda)/\delta^{**}$

n	r	0.02	0.05	0.10	0.25	0.40	0.50	0.60	0.75	0.90	0.95	0.98
3	3	-4.47	-2.54	-1.49	-0.52	-0.10	0.10	0.31	0.69	1.46	2.12	3.39
4	3	-6.92	-3.85	-2.32	-0.84	-0.29	-0.04	0.18	0.50	1.06	1.55	2.43
	4	-2.37	-1.50	-0.96	-0.37	-0.08	0.09	0.25	0.55	1.07	1.49	2.15
5	3	-9.35	-5.22	-3.04	-1.22	-0.50	-0.19	0.06	0.40	0.86	1.20	1.76
	4	-3.13	-1.94	-1.24	-0.50	-0.16	0.02	0.18	0.45	0.88	1.22	1.74
	5	-1.63	-1.08	-0.73	-0.31	-0.06	0.08	0.22	0.47	0.89	1.20	1.64
6	3	-10.54	-6.12	-3.72	-1.56	-0.69	-0.32	-0.04	0.33	0.75	1.02	1.39
	4	-3.69	-2.39	-1.59	-0.67	-0.25	-0.05	0.12	0.38	0.76	1.03	1.42
	5	-2.05	-1.36	-0.91	-0.38	-0.11	0.04	0.17	0.40	0.77	1.04	1.41
	6	-1.29	-0.91	-0.64	-0.28	-0.06	0.07	0.19	0.41	0.77	1.04	1.39
7	3	-13.00	-7.39	-4.45	-1.87	-0.89	-0.48	-0.16	0.26	0.68	0.90	1.20
	4	-4.67	-2.95	-1.94	-0.84	-0.36	-0.13	0.05	0.32	0.66	0.89	1.20
	5	-2.48	-1.59	-1.10	-0.48	-0.17	-0.02	0.12	0.34	0.66	0.89	1.21
	6	-1.54	-1.04	-0.73	-0.32	-0.10	0.03	0.15	0.35	0.67	0.90	1.20
	7	-1.09	-0.79	-0.56	-0.26	-0.06	0.05	0.17	0.36	0.68	0.90	1.18
8	3	-14.36	-8.15	-5.01	-2.14	-1.04	-0.58	-0.21	0.24	0.67	0.88	1.12
	4	-5.34	-3.30	-2.18	-0.99	-0.43	-0.19	0.02	0.30	0.64	0.83	1.07
	5	-2.78	-1.86	-1.25	-0.56	-0.22	-0.05	0.10	0.32	0.62	0.82	1.07
	6	-1.80	-1.20	-0.83	-0.36	-0.12	0.01	0.13	0.33	0.63	0.82	1.08
	7	-1.28	-0.88	-0.61	-0.27	-0.07	0.04	0.15	0.33	0.63	0.82	1.08
	8	-0.97	-0.70	-0.50	-0.22	-0.05	0.06	0.16	0.34	0.63	0.82	1.07
9	3	-15.68	-9.12	-5.64	-2.38	-1.17	-0.66	-0.28	0.20	0.66	0.86	1.06
	4	-6.31	-3.78	-2.47	-1.08	-0.50	-0.24	-0.01	0.28	0.61	0.79	1.00
	5	-3.19	-2.10	-1.40	-0.63	-0.26	-0.08	0.08	0.30	0.58	0.76	0.98
	6	-2.01	-1.38	-0.94	-0.41	-0.15	-0.01	0.11	0.30	0.57	0.76	0.99
	7	-1.43	-0.99	-0.70	-0.31	-0.10	0.02	0.13	0.31	0.57	0.76	0.99
	8	-1.08	-0.76	-0.55	-0.25	-0.07	0.04	0.14	0.31	0.58	0.76	0.99
	9	-0.87	-0.64	-0.47	-0.21	-0.05	0.05	0.15	0.32	0.58	0.76	0.98
10	3	-17.45	-9.98	-6.05	-2.58	-1.29	-0.76	-0.34	0.17	0.66	0.87	1.07
	4	-6.54	-4.17	-2.70	-1.22	-0.58	-0.28	-0.04	0.27	0.60	0.77	0.96
	5	-3.56	-2.37	-1.56	-0.73	-0.31	-0.12	0.05	0.28	0.56	0.72	0.93
	6	-2.21	-1.51	-1.03	-0.48	-0.19	-0.04	0.09	0.28	0.54	0.71	0.92
	7	-1.56	-1.08	-0.77	-0.35	-0.12	-0.00	0.11	0.28	0.54	0.70	0.93
	8	-1.20	-0.86	-0.62	-0.27	-0.08	0.02	0.12	0.28	0.53	0.71	0.93
	9	-0.97	-0.70	-0.50	-0.23	-0.06	0.04	0.13	0.29	0.54	0.71	0.93
	10	-0.80	-0.60	-0.44	-0.20	-0.04	0.04	0.14	0.29	0.54	0.71	0.92
11	3	-18.52	-10.68	-6.42	-2.76	-1.41	-0.85	-0.42	0.13	0.65	0.87	1.07
	4	-7.26	-4.57	-2.95	-1.37	-0.66	-0.36	-0.10	0.24	0.58	0.75	0.92
	5	-4.00	-2.58	-1.75	-0.81	-0.37	-0.16	0.01	0.26	0.54	0.69	0.88
	6	-2.45	-1.67	-1.16	-0.53	-0.22	-0.07	0.06	0.26	0.52	0.66	0.85
	7	-1.70	-1.21	-0.85	-0.40	-0.15	-0.02	0.09	0.26	0.50	0.65	0.86
	8	-1.30	-0.92	-0.66	-0.30	-0.11	0.00	0.10	0.26	0.50	0.65	0.86
	9	-1.06	-0.76	-0.54	-0.25	-0.08	0.02	0.11	0.26	0.50	0.65	0.86
	10	-0.87	-0.63	-0.46	-0.21	-0.06	0.03	0.12	0.27	0.50	0.65	0.86
	11	-0.75	-0.55	-0.42	-0.19	-0.05	0.03	0.12	0.27	0.50	0.65	0.85
12	3	-19.08	-11.23	-6.92	-3.03	-1.58	-0.97	-0.49	0.10	0.64	0.88	1.10
	4	-7.44	-4.81	-3.17	-1.47	-0.74	-0.40	-0.14	0.21	0.58	0.75	0.92
	5	-4.17	-2.72	-1.88	-0.89	-0.42	-0.20	-0.01	0.24	0.53	0.68	0.84
	6	-2.63	-1.83	-1.27	-0.60	-0.26	-0.10	0.05	0.25	0.50	0.64	0.81
	7	-1.91	-1.32	-0.92	-0.42	-0.17	-0.04	0.08	0.25	0.48	0.62	0.80
	8	-1.41	-1.00	-0.71	-0.33	-0.12	-0.01	0.09	0.25	0.48	0.62	0.79
	9	-1.15	-0.80	-0.58	-0.27	-0.09	0.01	0.10	0.25	0.47	0.62	0.80
	10	-0.91	-0.67	-0.48	-0.23	-0.07	0.02	0.11	0.25	0.47	0.62	0.80
	11	-0.78	-0.58	-0.43	-0.20	-0.06	0.03	0.11	0.25	0.47	0.62	0.80
	12	-0.69	-0.53	-0.39	-0.19	-0.05	0.03	0.11	0.25	0.47	0.62	0.79

From N. R. Mann, R. E. Schafer, and N. D. Singpurawalla, *Methods for Statistical Analysis of Reliability and Life Data*, Wiley, New York, 1974, pp. 226–229.

Corresponding limits for the Weibull scale parameter are

$$\underset{\sim}{\alpha}=\exp(\underset{\sim}{\lambda}), \qquad \tilde{\alpha}=\exp(\tilde{\lambda}).$$

The limits above yield one-sided $100\gamma\%$ confidence limits when γ replaces $(1+\gamma)/2$ or $1-\gamma$ replaces $(1-\gamma)/2$. The limits are exact (approximate) for failure (time) censored data.

Insulating fluid example. For 30 kV, $\lambda^*=8.583$, $\delta^*=1.984$, $B=0.08175$, and $C=-0.00228$. For a two-sided 90% confidence interval, $\underset{\sim}{\lambda}=8.583-\{-0.00228 + t^{**}[(1 + 0.90)/2; 12, 10]\}(1 + 0.08175)^{-1}1.984 = 8.583-(-0.00228+0.62)(1.08175)^{-1}1.984=7.405$ and $\tilde{\lambda}=8.583-(-0.00228-0.67)(1.08175)^{-1}1.984=9.816$. Corresponding limits for the Weibull scale parameter are $\underset{\sim}{\alpha}=\exp(7.405)=1640$ and $\tilde{\alpha}=\exp(9.816)=18{,}300$ seconds.

Approximate limits. Two-sided approximate $100\gamma\%$ confidence limits for λ are

$$\underset{\sim}{\lambda}\cong\lambda^*-K_\gamma A^{1/2}\delta^*, \qquad \tilde{\lambda}\cong\lambda^*+K_\gamma A^{1/2}\delta^*, \tag{4.33}$$

where $\mathrm{Var}(\lambda^*)=A\delta^2$ and K_γ is the $[100(1+\gamma)/2]$th standard normal percentile. The limits are more exact the larger r is, and the approximate interval tends to be too narrow. δ^* may be a pooled estimate (Section 5.5) for δ.

Insulating fluid example. For 30 kV, $\lambda^*=8.583$, $\delta^*=1.984$, and $A=0.10574$. Two-sided approximate 90% confidence limits for λ are $\underset{\sim}{\lambda}=8.583-1.645(0.10574)^{1/2}1.984=7.522$ and $\tilde{\lambda}=8.583+1.061=9.644$. Corresponding limits for the Weibull scale parameter are $\underset{\sim}{\alpha}=\exp(7.522)=1850$ and $\tilde{\alpha}=\exp(9.644)=15{,}400$ seconds.

Confidence Limits for Percentiles

Exact limits. Two-sided exact $100\gamma\%$ confidence limits for the $100P$th extreme value percentile $y_P=\lambda+u_P\delta$ are

$$\underset{\sim}{y_P}=\lambda^{**}-t^{**}\big[(1+\gamma)/2; P, n, r\big]\delta^{**},$$
$$\tilde{y}_P=\lambda^{**}-t^{**}\big[(1-\gamma)/2; P, n, r\big]\delta^{**}, \tag{4.34}$$

where λ^{**} and δ^{**} are the BLIEs and $t^{**}(\varepsilon; P, n, r)$ is the 100εth percentile of the distribution of $t^{**}=(\lambda^{**}-y_P)/\delta^{**}$. These percentiles appear in Table 4.4 for $P=0.10$, $n=3(1)12$, $r=3(1)n$, and $1-\varepsilon$ and $\varepsilon=.02, .05, .10$,

Table 4.4. Percentiles of the Distribution of $t_{.10}^{**} = (\lambda^{**} - y_{.10})/\delta^{**}$

n	r	0.02	0.05	0.10	0.25	0.40	0.50	0.60	0.75	0.90	0.95	0.98
3	3	0.75	1.10	1.43	2.18	2.88	3.40	4.06	5.50	8.99	13.16	20.93
4	3	0.78	1.16	1.49	2.18	2.82	3.33	3.96	5.38	9.03	13.07	20.23
	4	0.87	1.16	1.46	2.06	2.60	2.99	3.45	4.40	6.47	8.39	11.66
5	3	0.78	1.18	1.51	2.17	2.79	3.27	3.87	5.24	8.78	12.58	20.38
	4	0.97	1.23	1.51	2.09	2.61	2.99	3.44	4.40	6.49	8.48	11.73
	5	0.97	1.23	1.49	2.02	2.49	2.82	3.20	3.93	5.48	6.73	8.66
6	3	0.73	1.18	1.53	2.15	2.73	3.18	3.74	4.98	8.24	11.74	18.65
	4	1.00	1.28	1.55	2.10	2.60	2.98	3.41	4.30	6.33	8.18	11.39
	5	1.02	1.29	1.54	2.05	2.50	2.82	3.21	3.94	5.42	6.73	8.89
	6	1.02	1.27	1.53	2.01	2.42	2.70	3.04	3.67	4.86	5.83	7.31
7	3	0.64	1.18	1.53	2.13	2.66	3.08	3.60	4.79	7.80	11.12	17.54
	4	1.04	1.31	1.58	2.10	2.57	2.91	3.33	4.21	6.16	7.89	10.90
	5	1.08	1.33	1.57	2.06	2.49	2.80	3.15	3.87	5.36	6.68	8.44
	6	1.08	1.32	1.56	2.03	2.42	2.70	3.01	3.63	4.86	5.82	7.23
	7	1.08	1.32	1.55	2.00	2.37	2.62	2.90	3.44	4.46	5.25	6.37
8	3	0.49	1.13	1.52	2.11	2.62	3.01	3.48	4.62	7.51	10.67	16.36
	4	1.04	1.33	1.60	2.10	2.56	2.88	3.27	4.10	5.96	7.79	10.75
	5	1.11	1.36	1.60	2.08	2.49	2.78	3.12	3.82	5.28	6.50	8.62
	6	1.13	1.36	1.59	2.05	2.43	2.71	3.02	3.62	4.83	5.83	7.18
	7	1.12	1.36	1.58	2.03	2.38	2.64	2.93	3.46	4.49	5.31	6.40
	8	1.12	1.36	1.58	2.01	2.34	2.57	2.83	3.32	4.21	4.90	5.84
9	3	0.42	1.12	1.51	2.09	2.57	2.95	3.40	4.43	7.14	10.21	15.61
	4	1.06	1.36	1.61	2.10	2.52	2.84	3.21	4.00	5.77	7.39	10.26
	5	1.17	1.41	1.63	2.08	2.47	2.76	3.08	3.76	5.13	6.34	8.13
	6	1.19	1.41	1.62	2.06	2.43	2.70	2.99	3.59	4.74	5.67	7.06
	7	1.19	1.41	1.62	2.04	2.39	2.64	2.91	3.45	4.48	5.28	6.46
	8	1.19	1.40	1.61	2.02	2.36	2.59	2.84	3.34	4.26	4.95	5.94
	9	1.19	1.40	1.60	2.00	2.33	2.55	2.78	3.22	4.04	4.66	5.50
10	3	0.09	0.99	1.46	2.05	2.51	2.84	3.27	4.25	6.75	9.36	14.88
	4	0.99	1.34	1.62	2.08	2.48	2.77	3.13	3.90	5.56	7.17	9.60
	5	1.17	1.42	1.64	2.07	2.45	2.71	3.02	3.67	5.00	6.13	8.02
	6	1.20	1.43	1.64	2.05	2.41	2.66	2.94	3.53	4.67	5.59	6.99
	7	1.21	1.43	1.64	2.04	2.38	2.62	2.88	3.41	4.41	5.18	6.29
	8	1.21	1.43	1.63	2.02	2.35	2.58	2.83	3.31	4.22	4.91	5.83
	9	1.21	1.42	1.63	2.01	2.32	2.54	2.77	3.22	4.03	4.63	5.51
	10	1.21	1.42	1.62	1.99	2.30	2.50	2.72	3.13	3.86	4.41	5.16
11	3	-0.09	0.90	1.42	2.01	2.45	2.77	3.17	4.07	6.41	9.11	14.47
	4	0.97	1.35	1.61	2.06	2.44	2.73	3.06	3.79	5.46	7.04	9.98
	5	1.18	1.43	1.64	2.05	2.41	2.68	2.98	3.60	4.90	6.07	7.83
	6	1.24	1.45	1.64	2.04	2.38	2.63	2.91	3.46	4.58	5.52	6.96
	7	1.25	1.45	1.64	2.03	2.35	2.59	2.86	3.36	4.36	5.16	6.34
	8	1.25	1.45	1.64	2.01	2.33	2.56	2.80	3.28	4.15	4.87	5.82
	9	1.25	1.44	1.64	2.00	2.31	2.53	2.76	3.21	4.01	4.63	5.54
	10	1.25	1.44	1.64	1.99	2.29	2.49	2.71	3.14	3.87	4.44	5.23
	11	1.25	1.45	1.63	1.98	2.28	2.46	2.67	3.06	3.76	4.26	4.94
12	3	-0.38	0.75	1.37	1.98	2.41	2.71	3.08	3.89	6.00	8.40	12.96
	4	0.95	1.34	1.60	2.05	2.42	2.69	3.00	3.67	5.17	6.60	9.07
	5	1.20	1.44	1.66	2.05	2.40	2.65	2.93	3.52	4.72	5.79	7.35
	6	1.26	1.46	1.67	2.04	2.38	2.62	2.88	3.39	4.41	5.31	6.61
	7	1.28	1.47	1.67	2.03	2.36	2.58	2.82	3.30	4.21	4.98	5.09
	8	1.28	1.47	1.66	2.02	2.34	2.54	2.78	3.22	4.06	4.75	5.71
	9	1.27	1.46	1.66	2.01	2.31	2.52	2.74	3.16	3.94	4.53	5.40
	10	1.27	1.47	1.65	2.00	2.30	2.49	2.70	3.11	3.82	4.37	5.11
	11	1.27	1.46	1.64	2.00	2.28	2.47	2.67	3.05	3.72	4.23	4.88
	12	1.28	1.47	1.64	1.99	2.27	2.44	2.63	3.00	3.62	4.07	4.68

From N. R. Mann, R. E. Schafer, and N. D. Singpurawalla, *Methods for Statistical Analysis of Reliability and Life Data*, Wiley, New York, 1974, pp. 230–233.

.25, .40, .50. Further tables for $P=0.01$ and 0.05 are given by Mann and Fertig (1973) for n up to 16 and by Mann, Fertig, and Scheuer (1971) for n up to 25 and the same r and ε values. The confidence limits (4.31) for λ are confidence limits for $y_{.632}$.

Such limits in terms of the BLUEs λ^* and δ^* are

$$\underline{y}_P=\lambda^*-\left\{C+t^{**}\left[(1+\gamma)/2;\,P,n,r\right]\right\}(1+B)^{-1}\delta^*,$$

$$\tilde{y}_P=\lambda^*-\left\{C+t^{**}\left[(1-\gamma)/2;\,P,n,r\right]\right\}(1+B)^{-1}\delta^*,$$

(4.35)

where $\mathrm{Var}(\lambda^*)=B\delta^2$ and $\mathrm{Cov}(\lambda^*,\delta^*)=C\delta^2$.

Corresponding limits for the $100P$th Weibull percentile $t_P=\exp(y_P)$ are

$$\underline{t}_P=\exp(\underline{y}_P),\qquad \tilde{t}_P=\exp(\tilde{y}_P).\tag{4.36}$$

The limits above yield one-sided $100\gamma\%$ confidence limits when γ replaces $(1+\gamma)/2$ or $1-\gamma$ replaces $(1-\gamma)/2$. The limits are exact (approximate) for failure (time) censored data.

Insulating fluid example.   For 30 kV, $\lambda^*=8.583$, $\delta^*=1.984$, $B=0.08175$, $C=-0.00228$, and $y^*_{10}=4.118$. For a two-sided 90% confidence interval, $\underline{y}_{.10}=8.583-\{-0.00228+t^{**}[(1+0.90)/2;0.10,12,10]\}(1+0.08175)^{-1}$ $1.984=8.583-(-0.00228+4.37)(1.08175)^{-1}1.984=0.572$ and $\tilde{y}_{10}=8.583-(-0.00228+1.47)(1.08175)^{-1}1.984=5.891$. The corresponding limits for the Weibull percentile are $\underline{t}_{.10}=\exp(0.572)=1.8$ and $\tilde{t}_{.10}=\exp(5.891)=362$ seconds.

Approximate limits for low percentiles.   Suppose that λ^* and δ^* are the BLUEs for λ and δ, and $\mathrm{Var}(\lambda^*)=A\delta^2$, $\mathrm{Var}(\delta^*)=B\delta^2$, and $\mathrm{Cov}(\lambda^*,\delta^*)=C\delta^2$. Then a one-sided lower approximate $100\gamma\%$ confidence limit for the $100P$th extreme value percentile is

$$\underline{y}_P\cong y^*_P+\delta^*\left[F(\gamma;a,b)-1\right]\left[(C/B)+u_P\right],\tag{4.37}$$

where $u_P=\ln[-\ln(1-P)]$ and $F(\gamma;a,b)$ is the 100γth F percentile with $a=2[u_P+(C/B)]^2/[A-(C^2/B)]$ degrees of freedom above and $b=2/B$ below. Mann, Schafer, and Singpurwalla (1974) state that this approximation is satisfactory for large r (say, greater than 10) and $P<0.10$. P must be small to insure that $[(C/B)+u_P]$ is negative. The lower limit for the $100P$th Weibull percentile is $\underline{t}_P\cong\exp(\underline{y}_P)$.

For percentiles that are not low or high, approximate limits like (4.33) can be used. Then y_p^* replaces λ^*, and D of (4.13) replaces A in (4.33).

Insulating fluid example.   For 30 kV, $\lambda^* = 8.583$, $\delta^* = 1.984$, $A = 0.10574$, $B = 0.08175$, and $C = -0.00228$. The estimate of the 10th percentile is $y_{.10}^* = 4.118$. A lower approximate 95% confidence limit for $y_{.10}$ is calculated as follows:

$$a = 2\left[-2.2504 + (-0.00228/0.08175)\right]^2 /$$

$$\left[0.10574 - \left((-0.00228)^2/(0.08175)\right)\right] = 98.2,$$

$$b = 2/0.08175 = 24.46, \qquad F(0.95; 98.2, 24.46) \cong 1.792,$$

$$\underset{\sim}{y}_{.10} \cong 4.118 + 1.984(1.792 - 1)\left[(-0.00228/0.08175) + (-2.2504)\right] = 0.538.$$

The lower limit for the 10th Weibull percentile is $\underset{\sim}{t}_{.10} \cong \exp(0.538) = 1.7$ seconds.

Confidence Limits for Reliability

Exact limits.   Tables for exact limits for reliability based on BLUEs or BLIEs have not been developed. The following method could be used in principle to obtain limits for the proportion $F(y)$ below y for an extreme value distribution. Two-sided $100\gamma\%$ confidence limits $\underset{\sim}{F}$ and $\tilde{F}$ for $F(y)$ are the solutions of

$$y = y_{\tilde{F}} = \lambda^{**} - t^{**}\left[(1+\gamma)/2; \tilde{F}, n, r\right]\delta^{**},$$
$$\hspace{6cm} (4.38)$$
$$y = y_{\underset{\sim}{F}} = \lambda^{**} - t^{**}\left[(1-\gamma)/2; \underset{\sim}{F}, n, r\right]\delta^{**};$$

the t^{**} percentiles are tabulated for a few values $F = .01, .05, .10, .632$. In practice, these F values are too widely spaced for accurate inverse interpolation to obtain $\underset{\sim}{F}$ and $\tilde{F}$; moreover, limits outside the range .01 to .632 require extrapolation in F and would be inaccurate. Direct tables are needed.

Corresponding limits for the proportion below t for a Weibull distribution are those above where $y = \ln(t)$.

The limits above yield one-sided $100\gamma\%$ confidence limits when γ replaces $(1+\gamma)/2$ or $1-\gamma$ replaces $(1-\gamma)/2$. The limits are exact (approximate) for failure (time) censored data.

Approximate limits.   Two-sided approximate $100\gamma\%$ confidence limits for reliability $R(y)$ at (ln) age y are calculated as follows. Estimate the standardized deviate $u=(y-\lambda)/\delta$ with

$$u^*=(y-\lambda^*)/\delta^*, \qquad (4.39)$$

where $\mathrm{Var}(\lambda^*)=A\delta^2$, $\mathrm{Var}(\delta^*)=B\delta^2$, and $\mathrm{Cov}(\lambda^*,\delta^*)=C\delta^2$. λ^* and δ^* may be pooled estimates (Section 5.5). Calculate

$$D^*=A+u^{*2}B+2u^*C, \qquad (4.40)$$

$$\underset{\sim}{u}=u^*-K_\gamma D^{*1/2}, \qquad \tilde{u}=u^*+K_\gamma D^{*1/2}, \qquad (4.41)$$

where K_γ is the standard normal $[100(1+\gamma)/2]$th percentile. The two-sided limits for $R(y)$ are

$$R(y)\cong\exp[-\exp(\tilde{u})], \qquad \tilde{R}(y)\cong\exp[-\exp(\underset{\sim}{u})]. \qquad (4.42)$$

These limits are more accurate the larger r is. The interval is shorter than an exact one. For a one-sided $100\gamma\%$ confidence limit, replace K_γ above by z_γ, the standard normal 100γth percentile. Corresponding limits for the fraction failed $F(y)$ are $\tilde{F}(y)=1-\tilde{R}(y)$ and $\tilde{F}(y)=1-R(y)$. Mann, Schafer, and Singpurwalla (1974) present more accurate approximations.

Insulating fluid example.   For 30 kV, $\lambda^*=8.583$, $\delta^*=1.984$, $A=0.10574$, $B=0.08175$, and $C=-0.00228$. A lower one-sided 95% confidence limit for reliability at five seconds is calculated as follows:

$$y=\ln(5)=1.609, \qquad u^*=(1.609-8.583)/1.984=-3.515,$$

$$D^*=0.10574+(-3.515)^2 0.08175+2(-3.515)(-0.00228)=1.132,$$

$$\tilde{u}=-3.515+1.645(1.132)^{1/2}=-1.765,$$

$$R(5)=\exp[-\exp(-1.765)]=0.843.$$

PREDICTION

Exact prediction limits based on a singly censored sample from a Weibull or extreme value distribution have been developed for the smallest observation of an independent future sample by Mann and Saunders (1969), Mann

(1970b), and Lawless (1973). The following approximate limits for the jth observation of the same sample come from Mann, Schafer, and Singpurwalla (1974).

jth Observation of the Same Sample

Predictor. A simple linear unbiased predictor for the jth (ln) order statistic $Y_{(j)}$ of the same sample of size n is

$$\hat{Y}_{(j)} = Y_{(r)} + \left(EU_{(j)} - EU_{(r)} \right) \delta^*, \tag{4.43}$$

where $EU_{(j)}$ and $EU_{(r)}$ are the expectations of the jth and rth standard extreme value order statistics in a sample of size n, and δ^* is the BLUE for δ. Tables of these expectations are given by White (1967) for $n = 1(1)50(5)100$, and David (1970) lists other tables. The predictor for the jth Weibull order statistic is $\hat{T}_{(j)} = \exp(\hat{Y}_{(j)})$. $Y_{(n)}$ is usually of greatest interest, as it is the time required for all sample units to fail.

Insulating fluid example. For 30 kV, the simple unbiased predictor for the last (ln) observation $Y_{(12)}$ is $\hat{Y}_{(12)} = 9.668 + (1.057 - 0.411)1.984 = 10.950$. The corresponding prediction of the last Weibull order statistic is $\hat{T}_{(12)} = \exp(10.950) = 56,900$ seconds.

Prediction limits. Two-sided approximate $100\gamma\%$ prediction limits for $Y_{(j)}$ are

$$\underline{Y}_{(j)} \simeq Y_{(r)} + \left\{ 1/F\left[(1+\gamma)/2; b, a \right] \right\} \left(EU_{(j)} - EU_{(r)} \right) \delta^*,$$
$$\tag{4.44}$$
$$\tilde{Y}_{(j)} \simeq Y_{(r)} + F\left[(1+\gamma)/2; a, b \right] \left(EU_{(j)} - EU_{(r)} \right) \delta^*,$$

where $F(\epsilon; a, b)$ is the 100ϵth F percentile with $a = 2(EU_{(j)} - EU_{(r)})^2/$ $[\text{Var}(U_{(j)}) + \text{Var}(U_{(r)}) - 2\text{Cov}(U_{(j)}, U_{(r)})]$ degrees of freedom above and $b = 2/B_{n,r}$ below, where $\text{Var}(\delta^*) = B_{n,r}\delta^2$. David (1970) lists tables of $EU_{(j)}$, $\text{Var}(U_{(j)})$, and $\text{Cov}(U_{(j)}, U_{(r)})$. Prediction limits for the jth Weibull order statistic are $\exp(\underline{Y}_{(j)})$ and $\exp(\tilde{Y}_{(j)})$. The limits above yield one-sided $100\gamma\%$ prediction limits when γ replaces $(1+\gamma)/2$. The limits should be a better approximation for failure censored data than for time censored data. In the limits, a and b are reversed.

An upper prediction limit for the nth observation is related to an outlier test. If the nth observation is above a $100\gamma\%$ limit, it is a statistically significant outlier at the $[100(1-\gamma)]\%$ level.

Insulating fluid example. For 30 kV, a one-sided upper approximate 90% prediction limit for $Y_{(12)}$ is calculated as follows:

$$a = 2(1.057 - 0.411)^2/(0.1525 + 0.1249 - 2 \times 0.0687) = 5.96,$$

$$b = 2/0.08175 = 24.5, \qquad F(0.90; 5.96, 24.5) = 2.022,$$

$$\tilde{Y}_{(12)} = 9.668 + 2.022(1.057 - 0.411)1.984 = 12.260,$$

$$\tilde{T}_{(12)} = \exp(12.260) = 211{,}000 \text{ seconds.}$$

Under the assumption of a Weibull distribution, the largest censored (ln) value, $11.363+$, does not exceed the prediction limit and is not an outlier, but it would be one if observed and greater than 12.260.

First Observation of a Future Sample

Predictor. The best linear unbiased predictor for the first (ln) order statistic $X_{(1)}$ of an independent future sample of size m is

$$\hat{X}_{(1)} = \lambda^* - \delta^* u_m, \tag{4.45}$$

where λ^* and δ^* are the BLUEs for λ and δ, $u_m = \ln(m) + 0.5772$, and $0.5772 \cdots$ is Euler's constant. The variance of its prediction error $(\hat{X}_{(1)} - X_{(1)})$ is

$$\mathrm{Var}\left(\hat{X}_{(1)} - X_{(1)}\right) = \delta^2 \left(A_{n,r} + u_m^2 B_{n,r} - 2u_m C_{n,r} + 1.6449\right), \tag{4.46}$$

where $\mathrm{Var}(\lambda^*) = A_{n,r}\delta^2$, $\mathrm{Var}(\delta^*) = B_{n,r}\delta^2$, $\mathrm{Cov}(\lambda^*, \delta^*) = C_{n,r}\delta^2$, and $1.6449 \cdots = \pi^2/6$. The predictor for the first Weibull order statistic is $\exp(\hat{X}_{(1)})$.

Insulating fluid example. For 30 kV, the prediction of the first ln order statistic in a future sample of $m = 12$ is $\hat{X}_{(1)} = 8.583 - 1.984[\ln(12) + 0.5772] = 2.508$. The prediction of the Weibull observation is $\exp(2.508) = 12.3$ seconds.

Prediction limits. A one-sided lower approximate $100\gamma\%$ prediction limit for $X_{(1)}$ is

$$\underline{X}_{(1)} \cong \hat{X}_{(1)} - \delta^* \left[F(\gamma; a, b) - 1\right]\left[u_m - (C_{n,r}/B_{n,r})\right], \tag{4.47}$$

where $F(\gamma; a, b)$ is the 100γth F percentile with $a = 2[\ln(m) + 0.5772 - (C_{n,r}/B_{n,r})]^2 / [A_{n,r} - (C^2_{n,r}/B_{n,r}) + 1.6449]$ degrees of freedom above and $b = 2/B_{n,r}$ below. A lower prediction limit for the first Weibull order statistic is $\exp(\underset{\sim}{X}_{(1)})$. Mann, Schafer, and Singpurwalla (1974) present this approximate limit, which they call a "$100\gamma\%$ **warranty period**" for a future lot of size m. Fertig, Meyer, and Mann (1980) table exact prediction limits for $m = 1$, $n = 5(5)25$, $r = 3, 5(5)n$, based on BLIEs.

Insulating fluid example. For 30 kV, the calculation of the lower approximate 90% prediction limit is $a = 2[\ln(12) + 0.5772 - (-0.00228/0.08175)]^2 / \{0.10574 - [(-0.00228)^2/0.08175] + 1.6449\} = 181$, $b = 2/0.01875 = 24.46$, $F(0.90; 181, 24.46) = 1.552$, $\underset{\sim}{X}_{(1)} = 2.508 - 1.984(1.552 - 1)[\ln(12) + 0.5772 - (-0.00228/0.08175)] = -0.876$. The lower limit for the first Weibull observation is $\exp(-0.876) = 0.42$ second.

METHODS FOR THE WEIBULL DISTRIBUTION WITH A KNOWN SHAPE PARAMETER

Sometimes data come from a Weibull distribution with a shape parameter β whose value is assumed to be known. For example, the distribution may be exponential ($\beta = 1$) or Rayleigh ($\beta = 2$). Then one needs to estimate the Weibull scale parameter α. Methods for the exponential distribution can be applied to such data. Transform each Weibull observation Y to $U = Y^\beta$, which comes from an exponential distribution with mean $\theta = \alpha^\beta$. Estimates and confidence limits for θ and functions of it can then be transformed into ones for the Weibull distribution. The same approach can be applied to data from an extreme value distribution with a known scale parameter; however, that situation seldom occurs in practice. The β value may come from previous or related data or may be a widely accepted value for the product, or even an engineering guess. Of course, the accuracy of this approach depends on the accuracy of the β value. Different β values may be tried to assess their effects on the results.

5. ORDER STATISTICS

This **advanced** section derives theory for order statistics from a continuous parent distribution. It includes (1) the distribution and moments of an order statistic, (2) the joint distribution and moments of pairs or any number of order statistics, (3) the distribution of the random fraction of the population below an order statistic, (4) a theoretical basis for probability and hazard plotting, and (5) the theory for best linear unbiased estimators.

Suppose that the n independent observations come from a continuous cumulative distribution $F( )$. Denote the ordered observations by $Y_1 \leqslant Y_2$

$\leqslant \cdots \leqslant Y_n$. Although stated in terms of complete samples, the following theory applies to samples singly **failure** censored on the right or left or both sides. Theory for multiply censored samples is more complicated and is not well developed; an exception, Mann (1970a) presents theory and tables for the Weibull distribution and all multiple censorings for samples up through size $n = 9$. Further information on order statistics is given by David (1970), Sarhan and Greenberg (1962), and Harter (1977b).

5.1. Distribution of an Order Statistic

Suppose random samples of size n are repeatedly taken from a continuous cumulative distribution $F(y)$. In each sample, the ith order statistic Y_i has a different value. So Y_i has a distribution. This distribution depends on $F(y)$ and n and i; it is given below. Later it is used to justify probability and hazard plotting and best linear unbiased estimates.

Cumulative distribution. This paragraph derives $G_i(y) = P\{Y_i \leqslant y\}$, the cumulative distribution of Y_i. The event $\{Y_i \leqslant y\}$ is equivalent to the event that at least i observations fall below y. There are n independent observations and each has a probability $F(y)$ of falling below y. So $P\{Y_i \leqslant y\}$ is the binomial probability of i or more observations below y, namely,

$$G_i(y) = \sum_{k=i}^{n} \binom{n}{k} [F(y)]^k [1 - F(y)]^{n-k}$$

$$= 1 - \sum_{k=0}^{i-1} \binom{n}{k} [F(y)]^k [1 - F(y)]^{n-k}. \tag{5.1}$$

Probability density. Suppose the parent distribution has a probability density $f(y)$. Then the probability density of Y_i is

$$g_i(y) = \frac{n!}{(i-1)!1!(n-i)!} [F(y)]^{i-1} f(y) [1 - F(y)]^{n-i}. \tag{5.2}$$

This comes from differentiating $G_i(y)$. Figure 5.1 motivates (5.2) as follows. The event that Y_i is in the infinitesimal interval y to $y + dy$ corresponds to the event that $i - 1$ observations fall below y, one falls between y and $y + dy$, and $n - i$ fall above $y + dy$. Each of the n independent observations can fall into one of those three intervals with approximate probabilities $F(y)$, $f(y) dy$, and $1 - F(y)$. The trinomial probability of the above event is

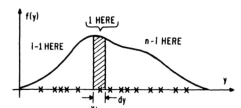

Figure 5.1. Motivation for the density of order statistic Y_i.

approximately

$$g_i(y)\,dy \simeq \frac{n!}{(i-1)!\,1!\,(n-i)!}\,[F(y)]^{i-1}[f(y)\,dy]^1[1-F(y)]^{n-i}.$$

This motivates (5.2).

Figure 5.2 depicts the density of a parent distribution and those of the order statistics of a sample of size 4.

Exponential. Y_i from an exponential distribution with failure rate λ has

$$G_i(y) = \sum_{k=i}^{n} \binom{n}{k}(1-e^{-\lambda y})^k e^{-(n-k)\lambda y}, \qquad y \geqslant 0,$$

$$g_i(y) = \frac{n!}{(i-1)!\,(n-i)!}(1-e^{-\lambda y})^{i-1}\lambda e^{-\lambda y}e^{-\lambda(n-i)y}. \qquad (5.3)$$

The first-order statistic has an exponential distribution

$$g_1(y) = (n\lambda)e^{-(n\lambda)y}, \qquad y \geqslant 0.$$

Its failure rate is $n\lambda$; that is, the failure rates of the n sample units add.

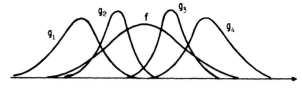

Figure 5.2. Parent f and order statistic densities g_i.

Normal. Y_i from a normal distribution with mean μ and standard deviation σ has

$$G_i(y) = \sum_{k=i}^{n} \binom{n}{k} \{\Phi[(y-\mu)/\sigma]\}^k \{1 - \Phi[(y-\mu)/\sigma]\}^{n-k},$$

$$g_i(y) = \frac{n!}{(i-1)!(n-i)!} [\Phi((y-\mu)/\sigma)]^{i-1}$$

$$\times \frac{1}{\sigma} \phi((y-\mu)/\sigma) [1 - \Phi((y-\mu)/\sigma)]^{n-i}, \tag{5.4}$$

where $\Phi(\ )$ and $\phi(\ )$ are the standard normal cumulative distribution and probability density.

Extreme value. Y_i from an extreme value distribution with location and scale parameters λ and δ has

$$G_i(y) = \sum_{k=i}^{n} \binom{n}{k} [1 - \exp(-e^{(y-\lambda)/\delta})]^k \exp[-(n-k)e^{(y-\lambda)/\delta}],$$

$$g_i(y) = \frac{n!}{(i-1)!(n-i)!} [1 - \exp(-e^{(y-\lambda)/\delta})]^{i-1}$$

$$\times \frac{1}{\delta} e^{(y-\lambda)/\delta} \exp[-(n-i+1)e^{(y-\lambda)/\delta}]. \tag{5.5}$$

Moments. Moments of an order statistic are defined like any other moments. For example, the expectation (or mean) of Y_i is

$$EY_i = \int_{-\infty}^{\infty} y g_i(y) \, dy. \tag{5.6}$$

The variance of Y_i is

$$\text{Var}(Y_i) = \int_{-\infty}^{\infty} (y - EY_i)^2 g_i(y) \, dy. \tag{5.7}$$

Suppose a distribution has location and scale parameters μ and σ. Let $U_i = (Y_i - \mu)/\sigma$ denote the standardized ith order statistic; it comes from the standardized parent distribution with $\mu = 0$ and $\sigma = 1$. Then

$$EY_i = \mu + (EU_i)\sigma, \qquad \text{Var}(Y_i) = \sigma^2 \text{Var}(U_i). \tag{5.8}$$

The relationships are useful, since there are tables of moments of the U_i for various distributions; such tables are listed by David (1970). These standardized moments depend on just i, n, and the parent distribution. For a symmetric parent distribution, such as the normal distribution,

$$EU_{n-i+1} = -EU_i, \qquad \text{Var}(U_{n-i+1}) = \text{Var}(U_i). \tag{5.9}$$

So only half of the expectations and variances need to be tabulated.

Example.   This example derives the moments of the order statistics of a sample of $n=2$ from a standard normal distribution. For U_1,

$$EU_1 = \int_{-\infty}^{\infty} u \frac{2!}{(2-1)!} \phi(u)[1-\Phi(u)] \, du = -2 \int_{-\infty}^{\infty} u\phi(u)\Phi(u) \, du,$$

since $\int_{-\infty}^{\infty} u\phi(u) \, du = 0$, the population mean. Integrate by parts to get

$$EU_1 = -2\left\{ \left[-\phi(u)\Phi(u) \right]_{-\infty}^{\infty} - \int_{-\infty}^{\infty} -[\phi(u)]^2 \, du \right\}$$

$$= -(1/\pi)^{1/2} \approx -0.564,$$

where $\pi^{1/2}[\phi(u)]^2$ is the normal density for a mean of 0 and a variance of 2. By symmetry, $EU_2 = -EU_1 = (1/\pi)^{1/2}$. Similarly,

$$EU_1^2 = \int_{-\infty}^{\infty} u^2 2\phi(u)[1-\Phi(u)] \, du$$

$$= \int_{-\infty}^{\infty} u^2\phi(u) \, du + 2\int_{-\infty}^{\infty} u^2\phi(u)\left[\tfrac{1}{2}-\Phi(u)\right] \, du = 1,$$

since the first integral is 1 (the population variance) and the second integral is zero (the integrand is an odd function). Then

$$\text{Var}(U_1) = EU_1^2 - (EU_1)^2 = 1 - (1/\pi) \approx 0.682.$$

By symmetry, $\text{Var}(U_2) = \text{Var}(U_1)$.

Exponential.   For an exponential distribution with mean θ,

$$EY_i = \theta\left(\frac{1}{n} + \frac{1}{n-1} + \cdots + \frac{1}{n-i+1} \right),$$

$$\text{Var}(Y_i) = \theta^2\left(\frac{1}{n^2} + \frac{1}{(n-1)^2} + \cdots + \frac{1}{(n-i+1)^2} \right). \tag{5.10}$$

Each sum has i terms. Sarhan and Greenberg (1962) tabulate these moments.

Normal. For a normal distribution with mean μ and standard deviation σ, (5.8) gives the mean and variance of Y_i in terms of those of the standardized U_i. David (1970) lists tables of EU_i and $\text{Var}(U_i)$.

Extreme value. For an extreme value distribution with location and scale parameters λ and δ, (5.8) gives the mean and variance of Y_i in terms of those of the standardized U_i. David (1970) lists tables of EU_i and $\text{Var}(U_i)$.

Other distributions. David (1970) and Sarhan and Greenberg (1962) give distributions and moments of order statistics from a number of other distributions, including the logistic, gamma, rectangular, and right triangular distributions.

Distribution of $P_i = F(Y_i)$. Suppose that Y_i is the ith order statistic of a sample of size n from a continuous cumulative distribution $F(y)$. The distribution of $P_i = F(Y_i)$, the random proportion of the distribution below Y_i, is used for probability plotting positions and nonparametric statistical methods. The cumulative distribution of P_i is

$$V_i(p) = P\{F(Y_i) \leqslant p\} = \sum_{k=i}^{n} \binom{n}{k} p^k (1-p)^{n-k}. \qquad (5.11)$$

This binomial probability comes from the fact that there are n independent observations, each of which falls below the $100p$th percentile $F^{-1}(p)$ with probability p. Then Y_i falls below $F^{-1}(p)$ if i or more observations fall below $F^{-1}(p)$. (5.11) is the binomial probability of this event.

For $n=1$, $P=F(Y)$ has a standard $(0,1)$ uniform distribution, and $V(p) = p, 0 \leqslant p \leqslant 1$. Thus, for any n, the P_i are the order statistics of a sample from a uniform distribution. Suppose that Y_i is used as a one-sided lower confidence limit for the $100p$th percentile of the parent distribution. Then (5.11) gives the confidence level of the interval.

The probability density of P_i is

$$v_i(p) = \frac{n!}{(i-1)!1!(n-i)!} p^{i-1}(1-p)^{n-i}, \qquad 0 \leqslant p \leqslant 1, \qquad (5.12)$$

a beta distribution. This comes from differentiating (5.11) with respect to p or from (5.2) evaluated for a uniform parent distribution with density $v(p) = 1, 0 \leqslant p \leqslant 1$.

The mean and variance of P_i are

$$EP_i = i/(n+1), \qquad \text{Var}(P_i) = i(n-i+1)/\left[(n+1)^2(n+2)\right]. \quad (5.13)$$

The covariance of P_i and P_j $(i<j)$ is

$$\text{Cov}(P_i, P_j) = i(n-j+1)/\left[(n+1)^2(n+2)\right]. \qquad (5.14)$$

The joint distribution of P_i and P_j is given by (5.17), where the parent distribution is uniform. Suppose Y_i and Y_j are used as two-sided confidence limits for the $100p$th percentile of the parent distribution. Then the joint distribution of P_i and P_j can be used to obtain the confidence level of the interval, given by (7.10) of Chapter 6.

Probability plotting positions. The mean $EP_i = i/(n+1)$ is a commonly used plotting position for Y_i on probability paper. Another one is the distribution median $\bar{p}_i$, that is, the solution of $V_i(\bar{p}_i) = 0.50$. Johnson (1964) tabulates such $\bar{p}_i$.

Exponential order statistics. Certain properties of order statistics from an exponential distribution are useful. Suppose $Y_1 \leqslant \cdots \leqslant Y_n$ are the order statistics from an exponential distribution with mean θ. The differences $D_i = Y_i - Y_{i-1}$ $(i=1,\ldots,n$ and $Y_0 \equiv 0)$ are all statistically independent. Also, D_i has an exponential distribution with mean $\theta/(n-i+1)$. A motivation for this follows. At the time Y_{i-1} of the $(i-1)$th failure, the $(n-i+1)$ running units each have a conditional exponential distribution of further time to failure with mean θ. It follows that the earliest further time to failure, $D_i = Y_i - Y_{i-1}$, has an exponential distribution with mean $\theta/(n-i+1)$. Each D_i is independent of $D_1,\ldots,D_{i-1}$, since the distribution of D_i does not depend on them. So all the D_i are independent of each other.

Hazard plotting positions. The properties of exponential order statistics provide a basis for hazard plotting positions. The following proof for complete samples extends to multiply censored samples (Nelson, 1972b). The distribution of $P = F(Y)$ for a single observation is uniform on $[0,1]$. Then one can show that the distribution of $H = -\ln[1 - F(Y)]$ is standard exponential (with mean 1). So the corresponding cumulative hazard values $H_i = -\ln[1 - F(Y_i)]$ are order statistics from a standard exponential distribution. By (5.10), $EH_i = (1/n) + [1/(n-1)] + \cdots + [1/(n-i+1)]$; this expected value is the cumulative hazard plotting position for Y_i.

Asymptotic distributions. The following material presents simple approximations to the distributions and moments of order statistics from large samples. These approximations are useful for obtaining approximate confidence intervals and approximate best linear unbiased estimators. Suppose a large sample of size n comes from a continuous cumulative distribution $F(y)$.

The cumulative distribution function $G_i(y)$ of the ith order statistic Y_i is close to a normal one with mean EY_i and variance $\text{Var}(Y_i)$ when i and $n-i$ are both large. Also, then approximations for the mean and variance by propagation of error (Hahn and Shapiro, 1967, Chap. 7) are

$$EY_i \simeq F^{-1}[i/(n+1)], \qquad \text{Var}(Y_i) \simeq [f(EY_i)]^{-2} i(n-i)/n^3, \quad (5.15)$$

where $f(y)$ is the probability density of the parent distribution.

For example, consider the ith order statistic from an exponential cumulative distribution $F(y) = 1 - \exp(-y/\theta)$. Here $F^{-1}(P) = -\theta \ln(1-P)$ and $f(y) = (1/\theta)\exp(-y/\theta)$. So, using $n \simeq n+1$,

$$EY_i \simeq -\theta \ln[(n-i+1)/(n+1)], \tag{5.16}$$

$$\text{Var}(Y_i) \simeq \{\tfrac{1}{\theta}\exp[\theta \ln((n-i)/n)/\theta]\}^{-2} i(n-i)/n^3 = \theta^2 i/[(n-i)n].$$

When i and $n-i$ are large, the differences $D_i = Y_i - Y_{i-1}$ of adjoining order statistics are approximately statistically independent. Also, then the distribution of D_i is approximately exponential with mean $EY_i - EY_{i-1} \approx 1/f(EY_{i-1})$. For example, consider order statistics from an exponential distribution with mean θ. Then D_i has an approximate exponential distribution with mean $1/\{(1/\theta)\exp[\theta \ln[(n-i)/(n+1)]/\theta]\} = \theta(n+1)/(n-i)$. The properties of the D_i provide approximate confidence limits (Chapter 7) for a distribution scale parameter as shown by Mann, Schafer, and Singpurwalla (1974, Sec. 5.2.3c1).

When n is large, the extreme order statistics Y_1 and Y_n have one of the three types of extreme value distributions: (1) the smallest (largest) extreme value distribution for certain parent distributions with an infinitely long tail on the left (right), (2) the Weibull (or exponential) distribution when the parent distribution is bounded on the left (right), and (3) the Cauchy-type extreme value distribution for certain parent distributions with an infinitely long tail on the left (right). Gumbel (1958) and Galambos (1978) describe these distributions and asymptotic theory in detail.

5.2. Joint Distribution of a Pair of Order Statistics

Suppose that random samples of size n are repeatedly taken from a continuous cumulative distribution $F(y)$. From sample to sample, the ith and jth order statistics Y_i and Y_j have different values. So Y_i and Y_j have a joint distribution. This distribution depends on $F(y)$ and n, i, and j; it is given below. Later, joint moments are used to justify best linear unbiased estimates of location and scale parameters.

Joint density.   This paragraph derives the joint density of Y_i and Y_j, $Y_i \leqslant Y_j$. Suppose the parent distribution has a probability density $f(y)$. Then the joint density is

$$g_{ij}(y_i, y_j) = \frac{n!}{(i-1)!1!(j-i-1)!1!(n-j)!} [F(y_i)]^{i-1} f(y_i)$$

$$[F(y_j) - F(y_i)]^{j-i-1} f(y_j)[1 - F(y_j)]^{n-j}, \qquad y_i \leqslant y_j. \quad (5.17)$$

Figure 5.3 motivates (5.17). The event that Y_i and Y_j are in the infinitesimal intervals from y_i to $y_i + dy_i$ and from y_j to $y_j + dy_j$ has approximate probability $g_{ij}(y_i, y_j) \, dy_i \, dy_j$. Figure 5.3 shows that this event is equivalent to the event that $i-1$ observations fall below y_i, one falls between y_i and $y_i + dy_i$, $j-i-1$ fall between $y_i + dy_i$ and y_j, one falls between y_j and $y_j + dy_j$, and $n-j$ fall above $y_j + dy_j$. Each of the n independent observations can fall into one of those five intervals, with approximate probabilities $F(y_i), f(y_i) \, dy_i$, $F(y_j) - F(y_i), f(y_j) \, dy_j$, and $1 - F(y_j)$. The multinomial probability of the event is approximately

$$g_{ij}(y_i, y_j) \, dy_i \, dy_j \simeq \frac{n!}{(i-1)!1!(j-i-1)!1!(n-j)!} [F(y_i)]^{i-1} [f(y_i) \, dy_i]^1$$

$$\times [F(y_j) - F(y_i)]^{j-i-1} [f(y_j) \, dy_j]^1 [1 - F(y_j)]^{n-j}.$$

This motivates (5.17).

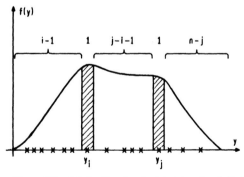

Figure 5.3.   Motivation for the joint density of Y_i, Y_j.

Joint moments. Joint moments of order statistics are defined like any other joint moments. In particular, for $i<j$, the covariance of Y_i and Y_j is

$$\mathrm{Cov}(Y_i, Y_j) = \int_{-\infty}^{\infty} \int_{-\infty}^{y_j} (y_i - EY_i)(y_j - EY_j) g_{ij}(y_i, y_j)\, dy_i\, dy_j. \quad (5.18)$$

These covariances are used to obtain best linear unbiased estimates. Suppose that the parent distribution has location and scale parameters μ and σ. Let $U_i = (Y_i - \mu)/\sigma$ denote the standardized ith order statistic; it comes from the standardized parent distribution with $\mu = 0$ and $\sigma = 1$. Then

$$\mathrm{Cov}(Y_i, Y_j) = \sigma^2 \mathrm{Cov}(U_i, U_j). \quad (5.19)$$

This gives any covariance in terms of a standardized one. For a symmetric distribution, such as the normal and logistic distributions,

$$\mathrm{Cov}(U_{n-i+1}, U_{n-j+1}) = -\mathrm{Cov}(U_i, U_j). \quad (5.20)$$

The standardized covariances depend on n, i, j, and the parent distribution. They are tabulated for various distributions as described below.

Example. This example shows the calculation of the covariance of the order statistics of a sample of $n=2$ from a standard normal distribution. First,

$$E(U_1 U_2) = \int_{-\infty}^{\infty} \int_{-\infty}^{u_2} u_1 u_2 2\phi(u_1)\phi(u_2)\, du_1\, du_2$$

$$= 2\int_{-\infty}^{\infty} u_2 \left[-\phi(u_1) \right]_{-\infty}^{u_2} \phi(u_2)\, du_2$$

$$= -2\int_{-\infty}^{\infty} u_2 \left[\phi(u_2) \right]^2 du_2 = 0,$$

since the last integrand is an odd function. Then

$$\mathrm{Cov}(U_1, U_2) = E(U_1 U_2) - (EU_1)(EU_2) = 0 - \left[-(1/\pi)^{1/2} \right](1/\pi)^{1/2} = 1/\pi.$$

Exponential. For an exponential distribution with mean θ,

$$\mathrm{Cov}(Y_i, Y_j) = \theta^2 \left(\frac{1}{n^2} + \frac{1}{(n-1)^2} + \cdots + \frac{1}{(n-i+1)^2} \right), \quad i<j. \quad (5.21)$$

Sarhan and Greenberg (1962) tabulate the standardized covariances ($\theta = 1$).

Normal. For a normal distribution with mean μ and standard deviation σ, (5.19) gives the covariance of Y_i and Y_j in terms of those of the standardized U_i and U_j. David (1970) lists tables of $\text{Cov}(U_i, U_j)$.

Extreme value. For an extreme value distribution with location and scale parameters λ and δ, (5.19) gives the covariance of Y_i and Y_j in terms of those of the standardized U_i and U_j. David (1970) lists tables of $\text{Cov}(U_i, U_j)$.

Other distributions. David (1970) and Sarhan and Greenberg (1962) give covariances of order statistics from a number of other distributions, including the logistic, gamma, rectangular, and right triangular distributions.

5.3. Joint Distribution of Any Order Statistics

Suppose that random samples of size n are repeatedly taken from a continuous cumulative distribution $F(y)$. From sample to sample, K chosen order statistics $Y_{i_1} \leqslant Y_{i_2} \leqslant \cdots \leqslant Y_{i_K}$ have different values. So the chosen order statistics have a joint distribution.

Joint density. The joint probability density of K chosen order statistics, $y_{i_1} \leqslant \cdots \leqslant y_{i_K}$ is

$$g_{i_1 \cdots i_K}(y_{i_1}, \ldots, y_{i_K}) = n! \left(\prod_{k=1}^{K} \frac{1}{(i_k - i_{k-1})!} \left[F(y_{i_k}) - F(y_{i_{k-1}}) \right]^{i_k - i_{k-1} - 1} \right.$$

$$\left. \times f(y_{i_k}) \right) \left[1 - F(y_{i_K}) \right]^{n - i_K} \tag{5.22}$$

for $y_{i_1} \leqslant \cdots \leqslant y_{i_K}$, and 0 elsewhere. In particular, the joint density of all n order statistics is

$$g_{1 \cdots n}(y_1, \ldots, y_n) = n! f(y_1) f(y_2) \cdots f(y_n), \tag{5.23}$$

where $y_1 \leqslant \cdots \leqslant y_n$, and the density is 0 elsewhere.

Asymptotic distribution. The following material presents a simple approximation to the joint distribution and moments of any number of order statistics from large samples. The approximation is useful for obtaining approximate confidence intervals and approximate best linear unbiased estimators. Suppose a large sample of size n comes from a continuous cumulative distribution $F(y)$. The joint distribution of K order statistics $Y_{i_1} \leqslant \cdots \leqslant Y_{i_K}$ is approximately K-variate normal with means EY_{i_k}, variances $\text{Var}(Y_{i_k})$, and covariances $\text{Cov}(Y_{i_k}, Y_{i_k})$. This approximation holds if all $i_k - i_{k-1}$ are large; that is, the order statistics are well separated.

Approximate means and variances are given by (5.15). An approximate covariance is

$$\text{Cov}(Y_{i_k}, Y_{i_{k'}}) \simeq [f(EY_{i_k})f(EY_{i_{k'}})]^{-1}\frac{i_k}{n}\left(1 - \frac{i_{k'}}{n}\right)\frac{1}{n}, \qquad (5.24)$$

where $f(\ )$ is the probability density.

5.4. Best Linear Unbiased Estimators

The following presentation gives the best (minimum variance) linear unbiased estimators (BLUEs) from order statistics. The derivation involves a linear regression model for a distribution location parameter μ and scale parameter σ. Regression theory (the Gauss–Markov Theorem) then provides the best linear unbiased estimators for μ and σ.

The model. $Y_{i_1} \leqslant \cdots \leqslant Y_{i_K}$ are K selected order statistics of a sample of size n from a distribution with location and scale parameters μ and σ. In this chapter, they are usually the first r order statistics $Y_1, Y_2, \ldots, Y_r$, and the corresponding theory for their moments is simple and a special case of that in Section 5.3. They may be order statistics of a multiply (progressively) failure censored sample; the corresponding theory for moments of such order statistics is more complex and is not a special case of that in Section 5.3. Then, by (5.8),

$$EY_{i_k} = \mu + \alpha_{i_k}\sigma, \qquad (5.25)$$

where $\alpha_{i_k} = E[(Y_{i_k} - \mu)/\sigma]$ is the expectation of the i_kth standardized order statistic ($\mu = 0$ and $\sigma = 1$). Similarly, for $k, k' = 1, \ldots, K$, by (5.8) and (5.19)

$$\text{Var}(Y_{i_k}) = V_{i_k i_k}\sigma^2, \qquad \text{Cov}(Y_{i_k}, Y_{i_{k'}}) = V_{i_k i_{k'}}\sigma^2, \qquad (5.26)$$

where $V_{i_k i_k} = E[(Y_{i_k} - EY_{i_k})^2/\sigma^2]$ and $V_{i_k i_{k'}} = E[(Y_{i_k} - EY_{i_k})(Y_{i_{k'}} - EY_{i_{k'}})/\sigma^2]$ are the variance and covariance of the standardized order statistics ($\mu = 0$ and $\sigma = 1$).

The estimators. The preceding equations can be expressed in matrix form as a linear regression model. Let $\mathbf{Y} = (Y_{i_1}, \ldots, Y_{i_K})'$, $e_{i_k} = Y_{i_k} - EY_{i_k}$, $\mathbf{e} = (e_{i_1}, \ldots, e_{i_K})'$, and

$$\mathbf{X} = \begin{bmatrix} 1 & \alpha_{i_1} \\ \vdots & \vdots \\ 1 & \alpha_{i_K} \end{bmatrix}. \qquad (5.27)$$

Throughout prime ($'$) denotes transpose. Then the linear regression model based on (5.25) is

$$Y = X(\mu\,\sigma)' + e, \tag{5.28}$$

where the covariance matrix of e is $\Sigma = V\sigma^2$; and $V = \{V_{i_k i_k'}\}$, $k, k' = 1, \ldots, K$, is given by (5.7) and (5.18). The Gauss–Markov Theorem (Rao, 1973) gives the best (minimum variance) linear unbiased estimators μ^* and σ^* for μ and σ; namely,

$$[\mu^*\ \sigma^*]' = X'(X'V^{-1}X)^{-1}X'V^{-1}Y = \begin{bmatrix} a_1 & \cdots & a_K \\ b_1 & \cdots & b_K \end{bmatrix} Y, \tag{5.29}$$

where the a_k and b_k are the coefficients for μ^* and σ^*, respectively. The covariance matrix of $[\mu^*\ \sigma^*]'$ is

$$\begin{bmatrix} \text{Var}(\mu^*) & \text{Cov}(\mu^*, \sigma^*) \\ \text{Cov}(\mu^*, \sigma^*) & \text{Var}(\sigma^*) \end{bmatrix} = \sigma^2(X'V^{-1}X)^{-1} = \sigma^2 \begin{bmatrix} A & C \\ C & B \end{bmatrix}. \tag{5.30}$$

David (1970) references tables of the coefficients $a_1, \ldots, a_K$ and $b_1, \ldots, b_K$ and the variance factors A and B and the covariance factor C for various distributions, sample sizes, and selected order statistics. The coefficients satisfy $a_1 + \cdots + a_K = 1$ and $b_1 + \cdots + b_K = 0$.

For samples with a large number of observed order statistics, μ^* and σ^* are approximately normally distributed, with means μ and σ and covariance matrix (5.30).

Example. This example presents the BLUEs for the mean μ and standard deviation σ of a normal distribution from a complete sample of size 2. As shown above,

$$X = \begin{bmatrix} 1 & -1/\pi^{1/2} \\ 1 & 1/\pi^{1/2} \end{bmatrix}, \qquad V = \begin{bmatrix} 1-(1/\pi) & 1/\pi \\ 1/\pi & 1-(1/\pi) \end{bmatrix}.$$

First

$$\begin{bmatrix} A & C \\ C & B \end{bmatrix} = \left[\begin{bmatrix} 1 & -1/\pi^{1/2} \\ 1 & 1/\pi^{1/2} \end{bmatrix}' \begin{bmatrix} 1-(1/\pi) & 1/\pi \\ 1/\pi & 1-(1/\pi) \end{bmatrix}^{-1} \begin{bmatrix} 1 & -1/\pi^{1/2} \\ 1 & 1/\pi^{1/2} \end{bmatrix} \right]^{-1}$$

$$= \begin{bmatrix} \frac{1}{2} & 0 \\ 0 & (\pi-2)/2 \end{bmatrix} = \begin{bmatrix} 0.500 & 0 \\ 0 & 0.571 \end{bmatrix}.$$

That is, $\text{Var}(\mu^*)=\sigma^2/2$, $\text{Var}(\sigma^*)=\sigma^2(\pi-2)/2\approx0.571\ \sigma^2$, and $\text{Cov}(\mu^*,\sigma^*)$ $=0$. Then

$$\begin{bmatrix} a_1 & a_2 \\ b_1 & b_2 \end{bmatrix}=\begin{bmatrix} 1 & -1/\pi^{1/2} \\ 1 & 1/\pi^{1/2} \end{bmatrix}\begin{bmatrix} \frac{1}{2} & 0 \\ 0 & (\pi-2)/2 \end{bmatrix}\begin{bmatrix} 1-(1/\pi) & 1/\pi \\ 1/\pi & 1-(1/\pi) \end{bmatrix}^{-1}$$

$$=\begin{bmatrix} \frac{1}{2} & \frac{1}{2} \\ -\pi^{1/2}/2 & \pi^{1/2}/2 \end{bmatrix}.$$

That is, $\mu^*=(Y_1+Y_2)/2$, the sample average (as expected), and $\sigma^*=(Y_2-Y_1)\pi^{1/2}/2\approx(Y_2-Y_1)0.886$.

Prediction. Kaminsky and Nelson (1975) and Goldberger (1962) give general theory for linear **prediction** of order statistics in a sample from earlier order statistics.

5.5. Other Linear Estimators

Many linear estimators have been proposed for various distributions, particularly the Weibull distribution. For example, Mann, Schafer, and Singpurwalla (1974, Chap. 4) and Bain (1978) present a number of such estimators for parameters of the Weibull distribution. Also, various authors have presented linear estimators based on least squares fitting of a straight line to a (possibly censored) sample plotted on paper of a distribution. For example, STATPAC of Nelson and others (1978) does this for exponential, Weibull, normal, lognormal, and extreme value hazard plots; Thomas and Wilson (1972) studied such estimates for the Weibull distribution and found that they compare favorably with BLUEs. BLIEs and pooled estimates are presented below.

BLIEs. Best linear invariant estimators (BLIEs) are presented here because available confidence limits for the Weibull distribution are given in terms of BLIEs. An estimator is called **invariant** if its mean square error does not depend on the location parameter μ. Such an estimator is "best" if it has the minimum mean squared error. Most best linear invariant estimators are biased. They are related to the best linear unbiased estimators as follows. Suppose that μ^* and σ^* are the BLUEs for μ and σ, where $\text{Var}(\mu^*)=A\sigma^2$, $\text{Var}(\sigma^*)=B\sigma^2$, and $\text{Cov}(\mu^*,\sigma^*)=C\sigma^2$, where A, B, and C depend on n and r but not on μ or σ. Then the BLIEs for μ and σ are

$$\mu^{**}=\mu^*-\sigma^*[C/(1+B)],\qquad \sigma^{**}=\sigma^*/(1+B).\qquad(5.31)$$

The mean squared errors of these estimators are

$$E(\mu^{**}-\mu)^2=\sigma^2\left[A-C^2(1+B)^{-1}\right], \qquad E(\sigma^{**}-\sigma)^2=\sigma^2B/(1+B),$$

$$E(\mu^{**}-\mu)(\sigma^{**}-\sigma)=\sigma^2C/(1+B). \tag{5.32}$$

Mann, Schafer, and Singpurwalla (1974) tabulate coefficients of BLIEs for the extreme value distribution; so these estimates can be calculated directly from the order statistics. For practical purposes, the BLUEs and BLIEs are almost the same unless the observed number r of order statistics is small. Moreover, for large r (and n), BLUEs and BLIEs and their mean squared errors are asymptotically equal. There is no compelling reason to choose either minimum mean squared error or unbiasedness as essential for a good estimator.

Pooled estimates.   Tables for linear estimators cover samples up to size 25. Larger samples can be handled as follows, and the same method is used to combine independent linear estimates. Randomly divide the sample into K smaller subsamples that can be handled with available tables. The subsamples should be as large as possible and nearly equal in size. Obtain the linear estimates of the distribution parameters from each subsample. Suppose the BLUEs are μ_k^* and σ_k^*, with respective variances $A_k\sigma^2$ and $B_k\sigma^2$ and covariance $C_k\sigma^2$ for $k=1,\ldots,K$. Then the best (minimum variance unbiased) pooled linear estimates are

$$\mu^*=\left[(\mu_1^*/A_1)+\cdots+(\mu_K^*/A_K)\right]A,$$

$$\sigma^*=\left[(\sigma_1^*/B_1)+\cdots+(\sigma_K^*/B_K)\right]B, \tag{5.33}$$

where

$$A=1/\left[(1/A_1)+\cdots+(1/A_K)\right], \qquad B=1/\left[(1/B_1)+\cdots+(1/B_K)\right]. \tag{5.34}$$

The variances and covariance of these estimates are

$$\text{Var}(\mu^*)=A\sigma^2, \qquad \text{Var}(\sigma^*)=B\sigma^2, \qquad \text{Cov}(\mu^*,\sigma^*)=C\sigma^2, \tag{5.35}$$

where

$$C=AB\left[C_1(A_1B_1)^{-1}+\cdots+C_K(A_KB_K)^{-1}\right]. \tag{5.36}$$

The same formulas are used to pool BLUEs from independent samples from different populations with common parameter values.

Formulas for combining BLIEs are more complex and are not given here.

Class B insulation example.   Section 3 presents life data in Table 3.1 on insulation assumed to have a lognormal life distribution at any temperature. The log standard deviation σ is assumed to be the same at all temperatures. The BLUEs of σ are $\sigma_1^* = 0.2265$ for 170°C, $\sigma_2^* = 0.4110$ for 190°C, and $\sigma_3^* = 0.0677$ for 220°C. Their variances are $Var(\sigma_1^*) = 0.0989 \, \sigma^2$ and $Var(\sigma_2^*) = Var(\sigma_3^*) = 0.1613 \, \sigma^2$. The pooled estimate of σ is obtained from

$$B = 1/\left[(1/0.0989) + (1/0.1613) + (1/0.1613)\right] = 0.04442,$$

$$\sigma^* = \left[(0.2265/0.0989) + (0.4110/0.1613) + (0.0677/0.1613)\right]0.04442$$

$$= 0.2336.$$

Then $Var(\sigma^*) = 0.04442 \, \sigma^2$.

The pooled estimates differ slightly when different subsamples are randomly chosen. Different people get slightly different estimates from the same data. McCool (1965) suggests appropriately averaging the estimates over all possible subsamples when the sample is complete. Attractive in principle, his method is unfortunately laborious in practice.

PROBLEMS

7.1.   Fluid breakdown.   Use the complete data on seconds (ln seconds) to insulating fluid breakdown at 35 kV in Tables 2.1 and 4.1.

(a)   Assuming that the life distribution is exponential, calculate the best linear unbiased estimate of the mean time to failure.

(b)   Calculate two-sided 95% confidence limits for the true mean.

(c)   Calculate the linear estimate of the 10% point of the exponential distribution.

(d)   Calculate two-sided 95% confidence limits for the 10% point.

(e)   Assuming that the life distribution is Weibull, calculate the best linear unbiased estimates of the parameters of the corresponding extreme value distribution.

(f)   Calculate two-sided exact 90% confidence limits for the extreme value location parameter λ.

(g)   Calculate two-sided exact 90% confidence limits for the extreme value scale parameter δ. Also, use the (Wilson–Hilferty) approximation.

(h) Calculate the corresponding estimates of the Weibull parameters.

(i) Calculate the corresponding confidence limits for the Weibull parameters.

(j) Calculate the linear estimate of the 10% point of the extreme value distribution.

(k) Calculate the theoretical standard error of the estimate of the 10% point in terms of δ.

(l) Calculate two-sided exact 90% confidence limits for the 10% point of the extreme value distribution.

(m) Calculate two-sided approximate 90% confidence limits for (l).

(n) Calculate the estimate and confidence limits for the corresponding Weibull 10% point.

(o) Plot the data on Weibull probability paper.

7.2. Pooled fluid breakdown.   Use the 30- and 35-kV samples in Table 4.1 and the results of Problem 7.1. Assume the true Weibull shape parameter β is the same at both voltages.

(a) Using the two estimates of the extreme value scale parameter, calculate a pooled estimate of the common true value.

(b) Give the theoretical variance of this estimate in terms of the unknown true value δ.

(c) Calculate the corresponding estimate of the common Weibull shape parameter.

(d) Calculate two-sided approximate 90% confidence limits for δ, using the (Wilson–Hilferty) approximation.

(e) Calculate the corresponding limits for the common Weibull shape parameter.

7.3. Insulating fluid.   Use the first nine times to breakdown of the 30-kV data of Table 2.1.

(a) Calculate an upper 90% prediction limit for the 10th failure time, assuming that time to breakdown has an exponential distribution.

(b) Do (a) assuming that time to breakdown has a Weibull distribution.

(c) Plot the 10 times to breakdown on Weibull or exponential probability paper.

(d) Comment on an appropriate analysis of the 30-kV data.

7.4. Class B insulation.   Use Table 3.1 and do the corresponding calculations of all estimates and confidence limits appearing in Section 3 for the following.

(a) The 190°C data.

(b) The 220°C data.

Do corresponding calculations of all predictions and prediction limits appearing in Section 3 for the following.

(c) The 190°C data.

(d) The 220°C data.

7.5. Class B insulation.   Use the data in Table 3.1, but fit Weibull distributions.

(a) Plot samples from each temperature on the same Weibull probability paper.

(b) Use natural logs of the data and calculate the BLUEs of the corresponding extreme value parameters λ and δ for a temperature.

(c) Calculate the estimate of the 50th Weibull percentile for that temperature.

(d) Use the chi-square approximation to get approximate 95% confidence limits for the Weibull shape parameter at that temperature.

(e) Calculate a pooled estimate of the shape parameter using the BLUEs from the four temperatures.

(f) Use the chi-square approximation to get approximate 95% confidence limits for the common shape parameter. Is the failure rate significantly increasing or decreasing?

(g) From the Weibull plot, assess whether the true Weibull shape parameter has the same value at all temperatures.

7.6.* The two-parameter exponential.   The cumulative distribution of the two-parameter exponential distribution can be written as

$$F(y) = 1 - \exp\left[-(y-\mu)/\sigma\right], \qquad y \geq \mu,$$

where μ is a location parameter and σ is a scale parameter. Derive the best linear estimates of μ and σ using the first r order statistics $Y_1, \ldots, Y_r$ from a sample of size n as follows.

*Asterisk denotes laborious or difficult.

(a) Give a formula for the expectation of Y_i in terms of that for a standard exponential distribution.

(b) Give a formula for the variance of Y_i in terms of that for a standard exponential distribution.

(c) Give a formula for the covariance of Y_i and Y_j in terms of that for a standard exponential distribution.

For $n=3$ and $r=2$:

(d) Numerically calculate the coefficients for the BLUEs μ^* and σ^*.

(e) Numerically calculate the factors for $Var(\mu^*)$, $Var(\sigma^*)$, and $Cov(\mu^*, \sigma^*)$.

7.7. Insulating fluid. The time to breakdown data in Table 6.1 of Chapter 6 yield the seven sets of linear estimates and variance and covariance factors below for parameters of an extreme value distribution for the ln data. According to theory for such data, time to breakdown has an exponential distribution at any test voltage (kV). Assess this with the following, assuming that the seven distributions are Weibull with a common shape parameter.

kV	n	λ^*	δ^*	A	B	C
26	3	7.125	2.345	0.40286	0.34471	−0.02477
28	5	5.857	1.224	0.23140	0.16665	−0.03399
30	11	4.373	0.987	0.10251	0.06417	−0.02033
32	15	3.310	1.898	0.07481	0.04534	−0.01556
34	19	2.531	1.353	0.05890	0.03502	−0.01256
36	15	1.473	1.154	0.07481	0.04534	−0.01556
38	8	0.0542	0.836	0.14198	0.09292	−0.02608

(a) Calculate the variance factor B for the pooled linear estimate δ^* for the extreme value scale parameter.

(b) Calculate the pooled linear estimate δ^* and the corresponding estimate of the common Weibull shape parameter β.

(c) Calculate two-sided 95% confidence limits for δ and β, using the normal approximation.

(d) Do (c), using the Wilson–Hilferty chi–square approximation. Compare with (c).

(e) Are the data (the pooled estimate) consistent with $\beta=1$ (exponential distribution)? If test conditions at a test voltage are not consistent, the β estimate tends to be lower. Are the data consistent with this possibility?

8

Maximum Likelihood Analyses of Multiply Censored Data

Introduction

Purpose. This advanced chapter shows how to use maximum likelihood (ML) methods to estimate distributions from **multiply censored data**. Such data consist of intermixed failure and running times and are also called "progressively," "hyper-," and "arbitrarily censored." The methods apply to multiply **time** censored data (Type I); such data are common in practice and are treated in detail here. The methods also apply to multiply **failure** censored data (Type II), to Types I and II singly censored data, and to complete data. This chapter is applied, but it is not easy.

Properties of ML methods. While difficult without sophisticated computer programs, ML methods are very important in life data analysis and elsewhere because they are very versatile. That is, they apply to most theoretical distributions and kinds of censored data. Also, there are computer programs that do ML calculations. Moreover, most ML estimators have good statistical properties. For example, under certain conditions (usually met in practice) on the distribution and data and for "large" sample sizes, the cumulative distribution function of a ML estimator is close to a normal one whose mean equals the quantity being estimated and whose variance is no greater than that of any other estimator. A ML estimator also usually has good properties for small samples.

The method. In principle, the ML method is simple. One first writes the sample likelihood (or its logarithm, the log likelihood). It is a function of the assumed distribution, the distribution parameters, and the data (including the censoring or other form of the data). The ML estimates of the parameters are the parameter values that maximize the sample likelihood or, equivalently, the log likelihood. The exact distributions of many ML estimators and confidence limits are not known. However, they are given approximately by the asymptotic (large-sample) theory, which involves the asymptotic covariance and Fisher information matrices of the ML estimates. The asymptotic theory is mathematically and conceptually advanced.

Asymptotic theory.   This chapter presents the asymptotic (large-sample) theory for ML estimators and confidence limits. For small samples, such intervals tend to be narrower than exact ones. Exact intervals from small samples are referenced; they have been developed for few distributions and only single Type II (failure) censoring. For the asymptotic theory to be a good approximation, the number of failures in the sample should be **large**. How large depends on the distribution, what is being estimated, the confidence level of limits, etc. For practical purposes, the asymptotic methods are applied to small samples, since crude theory is better than no theory. Shenton and Bowman (1977) give theory for higher-order terms for greater accuracy of the asymptotic theory. The theory distinguishes between three values of a parameter: the true population value θ_0, its ML estimate $\hat{\theta}$, and an arbitrary value θ.

Overview.   Chapter sections present ML methods for estimating distributions: exponential (Section 1), normal and lognormal (Section 2), and Weibull and extreme value (Section 3). The section for each distribution explains how to calculate the sample likelihood, the likelihood equations, the ML estimates of distribution parameters, their Fisher information and covariance matrices and confidence limits, and ML estimates and confidence limits for percentiles and reliabilities. Section 1 is much easier than the rest. Section 4 presents ML methods for data with competing failure modes. Section 5 presents general theory for ML methods, which apply to estimating other quantities and to other distributions. Section 6 describes numerical methods for ML calculations. Readers may need to write their own programs that do such calculations.

Those who have computer programs for the ML calculations may wish to read only Sections 1 through 3 or 4. Those who wish to understand the theory need to read Section 5, while reading earlier sections. Those who wish to write computer programs need to read Section 6. Such computer programs are essential for data analysis, as the computations are complicated and laborious.

Needed background. Needed background for this chapter includes (1) basic knowledge of the life distributions described in Chapter 2, (2) basic statistical concepts on estimates and confidence limits in Chapter 6, and for advanced material (3) partial differentiation and simple matrix algebra. Those who wish only to use computer programs for ML estimation do not need item (3).

Further methods. The bibliographies of Buckland (1964), Mendenhall (1958), and Govindarajulu (1964) list many references on ML theory for such data and various distributions. Cohen (1963, 1965, 1966), for example, presents the basic ML methods for the distributions in this chapter. Also, Gross and Clark (1975) and Mann, Schafer, and Singpurwalla (1974) present ML fitting of the gamma distribution to censored data. Prentice (1974) presents ML fitting of a three-parameter log gamma distribution that includes the Weibull (and extreme value) and lognormal (and normal) distributions as special cases; it can help one decide between those distributions. The CENSOR program of Meeker and Duke (1979) does ML fitting of the logistic and log-logistic distributions to multiply censored data. Kaplan and Meier (1958) give a nonparametric ML estimate (product-limit estimate) of the reliability function from multiply censored data; Kalbfleisch and Prentice (1980) and Gross and Clark (1975) present it in detail and give other nonparametric methods for such data. The BMDP routine PL1 of Dixon and Brown (1977) calculates the product-limit estimate and its standard error at each failure time.

Multiply censored data can also be graphically analyzed with probability or hazard plots (Chapter 4). In practice, both graphical and ML methods can be employed to extract maximum information from such data. Each method yields information not provided by the other. For example, a plot helps one to assess the validity of the assumed distribution and of the data, and ML analysis provides confidence limits and objective estimates. Chapters 9 and 12 present further ML methods.

Basic assumption. Like other methods (e.g., Chapter 4) for analysis of multiply censored data, ML methods depend on a basic assumption. It is assumed that units censored at any specific time come from the same life distribution as the units that run beyond that time. This assumption does not hold, for example, if units are removed from service unfailed when they look like they are about to fail. Lagakos (1979) discusses in detail this assumption, which he calls noninformative censoring, and alternative assumptions about the censoring.

Notation. The following notation is used throughout this chapter. In particular, note that this chapter does not use capital letters for random variables, following the custom for ML notation.

z_γ	100γth standard normal percentile.
K_γ	$[100(1 + \gamma)/2]$th standard normal percentile.
log	Base 10 logarithm.
ln	Base e logarithm.
n	Sample size.
r	Number of failure times in a sample.
$\Sigma, \Sigma', \Sigma''$	Sums that run over all, failed, and unfailed units, respectively.
Var( )	True asymptotic variance.
Vâr( )	ML estimate of Var( ).
var( )	Local estimate of Var( ).
$\theta_0, \theta, \hat{\theta}$	True, arbitrary, and ML values of a parameter, particularly the exponential mean.
$\mu_0, \mu, \hat{\mu}$	True, arbitrary, and ML values of the normal mean or lognormal μ parameter.
$\sigma_0, \sigma, \hat{\sigma}$	True, arbitrary, and ML values of the normal standard deviation or lognormal σ parameter.
$\alpha_0, \alpha, \hat{\alpha}$	True, arbitrary, and ML values of the Weibull scale parameter.
$\beta_0, \beta, \hat{\beta}$	True, arbitrary, and ML values of the Weibull shape parameter.
$y_1, y_2, \ldots$	Failure and running times, usually taken to be fixed given data values rather than random variables (which are denoted by capital letters in other chapters).

1. EXPONENTIAL DISTRIBUTION

This section presents maximum likelihood (ML) methods for fitting an exponential distribution to multiply censored data. These methods also apply to multiply time and failure censored data, to singly censored data of both types, and to complete data. The section presents

1. Log likelihood.
2. ML estimate $\hat{\theta}$ of the true mean θ_0.
3. Asymptotic variance of $\hat{\theta}$.
4. Confidence limits for θ_0.

5. Estimates and confidence limits for the failure rate $\dot{\lambda}_0 = 1/\theta_0$, percentiles, and reliability.

6. Computer programs for ML fitting.

Cohen (1963), Mann, Schafer, and Singpurwalla (1974), Bain (1978), Lawless (1982), and others give such results. General ML theory and motivation appear in Sections 5.1 and 5.2.

Here y_i denotes the failure or running time on sample unit i, n denotes the number of units in the sample, and r denotes the number of failures. Below, θ_0 denotes the true population mean, $\hat{\theta}$ its ML estimate, and θ an arbitrary value. The exponential distribution is described in Chapter 2.

Exponential Mean θ_0

Log likelihood. The sample log likelihood for multiply censored data from an exponential distribution is (as shown in Sections 5.1 and 5.2)

$$\mathcal{L}(\theta) = \sum_i{}' [-\ln(\theta) - (y_i/\theta)] + \sum_i{}'' (-y_i/\theta),$$

where the first sum runs over the failure times, and the second sum runs over the running times. For a given set of sample times, $\mathcal{L}(\theta)$ is regarded as a function of θ, an arbitrary value of the parameter over the allowed range $0 < \theta < \infty$. The θ value that maximizes $\mathcal{L}(\theta)$ for a sample is the sample ML estimate $\hat{\theta}$ of the true θ_0. The following results are derived in Sections 5.1 and 5.2.

ML estimate. For both time and failure censored samples, the ML estimate $\hat{\theta}$ for θ_0 is

$$\hat{\theta} = \sum_i y_i/r. \tag{1.1}$$

This is the total time on all n units divided by the number r of failures. For a **failure** censored sample, $\hat{\theta}$ equals the BLUE θ^* of Chapter 7; thus, results of Chapter 7 apply to $\hat{\theta}$. For a **time** censored sample, Bartholomew (1963) derives the exact distribution of $\hat{\theta}$. For large r (say, over 15), the asymptotic cumulative distribution function of $\hat{\theta}$ is close to a normal one with mean θ_0 and variance (1.2) for time censoring and variance (1.5) for failure censoring.

Engine fan example. Table 1.1 shows life data on $n = 70$ diesel engine fans that accumulated 344,440 hours in service while $r = 12$ failed. Management wanted an estimate and confidence limits (given later) for the fraction

Table 1.1. Fan Failure Data (Hours)

450	1850 +	2200 +	3750 +	4300 +	6100 +	7800 +	8750
460 +	1850 +	3000 +	4150 +	4600	6100	7800 +	8750 +
1150	1850 +	3000 +	4150 +	4850 +	6100 +	8100 +	9400 +
1150	2030 +	3000 +	4150 +	4850 +	6100 +	8100 +	9900 +
1560 +	2030 +	3000 +	4150 +	4850 +	6300 +	8200 +	10100 +
1600	2030 +	3100	4300 +	4850 +	6450 +	8500 +	10100 +
1660 +	2070	3200 +	4300 +	5000 +	6450 +	8500 +	10100 +
1850 +	2070	3450	4300 +	5000 +	6700 +	8500 +	11500 +
1850 +	2080	3750 +		5000 +	7450 +	8750 +	

+ Denotes running time.

of such fans failing on an 8000-hour warranty. This information was to be used to determine whether unfailed fans should be replaced with a new design. The ML estimate of θ_0 is $\hat{\theta} = 344,440/12 = 28,700$ hours. Figure 1.1 is a Weibull hazard plot of the data. The ML fit of the exponential distribution is the center straight line on the plot. The other two straight lines are 95% confidence limits for percentiles (and reliabilities), as described later. This plot conveniently presents the data, fitted distribution, and confidence limits.

Variance. The (asymptotic) variance of $\hat{\theta}$ is needed later to obtain approximate confidence limits for θ_0, $\lambda_0 = 1/\theta_0$, percentiles, and reliabilities. For a **time** censored sample, the true theoretical asymptotic variance of $\hat{\theta}$ is

$$\text{Var}(\hat{\theta}) = \theta_0^2 \Big/ \sum_i [1 - \exp(-\eta_i/\theta_0)], \tag{1.2}$$

where η_i is the planned censoring time for unit i. The **ML estimate** of $\text{Var}(\hat{\theta})$ is

$$\text{V\^ar}(\hat{\theta}) = \hat{\theta}^2 \Big/ \sum_i [1 - \exp(-\eta_i/\hat{\theta})]; \tag{1.3}$$

that is, $\hat{\theta}$ replaces θ_0 in (1.2). This estimate requires a planned censoring time η_i for each sample unit. These times may not be known. However, they are not needed for the **local estimate**, which is

$$\text{var}(\hat{\theta}) = \hat{\theta}^2/r. \tag{1.4}$$

Where available, $\text{V\^ar}(\hat{\theta})$ is generally thought to be more accurate than $\text{var}(\hat{\theta})$. Var, V\^ar, and var respectively denote the true value, ML estimate, and local estimate of the asymptotic variance.

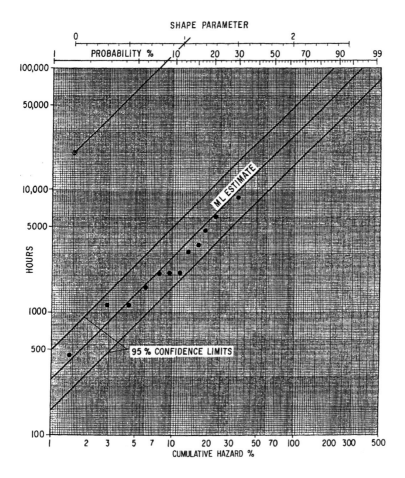

Figure 1.1. Weibull hazard plot of fan data and exponential fit.

For a **failure** censored sample, the true asymptotic variance of $\hat{\theta}$ is

$$\mathrm{Var}(\hat{\theta}) = \theta_0^2/r. \qquad (1.5)$$

This is also the exact variance for any sample size. Its ML (and local) estimate is

$$\mathrm{Var}(\hat{\theta}) = \mathrm{V\hat{a}r}(\hat{\theta}) = \hat{\theta}^2/r. \qquad (1.6)$$

The square root of (1.2) or (1.5) is the true asymptotic **standard error** of $\hat{\theta}$, that is, the standard deviation of the asymptotic distribution of $\hat{\theta}$.

Approximate limits. Approximate two-sided $100\gamma\%$ confidence limits for θ_0 are

$$\underset{\sim}{\theta} = \hat{\theta}/\exp\left\{K_\gamma\left[\text{Var}(\hat{\theta})\right]^{1/2}/\hat{\theta}\right\}, \quad \tilde{\theta} = \hat{\theta}\cdot\exp\left\{K_\gamma\left[\text{Var}(\hat{\theta})\right]^{1/2}/\hat{\theta}\right\}, \quad (1.7)$$

where $\text{Var}(\hat{\theta})$ is estimated with (1.3), (1.4), or (1.6). For a one-sided $100\gamma\%$ confidence limit replace K_γ in (1.7) with z_γ. The limits are more accurate the larger r is. The approximate limits (1.7) apply to time and failure censored data. For **singly** time censored data, Bartholomew (1963) gives complex exact limits.

Engine fan example. For the fans, the local estimate is $\text{var}(\hat{\theta}) = \hat{\theta}^2/12$. Approximate 95% confidence limits for θ_0 are $\underset{\sim}{\theta} \cong 28,700/\exp[1.960(\hat{\theta}^2/12)^{1/2}/\hat{\theta}] = 16,300$ hours and $\tilde{\theta} \cong 28,700(1.761) = 50,500$ hours. Each limit is a one-sided 97.5% confidence limit.

χ^2 **limits.** For multiply **failure** censored data, exact two-sided $100\gamma\%$ confidence limits for θ_0 are

$$\underset{\sim}{\theta} = \hat{\theta}2r/\chi^2\left[(1+\gamma)/2; 2r\right], \quad \tilde{\theta} = \hat{\theta}2r/\chi^2\left[(1-\gamma)/2; 2r\right], \quad (1.8)$$

where $\chi^2(\delta; 2r)$ is the 100δth chi-square percentile with $2r$ degrees of freedom. For time censoring, one can use (1.8) as an approximation, which may be more accurate than (1.7), particularly for small r (say, less than 15). For a one-sided $100\gamma\%$ confidence limit, replace $(1+\gamma)/2$ by γ or $(1-\gamma)/2$ by $1-\gamma$ in (1.8).

Engine fan. Approximate two-sided 95% confidence limits (1.8) for the time censored fan data are

$$\underset{\sim}{\theta} \cong 28,700\left[2\times12/\chi^2(0.975; 2\times12)\right] = 28,700(24/39.36) = 17,500,$$

$$\tilde{\theta} \cong 28,700\left[2\times12/\chi^2(0.025; 2\times12)\right] = 28,700(24/12.40) = 55,500.$$

Samples with no failures. For some samples, there are no failures when the data must be analyzed (time censored). Then (2.1) does not yield an estimate, and (2.3) does not yield confidence limits. A commonly used **incorrect** one-sided lower $100\gamma\%$ confidence limit for θ_0 is

$$\underset{\sim}{\theta} = 2(y_1 + \cdots + y_n)/\chi^2(\gamma; 2) = -(y_1 + \cdots + y_n)/\ln(1-\gamma).$$

This is incorrectly used to obtain one-sided limits for the failure rate, percentiles, and reliabilities. For an estimate, some use a 50% confidence limit; this method of estimation has no theoretical basis and can be seriously misleading. The confidence limit has a difficulty. As stated, it applies only when there are no failures. The limit must also be defined when there is one failure or two or three, etc.

Sample size. One may wish to choose a sample size to achieve a confidence interval of a desired length for θ_0. One measure of length is the ratio $\tilde{\theta}/\theta$, which is a function [(1.7) or (1.8)] only of the number of failures r, not the sample size n. The ratio can be calculated for a number of r values and a suitable r chosen.

Exponential Failure Rate λ_0

Estimate and confidence limits. The true failure rate is $\lambda_0 = 1/\theta_0$, and its ML estimate is

$$\hat{\lambda} = 1/\hat{\theta} = r \Big/ \sum_i y_i. \tag{1.9}$$

This is the number r of failures divided by the total accumulated time, the "sample failure rate." One- or two-sided $100\gamma\%$ confidence limits for λ_0 are

$$\underset{\sim}{\lambda} = 1/\tilde{\theta}, \qquad \tilde{\lambda} = 1/\underset{\sim}{\theta}, \tag{1.10}$$

where $\underset{\sim}{\theta}$ and $\tilde{\theta}$ are the one- or two-sided limits in (1.7) or (1.8).

Engine fan. For the fans, $\hat{\lambda} = 12/344,440 = 34.8$ failures per million hours. Approximate 95% confidence limits from (1.7) are $\underset{\sim}{\lambda} \approx 10^6/50,500 = 19.8$ and $\tilde{\lambda} \approx 10^6/16,300 = 61.3$ failures per million hours. Each limit is a one-sided 97.5% confidence limit.

Exponential Percentile

Estimate and confidence limits. The $100P$th exponential percentile is $y_P = -\theta_0 \ln(1-P)$, and its ML estimate is

$$\hat{y}_P = -\hat{\theta} \ln(1-P), \tag{1.11}$$

where (1.1) gives $\hat{\theta}$. One- or two-sided $100\gamma\%$ confidence limits for y_P are

$$\underset{\sim}{y}_P \approx -\underset{\sim}{\theta} \ln(1-P), \qquad \tilde{y}_P \approx -\tilde{\theta} \ln(1-P), \tag{1.12}$$

where $\underset{\sim}{\theta}$ and $\tilde{\theta}$ are the one- or two-sided limits in (1.7) or (1.8).

Engine fan. The ML estimate of the 10th percentile of the engine fan life is $\hat{y}_{.10} = -28,700 \ln(1 - 0.10) = 3020$ hours. Approximate 95% confidence limits from (1.7) are $y_{.10} \cong -16,300 \ln(1 - 0.10) = 1720$ and $\tilde{y}_{.10} \cong -50,500 \ln(1 - 0.10) = 5320$ hours. The ML line in Figure 1.1 gives the ML estimates of the exponential percentiles. On Weibull paper, the two-sided confidence limits for the exponential percentiles are given by straight lines parallel to the ML line.

Exponential Reliability

Estimate and confidence limits. The exponential reliability for an age y is $R(y) = \exp(-y/\theta_0)$, and its ML estimate is

$$\hat{R}(y) = \exp(-y/\hat{\theta}), \tag{1.13}$$

where (1.1) gives $\hat{\theta}$. One- or two-sided $100\gamma\%$ confidence limits for $R(y)$ are

$$\underset{\sim}{R}(y) = \exp(-y/\underset{\sim}{\theta}), \qquad \tilde{R}(y) = \exp(-y/\tilde{\theta}), \tag{1.14}$$

where $\tilde{\theta}$ and $\underset{\sim}{\theta}$ are the one- or two-sided limits in (1.7) or (1.8).

The corresponding estimate and limits for the fraction failing, $F(y) = 1 - R(y)$, are

$$\hat{F}(y) = 1 - \hat{R}(y), \qquad \underset{\sim}{F}(y) = 1 - \tilde{R}(y), \qquad \tilde{F}(y) = 1 - \underset{\sim}{R}(y). \tag{1.15}$$

Engine fan. The ML estimate of the fan reliability on an 8000-hour warranty is $\hat{R}(8000) = \exp(-8000/28,700) = 0.76$. Approximate 95% confidence limits from (1.7) are $\underset{\sim}{R}(8000) \cong \exp(-8000/16,300) = 0.61$ and $\tilde{R}(8000) \cong \exp(-8000/50,500) = 0.85$. The estimate and limits for the fraction failing are $\hat{F}(8000) = 0.24$, $\underset{\sim}{F}(8000) \cong 0.15$, and $\tilde{F}(8000) \cong 0.39$. Management used this information to decide whether to replace unfailed fans with an improved fan. In the Weibull plot in Figure 1.1, the straight-line confidence limits for the exponential percentiles are also the confidence limits for failure probabilities.

Exponential Prediction

Exponential prediction methods surveyed by Hahn and Nelson (1973) apply to predictions and prediction limits for a future sample. The formulas there are exact if the past sample is multiply **failure** censored. Then the observed number r of failures must be used in place of the sample size in formulas of their paper. For a time censored past sample, those formulas are approximate.

Computer Programs

Computer programs that ML fit an exponential distribution to multiply censored data are not necessary; the calculations are easy to do by hand. Programs that do the calculations are STATPAC by Nelson and others (1978) and SURVREG by Preston and Clarkson (1980). Each gives the ML estimates of the exponential mean, percentiles, and probabilities (including reliabilities) plus quantities programmed by the user. STATPAC gives approximate asymptotic confidence limits for these quantities based on the local estimate of the variance (1.4) and the normal approximation for $\ln(\hat{\theta})$.

Two-Parameter Exponential

ML fitting of the two-parameter exponential distribution is presented by Mann, Schafer, and Singpurwalla (1974, Secs. 5.1.2. and 5.1.3) and Bain (1978). This distribution is used much less frequently than the (one-parameter) exponential distribution, since most products do not have an initial period that is failure free.

2. NORMAL AND LOGNORMAL DISTRIBUTIONS

This section presents maximum likelihood (ML) methods for fitting normal and lognormal distributions to multiply time censored data (Type I). These methods also apply to multiply failure censored data (Type II), to singly censored data of both types, and to complete data.

This section first presents an example based on computer output, as most readers will simply use a computer program for such analyses. Then the section surveys some computer programs for such analyses. Next the section summarizes special (simpler) ML methods for complete and singly censored data. Lastly the section presents ML methods for fitting a (log) normal distribution to multiply censored data; these methods are difficult and will interest only advanced readers. Results include normal and lognormal parameter estimates, their information and covariance matrices and approximate confidence limits, and estimates and approximate confidence limits for percentiles and reliabilities. Cohen (1963) gives such results. Exact confidence limits have not been tabulated for multiply censored data. General ML theory and motivation appear in Sections 5.3 and 5.4.

Here t_i denotes the failure or running time on sample unit i from a lognormal distribution. Similarly, y_i denotes the failure or running time on sample unit i from a normal distribution. Below, μ_0 and σ_0 denote true population parameter values, and μ and σ denote arbitrary values.

Analyses of lognormal data t_i are done in terms of base 10 logs $y_i = \log(t_i)$; such y_i come from a normal distribution with μ_0 and σ_0 equal to the

Table 2.1. Thousands of Miles to Failure for Locomotive Controls

22.5	57.5	78.5	91.5	113.5	122.5
37.5	66.5	80.0	93.5	116.0	123.0
46.0	68.0	81.5	102.5	117.0	127.5
48.5	69.5	82.0	107.0	118.5	131.0
51.5	76.5	83.0	108.5	119.0	132.5
53.0	77.0	84.0	112.5	120.0	134.0
54.5					

59 controls ran 135,000 miles without failure.

lognormal parameters μ_0 and σ_0. The normal and lognormal distributions and their relationship are described in Chapter 2.

Locomotive Control Example

Table 2.1 shows singly time censored life data on 96 locomotive controls. Management wanted an estimate and confidence limits (given later) for the fraction of controls failing on an 80-thousand-mile warranty. Figure 2.1 is a lognormal probability plot (Chapter 3) of the data. The ML fit of the lognormal distribution is the straight line on the plot; the curved lines are 95% confidence limits for the distribution percentiles (and reliabilities).

Figure 2.2 shows computer output from STATPAC of Nelson and others (1978). The output gives ML estimates and approximate confidences limits for the parameters, selected percentiles, and the fraction failing on warranty. Also, the output gives the **local** estimate of the covariance matrix of the parameter estimates. The ML estimates of the lognormal parameters are $\hat{\mu} = 2.2223$ and $\hat{\sigma} = 0.3064$ (base 10 logs). The corresponding estimate of the

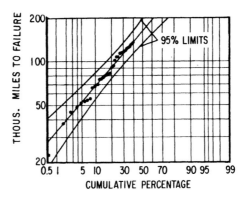

Figure 2.1. Lognormal probability plot of locomotive control failures.

* MAXIMUM LIKELIHOOD ESTIMATES FOR DIST. PARAMETERS
  WITH APPROXIMATE 95% CONFIDENCE LIMITS

PARAMETERS	ESTIMATE	LOWER LIMIT	UPPER LIMIT
CENTER μ	2.222269	2.133593	2.310946
SPREAD σ	0.3064140	0.2365178	0.3969661

* COVARIANCE MATRIX

PARAMETERS	CENTER $\hat{\mu}$	SPREAD $\hat{\sigma}$
CENTER $\hat{\mu}$	0.2046921E-02	
SPREAD $\hat{\sigma}$	0.1080997E-02	0.1638384E-02

PCTILES

* MAXIMUM LIKELIHOOD ESTIMATES FOR DIST. PCTILES
  WITH APPROXIMATE 95% CONFIDENCE LIMITS

PCT.	ESTIMATE	LOWER LIMIT	UPPER LIMIT
0.1	18.84908	11.73800	30.26819
0.5	27.09410	18.40041	39.89532
1	32.30799	22.85237	45.67604
5	52.25811	40.94096	66.70360
10	67.53515	55.28316	82.50246
20	92.13728	77.87900	109.0060
50	166.8282	136.0171	204.6188
80	302.0672	219.3552	415.9673
90	412.1062	278.7152	609.3370
95	532.5805	338.8504	837.0715
99	861.4479	487.2066	1523.157

PERCENT(LIMIT 80.) F(80)

* MAXIMUM LIKELIHOOD ESTIMATES FOR % WITHIN LIMITS
  WITH APPROXIMATE 95% CONFIDENCE LIMITS

	ESTIMATE	LOWER LIMIT	UPPER LIMIT
PCT	14.87503	9.895509	21.75530

Figure 2.2. STATPAC output on lognormal fit to the locomotive control data.

median of the lognormal distribution is antilog(2.2223) = 167 thousand miles; this is a "typical" life.

Figure 2.2 shows that the ML estimate of the percentage failing on warranty is $\hat{F}(80) = 15\%$. Approximate 95% confidence limits are $\underline{F}(80) = 10\%$ and $\tilde{F}(80) = 22\%$. Management decided that the control must be redesigned, as warranty costs would be too high.

Computer programs

Some computer programs that ML fit the normal and lognormal distributions to multiply censored data are the following.

1. STATPAC by Nelson and others (1978). It gives the ML estimates of normal and lognormal parameters (μ_0, σ_0), percentiles, and probabilities (including reliabilities) plus quantities programmed by users. It also gives approximate asymptotic confidence limits for these quantities. To do this it estimates the covariance matrix of the parameter estimates from the local Fisher information matrix. Figure 2.2 shows STATPAC output.

2. CENSOR by Meeker and Duke (1979). It is similar to STATPAC, costs much less, but is less easy to learn.

3. CENS by Hahn and Miller (1968) on General Electric Time-Sharing. It gives the ML estimates of the parameters and of their covariance matrix for Type I and II censoring.

4. Glasser's (1965) program. It is similar to that of Hahn and Miller (1968).

5. IMSL (1975) OTMLNR Program.

6. SURVREG by Preston and Clarkson (1980).

Programs for ML fitting of a normal distribution are used to fit a lognormal distribution to data. Then one transforms each lognormal data value t to a normal one $y = \log(t)$ and fits a normal distribution to the transformed values. Estimates and confidence limits for lognormal quantities are obtained in the obvious way from the normal results.

Complete Data

For a complete sample of n observations, the ML estimates are

$$\hat{\mu} = \bar{y}, \qquad \hat{\sigma} = s\left[(n-1)/n\right]^{1/2},$$

where $\bar{y}$ and s are the sample mean and standard deviation. The large-sample covariance matrix of $\hat{\mu}$ and $\hat{\sigma}$ has

$$\mathrm{Var}(\hat{\mu}) \cong \sigma_0^2/n, \qquad \mathrm{Var}(\hat{\sigma}) \cong \sigma_0^2/(2n), \qquad \mathrm{Cov}(\hat{\mu}, \hat{\sigma}) \cong 0.$$

The sampling distribution of $\hat{\mu}$ is exactly normal, with a mean of μ_0 and variance of σ_0^2/n. $n\hat{\sigma}^2/\sigma_0^2$ has a chi-square sampling distribution with $(n-1)$ degrees of freedom. Chapter 6 provides exact statistical methods based on the statistics $\bar{y} = \hat{\mu}$ and $s = \hat{\sigma}\,[n/(n-1)]^{1/2}$; these methods can be expressed in terms of $\hat{\mu}$ and $\hat{\sigma}$ with the obvious substitutions.

Singly Censored Data

The large-sample methods below apply to singly censored samples and provide ML estimates and approximate confidence limits.

For a singly Type I or II censored sample, the ML estimates $\hat{\mu}$ and $\hat{\sigma}$ are easily calculated with tables of Cohen (1961) and Schmee and Nelson (1977). The tables circumvent iterative solution of the likelihood equations and directly yield the ML estimates. Cohen (1961) and Harter and Moore (1966) tabulate the asymptotic covariance matrix, which is the same for Types I and II single censoring; for censoring at (log) time η, the matrix is a function of $\zeta = (\eta - \mu_0)/\sigma_0$. Evaluation of the matrix at $\hat{\zeta} = (\eta - \hat{\mu})/\hat{\sigma}$ with the Cohen table (also given by Schmee and Nelson) yields the ML estimate of the covariance matrix. Harter and Moore (1966) give Monte Carlo estimates of the exact means and variances of $\hat{\mu}$ and $\hat{\sigma}$ for small samples.

For a singly Type II censored sample, Schmee and Nelson (1976) tabulate factors for exact confidence limits for μ_0 and σ_0, using ML estimates for $n = 2(1)10, r = 2(1)n$. Bain (1978, p. 399) tabulates confidence limits for μ_0 and σ_0 for $n = 20, 50, 100, 200, \infty$, $r/n = .3(.2).9$, and $1 - \delta$ and $\delta = .02, .05$, .10, .20. The approximate limits below can be used for such data, too. For exact confidence limits for percentiles and reliabilities, only the linear methods of Chapter 7 are available.

One can use exact or approximate limits to choose n and r to get confidence intervals of desired length.

Other Methods

Prediction methods have not been developed for multiply censored (log) normal samples. For multiple failure censoring, exact prediction limits could be obtained by simulation methods like those of Thoman, Bain, and Antle (1970). Tabulations of such limits would be large, because a sample of size n can be Type II multiply censored in 2^{n-1} ways, a large number. In general, tests of fit and other methods have not been developed for multiply censored data.

Approximate best linear unbiased estimators for the parameters from multiply failure censored data can be derived with the methods of Thomas and Wilson (1972). Their simple methods also give approximate variances and covariances of the estimates.

Three-Parameter Lognormal

ML fitting of the three-parameter lognormal distribution is presented by Cohen (1976), who treats multiply censored data, by Mann, Schafer, and Singpurwalla (1974, Sec. 5.4.3), who treat singly censored data, and by their references.

Normal and Lognormal Parameters

The following advanced material presents ML methods for fitting a (log) normal distribution to multiply (right) censored data. It is intended for those who desire a deeper understanding or for those who wish to write a computer program for such calculations. Section 5 presents detailed theory underlying the methods.

ML estimates. The log likelihood for a sample of n units with r failures is

$$\mathcal{L} = \sum_i{}' \ln\{(1/\sigma)\phi[(y_i - \mu)/\sigma]\} + \sum_i{}'' \ln\{1 - \Phi[(y_i - \mu)/\sigma]\}, \quad (2.1)$$

where $\phi(\ )$ and $\Phi(\ )$ are the standard normal probability density and cumulative distribution. The sum $\sum_i'$ runs over the r failures times and the sum $\sum_i''$ runs over $n - r$ running times. The ML estimates $\hat{\mu}$ and $\hat{\sigma}$ for μ_0 and σ_0 are unique and are the μ and σ values that maximize (2.1); they are also the solutions of the likelihood equations

$$0 = \partial\mathcal{L}/\partial\mu = (1/\sigma)\sum_i{}'' h[(y_i - \mu)/\sigma] + (r/\sigma)[(\bar{y} - \mu)/\sigma],$$

$$0 = \partial\mathcal{L}/\partial\sigma = (1/\sigma)\sum_i{}'' [(y_i - \mu)/\sigma] h[(y_i - \mu)/\sigma] - (r/\sigma)$$

$$+ (r/\sigma^3)[s'^2 + (\bar{y} - \mu)^2], \quad (2.2)$$

where $h(z) = \phi(z)/[1 - \Phi(z)]$ is the standard normal hazard function, $\bar{y}$ is the average of the r failures, and $s' = [\sum_i'(y_i - \bar{y})^2/r]^{1/2}$ is their standard deviation. The nonlinear equations (2.2) must be iteratively solved by computer to obtain $\hat{\mu}$ and $\hat{\sigma}$, or (2.1) must be numerically maximized to obtain them. Section 6 describes such iterative methods. Computer programs that do the calculations are listed above. Readers may wish to write their own such programs.

For samples with many failures, the joint cumulative distribution function of $\hat{\mu}$ and $\hat{\sigma}$ is close to a joint normal one with means μ_0 and σ_0 and the covariance matrix (2.5).

Information matrix. Needed for confidence limits, the negative second partial derivatives of the log likelihood are

$$-\partial^2 \mathcal{L}/\partial\mu^2 = (1/\sigma^2)\left\{r + \sum_i{}''\left[h^2(z_i) - z_ih(z_i)\right]\right\},$$

$$-\partial^2 \mathcal{L}/\partial\sigma^2 = (2/\sigma^2)\sum_i{}''\left[z_ih(z_i) + z_i^2h^2(z_i) - z_i^3h(z_i)\right], \qquad (2.3)$$

$$-\partial^2 \mathcal{L}/\partial\mu\,\partial\sigma = -\partial^2 \mathcal{L}/\partial\sigma\,\partial\mu$$

$$= (1/\sigma^2)\sum_i{}''\left[h(z_i) - z_ih^2(z_i) + z_i^2h(z_i)\right] - (2r/\sigma^3)(\bar{y} - \mu),$$

where $z_i = (y_i - \mu)/\sigma$ is the standard deviate. The expectations of (2.3) for $\mu = \mu_0$ and $\sigma = \sigma_0$ are the elements of the true Fisher information matrix $\mathbf{F}_0$; that is,

$$\mathbf{F}_0 = \begin{bmatrix} E_0\{-\partial^2 \mathcal{L}/\partial\mu^2\}_0 & E_0\{-\partial^2 \mathcal{L}/\partial\mu\,\partial\sigma\}_0 \\ E_0\{-\partial^2 \mathcal{L}/\partial\sigma\,\partial\mu\}_0 & E_0\{-\partial^2 \mathcal{L}/\partial\sigma^2\}_0 \end{bmatrix}.$$

For a time censored sample,

$$E_0\{-\partial^2 \mathcal{L}/\partial\mu^2\}_0 = (1/\sigma_0^2)\sum_i\left\{\Phi(\zeta_i) - \phi(\zeta_i)[\zeta_i - h(\zeta_i)]\right\},$$

$$E_0\{-\partial^2 \mathcal{L}/\partial\sigma^2\}_0 = (1/\sigma_0^2)\sum_i\left\{2\Phi(\zeta_i) - \zeta_i\phi(\zeta_i)[1 + \zeta_i - \zeta_ih(\zeta_i)]\right\},$$

$$E_0\{-\partial^2 \mathcal{L}/\partial\mu\,\partial\sigma\}_0 = E_0\{-\partial^2 \mathcal{L}/\partial\sigma\,\partial\mu\}_0 \qquad (2.4)$$

$$= (1/\sigma_0^2)\sum_i\left(-\phi(\zeta_i)\{1 + \zeta_i[\zeta_i - h(\zeta_i)]\}\right).$$

These are functions of the true standard deviates $\zeta_i = (\eta_i - \mu_0)/\sigma_0$ of the **planned** censoring times η_i of all sample units; Σ_i runs over all n units. For the locomotive control, all $\eta_i = \log(135)$, the common log censoring time.

Locomotive control. The ML estimate of the Fisher information matrix for the locomotive control data has $E_0\{-\partial^2 \mathcal{L}/\partial\mu^2\}_0 = 753$, $E_0\{-\partial^2 \mathcal{L}/\partial\sigma^2\}_0 = 939$, and $E_0\{-\partial^2 \mathcal{L}/\partial\mu\,\partial\sigma\}_0 = -504$, where $\mu_0 = \hat{\mu}$ and $\sigma_0 = \hat{\sigma}$, from the CENS program of Hahn and Miller (1968).

Covariance matrix.   The inverse of the true Fisher matrix (2.4) is the true large-sample covariance matrix of $\hat{\mu}$ and $\hat{\sigma}$; namely,

$$\begin{bmatrix} \text{Var}(\hat{\mu}) & \text{Cov}(\hat{\mu},\hat{\sigma}) \\ \text{Cov}(\hat{\mu},\hat{\sigma}) & \text{Var}(\hat{\sigma}) \end{bmatrix} = \begin{bmatrix} E_0\{-\partial^2\mathcal{L}/\partial\mu^2\}_0 & E_0\{-\partial^2\mathcal{L}/\partial\mu\,\partial\sigma\}_0 \\ E_0\{-\partial^2\mathcal{L}/\partial\sigma\,\partial\mu\}_0 & E_0\{-\partial^2\mathcal{L}/\partial\sigma^2\}_0 \end{bmatrix}^{-1}$$

(2.5)

It is in terms of standardized planned censoring times ζ_i for **all** sample units. When the ζ_i are all known, one can obtain the ML estimate of the covariance matrix for a time censored sample. Replace ζ_i in (2.4) by $\hat{\zeta}_i = (\eta_i - \hat{\mu})/\hat{\sigma}$ to estimate the Fisher information matrix. Invert this estimate to get the **ML estimate** of the covariance matrix.

Also, one can obtain the sample **local estimate** of the covariance matrix, even when censoring times for all sample units are not known. Replace μ and σ in (2.3) by $\hat{\mu}$ and $\hat{\sigma}$ to get elements of the local Fisher information matrix. Invert that matrix to get the local estimate of the covariance matrix. This estimate applies to both time and failure censored samples.

Locomotive control.   The ML estimate of the covariance matrix has $\hat{\text{Var}}(\hat{\mu}) \cong 2.07 \times 10^{-3}$, $\hat{\text{Var}}(\hat{\sigma}) \cong 1.66 \times 10^{-3}$, and $\hat{\text{Cov}}(\hat{\mu},\hat{\sigma}) \cong 1.11 \times 10^{-3}$; this was obtained with the CENS program of Hahn and Miller (1968) by inverting the ML estimate of the Fisher information matrix (2.5), namely,

$$\begin{bmatrix} \hat{\text{Var}}(\hat{\mu}) & \hat{\text{Cov}}(\hat{\mu},\hat{\sigma}) \\ \hat{\text{Cov}}(\hat{\mu},\hat{\sigma}) & \hat{\text{Var}}(\hat{\sigma}) \end{bmatrix} = \begin{bmatrix} 753 & -504 \\ -504 & 939 \end{bmatrix}^{-1}$$

$$= \begin{bmatrix} 2.07 \times 10^{-3} & 1.11 \times 10^{-3} \\ 1.11 \times 10^{-3} & 1.66 \times 10^{-3} \end{bmatrix}.$$

Figure 2.2 shows the local estimate, which differs.

Confidence limits.   Two-sided approximate $100\gamma\%$ confidence limits for μ_0 and σ_0 are

$$\underset{\sim}{\mu} \cong \hat{\mu} - K_\gamma[\text{Var}(\hat{\mu})]^{1/2}, \qquad \tilde{\mu} \cong \hat{\mu} + K_\gamma[\text{Var}(\hat{\mu})]^{1/2}, \qquad (2.6)$$

$$\underset{\sim}{\sigma} \cong \hat{\sigma}/\exp\{K_\gamma[\text{Var}(\hat{\sigma})]^{1/2}/\hat{\sigma}\}, \qquad \tilde{\sigma} \cong \hat{\sigma}\cdot\exp\{K_\gamma[\text{Var}(\hat{\sigma})]^{1/2}/\hat{\sigma}\},$$

here $\text{Var}(\hat{\mu})$ and $\text{Var}(\hat{\sigma})$ must be estimated as described above. For one-sided

$100\gamma\%$ confidence limits, replace K_γ by z_γ. Exact limits for singly censored data are referenced above.

Locomotive control. Two-sided approximate 95% confidence limits are

$$\underset{\sim}{\mu} \cong 2.2223 - 1.960(2.07 \times 10^{-3})^{1/2} = 2.133,$$

$$\tilde{\mu} \cong 2.2223 + 1.960(2.07 \times 10^{-3})^{1/2} = 2.311,$$

$$\underset{\sim}{\sigma} \cong 0.3064/\exp\left[1.960(1.66 \times 10^{-3})^{1/2}/0.3064\right] = 0.236,$$

$$\tilde{\sigma} \cong 0.3064 \times 1.298 = 0.398.$$

Corresponding limits for the median of the lognormal distribution are antilog(2.133) = 136 and antilog(2.311) = 205 thousand miles. Each limit is a one-sided 97.5% confidence limit.

Percentiles

ML estimates. The $100P$th percentile of a normal distribution is $y_P = \mu_0 + z_P\sigma_0$, where z_P is the $100P$th standard normal percentile. Its ML likelihood estimate is

$$\hat{y}_P = \hat{\mu} + z_P\hat{\sigma}. \tag{2.7}$$

For large samples, the cumulative distribution function of $\hat{y}_P$ is close to a normal one, with a mean equal to y_P and a variance

$$\mathrm{Var}(\hat{y}_P) \cong \mathrm{Var}(\hat{\mu}) + z_P^2\mathrm{Var}(\hat{\sigma}) + 2z_P\mathrm{Cov}(\hat{\mu},\hat{\sigma}). \tag{2.8}$$

The terms of (2.8) are estimated from (2.3) or (2.4), as described above.

Confidence limits. Two-sided approximate $100\gamma\%$ confidence limits for y_P are

$$\underset{\sim}{y_P} \cong \hat{y}_P - K_\gamma[\mathrm{Var}(\hat{y}_P)]^{1/2}, \qquad \tilde{y}_P \cong \hat{y}_P + K_\gamma[\mathrm{Var}(\hat{y}_P)]^{1/2}, \tag{2.9}$$

where $\mathrm{Var}(\hat{y}_P)$ is estimated from (2.8). For a one-sided $100\gamma\%$ confidence limit, replace K_γ by z_γ.

The estimate and limits for the corresponding lognormal percentile $t_P = $ antilog(y_P) are $\hat{t}_P = $ antilog($\hat{y}_P$), $\underset{\sim}{t_P} \cong$ antilog($\underset{\sim}{y_P}$), and $\tilde{t}_P \cong$ antilog($\tilde{y}_P$).

Locomotive control. The ML estimate of the first percentile of log life is $\hat{y}_{.01} = 2.2223 + (-2.326)0.3064 = 1.5185$. The ML estimate of the first percentile of the lognormal life distribution is antilog(1.5185) = 33.0 thousand miles. The ML estimate of $\text{Var}(\hat{y}_{.01})$ is $\text{Vâr}(\hat{y}_{.01}) = 10^{-3}[2.07 + (-2.326)^2 1.66 + 2(-2.326)1.11] = 5.887 \times 10^{-3}$. Two-sided approximate 95% confidence limits are

$$\underset{\sim}{y}_{.01} = 1.5185 - 1.960(5.887 \times 10^{-3})^{1/2} = 1.3681,$$

$$\tilde{y}_{.01} = 1.5185 + 0.1504 = 1.6689.$$

Limits for the corresponding lognormal percentile are $\underset{\sim}{t}_{.01} = \text{antilog}(1.3681) = 23.3$ and $\tilde{t}_{.01} = \text{antilog}(1.6689) = 47.7$ thousand miles. Each limit is a one-sided 97.5% confidence limit. The ML line in Figure 2.1 gives the ML estimates of the lognormal percentiles. The curves on the plot give the two-sided confidence limits for the percentiles. The limits are narrowest near the "center" of the data.

Reliability

ML estimate. The fraction of a (log) normal distribution failing by a (log) age y is $F(y) = \Phi\{(y - \mu_0)/\sigma_0\}$. Its ML estimate is

$$\hat{F}(y) = \Phi[(y - \hat{\mu})/\hat{\sigma}]. \tag{2.10}$$

For large samples, the cumulative distribution function of $\hat{z} = (y - \hat{\mu})/\hat{\sigma}$ is close to a normal one, with a mean equal to $z = (y - \mu_0)/\sigma_0$ and a variance

$$\text{Var}(\hat{z}) \cong (1/\sigma_0^2)[\text{Var}(\hat{\mu}) + z^2 \text{Var}(\hat{\sigma}) + 2z\,\text{Cov}(\hat{\mu}, \hat{\sigma})]. \tag{2.11}$$

The terms of (2.11) are estimated from (2.3) or (2.4), as described above.

Confidence limits. For two-sided approximate $100\gamma\%$ confidence limits for $F(y)$, calculate

$$\underset{\sim}{z} \cong \hat{z} - K_\gamma[\text{Var}(\hat{z})]^{1/2}, \qquad \tilde{z} \cong \hat{z} + K_\gamma[\text{Var}(\hat{z})]^{1/2}, \tag{2.12}$$

where $\text{Var}(\hat{z})$ is estimated from (2.11). Then

$$\underset{\sim}{F}(y) \cong \Phi(\underset{\sim}{z}), \qquad \tilde{F}(y) \cong \Phi(\tilde{z}). \tag{2.13}$$

For a one-sided $100\gamma\%$ confidence limit, replace K_γ by z_γ.

The corresponding reliability estimate and limits are

$$\hat{R}(y) = 1 - \hat{F}(y), \qquad \underset{\sim}{R}(y) \cong 1 - \tilde{F}(y), \qquad \tilde{R}(y) \cong 1 - \underset{\sim}{F}(y). \quad (2.14)$$

Locomotive control. The ML estimate of the fraction failing on an 80 thousand mile warranty is calculated as $\hat{z} = [\log(80) - 2.2223]/0.3064 = -1.0418$ and $\hat{F}(80) = \Phi(-1.0418) = 0.149$. Two-sided approximate 95% confidence limits are calculated as

$$\text{Vâr}(\hat{z}) \cong (1/0.3064)^2 10^{-3} \big[2.07 + (-1.0418)^2 1.66$$

$$+ 2(-1.0418)1.11 \big] = 0.01660,$$

$$\underset{\sim}{z} \cong -1.0418 - 1.960(0.01660)^{1/2} = -1.2943,$$

$$\tilde{z} \cong -1.0418 + 1.960(0.01660)^{1/2} = -0.7893,$$

$$\underset{\sim}{F}(80) \cong \Phi(-1.2943) = 0.098, \qquad \tilde{F}(80) \cong \Phi(-0.7893) = 0.215.$$

Each limit is a one-sided approximate 97.5% confidence limit. In Figure 2.1, the ML line gives the ML estimates of failure probabilities. The curves for the confidence limits for the percentiles are also approximately the confidence limits for failure probabilities.

3. WEIBULL AND EXTREME VALUE DISTRIBUTIONS

This section presents maximum likelihood (ML) methods for fitting Weibull and extreme value distributions to multiply time censored data (Type I). The methods also apply to multiply failure censored data (Type II), to singly censored data of both types, and to complete data.

Most readers will use a standard computer program for ML fitting of a Weibull (or extreme value) distribution to data. For this reason, this section first presents an example with such a computer analysis. Then the section surveys some available computer programs. Next it presents special results for complete and singly censored data. Finally, it presents the advanced computational methods for ML fitting of a Weibull (or extreme value) distribution to multiply censored data. The last topic will interest only those who desire a deeper understanding of ML methods or wish to write their own programs for ML fitting. The computational methods cover Weibull and extreme value parameter estimates, their information and covariance matrices and approximate confidence limits, and estimates and approximate confidence limits for percentiles and reliability. Exact confidence limits have not been tabulated for multiply censored data, only for complete and singly censored data. General ML theory and motivation appear in Sections 5.3 and 5.4.

Figure 3.1 is a Weibull hazard plot (Chapter 4) of the fan data. The ML fit of the Weibull distribution is the straight line in the plot, and the curves are approximate 95% confidence limits. In Figure 1.1 for the exponential fit, the confidence limits are closer together and are straight lines. Such plots are convenient and informative presentations of the data, the fitted distribution, and the confidence limits.

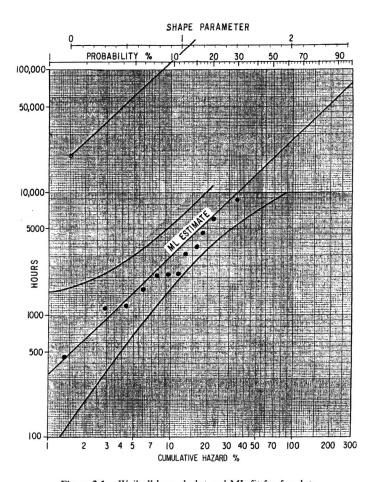

Figure 3.1. Weibull hazard plot and ML fit for fan data.

Figure 3.2 shows ML output of the WEIB program of the General Electric Information Services Company (1979)

The ML estimates of the parameters are $\hat{\alpha} = 26.3$ thousand hours and $\hat{\beta} = 1.06$. The shape parameter estimate is close to unity; this indicates that the failure rate is essentially constant. That is, new and old fans are about equally prone to failure. The approximate 95% confidence limits are $\beta = 0.64$ and $\tilde{\beta} = 1.74$. This interval encloses 1. Thus the data are consistent with $\beta_0 = 1$, corresponding to the exponential distribution and a constant failure rate. So management had no convincing indication that it was necessary to replace high-mileage fans before low-mileage ones. Replacements could be made in the most convenient and economical way.

The estimate of the fraction failing on an 8-thousand-hour warranty is $\hat{F}(8) = 1 - 0.75 = 0.25$. Approximate 95% confidence limits are $F(8) = 1 - 0.86 = 0.14$ and $\tilde{F}(8) = 1 - 0.61 = 0.39$. This helped management to decide to replace unfailed fans with a more durable fan. In Figure 3.1, the ML line gives the ML estimates of the failure probabilities. The curves for the confidence limits for the percentiles are also approximately the confidence limits for failure probabilities.

Computer Programs

Some computer programs that ML fit the Weibull and extreme value distributions to multiply censored data are the following.

1. STATPAC by Nelson and others (1978). It gives the ML estimates of Weibull (α, β) and extreme value (λ, δ) parameters, percentiles, probabilities (including reliabilities), plus quantities programmed by users. It also gives approximate asymptotic confidence limits for these quantities. For limits it uses the local estimate of the covariance matrix of the parameter estimates.

2. CENSOR by Meeker and Duke (1979) is similar to STATPAC, costs much less, but is less easy to learn.

3. SURVREG by Preston and Clarkson (1980) is similar to STATPAC.

4. WEIB of the General Electric Information Service Company (1979) gives the same results as STATPAC, but only for the Weibull distribution. Figure 3.2 shows output from WEIB.

5. Many organizations have simple computer programs that merely calculate the ML estimates of the Weibull parameters but not confidence intervals.

Programs for ML fitting of a Weibull distribution are widely available and can be used to fit an extreme value distribution to data. Then one first transforms each extreme value data value y to a Weibull one $t = \exp(y)$ and

```
ESTIMATES FOR THE CUMULATIVE WEIBULL DISTRIBUTION:
            F.(T) = 1-EXP[-(T/A)^B]


       ESTIMATE AND TWO-SIDED 95.00% CONFIDENCE
       INTERVALS FOR DISTRIBUTION PARAMETERS

    SHAPE (BETA) PARAMETER:              1.0584
    LOWER LIMIT:                         0.64408
    UPPER LIMIT:                         1.7394


    SCALE (63.2 PCTILE) PARAMETER:       26.297
    LOWER LIMIT:                         10.552
    UPPER LIMIT:                         65.535


      ESTIMATED COVARIANCE MATRIX OF PARAMETER ESTIMATES:

                       SCALE             SHAPE
            SCALE      150.10           -2.6645
            SHAPE     -2.6645         0.71958E-01


         ESTIMATE AND TWO-SIDED 95.00% CONFIDENCE
         INTERVALS FOR DISTRIBUTION PERCENTILES

    PERCEN        PERCENTILE        LOWER          UPPER
     TAGE          ESTIMATE        LIMIT          LIMIT

     0.10      0.38526E-01    0.29849E-02        0.49726
     0.50        0.17659      0.28467E-01        1.0954
     1.00        0.34072      0.74823E-01        1.5515
     5.00        1.5892         0.68310          3.6974
    10.00        3.1372         1.6862           5.8369

    20.00        6.3747         3.7305           10.893
    50.00        18.600         8.5248           40.584
    70.00        31.338         11.702           83.924
    90.00        57.826         16.541           202.16
    95.00        74.148         18.949           290.14

    99.00        111.31         23.578           525.47
    99.50        127.08         25.301           638.23
    99.90        163.27         28.891           922.66
```

WANT AN ESTIMATE OF PROBABILITY OF SURVIVAL BEYOND A SPECIFIC TIME?
IF SO, TYPE A DESIRED TIME; OTHERWISE, TYPE 0 --? **8**

```
        ESTIMATE AND TWO-SIDED 95.00% CONFIDENCE
    INTERVAL FOR THE DISTRIBUTION PERCENTAGE ABOVE      8.000

          ESTIMATE:                    0.75293
          LOWER LIMIT:                 0.60817
          UPPER LIMIT:                 0.85681
```

Figure 3.2. WEIB output for fan failure data.

fits a Weibull distribution to the transformed values. The ML estimates $\hat{\alpha}$ and $\hat{\beta}$ of the Weibull distribution fitted to the t values are then converted to $\hat{\lambda} = \ln(\hat{\alpha})$ and $\hat{\delta} = 1/\hat{\beta}$. Estimates and confidence limits for these and other quantities come in the obvious way from the Weibull results.

Complete Data

The ML methods below apply to complete samples and provide ML estimates and approximate and exact confidence limits for complete data.

For a complete sample of n observations, the ML estimates of the extreme value or Weibull percentiles still entail an iterative solution of the likelihood equations. The asymptotic covariance matrix for the extreme value parameter estimates has (Harter and Moore, 1968)

$$\text{Var}(\hat{\lambda}) \cong \delta_0^2 1.1087/n, \qquad \text{Var}(\hat{\delta}) \cong \delta_0^2 0.6079/n,$$

$$\text{Cov}(\hat{\lambda}, \hat{\delta}) \cong -\delta_0^2 0.2570/n.$$

Similarly, for the Weibull parameter estimates (Bain, 1978, p. 215),

$$\text{Var}(\hat{\alpha}) \cong (\alpha_0/\beta_0)^2 1.1087/n, \qquad \text{Var}(\hat{\beta}) \cong \beta_0^2 0.6079/n,$$

$$\text{Cov}(\hat{\alpha}, \hat{\beta}) \cong \alpha_0 0.2570/n.$$

These can be used to obtain approximate confidence limits for the parameters and functions of them as described below. These asymptotic variances are smaller than those for other estimates. That is, ML estimates are asymptotically at least as good as any others, for example, the moment estimators of Chapter 6.

There are tabulations of factors for exact $100\gamma\%$ confidence limits based on ML estimates for the following:

1. **Distribution parameters.** One- or two-sided limits for $n = 5(1)20(2)80(5)100(10)120$ and $\gamma = .02, .05, .25, .40(.10).70(.05).95, .98$ by Thoman, Bain, and Antle (1969). Lemon and Wattier (1976) tabulate lower limits for $n = 2(1)30$ and $\gamma = .80, .90, .95$. Bain (1978) gives tables for $n = 5(1)16(2)24(4)40(5)60, 70, 80, 100, 120$ and $1 - \gamma$ and $\gamma = .02, .05, .10, .50$.

2. **Reliabilities, $R(y)$.** One-sided lower limits for $n = 8$, 10, 12, 15, 18, 20, 25, 30, 40, 50, 75, 100, $\gamma = .75, .90, .95, .98$, and $\hat{R}(y) = .50(.02).98$ by Thoman, Bain, and Antle (1970) and Bain (1978).

3. **Percentiles, y_P.** One-sided lower limits can be obtained from the above table of limits for reliabilities as described by Thoman, Bain, and Antle (1970). Lemon and Wattier (1976) tabulate lower limits for the 1st

and 10th percentiles for $n = 2(1)30$ and $\gamma = .80, .90, .95$. Bain (1978) gives tables for his n and γ above and $1 - P = .02, .05, .10, .135, .25, .333, .368, .407, .50, .60, .80, .90, .95, .98$.

Lawless (1972b, 1975, 1978) gives a method for exact conditional confidence limits for parameters, percentiles, and reliabilities from complete data. His limits require a special computer program for the laborious calculations.

Sample size can be chosen to yield confidence intervals of desired length.

Singly Censored Data

The large-sample methods below apply to singly censored samples and provide ML estimates and approximate confidence limits.

For a singly Type I or II censored sample, the ML estimates of extreme value or Weibull parameters still entail an iterative solution of the likelihood equations. Meeker and Nelson (1974, 1977) tabulate the asymptotic covariance matrices of the ML estimates of the distribution parameters. The matrix is the same for censoring Types I and II; it is a function of the standardized censoring time $\zeta = (\eta - \lambda_0)/\delta_0$, where η is the (ln) censoring time. Evaluation of the matrix at $\hat{\zeta} = (\eta - \hat{\lambda})/\hat{\delta}$ with the aid of the table yields the ML estimate of the covariance matrix. Harter and Moore (1968) briefly tabulate the asymptotic Fisher information and covariance matrices and give Monte Carlo means and variances of ML parameter estimates for selected sample sizes and censorings.

For Type II single censoring, there are Monte Carlo tabulations of exact factors for confidence limits, which are used for approximate limits for Type I single censoring, namely,

1. Distribution parameters. Both parameters for $n = 40(20)120, r = .50n, .75n, \gamma = .01, .05, .10, .90, .95, .99$ by Billman, Antle, and Bain (1972). McCool (1970a) gives limits for the Weibull shape parameter for $\gamma = .10, .50, .90, .95$ and for n and r values in (3) below. Bain (1978) gives tables for both Weibull parameters for $n = 5, 10, 20, 40, 120, \infty, r/n = .5, .75$, and $1 - \gamma$ and $\gamma = .01, .05, .10$. McCool (1974) tabulates limits for both parameters for γ and $1 - \gamma = .01, .02, .025, .05(.05).30, .40, .50$, where $n = 5(r = 3, 5)$, $n = 10(r = 3, 5, 10), n = 15(r = 5, 10, 15), n = 20[r = 5(5)20], n = 30(r = 5, 10, 15, 20, 30)$.

2. Reliabilities $R(y)$. One-sided lower limits for $n = 40(20)120, \gamma = .90, .95, .98, .99$, and $\hat{R}(y) = .70(.02).94(.01).99(.0025).995(.001).998, .9985, .999$ by Billman, Antle, and Bain (1972). Bain (1978) gives one-sided limits for $\hat{R} = .70(.02).94(.01).99, .9925, .995(.001).998, .9985, .999, n = 40(20)100, r/n = .50, .75$, and $\gamma = .90, .95, .98, .99$.

3. Percentiles y_p. One-sided lower limits can be obtained from the above table of limits for reliabilities as described by Thoman, Bain, and Antle (1970). McCool (1970b) gives limits for $t_{.10}$ for $\gamma = .05, .10, .50, .90, .95$ and for $n = 10[r = 2(1)5, 7, 10]$, $n = 20[r = 3(1)7, 10, 15, 20]$, and $n = 30[r = 3(1)10, 15, 30]$. McCool (1974) gives limits for $t_{.01}, t_{.10}, t_{.50}$ for the γ, n, and r in (1) above.

Lawless (1972b, 1975, 1978) gives a method for exact conditional confidence limits for parameters, percentiles, and reliabilities from singly failure censored data. Also, Lawless (1973) gives a method for exact conditional prediction limits for the smallest order statistic of a future sample. His limits require a special computer program for the laborious calculations.

Sample size n and the number r of failures can be chosen to yield confidence intervals of desired length.

Other Methods

Predictions methods based on ML estimates have not been developed for multiply censored Weibull (or extreme value) samples. For multiple failure censoring, exact prediction limits could be obtained by simulation methods like those of Billman, Antle, and Bain (1972). Tabulations of such limits would be large, because a sample of size n can be multiply Type II censored in $2^{n-1} - 1$ ways, a large number. In general, tests of fit and other methods have not been developed for multiply censored data.

Mann (1971) tabulates coefficients for best linear invariant estimators of the extreme value parameters from all multiply failure censored samples up to size 6. She also tables the exact variances and covariances of these parameter estimates. She references tables for 90% confidence limits for the 10th percentile from multiply failure censored samples up to size 9. These results apply to the Weibull distribution in the usual way. Thomas and Wilson (1972) give simple approximately best linear unbiased estimators for the parameters from such data and their approximate variances and covariances.

The method of Lawless (1972b, 1975, 1978) can be extended to yield exact conditional confidence limits for parameters, percentiles, reliabilities, and other quantities from multiply failure censored data. His method requires a special computer program for the laborious calculations.

Three-Parameter Weibull

ML fitting of the three-parameter Weibull distribution is presented by Cohen (1975) and Wingo (1973) for multiply censored data, by Mann, Schafer, and Singpurwalla (1974, Sec. 5.2.1), by Bain (1978, Sec. 4.4) for singly censored data, and by their references.

Computational Methods

The rest of this section presents advanced computations for ML methods for Weibull and extreme value data. Section 5 provides more detail and theory. The following material will interest only those who will write computer programs for the computations or who seek a deeper understanding.

Distribution Parameters

Weibull parameter estimates. The Weibull log likelihood for a sample of n units with r failures is

$$\mathcal{L} = \sum_i \left[\ln(\beta) + (\beta - 1)\ln(t_i) - \beta\ln(\alpha) - (t_i/\alpha)^\beta \right] + \sum_i'' \left[-(t_i/\alpha)^\beta \right].$$

$$(3.1)$$

The sums $\sum_i, \sum_i'$, and $\sum_i''$ respectively run over all, failed, and unfailed units.

The ML estimates $\hat{\alpha}$ and $\hat{\beta}$ for α_0 and β_0 are the α and β values that maximize (3.1); $\hat{\alpha}$ and $\hat{\beta}$ are unique and are also the solutions of the likelihood equations

$$0 = \partial\mathcal{L}/\partial\alpha = \sum_i' \left[-(\beta/\alpha) + (\beta/\alpha)(t_i/\alpha)^\beta \right] + \sum_i'' (\beta/\alpha)(t_i/\alpha)^\beta,$$

$$0 = \partial\mathcal{L}/\partial\beta = \sum_i' \left[(1/\beta) + \ln(t_i/\alpha) - (t_i/\alpha)^\beta \ln(t_i/\alpha) \right]$$

$$- \sum_i'' (t_i/\alpha)^\beta \ln(t_i/\alpha).$$

$$(3.2)$$

These nonlinear equations must be iteratively solved by computer to obtain $\hat{\alpha}$ and $\hat{\beta}$. Otherwise, (3.1) must be numerically maximized to get them. Section 6 describes how to solve such equations iteratively. Alternatively, the likelihood equations (3.2) can be combined to eliminate α. This yields a single equation in β that is easier to solve, namely,

$$\sum_i' \ln(t_i)/r = \left(\sum_i t_i^\beta \ln(t_i) \right) \left(\sum_i t_i^\beta \right)^{-1} - (1/\beta).$$

$$(3.3)$$

The left-hand sum runs over just the failure times. It is easy to iteratively solve (3.3) to get $\hat{\beta}$, since the right-hand side is a monotone function of β.

Then calculate

$$\hat{\alpha} = \left(\sum_i t_i^{\hat{\beta}} / r \right)^{1/\hat{\beta}}. \tag{3.4}$$

Computer programs that do these calculations are listed above.

For large samples, the joint cumulative distribution function of $\hat{\alpha}$ and $\hat{\beta}$ is close to a joint normal one with means α_0 and β_0 and the covariance matrix given below.

The ML estimates for the parameters of the corresponding extreme value distribution are

$$\hat{\lambda} = \ln(\hat{\alpha}), \qquad \hat{\delta} = 1/\hat{\beta}. \tag{3.5}$$

Extreme value parameter estimates.   The extreme value log likelihood for a sample of n units with r failures is

$$\mathcal{L} = \sum_i \{ -\ln(\delta) - \exp[(y_i - \lambda)/\delta] + [(y_i - \lambda)/\delta] \} - \sum_i'' \exp[(y_i - \lambda)/\delta]. \tag{3.6}$$

The ML estimates $\hat{\lambda}$ and $\hat{\delta}$ for λ_0 and δ_0 are the λ and δ values that maximize (3.6); $\hat{\lambda}$ and $\hat{\delta}$ are also the solutions of the likelihood equations

$$0 = \partial \mathcal{L} / \partial \lambda = (1/\delta) \left[\sum_i \exp((y_i - \lambda)/\delta) - r \right], \tag{3.7}$$

$$0 = \partial \mathcal{L} / \partial \delta = (1/\delta) \left[\sum_i ((y_i - \lambda)/\delta) \exp((y_i - \lambda)/\delta) - \sum_i' ((y_i - \lambda)/\delta) - r \right].$$

These nonlinear equations in λ and δ must be iteratively solved by computer to obtain $\hat{\lambda}$ and $\hat{\delta}$, or (3.6) must be numerically maximized to get them. Alternatively, the likelihood equations can be combined to eliminate $\exp(-\lambda/\delta)$ and yield

$$\sum_i' y_i / r = \left(\sum_i y_i \exp(y_i/\delta) \right) \left(\sum_i \exp(y_i/\delta) \right)^{-1} - \delta. \tag{3.8}$$

The right-hand side is a monotone function of δ, so it is easy to solve iteratively for $\hat{\delta}$. Then calculate

$$\hat{\lambda} = \hat{\delta} \cdot \ln \left(\sum_i \exp(y_i/\hat{\delta})/r \right). \tag{3.9}$$

For large r, the joint cumulative distribution function of $\hat{\lambda}$ and $\hat{\delta}$ is close to a joint normal one with means λ_0 and δ_0 and the covariance matrix given below.

The ML estimates for the parameters of the corresponding Weibull distribution are

$$\hat{\alpha} = \exp(\hat{\lambda}), \qquad \hat{\beta} = 1/\hat{\delta}. \tag{3.10}$$

Information matrix.   The local Fisher information matrix for the extreme value distribution is

$$\mathbf{F} = \begin{bmatrix} -\partial^2 \mathcal{L}/\partial\lambda^2 & -\partial^2 \mathcal{L}/\partial\lambda\partial\delta \\ -\partial \mathcal{L}/\partial\delta\partial\lambda & -\partial^2 \mathcal{L}/\partial\delta^2 \end{bmatrix}.$$

These derivatives are

$$-\partial^2 \mathcal{L}/\partial\lambda^2 = (1/\delta^2)\sum_i \exp(z_i),$$

$$-\partial^2 \mathcal{L}/\partial\delta^2 = (1/\delta^2)\left[2(\partial\mathcal{L}/\partial\delta) + r + \sum_i z_i^2 \exp(z_i)\right], \tag{3.11}$$

$$-\partial^2 \mathcal{L}/\partial\lambda\partial\delta = -\partial^2 \mathcal{L}/\partial\delta\partial\lambda = (1/\delta^2)\left[\delta(\partial\mathcal{L}/\partial\lambda) + \sum_i z_i^2 \exp(z_i)\right],$$

where $z_i = (y_i - \lambda)/\delta$. Evaluated at $\lambda = \hat{\lambda}$ and $\delta = \hat{\delta}, \partial\mathcal{L}/\partial\lambda = \partial\mathcal{L}/\partial\mu = 0$ according to (3.7). The expectations of (3.11) for $\lambda = \lambda_0$ and $\delta = \delta_0$ are the terms of the true Fisher information matrix; namely, for a time censored sample, the terms are

$$E_0\{-\partial^2 \mathcal{L}/\partial\lambda^2\}_0 = (1/\delta_0^2)\sum_i \{1 - \exp[-\exp(\zeta_i)]\},$$

$$E_0\{-\partial^2 \mathcal{L}/\partial\delta^2\}_0 = (1/\delta_0^2)\sum_i \left\{1 + \int_0^{\exp(\zeta_i)}[\ln(w)]^2 w \cdot \exp(-w)\, dw\right.$$

$$\left. + \exp[-\exp(\zeta_i)]\zeta_i^2 \exp(\zeta_i)\right\}, \tag{3.12}$$

$$E_0\{-\partial^2 \mathcal{L}/\partial\lambda\partial\delta\}_0 = E_0\{-\partial^2 \mathcal{L}/\partial\delta\partial\lambda\}_0$$

$$= (1/\delta_0^2)\sum_i \left\{\int_0^{\exp(\zeta_i)}[\ln(w)] w \cdot \exp(-w)\, dw + \exp[-\exp(\zeta_i)]\zeta_i \exp(\zeta_i)\right\}.$$

Meeker and Nelson (1974) tabulate the terms in curly brackets as a function of the standard deviates $\zeta_i = (\eta_i - \lambda_0)/\delta_0$ of the planned censoring times η_i of all sample units.

The Weibull Fisher information matrix is not given here. It is more convenient to work in terms of the related extreme value Fisher information matrix (3.12) as described below.

Covariance matrix. The inverse of the Fisher information matrix (3.12) is the true large-sample covariance matrix of $\hat{\lambda}$ and $\hat{\delta}$; it is in terms of the planned censoring times ζ_i for all sample units. When planned censoring times of all sample units are **known**, one can obtain the ML estimate of the covariance matrix. Replace λ_0 and δ_0 in (3.12) by $\hat{\lambda}$ and $\hat{\delta}$, and invert this estimate of the Fisher matrix to get the ML estimate of the covariance matrix for a time censored sample.

Also, one can obtain the estimate based on the sample local Fisher information matrix, even when planned censoring times for all sample units are not known. Replace λ and δ in (3.11) by $\hat{\lambda}$ and $\hat{\delta}$, and invert the resulting estimate of the local Fisher information matrix to get the local estimate of the covariance matrix. This local estimate applies to both time and failure censored samples. Written out, the local estimate is

$$\begin{pmatrix} \text{var}(\hat{\lambda}) & \text{cov}(\hat{\lambda}, \hat{\delta}) \\ \text{cov}(\hat{\lambda}, \hat{\delta}) & \text{var}(\hat{\delta}) \end{pmatrix} = \begin{pmatrix} -\partial^2 \mathcal{L}/\partial\lambda^2 & -\partial^2 \mathcal{L}/\partial\lambda\partial\delta \\ -\partial^2 \mathcal{L}/\partial\delta\partial\lambda & -\partial^2 \mathcal{L}/\partial\delta^2 \end{pmatrix}^{-1}.$$

Relation between Weibull and extreme value covariance matrices. The asymptotic variances and covariances of the Weibull parameter estimates can be expressed in terms of those for the corresponding extreme value parameter estimates as

$$\text{Var}(\hat{\alpha}) \cong [\exp(\lambda_0)]^2 \text{Var}(\hat{\lambda}), \qquad \text{Var}(\hat{\beta}) \cong (1/\delta_0^4) \text{Var}(\hat{\delta}),$$

$$\text{Cov}(\hat{\alpha}, \hat{\beta}) \cong -(1/\delta_0^2)\exp(\lambda_0)\text{Cov}(\hat{\lambda}, \hat{\delta}). \qquad (3.13)$$

Similarly,

$$\text{Var}(\hat{\lambda}) \cong (1/\alpha_0)^2 \text{Var}(\hat{\alpha}), \qquad \text{Var}(\hat{\delta}) \cong (1/\beta_0^4) \text{Var}(\hat{\beta}),$$

$$\text{Cov}(\hat{\lambda}, \hat{\delta}) \cong -\beta_0^{-2}\alpha_0^{-1} \text{Cov}(\hat{\alpha}, \hat{\beta}). \qquad (3.14)$$

These formulas come from (5.3.16) and (5.3.17),

Engine fan. The local estimate of the Weibull covariance matrix is $\text{var}(\hat{\alpha}) = 150.1, \text{var}(\hat{\beta}) = 0.07196$ and $\text{cov}(\hat{\alpha}, \hat{\beta}) = -2.664$. This was obtained with the WEIB routine of the General Electric Information Service Company (1979); see Figure 3.2. The parameter estimates for the corresponding extreme value distribution of ln life are $\hat{\lambda} = \ln(26.30) = 3.270$ and $\hat{\delta} = 1/1.058 = 0.9452$. The local estimates of their variances and covariance are

$$\text{var}(\hat{\lambda}) = (1/26.3)^2 150.1 = 0.217,$$

$$\text{var}(\hat{\delta}) = (1/1.058)^4 0.07196 = 0.0574,$$

$$\text{cov}(\hat{\lambda}, \hat{\delta}) = -(1.058)^{-2} 26.3^{-1}(-2.664) = 0.0905.$$

Confidence limits for extreme value parameters. Two-sided approximate $100\gamma\%$ confidence limits for the true extreme value parameters λ_0 and δ_0 are obtained from (5.1.18) and (5.1.19) as

$$\underline{\lambda} = \hat{\lambda} - K_\gamma \left[\text{Var}(\hat{\lambda}) \right]^{1/2}, \quad \tilde{\lambda} = \hat{\lambda} + K_\gamma \left[\text{Var}(\hat{\lambda}) \right]^{1/2}, \quad (3.15)$$

$$\underline{\delta} = \hat{\delta}/\exp\left\{ K_\gamma \left[\text{Var}(\hat{\delta}) \right]^{1/2} /\delta \right\}, \quad \tilde{\delta} = \hat{\delta} \cdot \exp\left\{ K_\gamma \left[\text{Var}(\hat{\delta}) \right]^{1/2} /\delta \right\}, \quad (3.16)$$

where $\text{Var}(\hat{\lambda})$ and $\text{Var}(\hat{\delta})$ must be estimated as described above. For one-sided $100\gamma\%$ confidence limits, replace K_γ by z_γ. The limits for the corresponding Weibull parameters are

$$\underline{\alpha} = \exp(\underline{\lambda}), \quad \tilde{\alpha} = \exp(\tilde{\lambda}), \quad (3.17)$$

$$\underline{\beta} = 1/\tilde{\delta}, \quad \tilde{\beta} = 1/\underline{\delta}. \quad (3.18)$$

Confidence limits for Weibull parameters. Two-sided approximate $100\gamma\%$ confidence limits for the true Weibull parameters α_0 and β_0 are

$$\underline{\alpha} = \hat{\alpha}/\exp\left\{ K_\gamma \left[\text{Var}(\hat{\alpha}) \right]^{1/2} /\hat{\alpha} \right\}, \quad \tilde{\alpha} = \hat{\alpha} \cdot \exp\left\{ K_\gamma \left[\text{Var}(\hat{\alpha}) \right]^{1/2} /\hat{\alpha} \right\},$$

$$(3.19)$$

$$\underline{\beta} = \hat{\beta}/\exp\left\{ K_\gamma \left[\text{Var}(\hat{\beta}) \right]^{1/2} /\hat{\beta} \right\}, \quad \tilde{\beta} = \hat{\beta} \cdot \exp\left\{ K_\gamma \left[\text{Var}(\hat{\beta}) \right]^{1/2} /\hat{\beta} \right\}.$$

$$(3.20)$$

For one-sided approximate $100\gamma\%$ limits, replace K_γ by z_γ. The limits for the corresponding extreme value parameters are

$$\underset{\sim}{\lambda}=\ln(\underset{\sim}{\alpha}), \qquad \tilde{\lambda}=\ln(\tilde{\alpha}), \tag{3.21}$$

$$\underset{\sim}{\delta}=1/\hat{\beta}, \qquad \tilde{\delta}=1/\underset{\sim}{\beta}. \tag{3.22}$$

Engine fan. Two-sided approximate 95% confidence limits are

$$\underset{\sim}{\alpha}=26.30/\exp\left[1.960(150.10)^{1/2}/26.30\right]=10.6, \qquad \tilde{\alpha}=26.30\cdot2.492=65.5,$$

$$\underset{\sim}{\beta}=1.058/\exp\left[1.960(0.07196)^{1/2}/1.058\right]=0.64, \qquad \tilde{\beta}=1.058\cdot1.644=1.74.$$

The confidence limits appear in Figure 3.2.

Percentiles

ML estimates. The $100P$ th percentile of a smallest extreme value distribution is $y_P=\lambda_0+u_P\delta_0$, where $u_P=\ln[-\ln(1-P)]$. Its ML estimate is

$$\hat{y}_P=\hat{\lambda}+u_P\hat{\delta}. \tag{3.23}$$

For large samples, the cumulative distribution function of $\hat{y}_P$ is close to a normal one, with a mean equal to y_P and a variance

$$\text{Var}(\hat{y}_P)\simeq\text{Var}(\hat{\lambda})+u_P^2\text{Var}(\hat{\delta})+2u_P\text{Cov}(\hat{\lambda},\hat{\delta}). \tag{3.24}$$

The variances and covariance are estimated from (3.11) or (3.12). The ML estimate for the corresponding Weibull percentile $t_P=\exp(y_P)=\alpha_0[-\ln(1-P)]^{1/\beta_0}$ is

$$\hat{t}_P=\exp(\hat{y}_P)=\hat{\alpha}\left[-\ln(1-P)\right]^{1/\hat{\beta}}. \tag{3.25}$$

Confidence limits. Two-sided approximate $100\gamma\%$ confidence limits for y_P are

$$\underset{\sim}{y}_P=\hat{y}_P-K_\gamma\left[\text{Var}(\hat{y}_P)\right]^{1/2}, \qquad \tilde{y}_P=\hat{y}_P+K_\gamma\left[\text{Var}(\hat{y}_P)\right]^{1/2}, \tag{3.26}$$

where $\text{Var}(\hat{y}_P)$ is estimated from (3.24). For a one-sided approximate

$100\gamma\%$ confidence limit, replace K_γ by z_γ. The limits for the corresponding Weibull percentile $t_P = \exp(y_P)$ are

$$\underset{\sim}{t_P} \approx \exp(\underset{\sim}{y_P}), \qquad \tilde{t}_P \approx \exp(\tilde{y}_P). \tag{3.27}$$

Engine fan. The ML estimate of the 10th percentile of the ln life is $\hat{y}_{.10} = 3.270 + (-2.2504)0.9452 = 1.143$. The ML estimate of the 10th percentile of the Weibull life distribution is $\hat{t}_{.10} = \exp(1.143) = 3.14$ thousand hours. The local estimate of Var($\hat{y}_{.10}$) is var($\hat{y}_{.10}$) $= 0.217 + (-2.2504)^2$ $0.0574 + 2(-2.2504)0.0905 = 0.100$. Two-sided approximate 95% confidence limits are

$$\underset{\sim}{y}_{.10} \approx 1.143 - 1.960(0.100)^{1/2} = 0.523, \qquad \tilde{y}_{.10} \approx 1.143 + 0.620 = 1.763.$$

Limits for the corresponding Weibull percentile are $\underset{\sim}{t}_{.10} = \exp(0.523) = 1.69$ and $\tilde{t}_{.10} = \exp(1.763) = 5.83$ thousand hours. Each limit is a one-sided approximate 97.5% confidence limit. The ML line in Figure 3.1 gives the ML estimates of the Weibull percentiles. The curves on the plot give the confidence limits for the true percentiles. The limits are narrowest near the "center" of the data. The Weibull limits in Figure 3.1 are wider than the exponential limits in Figure 1.1 for the same data. This should be expected, since the Weibull distribution has an additional unknown shape parameter, resulting in greater uncertainty in estimates.

Reliability

ML estimate. The fraction of an extreme value distribution surviving an age y is $R(y) = \exp\{-\exp[(y - \lambda_0)/\delta_0]\}$. Its ML estimate is

$$\hat{R}(y) = \exp\{-\exp[(y - \hat{\lambda})/\hat{\delta}]\}. \tag{3.28}$$

For large samples, the cumulative distribution function of $\hat{u} = (y - \hat{\lambda})/\hat{\delta}$ is close to a normal one, with a mean of $u = (y - \lambda_0)/\delta_0$ and a variance

$$\text{Var}(\hat{u}) \approx (1/\delta_0^2)[\text{Var}(\hat{\lambda}) + u^2 \text{Var}(\hat{\delta}) + 2u\text{Cov}(\hat{\lambda}, \hat{\delta})]. \tag{3.29}$$

The fraction of a Weibull distribution surviving age t is $R(t) = \exp[-(t/\alpha_0)^{\beta_0}]$. Its ML estimate is

$$\hat{R}(t) = \exp[-(t/\hat{\alpha})^{\hat{\beta}}]. \tag{3.30}$$

Confidence limits. Two-sided approximate $100\gamma\%$ confidence limits for $R(y)$ are calculated as

$$\underset{\sim}{u} = \hat{u} - K_\gamma\left[\text{Var}(\hat{u})\right]^{1/2} \qquad \tilde{u} = \hat{u} + K_\gamma\left[\text{Var}(\hat{u})\right]^{1/2} \qquad (3.31)$$

where $\text{Var}(\hat{u})$ is estimated from (3.29) and (3.11) or (3.12), and

$$\underset{\sim}{R}(y) = \exp\left[-\exp(\tilde{u})\right], \qquad \tilde{R}(y) = \exp\left[-\exp(\underset{\sim}{u})\right]. \qquad (3.32)$$

For a one-sided approximate $100\gamma\%$ limit, replace K_γ by z_γ. The corresponding estimate and confidence limits for the fraction failing are

$$\hat{F}(y) = 1 - \hat{R}(y), \qquad \underset{\sim}{F}(y) = 1 - \tilde{R}(y), \qquad \tilde{F}(y) = 1 - \underset{\sim}{R}(y). \qquad (3.33)$$

Confidence limits for a Weibull reliability at age t are the same as those for the corresponding extreme value reliability $R(y)$, where $y = \ln(t)$.

Engine fan. The ML estimate of the fraction failing on an 8 thousand-hour warranty is calculated as $\hat{u} = [\ln(8) - 3.270]/0.945 = -1.260$ and $\hat{F}(8) = 1 - \exp[-\exp(-1.260)] = 0.25$. A one-sided upper approximate 95% confidence limit is calculated as

$$\text{Var}(\hat{u}) \cong (1/0.945^2)\left[0.217 + (-1.260)^2 0.0574 + 2(-1.260)0.0905\right]$$

$$= 0.0897,$$

$$\tilde{u} = -1.260 + 1.645(0.0897)^{1/2} = -0.767,$$

$$\tilde{F}(8) = 1 - \exp\left[-\exp(-0.767)\right] = 0.37.$$

4. DATA WITH COMPETING FAILURE MODES—MAXIMUM LIKELIHOOD ANALYSIS

4.1 Introduction

Life data with a mix of failure modes requires special methods. It is clearly wrong to use data on one failure mode to estimate the life distribution of another failure mode. This section describes maximum likelihood (ML) methods for analyzing such data where the mode (cause) of each failure is

known. Section 5.5 gives the theoretical basis of these ML methods. Chapter 5 presents graphical methods for such data.

This section presents ML methods for estimating (1) a separate life distribution for each failure mode, (2) the life distribution when all modes act, and (3) the life distribution that would result if certain modes were eliminated. The topics of this section are the following.

Illustrative data (Section 4.2).

The model (Section 4.3).

Analysis of data on a failure mode (Section 4.4).

The life distribution when all failure modes act (Section 4.5).

The life distribution with certain failure modes eliminated (Section 4.6).

Other methods (Section 4.7).

Theory for these analyses is in Section 5.5. The models are applied to an example where each sample unit fails from just one cause. However, the methods extend to data where each unit can keep running and accumulate failures from more than one cause; Nelson (1974b) presents this.

McCool (1974, 1976, 1978a) gives ML estimates and confidence limits for independent competing failure modes with Weibull distributions with a common (unknown) shape parameter value. Cox (1959) gives ML estimates for exponential distributions for (1) competing failure modes and (2) a mixture—both for data with the failure cause (a) identified and (b) not identified. Boardman and Kendell (1970) provide ML estimates for exponential distributions from time censored data with competing failure modes that are identified. Birnbaum (1979), David and Moeschberger (1979), and Kalbfleisch and Prentice (1980) present ML methods for such data; they emphasize biomedical applications and nonparametric methods.

4.2. Illustrative Data

The data of King (1971) in Table 4.1 illustrate the methods given here. The data are the breaking strengths of 20 wire connections. Such wires are bonded at one end to a semiconductor wafer and at the other end to a terminal post. Each failure results from breaking of the wire or a bond (whichever is weaker), as shown in Table 4.1. The main problem is to estimate the strength distributions of (1) wires, (2) bonds, and (3) connections. The specification for the connections requires that less than 1% have strength below 500 mg. Another problem is to estimate the strength distribution that would result from a redesign that eliminates (1) wire failures or else (2) bond failures. Strength takes the place of time to failure in this example.

Table 4.1.   Connection Strength Data

Strength mg.	Break of	Strength mg.	Break of
550	B	1250	B
750	W	1350	W
950	B	1450	B
950	W	1450	B
1150	W	1450	W
1150	B	1550	B
1150	B	1550	W
1150	W	1550	W
1150	W	1850	W
1250	B	2050	B

Figure 4.1 shows a normal hazard plot of the 20 strengths—ignoring cause of failure. Figure 4.2 is such a plot for the wire failures, and Figure 4.3 is such a plot for the bond failures. One makes and uses such plots to assess the validity of the data and the normal distribution as described in Chapter 5. Also, the plot is a convenient means of presenting the fitted distribution and confidence limits.

4.3.   The Model

The model for data with competing failure modes is that of Chapter 5. It is briefly presented here. It consists of

1.   A separate life distribution for each failure mode.

2.   The relationship between the failure times for each failure mode and the failure time for a test unit—the series-system model for independent competing failure modes here.

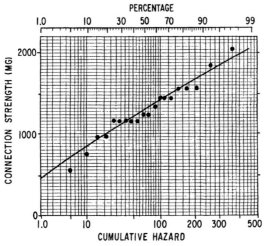

Figure 4.1.   Normal hazard plot of connection strengths.

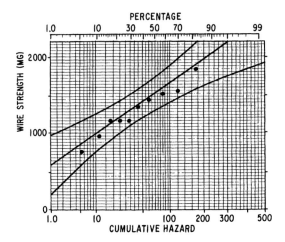

Figure 4.2. Normal hazard plot of wire strengths.

The series-system model. The series-system model for the life of units with independent competing failure modes assumes

1. Each unit has M potential times to failure, one from each mode, and the times are statistically **independent**.

2. The time to failure of a unit is the smallest of its M potential times.

Let $R_m(y)$ denote the reliability [probability that failure mode m does not occur by (log) time y] if only mode m were acting. The reliability function

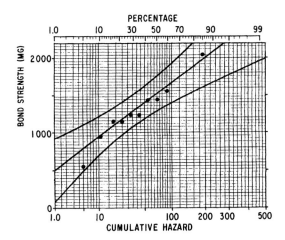

Figure 4.3. Normal hazard plot of bond strengths.

$R(y)$ of units with M statistically independent failure modes is given by the **product rule**:

$$R(y) = R_1(y)R_2(y) \cdots R_M(y). \tag{4.1}$$

This is used in Sections 4.5 and 4.6 to estimate life distributions from data with competing failure modes. Chapter 5 describes this model further.

4.4. Analysis of Data on a Single Failure Mode

Point of view. The following point of view simplifies analysis of data on a failure mode, say, mode 1. Each unit has a time to failure with mode 1 or else a time without failure mode 1. A time when a unit fails by another mode or when the unit has not failed is treated as a censoring time for mode 1, since the failure time for mode 1 is beyond the censoring time. Then the running and censoring times for mode 1 are a multiply censored sample, and the parameters of the distribution for mode 1 can be estimated with ML methods from that multiply censored sample. This is valid (1) if and only if a unit's failure time for mode 1 is statistically independent of its times for all other modes and (2) if a parameter value of the distribution of another mode has the same value, that fact not being used to estimate the parameter for mode 1. Restriction (1) holds if the series-system model applies (all failure modes are mutually independent). Then the ML estimates of the distribution parameters for different modes are close to statistically independent for large samples when the distributions do not have the same values of any parameters.

Maximum likelihood estimation. The following explains ML methods for estimating a separate life distribution for each failure mode. The methods also provide approximate confidence limits. Theory for these methods is in Section 5.5.

For ML analysis of a particular mode, failure and censoring times for that mode must be extracted from the data as explained above. Such data on the wire and bond failures are shown in Tables 4.2 and 4.3. Then a distribution is fitted to such multiply censored data on a mode by the ML method. The laborious ML calculations for such data must be done with computer programs such as STATPAC by Nelson and Hendrickson (1972) and Nelson and others (1978).

The ML fit of the normal distribution to the data for each failure mode was calculated by the STATPAC program of Nelson and others (1978). The fitted distributions for wire and bond failures appear in Figures 4.2 and 4.3. Figures 4.4 and 4.5 for wire and bond failures show the ML estimates and

Table 4.2. Wire Failure Data

550+	1150+	1250+	1550+
750	1150+	1350	1550
950+	1150	1450+	1550
950	1150	1450+	1850
1150	1250+	1450	2050+

* MAXIMUM LIKELIHOOD ESTIMATES FOR DIST. PARAMETERS
WITH APPROXIMATE 95% CONFIDENCE LIMITS

PARAMETERS	ESTIMATE	LOWER LIMIT	UPPER LIMIT
CENTER $\hat{\mu}_w=$ 1517.384		$\mu_w=$ 1298.909	$\tilde{\mu}_w=$ 1735.859
SPREAD $\hat{\sigma}_w=$ 398.8265		$\sigma_w=$ 256.3974	$\tilde{\sigma}_w=$ 620.3751

* COVARIANCE MATRIX

PARAMETERS	CENTER $\hat{\mu}_w$	SPREAD $\hat{\sigma}_w$
CENTER $\hat{\mu}_w$	12424.89	
SPREAD $\hat{\sigma}_w$	3606.373	8081.704

PCTILES

* MAXIMUM LIKELIHOOD ESTIMATES FOR DIST. PCTILES
WITH APPROXIMATE 95% CONFIDENCE LIMITS

	$\hat{y}_p$	y_p	$\tilde{y}_p$
PCT. P	ESTIMATE	LOWER LIMIT	UPPER LIMIT
0.1	284.8018	-223.7595	793.3631
0.5	489.9127	62.82445	917.0010
1	589.4003	200.3698	978.4307
5	861.2300	567.6657	1154.794
10	1006.197	754.7586	1257.634
20	1181.789	966.3855	1397.192
50	1517.384	1298.909	1735.859
80	1852.979	1547.971	2157.988
90	2028.571	1662.169	2394.973
95	2173.538	1752.405	2594.670
99	2445.367	1915.946	2974.789

PERCENT(500. LIMIT)

* MAXIMUM LIKELIHOOD ESTIMATES FOR % WITHIN LIMITS
WITH APPROXIMATE 95% CONFIDENCE LIMITS

	ESTIMATE	LOWER LIMIT	UPPER LIMIT
	$\hat{R}(500)$	$R(500)$	$\tilde{R}(500)$
PCT	99.46284%	89.66245	99.97471

Figure 4.4. STATPAC output on normal ML fit to wire strength.

Table 4.3.   Bond Failure Data

550	1150	1250	1550
750+	1150	1350+	1550+
950	1150+	1450	1550+
950+	1150+	1450	1850+
1150+	1250	1450+	2050

* MAXIMUM LIKELIHOOD ESTIMATES FOR DIST. PARAMETERS
WITH APPROXIMATE 95% CONFIDENCE LIMITS

PARAMETERS	ESTIMATE	LOWER LIMIT	UPPER LIMIT
CENTER $\hat{\mu}_B$=	1522.314	$\underset{\sim}{\mu}_B$=1283.975	$\tilde{\mu}_B$=1760.654
SPREAD $\hat{\sigma}_B$=	434.9267	$\underset{\sim}{\sigma}_B$=279.7427	$\tilde{\sigma}_B$=676.1973

* COVARIANCE MATRIX

PARAMETERS	CENTER $\hat{\mu}_B$	$\hat{\sigma}_B$ SPREAD
CENTER $\hat{\mu}_B$	14787.01	
SPREAD $\hat{\sigma}_B$	4466.143	9589.641

PCTILES

* MAXIMUM LIKELIHOOD ESTIMATES FOR DIST. PCTILES
WITH APPROXIMATE 95% CONFIDENCE LIMITS

PCT. P	$\hat{y}_P$ ESTIMATE	$\underset{\sim}{y}_P$ LOWER LIMIT	$\tilde{y}_P$ UPPER LIMIT
0.1	178.1635	-371.9489	728.2759
0.5	401.8403	-59.57930	863.2599
1	510.3331	90.32051	930.3456
5	806.7678	490.4355	1123.100
10	964.8561	694.0329	1235.679
20	1156.342	923.9285	1388.756
50	1522.314	1283.975	1760.654
80	1888.286	1553.975	2222.598
90	2079.772	1678.140	2481.405
95	2237.861	1776.367	2699.354
99	2534.295	1954.560	3114.030

PERCENT(500. LIMIT)

* MAXIMUM LIKELIHOOD ESTIMATES FOR % WITHIN LIMITS
WITH APPROXIMATE 95% CONFIDENCE LIMITS

	ESTIMATE $\hat{R}(500)$	LOWER LIMIT $\underset{\sim}{R}(500)$	UPPER LIMIT $\tilde{R}(500)$
PCT	99.06270	88.25481	99.93277

Figure 4.5.   STATPAC output on normal ML fit to bond strength.

confidence intervals for parameters and percentiles. The wire and bond strength distributions are very similar (see Figures 4.2 and 4.3).

4.5. The Life Distribution When All Failure Modes Act

When a product is in actual use, any mode can cause failure. This section presents the ML method for estimating the life distribution when all failure modes act.

ML estimate of the life distribution.   Suppose that the product reliability at a given age is to be estimated. First obtain the ML estimate of the reliability for that age for each failure mode as described in Section 4.4. Then the ML estimate of reliability when all failure modes act is the product (4.1) of the reliability estimates for each failure mode. Confidence limits for such a reliability are described in Section 5.5. The estimate and confidence limits can be calculated for any number of ages, as needed.

For example, for the connections, suppose that the reliability for a strength of 500 mg is to be estimated. The ML estimate of reliability at 500 mg is obtained as described above and is 0.995 for wires and 0.991 for bonds. So the ML estimate of reliability when all failure modes act is $0.995 \times 0.991 = 0.986$.

The estimate of the fraction failing from any cause below 500 mg is $1 - 0.986 = 0.014$. This point is on the curve in Figure 4.1. That curve is the ML estimate of the strength distribution of connections when all failure modes act. It was obtained by the above calculations for various strengths. Such calculations yield a distribution curve, **not** a straight line, since the calculated distribution of connection strength is not normal. The ML estimate of any percentile can be obtained from the plot or by numerical calculation.

Rough estimate.   A rough analysis of such data may do. This involves (1) using all failure times (ignoring cause) and running times of all unfailed units, and (2) ML fitting a single distribution to that data. This may be satisfactory in the range of the data. It may mislead in extrapolating into the tail above or below the data; then this approach is generally not recommended.

Hazard plot of the data with all failure modes.   Chapter 5 explains hazard plotting of a distribution of time to failure when all modes act. The hazard plot (Figure 4.1 for all modes) provides a check on the data and the fit of the model. The connection data and fitted curve appear valid.

4.6. The Life Distribution with Some Failure Modes Eliminated

For some products, some failure modes can be eliminated by design changes. This section shows how to estimate the resulting life distribution from the data on the remaining failure modes. The worth of a redesign can thus be assessed in advance.

ML estimate of the life distribution. Suppose that the reliability of redesigned units at a given age is to be estimated. First obtain the ML estimate of the reliability at that age for each remaining failure mode, as described in Section 4.4. Then the ML estimate of the redesign reliability for that age is the product (4.1) of those reliability estimates. This method assumes that the reliability for an eliminated failure mode is 1. In practice, a failure mode may not be completely eliminated; Section 3 of Chapter 5 describes how to handle such a situation.

For the connections, bond failures have a greater percentage below the 500-mg specification. Bond failures could be eliminated through a redesign. Suppose that the redesign reliability for 500 mg is to be estimated. The 500-mg reliability estimated for wire failure is 0.995. So the ML estimate of the redesign reliability is $0.955 \times 1 = 0.995$.

The estimate of the fraction below 500 mg is $1 - 0.995 = 0.005$. This point is off scale on normal hazard paper in Figure 4.2. There the line is the ML estimate of the life distribution of the redesign. Such a distribution curve is obtained through repeated calculation of the fraction failing by various strengths (ages). The estimate of the percentage below the 500-mg specification is 0.5%; this is marginal. So the connections need further improvement (particularly the wire). Confidence limits may be obtained as described in Section 5.3.

Rough estimate. A rough analysis of such data may do. This involves (1) using all failure times of the remaining failure modes, ignoring causes and treating all other times as running times, and (2) fitting a single distribution to that data. This approach is generally **not** recommended for extrapolating.

Sometimes a mode is not completely eliminated. An estimate of the new distribution for that mode can be combined with the estimates for the other remaining modes with (4.1), as shown by Nelson (1974b); see also Section 3 of Chapter 5.

Hazard plot with modes eliminated. Chapter 5 explains hazard plotting of a life distribution when certain failure modes are eliminated. The hazard plot (Figure 4.2 without bond failures) provides a check on the data and on the ML fit of the model. The connection data and the fitted model appear satisfactory.

4.7. Other Methods

Other analyses of such life data include checks on the life distribution, the data, and the independence of the failure modes. The series-system model assumes that the failure modes are statistically independent. Nelson (1974b) describes how to check independence (units must have a failure from each mode) and the validity of the distribution and the data.

Correlations between the failure times for dependent modes are usually positive. This means that long (short) failure times of one mode tend to go with long (short) failure times of the other mode. That is, a unit is generally strong or generally weak. The failure time of the mode of interest soon follows the failure (censoring time) from another mode when the modes are dependent. Thus, assuming independence, the ML estimate (Section 4.4) for the life distribution of, say, failure mode 1 is biased toward long life; so this estimate for the distribution of mode 1 tends to be on the high side. To obtain one on the low side, treat the failure times for the correlated failure modes as if they were from mode 1. If close, the two estimated bounds may serve practical purposes.

A graphical check for independence is given by Nadas (1969). It applies to normal and lognormal life distributions. Moeschberger and David (1971) and David and Moeschberger (1979) give ML estimates for the joint distribution of correlated failure modes.

In some applications, the cause of failure is not identified. Then the data are treated as coming from the many-parameter distribution (4.1). For example, Friedman and Gertsbakh (1980) present ML estimates for a three-parameter model based on two competing failure modes—one exponential and the other Weibull. See also David and Moeschberger (1979).

5.   GENERAL MAXIMUM LIKELIHOOD METHODS AND THEORY

This advanced technical section presents general maximum likelihood (ML) methods for fitting a distribution to data multiply censored on the right. These methods provide estimates and approximate confidence intervals. Results and methods are presented without regularity conditions and proofs, which are given by Wilks (1962), Rao (1973), and Hoadley (1971).

Section 5.1 presents general ML theory for a distribution with one unknown parameter and complete data; the theory is illustrated with the exponential distribution, and it also applies to the binomial, Poisson, and other one-parameter distributions. Section 5.2 presents general ML theory for such a distribution and multiply time censored data (on the right); the theory is illustrated with the exponential distribution. Section 5.3 presents general ML theory for a distribution with two (or more) unknown parameters and complete data; the theory is illustrated with the normal distribution and also applies to the Weibull, extreme value, and other distributions. Section 5.4 presents general ML theory for such a distribution and multiply time censored data (on the right). Section 5.5 presents ML theory for competing failure modes. Section 5.6 motivates the ML theory.

Each section covers (1) the (log) likelihood, (2) ML estimates of distribution parameters, (3) their Fisher information matrix, (4) their covariance matrix, (5) estimates of functions of the parameters, and (6) approximate confidence limits. Although the asymptotic ML theory applies to large samples, ML estimates and methods are usually good with small samples.

5.1 One-Parameter Distribution and Complete Data (Exponential Example)

1. LIKELIHOOD

For a continuous distribution, the **likelihood** $L(\theta)$ for a complete sample of n observations $y_1, \ldots, y_n$ is defined as the joint probability density $g(y_1, \ldots, y_n; \theta)$. The likelihood $L(\theta)$ is viewed as a function of θ, an arbitrary value of the distribution parameter. The (unknown) true value is denoted by θ_0. Usually the n observations are a random sample of **independent** observations from the same distribution with probability density $f(y; \theta)$. Then the **sample likelihood** is

$$L(\theta) = f(y_1; \theta) \times \cdots \times f(y_n; \theta). \tag{5.1.1}$$

For a discrete distribution, the probability mass function is used in place of the probability density.

The sample **log likelihood** is the natural log of the likelihood; namely,

$$\mathcal{L}(\theta) = \ln[L(\theta)] = \ln[f(y_1; \theta)] + \cdots + \ln[f(y_n; \theta)] \tag{5.1.2}$$

for **independent** observations from the same distribution. This function of θ has different shapes for different sets of sample $y_1, \ldots, y_n$. The log likelihood for just observation i is

$$\mathcal{L}_i(\theta) = \ln[f(y_i; \theta)]. \tag{5.1.3}$$

So the sample log likelihood is the sum of the n log likelihoods of the independent observations.

Exponential likelihood. For a complete sample of n independent observations $y_1, \ldots, y_n$ from an exponential distribution with mean θ, the sample likelihood is

$$L(\theta) = (1/\theta)\exp(-y_1/\theta) \times \cdots \times (1/\theta)\exp(-y_n/\theta)$$

$$= \theta^{-n}\exp[-(y_1 + \cdots + y_n)/\theta].$$

The sample log likelihood is

$$\mathcal{L}(\theta) = -n\ln(\theta) - [(y_1 + \cdots + y_n)/\theta]. \tag{5.1.4}$$

The log likelihood for just observation y_i is

$$\mathcal{L}_i(\theta) = -\ln(\theta) - (y_i/\theta).$$

2. ML ESTIMATE

The **ML estimate** $\hat{\theta}$ is the θ value that maximizes the sample likelihood $L(\theta)$. Equivalently, $\hat{\theta}$ maximizes the sample log likelihood $\mathcal{L}(\theta)$, since it is a monotone function of $L(\theta)$. The ML estimate is a function of the sample values $y_1, \ldots, y_n$; this function could theoretically be written as $\hat{\theta} = \hat{\theta}(y_1, \ldots, y_n)$. In practice, however, one may not be able to write $\hat{\theta}$ as an explicit function of $y_1, \ldots, y_n$.

The value $\hat{\theta}$ that maximizes $\mathcal{L}(\theta)$ can be found by the usual calculus method of setting the derivative of $\mathcal{L}(\theta)$ with respect to θ equal to zero and solving for $\hat{\theta}$; namely, solve the **likelihood equation:**

$$\partial\mathcal{L}(\theta)/\partial\theta = 0. \tag{5.1.5}$$

After finding a solution $\hat{\theta}$ of (5.1.5), one should confirm that $\hat{\theta}$ maximizes $\mathcal{L}(\theta)$. If $\partial^2\mathcal{L}(\theta)/\partial\theta^2$ evaluated at $\theta = \hat{\theta}$ is negative, then $\hat{\theta}$ corresponds to a local optimum (not necessarily a global one, which is desired).

For n large, the cumulative distribution of the ML estimator $\hat{\theta}$ is close to a normal one, with mean θ_0 and asymptotic variance given by (5.1.12), provided that the life distribution satisfies regularity conditions (Wilks, 1962; Rao, 1973; Hoadley, 1971). Also, under regularity conditions on the life distribution, no other estimator with an asymptotic normal distribution has smaller asymptotic variance.

The partial derivative $\partial\mathcal{L}_i(\theta)/\partial\theta$ is called the **score** for observation i. Evaluated at $\theta = \theta_0$, its expectation when the observation is from $f(y_i; \theta_0)$ satisfies

$$0 = E_0\{\partial\mathcal{L}_i(\theta)/\partial\theta\}_0 \equiv \int_{-\infty}^{\infty} \{\partial\mathcal{L}_i(\theta)/\partial\theta\}_0 f(y_i; \theta_0)\, dy_i.$$

The subscript zero on $\{\ \}_0$ indicates that the quantity inside is evaluated at $\theta = \theta_0$. The subscript zero on E_0 indicates that the expectation is taken with respect to the true distribution with θ_0. Consequently, for independent observations, the **sample score** $\partial\mathcal{L}(\theta)/\partial\theta$ satisfies

$$0 = E_0\{\partial\mathcal{L}(\theta)/\partial\theta\}_0.$$

These relationships can aid in the calculation of theoretical expectations, for example, for the Fisher information below.

ML estimate of exponential mean. The likelihood equation for a complete sample of n independent observations $y_1, \ldots, y_n$ is the derivative of (5.1.4):

$$\partial \mathfrak{L}(\theta)/\partial \theta = -(n/\theta) + ((y_1 + \cdots + y_n)/\theta^2) = 0. \qquad (5.1.6)$$

The solution of this equation is the ML estimate

$$\hat{\theta} = (y_1 + \cdots + y_n)/n = \bar{y}, \qquad (5.1.7)$$

the sample average. As seen in Chapter 6, the distribution of $\hat{\theta}$ is approximately normal, with mean θ_0 and variance (5.1.14).

3. FISHER INFORMATION

One must calculate the Fisher information to obtain the asymptotic variance of the ML estimator of the distribution parameter. The variance is used to obtain approximate confidence limits for the parameter and functions of it. The **Fisher information** is the expectation of the negative of the second partial derivative of the log likelihood with respect to the parameter; that is,

$$E_0\{-\partial^2 \mathfrak{L}(\theta)/\partial \theta^2\}_0 = \int_{-\infty}^{\infty} \cdots \int_{-\infty}^{\infty} \{-\partial^2 \mathfrak{L}(\theta)/\partial \theta^2\}_0$$
$$\times g(y_1, \ldots, y_n; \theta_0)\, dy_1 \cdots dy_n. \qquad (5.1.8)$$

The notation $\{\ \}_0$ denotes that the second derivative is evaluated at $\theta = \theta_0$ and the expectation E_0 is calculated assuming that the observations come from the joint distribution $g(y_1, \ldots, y_n; \theta_0)$, where the parameter has the true value θ_0. The second derivative is an implicit function of the observations, which are regarded as random quantities for the purposes of calculating the expectation. For n independent observations from the same distribution, the Fisher information (5.1.8) becomes

$$E_0\{-\partial^2 \mathfrak{L}(\theta)/\partial \theta^2\}_0 = \sum_{i=1}^{n} \int_{-\infty}^{\infty} \{-\partial^2 \mathfrak{L}_i(\theta)/\partial \theta^2\}_0 f(y_i; \theta_0)\, dy_i$$
$$= n \int_{-\infty}^{\infty} \{-\partial^2 \ln[f(y; \theta)]/\partial \theta^2\}_0 f(y; \theta_0)\, dy. \qquad (5.1.9)$$

If one knows the expectation of the quantity in braces $\{\ \}$, one need not evaluate the integral. Also, one can use the properties of expectation to evaluate the Fisher information. An equivalent formula (under regularity conditions) is

$$E_0\{-\partial^2\mathcal{L}(\theta)/\partial\theta^2\}_0 = E_0\{(\partial\mathcal{L}(\theta)/\partial\theta)^2\}_0 = nE_0\{(\partial\mathcal{L}_i(\theta)/\partial\theta)^2\}_0$$

$$= n\int_{-\infty}^{\infty} (\partial\ln[f(y;\theta)]/\partial\theta)_0^2 f(y;\theta_0)\,dy. \quad (5.1.10)$$

For a discrete distribution, the expectations above are obtained by replacing the integrals by sums running over all possible discrete outcomes.

Exponential Fisher information. For a complete sample of n independent observations from an exponential distribution with mean θ_0, the Fisher information is calculated as follows. First, for an arbitrary observation i,

$$\partial^2\mathcal{L}_i(\theta)/\partial\theta^2 = (1/\theta^2) - 2(y_i/\theta^3).$$

Then

$$E_0\{-\partial^2\mathcal{L}(\theta)/\partial\theta^2\}_0 = -nE_0\{(1/\theta_0^2) - 2(y_i/\theta_0^3)\}$$

$$= -n\{E_0(1/\theta_0^2) - 2(E_0 y_i/\theta_0^3)\}$$

$$= -n\{(1/\theta_0^2) - 2(\theta_0/\theta_0^3)\},$$

since $E_0 y_i = \theta_0$, the mean. Then

$$E_0\{-\partial^2\mathcal{L}(\theta)/\partial\theta^2\}_0 = n/\theta_0^2. \quad (5.1.11)$$

By the equivalent formula (5.1.10) and (5.1.6), one also obtains (5.1.11) as

$$E_0\{-\partial^2\mathcal{L}(\theta)/\partial\theta^2\}_0 = nE_0\{[-(1/\theta_0) + (y_i/\theta_0^2)]^2\}$$

$$= n\{E_0(1/\theta_0^2) - 2(E_0 y_i/\theta_0^3) + (E_0 y_i^2/\theta_0^4)\}$$

$$= n\{(1/\theta_0^2) - 2(1/\theta_0^2) + 2(1/\theta_0^2)\} = n/\theta_0^2.$$

4. ASYMPTOTIC VARIANCE

The true asymptotic (for large n) variance $\mathrm{Var}(\hat\theta)$ of the ML estimator $\hat\theta$ is the inverse of the true Fisher information; namely,

$$\mathrm{Var}(\hat\theta) = 1/E_0\{-\partial^2\mathcal{L}(\theta)/\partial\theta^2\}_0. \quad (5.1.12)$$

This variance is generally a function of θ_0. To obtain confidence limits, one must calculate one of the following two estimates of this variance. The **ML estimate** $\text{Vâr}(\hat{\theta})$ is (5.1.12), with θ_0 replaced by $\hat{\theta}$. The **local estimate** is

$$\text{var}(\hat{\theta}) = 1 / \left(-\partial^2 \mathcal{L}(\theta) / \partial \theta^2 \right)_{\theta = \hat{\theta}}; \qquad (5.1.13)$$

this is the reciprocal of the local Fisher information, that is, the negative of the second partial derivative of the log likelihood evaluated at $\theta = \hat{\theta}$. Here $\mathcal{L}(\theta)$ is calculated for the actual sample values $y_1, \ldots, y_n$ of the observations. The local estimate is generally easier to calculate.

Exponential asymptotic variance. By (5.1.12) and (5.1.11), the true asymptotic variance of the ML estimator $\hat{\theta}$ for the mean θ_0 of an exponential distribution is

$$\text{Var}(\hat{\theta}) = \theta_0^2 / n. \qquad (5.1.14)$$

The ML estimate of this variance is

$$\text{Vâr}(\hat{\theta}) = \hat{\theta}^2 / n = \bar{y}^2 / n. \qquad (5.1.15)$$

The local estimate of the variance is

$$\text{var}(\hat{\theta}) = 1 / \left(-\sum_{i=1}^{n} \left[(1/\theta^2) - 2(y_i / \theta^3) \right]_{\theta = \hat{\theta}} \right) = \hat{\theta}^2 / n = \bar{y}^2 / n. \quad (5.1.16)$$

For complete exponential data, the ML and local estimates of the asymptotic variance are equal. For most distributions and censored data, these estimates differ.

5. ML ESTIMATE OF A FUNCTION OF THE PARAMETER

Suppose that $h = h(\theta)$ is a continuous function of θ. For the exponential distribution, the failure rate $\lambda = 1/\theta$ and reliability $R(y) = \exp(-y/\theta)$ are such functions. The ML estimate for the true value $h_0 = h(\theta_0)$ is $\hat{h} = h(\hat{\theta})$; this is called the **invariance property** of ML estimators. For large sample size n, the cumulative distribution function of $\hat{h}$ is close to a normal one, with a mean equal to h_0 and a (true asymptotic) variance

$$\text{Var}(\hat{h}) = (\partial h / \partial \theta)_0^2 \text{Var}(\hat{\theta}), \qquad (5.1.17)$$

where the notation $(\ )_0$ denotes that the partial derivative is evaluated at $\theta = \theta_0$. The partial derivative $\partial h / \partial \theta$ must be a continuous function of θ in a neighborhood of the true value θ_0. The ML and local estimates of (5.1.17) are obtained by using respectively $\text{Vâr}(\hat{\theta})$ and $\text{var}(\hat{\theta})$ for $\text{Var}(\hat{\theta})$ and using $\hat{\theta}$ for θ_0 in the partial derivative.

ML estimate of the failure rate. For an exponential distribution, the failure rate $\lambda_0 = 1/\theta_0$ is a continuous function of θ_0. Its ML estimate is

$$\hat{\lambda} = 1/\hat{\theta} = 1/\bar{y}.$$

The calculation of its asymptotic variance is

$$\partial\lambda/\partial\theta = -1/\theta^2, \qquad \text{Var}(\hat{\lambda}) = \left(-1/\theta_0^2\right)^2\theta_0^2/n = \left(1/\theta_0^2\right)/n = \lambda_0^2/n.$$

For complete data, the ML and local estimates of this variance are equal and are

$$\text{Vâr}(\hat{\lambda}) = \text{var}(\hat{\lambda}) = \hat{\lambda}^2/n = 1/(n\bar{y}^2).$$

6. CONFIDENCE LIMITS

Two-sided approximate $100\gamma\%$ confidence limits for the true value h_0 of a function are

$$\underset{\sim}{h} = \hat{h} - K_\gamma\left[\text{var}(\hat{h})\right]^{1/2}, \qquad \tilde{h} = \hat{h} + K_\gamma\left[\text{var}(\hat{h})\right]^{1/2}, \qquad (5.1.18)$$

where K_γ is the $[100(1+\gamma)/2]$th standard normal percentile. Either the ML or local estimate of variance can be used in (5.1.18) and (5.1.19). This interval employs the large-sample normal distribution of $\hat{h}$. Of course, (5.1.18) is an interval for the parameter θ_0 itself; then

$$\underset{\sim}{\theta} = \hat{\theta} - K_\gamma\left[\text{var}(\hat{\theta})\right]^{1/2}, \cdot \qquad \tilde{\theta} = \hat{\theta} + K_\gamma\left[\text{var}(\hat{\theta})\right]^{1/2}.$$

If $\hat{h}$ must be positive, one treats $\ln(\hat{h})$ as normally distributed. Then positive two-sided approximate $100\gamma\%$ confidence limits for h_0 are

$$\underset{\sim}{h} = \hat{h}/\exp\left\{K_\gamma\left[\text{var}(\hat{h})\right]^{1/2}/\hat{h}\right\}, \quad \tilde{h} = \hat{h}\cdot\exp\left\{K_\gamma\left[\text{var}(\hat{h})\right]^{1/2}/\hat{h}\right\}. \quad (5.1.19)$$

A one-sided approximate $100\gamma\%$ confidence limit is obtained from a preceding limit when K_γ is replaced by z_γ, the 100γth standard normal percentile.

It may be possible to find a function such that the distribution of $\hat{h}$ is closer to normal than that of $\hat{\theta}$, yielding more accurate confidence limits.

Confidence limits for an exponential mean. The mean θ_0 of an exponential distribution must be positive. By (5.1.19), positive two-sided approximate $100\gamma\%$ confidence for θ_0 are

$$\underline{\theta} = \bar{y}/\exp\left[K_\gamma(\bar{y}^2/n)^{1/2}/\bar{y}\right] = \bar{y}/\exp\left(K_\gamma/n^{1/2}\right),$$

$$\tilde{\theta} = \bar{y}\cdot\exp\left(K_\gamma/n^{1/2}\right). \tag{5.1.20}$$

5.2. One-Parameter Distribution and Multiply Censored Data (Exponential Example)

ML theory for a one-parameter distribution and multiply time censored data is the same as that for complete data; however, the (log) likelihood functions differ to take censoring into account. This section uses the assumption that the lives $y_1, \ldots, y_n$ of the n sample units are statistically independent and come from the same distribution with probability density $f(y; \theta)$. Throughout θ denotes an arbitrary value of the distribution parameter and θ_0 denotes the true (unknown) value.

1. LIKELIHOOD

For a continuous distribution, the likelihood for unit i with an observed failure at time y_i is the probability density

$$L_i(\theta) = f(y_i; \theta). \tag{5.2.1}$$

The corresponding log likelihood is

$$\mathcal{L}_i(\theta) = \ln\left[f(y_i; \theta)\right]. \tag{5.2.2}$$

For unit i with a (Type I) censoring time y_i (censored on the right), the likelihood is the survival probability

$$L_i(\theta) = R(y_i; \theta). \tag{5.2.3}$$

The corresponding log likelihood is

$$\mathcal{L}_i(\theta) = \ln\left[R(y_i; \theta)\right]. \tag{5.2.4}$$

Because the sample units are statistically independent, the sample log likelihood is the product of the likelihoods of the n sample units; namely,

$$L(\theta)= \prod_i {}' f(y_i;\theta) \prod_i {}'' R(y_i; \theta),$$ (5.2.5)

where the first product runs over the r failure times, and the second product runs over the $n-r$ censoring times. The corresponding sample log likelihood is the sum of the log likelihoods of the n independent units; namely,

$$\mathcal{L}(\theta)= \sum_i {}' \ln[f(y_i;\theta)] + \sum_i {}'' \ln[R(y_i; \theta)],$$ (5.2.6)

where the first sum runs over the r failure times, and the second sum runs over the $n-r$ censoring times.

Exponential likelihood.   For a multiply censored sample of independent units, suppose that the r failure times are $y_1,\ldots, y_r$ and the $n-r$ censoring times are $y_{r+1},\ldots, y_n$. For an exponential life distribution, the sample likelihood is

$$L(\theta)= \prod_{i=1}^r (1/\theta)\exp(-y_i/\theta) \prod_{i=r+1}^n \exp(-y_i/\theta).$$

The corresponding log likelihood is

$$\mathcal{L}(\theta)= -r\ln(\theta)-(1/\theta) \sum_{i=1}^n y_i.$$ (5.2.7)

2. ML ESTIMATE

As in Section 5.1, the ML estimate $\hat{\theta}$ is the θ value that maximizes the sample likelihood $L(\theta)$ or log likelihood $\mathcal{L}(\theta)$. $\hat{\theta}$ also is a solution of the likelihood equation

$$\partial\mathcal{L}(\theta)/\partial\theta =0.$$ (5.2.8)

One should confirm that a solution of (5.2.8) maximizes $\mathcal{L}(\theta)$.

As in Section 5.1, for large r, the cumulative distribution of $\hat{\theta}$ is close to a normal one, with mean θ_0, the true value, and (asymptotic) variance given by (5.2.16). This is so, provided that the life distribution and censoring times satisfy various regularity conditions given by Wilks (1962), Rao (1973), Hoadley (1971), and especially Basu and Ghosh (1980). Also, under such regularity conditions, no other estimator with an asymptotic normal distribution has smaller asymptotic variance.

ML estimate of the exponential mean. From (5.2.7), the likelihood equation for the exponential mean is

$$0 = \partial \mathcal{L}(\theta)/\partial\theta = -(r/\theta) + (1/\theta^2) \sum_{i=1}^{n} y_i.$$

Its solution is the ML estimate

$$\hat{\theta} = \left(\sum_{i=1}^{n} y_i \right)/r. \tag{5.2.9}$$

This is the total running time on failed and censored units, divided by the number of failures.

3. FISHER INFORMATION

As before, one needs the Fisher information to obtain the asymptotic variance of the ML estimator. For Type I (time) censoring, we assume that y_i is the random failure time of unit i and η_i is the (known) planned censoring time. Often the η_i of the failed units are not known; this situation is discussed below. For unit i, we use the indicator function

$$I_i(y_i) = \begin{cases} 1 & \text{if } y_i < \eta_i \text{ (failure observed)}, \\ 0 & \text{if } y_i \geq \eta_i \text{ (censored)}. \end{cases} \tag{5.2.10}$$

For purposes of calculating expectations, y_i and $I_i(y_i)$ are regarded as random variables. For unit i, the log likelihood can be written

$$\mathcal{L}_i(\theta) = I_i(y_i)\ln[f(y_i; \theta)] + [1 - I_i(y_i)]\ln[R(\eta_i; \theta)]. \tag{5.2.11}$$

The Fisher information for unit i is

$$E_0\{-\partial^2\mathcal{L}_i(\theta)/\partial\theta^2\}_0 = \int_{-\infty}^{\infty} \{-\partial^2\mathcal{L}_i(\theta)/\partial\theta^2\}_0 f(y_i; \theta_0)dy_i$$

$$= \int_{-\infty}^{\eta_i} \{-\partial^2\ln[f(y_i; \theta)]/\partial\theta^2\}_0 f(y_i; \theta_0) \, dy_i$$

$$+ \{-\partial^2\ln[R(\eta_i; \theta)]/\partial\theta^2\}_0 R(\eta_i; \theta_0) \tag{5.2.12}$$

where the subscript 0 indicates that the quantity is evaluated at $\theta = \theta_0$, the true value. For a discrete distribution, sums replace the integrals above. For independent y_i, the true sample Fisher information is

$$E_0\{-\partial^2 \mathfrak{L}(\theta)/\partial\theta^2\}_0 = \sum_{i=1}^{n} E_0\{-\partial^2 \mathfrak{L}_i(\theta)/\partial\theta^2\}_0. \qquad (5.2.13)$$

The sample local Fisher information is the negative second partial derivative of the sample log likelihood

$$-\partial^2 \mathfrak{L}(\theta)/\partial\theta^2 = -\sum_i \partial^2 \ln[f(y_i;\theta)]/\partial\theta^2 - \sum_i \partial^2 \ln[R(y_i;\theta)]/\partial\theta^2,$$

$$(5.2.14)$$

evaluated at $\theta = \hat{\theta}$, the ML estimate, where the first (second) sum runs over the failed (censored) units.

Exponential Fisher information. For a Type I (time) censored unit i from an exponential distribution with mean θ_0,

$$\{-\partial^2 \ln[f(y_i;\theta)]/\partial\theta^2\} = -\partial^2[-\ln(\theta)-(y_i/\theta)]/\partial\theta^2$$

$$= \{(-1/\theta^2)+2(y_i/\theta^3)\},$$

$$\int_0^{\eta_i}\{(-1/\theta_0^2)+2(y_i/\theta_0^3)\}(1/\theta_0)\exp(-y_i/\theta_0)\,dy_i$$

$$= (1/\theta_0^2)[1-\exp(-\eta_i/\theta_0)-2(\eta_i/\theta_0)\exp(-\eta_i/\theta_0)],$$

$$\{-\partial^2 \ln[R(\eta_i;\theta)]/\partial\theta^2\}_0 = \{-\partial^2(-\eta_i/\theta)/\partial\theta^2\}_0 = (2\eta_i/\theta_0^3),$$

$$E_0\{-\partial^2 \mathfrak{L}_i(\theta)/\partial\theta^2\}_0 = (1/\theta_0^2)[1-\exp(-\eta_i/\theta_0)-2(\eta_i/\theta_0)\exp(-\eta_i/\theta_0)]$$

$$+ (2\eta_i/\theta_0^3)\exp(-\eta_i/\theta_0) = (1/\theta_0^2)[1-\exp(-\eta_i/\theta_0)].$$

For a Type I (time) censored sample of independent units, the true sample Fisher information is

$$E_0\{-\partial^2 \mathfrak{L}(\theta)/\partial\theta^2\}_0 = (1/\theta_0^2)\sum_{i=1}^{n}[1-\exp(-\eta_i/\theta_0)]. \qquad (5.2.15)$$

The local estimate of the Fisher information is [from (5.2.14)]

$$-\partial^2 \mathfrak{L}(\theta)/\partial\theta^2|_{\theta=\hat{\theta}} = \sum_i'[(-1/\hat{\theta}^2)+2(y_i/\hat{\theta}^3)] + \sum_i''(2y_i/\hat{\theta}^3) = r/\hat{\theta}^2,$$

where $\hat{\theta} = \sum_{i=1}^{n} y_i/r$.

4. ASYMPTOTIC VARIANCE

For a large number r of failures, the true asymptotic variance $\text{Var}(\hat{\theta})$ of the ML estimator $\hat{\theta}$ is the inverse of the true Fisher information; namely,

$$\text{Var}(\hat{\theta})=1/E_0\{-\partial^2\mathcal{L}(\theta)/\partial\theta^2\}_0. \qquad (5.2.16)$$

To obtain confidence limits, one must use one of the following two estimates of this variance. The ML estimate $\hat{\text{Var}}(\hat{\theta})$ is (5.2.16), where θ_0 is replaced by $\hat{\theta}$. The local estimate is the inverse of the sample local Fisher information, namely,

$$\text{var}(\hat{\theta})=1/\left(-\partial^2\mathcal{L}(\theta)/\partial\theta^2\right)_{\theta=\hat{\theta}}. \qquad (5.2.17)$$

Exponential asymptotic variance. By (5.2.16) and (5.2.15), the true asymptotic variance of the ML estimate of an exponential mean (Type I censoring) is

$$\text{Var}(\hat{\theta})=\theta_0^2\Big/\left(\sum_{i=1}^{n}[1-\exp(-\eta_i/\theta_0)]\right). \qquad (5.2.18)$$

The ML estimate of this variance is (5.2.18), with θ_0 replaced by $\hat{\theta}$. This variance estimate contains a planned censoring time η_i for each unit. In many applications, such times are unknown for units that have failed. Then one can use the local estimate of this variance, namely,

$$\text{var}(\hat{\theta})=\hat{\theta}^2/r. \qquad (5.2.19)$$

This estimate does not involve planned censoring times. For Type I (time) censored data, the ML and local estimates differ but are numerically close.

5. ML ESTIMATE OF A FUNCTION OF THE PARAMETER

Theory for estimating a function of the parameter is exactly the same as that in Section 5.1.

6. CONFIDENCE LIMITS

Theory for approximate confidence limits is exactly the same as that in Section 5.1.

5.3. Multiparameter Distribution and Complete Data (Normal Example)

ML theory for a multiparameter distribution is given here in terms of a continuous distribution with two parameters μ and σ and probability density $f(y; \mu, \sigma)$. As before, μ and σ denote arbitrary parameter values, and μ_0 and σ_0 denote the (unknown) true values. The ML theory is much like that for a one-parameter distribution, but calculation of the variances and covariances of the ML estimates must be generalized to any number of parameters. The theory involves some basic matrix algebra. The following theory readily extends to distributions with more than two parameters and to discrete distributions.

1. LIKELIHOOD

For a complete sample of n independent observations $y_1, \ldots, y_n$ each from a continuous distribution with probability density $f(y; \mu, \sigma)$, the **sample likelihood** is

$$L(\mu, \sigma) = f(y_1; \mu, \sigma) \times \cdots \times f(y_n; \mu, \sigma); \qquad (5.3.1)$$

this is the "probability" of obtaining the sample observations if μ and σ are the parameter values. Here $L(\mu, \sigma)$ is viewed as a function of μ and σ, but it is also a function of $y_1, \ldots, y_n$.

The sample **log likelihood** is

$$\mathcal{L}(\mu, \sigma) = \ln[L(\mu, \sigma)] = \ln[f(y_1; \mu, \sigma)] + \cdots + \ln[f(y_n; \mu, \sigma)], \quad (5.3.2)$$

where the observations are independent and from the same probability density. The log likelihood for just observation i is

$$\mathcal{L}_i(\mu, \sigma) = \ln[f(y_i; \mu, \sigma)]. \qquad (5.3.3)$$

Normal likelihood. For a complete sample of n independent observations $y_1, \ldots, y_n$ from a normal distribution with mean μ_0 and standard deviation σ_0, the sample likelihood is

$$L(\mu, \sigma) = (2\pi\sigma^2)^{-1/2} \exp[-(y_1 - \mu)^2/(2\sigma^2)] \times \cdots \times (2\pi\sigma^2)^{-1/2}$$

$$\times \exp[-(y_n - \mu)^2/(2\sigma^2)]$$

$$= (2\pi\sigma^2)^{-n/2} \exp\left(-\sum_{i=1}^{n} (y_i - \mu)^2/(2\sigma^2)\right).$$

The sample log likelihood is

$$\mathcal{L}(\mu, \sigma) = -(n/2)\ln(2\pi\sigma^2) - \left(\sum_{i=1}^{n} (y_i - \mu)^2/(2\sigma^2)\right). \qquad (5.3.4)$$

The log likelihood for just observation y_i is

$$\mathcal{L}_i(\mu, \sigma) = -\tfrac{1}{2}\left[\ln(2\pi) + \ln(\sigma^2)\right] - \left[(y_i - \mu)^2/(2\sigma^2)\right].$$

2. ML ESTIMATES

The **ML estimates** $\hat{\mu}$ and $\hat{\sigma}$ are the μ and σ values that maximize the sample likelihood $L(\mu, \sigma)$ or, equivalently, maximize the sample log likelihood. The values $\hat{\mu}$ and $\hat{\sigma}$ may be found with the usual calculus method. Namely, the μ and σ values that satisfy the **likelihood equations**

$$\partial\mathcal{L}(\mu, \sigma)/\partial\mu = 0, \qquad \partial\mathcal{L}(\mu, \sigma)/\partial\sigma = 0 \qquad (5.3.5)$$

are the ML estimates $\hat{\mu}$ and $\hat{\sigma}$. For most distributions and sample data, the solution of these equations is unique. These estimates are functions of the observations $y_1,\ldots, y_n$, say, $\hat{\mu} = \hat{\mu}(y_1,\ldots, y_n)$ and $\hat{\sigma} = \hat{\sigma}(y_1,\ldots, y_n)$; however, for many distributions it is not possible to solve for explicit functions. Instead one must obtain $\hat{\mu}$ and $\hat{\sigma}$ by numerical methods from the observed $y_1,\ldots, y_n$, as described in Section 6.

For large n, the joint cumulative distribution of $\hat{\mu}$ and $\hat{\sigma}$ is close to a joint normal one with means μ_0 and σ_0 and covariance matrix given by (5.3.14). This is so provided that the life distribution satisfies regularity conditions given by Wilks (1962), Rao (1973), and Hoadley (1971). Also, under regularity conditions on the life distribution, no other asymptotically joint normal estimators have smaller asymptotic variances.

The partial derivatives $\partial\mathcal{L}_i(\mu, \sigma)/\partial\mu$ and $\partial\mathcal{L}_i(\mu, \sigma)/\partial\sigma$ are called the **scores** for observation i. Evaluated at $\mu = \mu_0$ and $\sigma = \sigma_0$, their expectations when the observation is from $f(y_i; \mu_0, \sigma_0)$ satisfy

$$0 = E_0\{\partial\mathcal{L}_i(\mu, \sigma)/\partial\mu\}_0 = \int_{-\infty}^{\infty} \{\partial\mathcal{L}_i(\mu, \sigma)/\partial\mu\}_0 f(y_i; \mu_0, \sigma_0)\, dy_i,$$

$$0 = E_0\{\partial\mathcal{L}_i(\mu, \sigma)/\partial\sigma\}_0 = \int_{-\infty}^{\infty} \{\partial\mathcal{L}_i(\mu, \sigma)/\partial\sigma\}_0 f(y_i; \mu_0, \sigma_0)\, dy_i;$$

the subscript 0 on { } indicates that the quantity (partial derivative) inside is evaluated at $\mu = \mu_0$ and $\sigma = \sigma_0$, and the 0 on the expectation E_0 indicates that the observation y_i comes from $f(y_i; \mu_0, \sigma_0)$. Consequently, the **sample scores** $\partial\mathcal{L}(\mu, \sigma)/\partial\mu$ and $\partial\mathcal{L}(\mu, \sigma)/\partial\sigma$ satisfy

$$0 = E_0\{\partial\mathcal{L}(\mu, \sigma)/\partial\mu\}_0, \qquad 0 = E_0\{\partial\mathcal{L}(\mu, \sigma)/\partial\sigma\}_0$$

since $\mathcal{L}(\mu, \sigma) = \sum_{i=1}^{n}\mathcal{L}_i(\mu, \sigma)$. These relationships can aid in the calculation of theoretical expectations, for example, for the Fisher information matrix below.

ML estimates—normal.   The likelihood equations for a complete sample are

$$0 = \partial \mathcal{L}(\mu, \sigma)/\partial \mu = \sum_{i=1}^{n} 2(y_i - \mu)/(2\sigma^2),$$

$$\text{(5.3.6)}$$

$$0 = \partial \mathcal{L}(\mu, \sigma)/\partial \sigma = -(n/\sigma) + (1/\sigma^3)\left(\sum_{i=1}^{n}(y_i - \mu)^2\right).$$

Their solution is

$$\hat{\mu} = \bar{y} = (y_1 + \cdots + y_n)/n, \qquad \hat{\sigma} = \left(\sum_{i=1}^{n}(y_i - \bar{y})^2/n\right)^{1/2}. \quad \text{(5.3.7)}$$

Here $\hat{\sigma}$ differs slightly from the usual estimate S of the standard deviation, which has $(n-1)$ in the denominator rather than n.

3.   FISHER INFORMATION MATRIX

One must calculate the Fisher information matrix to obtain the asymptotic variances and covariances of the ML estimates of the distribution parameters. This matrix is used later to obtain approximate confidence limits for the parameters and functions of them. One uses the matrix of negative second partial derivatives of the sample log likelihood, namely,

$$\mathbf{F} = \begin{bmatrix} -\partial^2 \mathcal{L}(\mu, \sigma)/\partial \mu^2 & -\partial^2 \mathcal{L}(\mu, \sigma)/\partial \mu \partial \sigma \\ -\partial^2 \mathcal{L}(\mu, \sigma)/\partial \sigma \partial \mu & -\partial^2 \mathcal{L}(\mu, \sigma)/\partial \sigma^2 \end{bmatrix}. \quad \text{(5.3.8)}$$

This matrix is symmetric, since $\partial^2 \mathcal{L}/\partial \mu \partial \sigma = \partial^2 \mathcal{L}/\partial \sigma \partial \mu$. Evaluated at $\mu = \hat{\mu}$ and $\sigma = \hat{\sigma}$, (5.3.8) is the **local Fisher information matrix**. The true theoretical **Fisher information matrix** $\mathbf{F}_0$ is the expectation of $\mathbf{F}$ evaluated for $\mu = \mu_0$ and $\sigma = \sigma_0$, namely,

$$\mathbf{F}_0 = \begin{pmatrix} E_0\{-\partial^2 \mathcal{L}(\mu, \sigma)/\partial \mu^2\}_0 & E_0\{-\partial^2 \mathcal{L}(\mu, \sigma)/\partial \mu \partial \sigma\}_0 \\ E_0\{-\partial^2 \mathcal{L}(\mu, \sigma)/\partial \sigma \partial \mu\}_0 & E_0\{-\partial^2 \mathcal{L}(\mu, \sigma)/\partial \sigma^2\}_0 \end{pmatrix}; \quad \text{(5.3.9)}$$

as before the subscript 0 on { } indicates that the partial derivative is evaluated at $\mu = \mu_0$ and $\sigma = \sigma_0$, and that on the expectation E denotes that the independent random observations $y_1, \ldots, y_n$ come from $f(y; \mu_0, \sigma_0)$. That is, for observation i,

$$E_0\{-\partial^2 \mathcal{L}_i(\mu, \sigma)/\partial \mu^2\}_0 = \int_{-\infty}^{\infty} \{-\partial^2 \mathcal{L}_i(\mu, \sigma)/\partial \mu^2\}_0 f(y_i; \mu_0, \sigma_0)\, dy_i,$$

$$(5.3.10)$$

$$E_0\{-\partial^2 \mathcal{L}_i(\mu, \sigma)/\partial \mu \partial \sigma\}_0 = \int_{-\infty}^{\infty} \{-\partial^2 \mathcal{L}_i(\mu, \sigma)/\partial \mu \partial \sigma\}_0 f(y_i; \mu_0, \sigma_0)\, dy_i,$$

$$E_0\{-\partial^2 \mathcal{L}_i(\mu, \sigma)/\partial \sigma^2\}_0 = \int_{-\infty}^{\infty} \{-\partial^2 \mathcal{L}_i(\mu, \sigma)/\partial \sigma^2\}_0 f(y_i; \mu_0, \sigma_0)\, dy_i.$$

Such an expectation for the sample is the sum of the corresponding expectations for the n units, since $\mathcal{L}(\mu, \sigma) = \Sigma_{i=1}^n \mathcal{L}_i(\mu, \sigma)$.

Equivalent formulas for the expectations (5.3.10) are

$$E_0\{-\partial^2 \mathcal{L}_i(\mu, \sigma)/\partial \mu^2\}_0 = E_0\{(\partial \mathcal{L}_i(\mu, \sigma)/\partial \mu)_0^2\}$$

$$= \int_{-\infty}^{\infty} (\partial \mathcal{L}_i(\mu, \sigma)/\partial \mu)_0^2 f(y_i; \mu_0, \sigma_0)\, dy_i,$$

$$E_0\{-\partial^2 \mathcal{L}_i(\mu, \sigma)/\partial \sigma^2\}_0 = E_0\{(\partial \mathcal{L}_i(\mu, \sigma)/\partial \sigma)_0^2\} \qquad (5.3.11)$$

$$= \int_{-\infty}^{\infty} (\partial \mathcal{L}_i(\mu, \sigma)/\partial \sigma)_0^2 f(y_i; \mu_0, \sigma_0)\, dy_i,$$

$$E_0\{-\partial^2 \mathcal{L}_i(\mu, \sigma)/\partial \mu \partial \sigma\}_0 = E_0\{(\partial \mathcal{L}_i(\mu, \sigma)/\partial \mu)_0 (\partial \mathcal{L}_i(\mu, \sigma)/\partial \sigma)_0\}$$

$$= \int_{-\infty}^{\infty} (\partial \mathcal{L}_i/\partial \mu)_0 (\partial \mathcal{L}_i/\partial \sigma)_0 f(y_i; \mu_0, \sigma_0)\, dy_i.$$

These are the expections of the squares and product of the scores.

For a discrete distribution, sums over all possible outcomes replace the integrals in the formulas above.

Fisher information—normal. For a normal distribution, terms of the Fisher information matrix are calculated as follows. For observation i,

$$-\partial^2 \mathcal{L}_i(\mu, \sigma)/\partial\mu^2 = -\partial^2\left\{-\tfrac{1}{2}\left[\ln(2\pi)+2\ln(\sigma)\right]-\left[(y_i-\mu)^2/(2\sigma^2)\right]\right\}/\partial\mu^2$$

$$= 1/\sigma^2,$$

$$-\partial^2 \mathcal{L}_i(\mu, \sigma)/\partial\sigma^2 = (-1/\sigma^2)+(3/\sigma^4)(y_i-\mu)^2,$$

$$-\partial^2 \mathcal{L}_i(\mu, \sigma)/\partial\mu\partial\sigma = 2(y_i-\mu)/\sigma^3,$$

$$E_0\left\{-\partial^2 \mathcal{L}_i(\mu, \sigma)/\partial\mu^2\right\}_0 = E_0\left\{1/\sigma_0^2\right\} = 1/\sigma_0^2,$$

$$E_0\left\{-\partial^2 \mathcal{L}_i(\mu, \sigma)/\partial\sigma^2\right\}_0 = E_0\left\{(-1/\sigma_0^2)+(3/\sigma_0^4)(y_i-\mu_0)^2\right\}$$

$$= (-1/\sigma_0^2)+(3/\sigma_0^4)\sigma_0^2 = 2/\sigma_0^2,$$

$$E_0\left\{-\partial^2 \mathcal{L}_i(\mu, \sigma)/\partial\mu\partial\sigma\right\}_0 = E_0\left\{2(y_i-\mu_0)/\sigma_0^3\right\} = 0.$$

Then, for the sample,

$$E_0\left\{-\partial^2 \mathcal{L}(\mu, \sigma)/\partial\mu^2\right\}_0 = \sum_{i=1}^{n}(1/\sigma_0^2) = n/\sigma_0^2,$$

$$E_0\left\{-\partial^2 \mathcal{L}(\mu, \sigma)/\partial\sigma^2\right\}_0 = \sum_{i=1}^{n}(2/\sigma_0^2) = 2n/\sigma_0^2, \qquad (5.3.12)$$

$$E_0\left\{-\partial^2 \mathcal{L}(\mu, \sigma)/\partial\mu\partial\sigma\right\}_0 = \sum_{i=1}^{n}0 = 0.$$

The true theoretical Fisher information matrix is

$$\mathbf{F}_0 = \begin{bmatrix} n/\sigma_0^2 & 0 \\ 0 & 2n/\sigma_0^2 \end{bmatrix}. \qquad (5.3.13)$$

4. ASYMPTOTIC COVARIANCE MATRIX

The true asymptotic (for large n) covariance matrix Σ of the ML estimators $\hat{\mu}$ and $\hat{\sigma}$ is the inverse of the true Fisher information matrix; namely,

$$\Sigma \equiv \begin{bmatrix} \text{Var}(\hat{\mu}) & \text{Cov}(\hat{\mu}, \hat{\sigma}) \\ \text{Cov}(\hat{\mu}, \hat{\sigma}) & \text{Var}(\hat{\sigma}) \end{bmatrix} = \mathbf{F}_0^{-1}. \qquad (5.3.14)$$

The variances and covariance in this matrix are generally functions of μ_0

and σ_0. The variances and covariances of the parameter estimates appear in the same positions in the covariance matrix where the corresponding partial derivatives with respect to those parameters appear in the Fisher matrix. For example, $\text{Var}(\hat{\mu})$ is in the upper left corner of $\boldsymbol{\Sigma}$, since $E_0\{-\partial^2 \mathcal{L}/\partial\mu^2\}_0$ is in the upper left corner of $\mathbf{F}_0$. The variances all appear on the main diagonal of $\boldsymbol{\Sigma}$.

To obtain confidence limits, one uses one of the following two estimates of the covariance matrix. The ML estimate $\hat{\boldsymbol{\Sigma}}$ is obtained by substituting $\hat{\mu}$ for μ_0 and $\hat{\sigma}$ for σ_0 in (5.3.14); the corresponding terms of the matrix are denoted by $\text{Vâr}(\hat{\mu})$, $\text{Vâr}(\hat{\sigma})$, and $\text{Côv}(\hat{\mu}, \hat{\sigma})$. The local estimate $\mathbf{V}$ is obtained by inverting the local Fisher information matrix $\mathbf{F}$ in (5.3.8); that is,

$$\mathbf{V} = \begin{bmatrix} \text{var}(\hat{\mu}) & \text{cov}(\hat{\mu}, \hat{\sigma}) \\ \text{cov}(\hat{\mu}, \hat{\sigma}) & \text{var}(\hat{\sigma}) \end{bmatrix} = \mathbf{F}^{-1},$$

where $\hat{\mu}$ and $\hat{\sigma}$ are used in place of μ and σ. In general, the local and ML estimates of $\boldsymbol{\Sigma}$ are not equal.

Asymptotic covariance matrix — normal. For a complete sample from a normal distribution, the true asymptotic covariance matrix of the ML estimators $\hat{\mu}$ and $\hat{\sigma}$ is

$$\boldsymbol{\Sigma} = \begin{bmatrix} \text{Var}(\hat{\mu}) & \text{Cov}(\hat{\mu}, \hat{\sigma}) \\ \text{Cov}(\hat{\mu}, \hat{\sigma}) & \text{Var}(\hat{\sigma}) \end{bmatrix} = \begin{bmatrix} n/\sigma_0^2 & 0 \\ 0 & 2n/\sigma_0^2 \end{bmatrix}^{-1} = \begin{bmatrix} \sigma_0^2/n & 0 \\ 0 & \sigma_0^2/(2n) \end{bmatrix}.$$

$$(5.3.15)$$

The ML estimate of this is

$$\hat{\boldsymbol{\Sigma}} = \begin{bmatrix} \text{Vâr}(\hat{\mu}) & \text{Côv}(\hat{\mu}, \hat{\sigma}) \\ \text{Côv}(\hat{\mu}, \hat{\sigma}) & \text{Vâr}(\hat{\sigma}) \end{bmatrix} = \begin{bmatrix} \hat{\sigma}^2/n & 0 \\ 0 & \hat{\sigma}^2/(2n) \end{bmatrix}.$$

The local estimate is

$$\mathbf{V} = \begin{bmatrix} \text{var}(\hat{\mu}) & \text{cov}(\hat{\mu}, \hat{\sigma}) \\ \text{cov}(\hat{\mu}, \hat{\sigma}) & \text{var}(\hat{\sigma}) \end{bmatrix}$$

$$= \begin{bmatrix} \sum_{i=1}^{n}(1/\hat{\sigma}^2) & \sum_{i=1}^{n} 2(y_i - \hat{\mu})/\hat{\sigma}^3 \\ \sum_{i=1}^{n} 2(y_i - \hat{\mu})/\hat{\sigma}^3 & \sum_{i=1}^{n}\left[(-1/\hat{\sigma}^2)+(3/\hat{\sigma}^4)(y_i - \hat{\mu})^2\right] \end{bmatrix}^{-1}$$

$$= \begin{bmatrix} n/\hat{\sigma}^2 & 0 \\ 0 & 2n/\hat{\sigma}^2 \end{bmatrix}^{-1} = \begin{bmatrix} \hat{\sigma}^2/n & 0 \\ 0 & \hat{\sigma}^2/(2n) \end{bmatrix},$$

since $\hat{\sigma}^2 = \Sigma_{i=1}^n (y_i - \hat{\mu})^2/n$. Here the local and ML estimates of the asymptotic covariance matrix are equal.

5. ML ESTIMATES OF FUNCTIONS OF THE PARAMETERS

Suppose that $h = h(\mu, \sigma)$ is a continuous function of μ and σ. For the normal distribution, such a function is the $100P$th percentile $y_P = \mu + z_P\sigma$ (z_P is the standard normal $100P$th percentile), and so is the reliability $R(y) = 1 - \Phi[(y - \mu)/\sigma]$. The ML estimate for the true value $h_0 = h(\mu_0, \sigma_0)$ is $\hat{h} = h(\hat{\mu}, \hat{\sigma})$; this is called the **invariance property** of ML estimators. For large sample size n, the cumulative distribution function of $\hat{h}$ is close to a normal one, with mean h_0 and a (true asymptotic) variance

$$\mathrm{Var}(\hat{h}) = (\partial h/\partial\mu)_0^2 \mathrm{Var}(\hat{\mu}) + (\partial h/\partial\sigma)_0^2 \mathrm{Var}(\hat{\sigma})$$

$$+ 2(\partial h/\partial\mu)_0(\partial h/\partial\sigma)_0 \mathrm{Cov}(\hat{\mu}, \hat{\sigma}); \qquad (5.3.16)$$

as before, the 0 subscript on ( ) denotes that the partial derivative is evaluated at $\mu = \mu_0$ and $\sigma = \sigma_0$. The partial derivatives must be continuous functions in the neighborhood of (μ_0, σ_0). The ML and local estimates for $\mathrm{Var}(\hat{h})$ are obtained by using respectively the ML and local estimates of the variances and covariance in (5.3.16) and using the estimates $\hat{\mu}$ and $\hat{\sigma}$ for μ_0 and σ_0 in the partial derivatives. (5.3.16) and (5.3.17) are based on propagation of error (Taylor expansions); see Hahn and Shapiro (1967) and Rao (1973).

Suppose that $g = g(\mu, \sigma)$ is another function of μ and σ. Then the asymptotic covariance of $\hat{g} = g(\hat{\mu}, \hat{\sigma})$ and $\hat{h}$ is

$$\mathrm{Cov}(\hat{g}, \hat{h}) \simeq (\partial g/\partial\mu)_0(\partial h/\partial\mu)_0 \mathrm{Var}(\hat{\mu}) + (\partial g/\partial\sigma)_0(\partial h/\partial\sigma)_0 \mathrm{Var}(\hat{\sigma})$$

$$+ \left[(\partial g/\partial\mu)_0(\partial h/\partial\sigma)_0 + (\partial g/\partial\sigma)_0(\partial h/\partial\mu)_0 \right] \mathrm{Cov}(\hat{\mu}, \hat{\sigma}),$$

$$(5.3.17)$$

where the partial derivatives are evaluated at μ_0 and σ_0. For large samples, the joint cumulative distribution of $\hat{g}$ and $\hat{h}$ is approximately a joint normal one with means $g_0 = g(\mu_0, \sigma_0)$ and h_0 and variances from (5.3.16) and covariance (5.3.17). The preceding applies to the joint distribution of any number of such ML estimates.

Weibull parameters. The Weibull shape parameter is $\beta = 1/\delta$ in terms of the extreme value parameters δ and λ. The variance of $\hat{\beta} = 1/\hat{\delta}$ involves $\partial\beta/\partial\lambda = 0$ and $\partial\beta/\partial\delta = -1/\delta^2$. By (5.3.16), $\mathrm{Var}(\hat{\beta}) = 0^2 \mathrm{Var}(\hat{\lambda}) + (-1/\delta_0^2)^2 \mathrm{Var}(\hat{\delta}) + 2(0)(-1/\delta_0^2)\mathrm{Cov}(\hat{\lambda}, \hat{\delta}) = (1/\delta_0^4)\mathrm{Var}(\hat{\delta})$. This is (3.13).

In addition, the Weibull scale parameter is $\alpha = \exp(\lambda)$ in terms of the extreme value location parameter λ. $\mathrm{Cov}(\hat{\alpha}, \hat{\beta})$ involves the derivatives $\partial \alpha / \partial \lambda = \exp(\lambda)$, $\partial \alpha / \partial \delta = 0$, $\partial \beta / \partial \lambda = 0$, and $\partial \beta / \partial \delta = -1/\delta^2$. Then, by (5.3.17),

$$\mathrm{Cov}(\hat{\alpha}, \hat{\beta}) \simeq \exp(\lambda_0)(0)\mathrm{Var}(\hat{\lambda}) + 0(-1/\delta_0^2)\mathrm{Var}(\hat{\delta})$$

$$+ \left[\exp(\lambda_0)(-1/\delta_0^2) + 0(0)\right]\mathrm{Cov}(\hat{\lambda}, \hat{\delta})$$

$$= -\exp(\lambda_0)(1/\delta_0^2)\mathrm{Cov}(\hat{\lambda}, \hat{\delta}).$$

This is (3.13).

ML estimate of a normal percentile. The ML estimate of the $100P$ th normal percentile $y_P = \mu_0 + z_P \sigma_0$ is $\hat{y}_P = \hat{\mu} + z_P \hat{\sigma}$. The calculation of its asymptotic variance is

$$\partial y_P / \partial \mu = 1, \qquad \partial y_P / \partial \sigma = z_P,$$

$$\mathrm{Var}(\hat{y}_P) = (1)^2(\sigma_0^2/n) + (z_P)^2\left[\sigma_0^2/(2n)\right] + 2(1)(z_P)^2 0$$

$$= \left[1 + (z_P^2/2)\right]\sigma_0^2/n.$$

For complete data, the ML and local estimates are equal and are

$$\mathrm{V\hat{a}r}(\hat{y}_P) = \left[1 + (z_P^2/2)\right]\hat{\sigma}^2/n.$$

6. CONFIDENCE LIMITS

Theory for approximate confidence limits is exactly the same as that in Section 5.1.

5.4. Multiparameter Distribution and Multiply Censored Data

This advanced section presents general methods and theory for ML analysis of multiply censored data. The methods and theory are the same as those of Section 5.3, but the log likelihood and covariance matrix differ to take the censoring into account. Section 5.3 is necessary background.

As before, a distribution's probability density is denoted by $f(y; \mu, \sigma)$, and its reliability function is denoted by $R(y; \mu, \sigma)$. Here μ and σ denote arbitrary values of the distribution parameters, and μ_0 and σ_0 denote the true values. The following methods are given for a two-parameter distribution, but they extend to distributions with any number of parameters, with suitable changes in formulas.

1. LIKELIHOOD

If a unit fails at age y_i, its log likelihood $\mathcal{L}_i$ is the natural log of the probability density at y_i, namely,

$$\mathcal{L}_i = \ln\left[f(y_i; \mu, \sigma) \right]. \tag{5.4.1}$$

If a unit is running at age y_i (it is censored on the right and its failure time is above y_i), then its log likelihood is the natural log of the distribution probability above y_i (i.e., of the reliability at age y_i), namely,

$$\mathcal{L}_i = \ln\left[R(y_i; \mu, \sigma) \right]. \tag{5.4.2}$$

In general, the log likelihood of an observation is the log of its probability model.

Suppose the sample contains n statistically independent units. The sample log likelihood is

$$\mathcal{L} = \sum_i \mathcal{L}_i = \sum_i{}' \ln\left[f(y_i; \mu, \sigma) \right] + \sum_i{}'' \ln\left[R(y_i; \mu, \sigma) \right]. \tag{5.4.3}$$

The sum Σ runs over all n units, Σ' runs over the failures, and Σ'' runs over the censored units. $\mathcal{L}$ is a function of the failure and running times y_i and of the distribution parameters μ and σ.

The ML methods apply to discrete distributions. Then the discrete probability mass function takes the place of the probability density in (5.4.3).

2. ML ESTIMATES

The ML estimates of μ_0 and σ_0 are the values $\hat{\mu}$ and $\hat{\sigma}$ that maximize the sample log likelihood (5.4.3). Under certain conditions usually satisfied in practice (Wilks, 1962; Rao 1973; and Hoadley, 1971), the ML estimates are unique. Also, for samples with large numbers of failures, the cumulative sampling distribution of $\hat{\mu}$ and $\hat{\sigma}$ is close to a joint normal cumulative distribution with means μ_0 and σ_0 and covariance matrix (5.4.14). That is, the cumulative sampling distribution of $\hat{\mu}$ and $\hat{\sigma}$ converges "in law" ("in distribution") to a joint normal cumulative distribution. This does not necessarily mean that the means, variances, and covariance of $\hat{\mu}$ and $\hat{\sigma}$ converge to those of the asymptotic distribution. However, the asymptotic normal distribution is valid for calculating the approximate confidence limits for μ_0 and σ_0, since the limits employ the (normal) probability that a ML estimate is within a specified multiple of the standard error from the true value.

One can numerically obtain the ML estimates by iteratively optimizing the log likelihood (5.4.3) with respect to μ and σ. Alternatively, one can calculate the partial derivatives of $\mathcal{L}$ with respect to μ and σ and set them equal to zero to get the **likelihood equations:**

$$\partial \mathcal{L} / \partial \mu = 0, \qquad \partial \mathcal{L} / \partial \sigma = 0. \tag{5.4.4}$$

The solution of these simultaneous nonlinear equations in μ and σ are the ML estimates. These must usually be iteratively solved by numerical methods described in Section 6.

3. FISHER INFORMATION MATRIX

One needs the Fisher information matrix to obtain the asymptotic (large-sample) covariance matrix of the ML estimates and approximate confidence intervals. One uses the matrix of negative second partial derivatives,

$$\mathbf{F} = \begin{bmatrix} -\partial^2 \mathcal{L} / \partial \mu^2 & -\partial^2 \mathcal{L} / \partial \mu \partial \sigma \\ -\partial^2 \mathcal{L} / \partial \sigma \partial \mu & -\partial^2 \mathcal{L} / \partial \sigma^2 \end{bmatrix}. \tag{5.4.5}$$

This is the **local Fisher information matrix** when $\hat{\mu}$ and $\hat{\sigma}$ are used in place of μ and σ.

The theoretical **Fisher information matrix** $\mathbf{F}_0$ is the expectation of (5.4.5), where the subscript 0 denotes that μ_0 and σ_0 are used for μ and σ and the observations y_i are treated as random variables. That is,

$$\mathbf{F}_0 = \begin{bmatrix} E_0\{-\partial^2 \mathcal{L} / \partial \mu^2\}_0 & E_0\{-\partial^2 \mathcal{L} / \partial \mu \partial \sigma\}_0 \\ E_0\{-\partial^2 \mathcal{L} / \partial \sigma \partial \mu\}_0 & E_0\{-\partial^2 \mathcal{L} / \partial \sigma^2\}_0 \end{bmatrix}. \tag{5.4.6}$$

The expectations are calculated as follows. In general, such an expectation of any function $g(y_i; \mu, \sigma)$ of an observation y_i is

$$E_0\{g(y_i; \mu, \sigma)\}_0 = \int_{-\infty}^{\infty} g(y_i; \mu_0, \sigma_0) f(y_i; \mu_0, \sigma_0) \, dy_i. \tag{5.4.7}$$

For a discrete distribution,

$$E_0\{g(y_i; \mu, \sigma)\}_0 = \sum_i g(y_i; \mu_0, \sigma_0) f(y_i; \mu_0, \sigma_0). \tag{5.4.8}$$

An evaluation of (5.4.6) follows. The log likelihood for the ith sample unit is reexpressed as

$$\mathcal{L}_i = I_i(y_i) \ln[f(y_i; \mu, \sigma)] + [1 - I_i(y_i)] \ln[R(\eta_i; \mu, \sigma)], \tag{5.4.9}$$

where $I_i(y_i)$ is an indicator variable such that $I_i(y_i)=1$ if $y_i < \eta_i$ (a failure is observed), and $I_i(y_i)=0$ if $y_i \geq \eta_i$ (the observation is censored). Here I_i and y_i are regarded as random variables, and the censoring time η_i may differ from unit to unit. Then, for unit i,

$$E_0\{-\partial^2 \mathcal{L}_i/\partial\mu^2\}_0 = \int_{-\infty}^{\infty} \{-\partial^2 \mathcal{L}_i/\partial\mu^2\}_0 f(y_i;\mu_0,\sigma_0)\,dy_i$$

$$= \int_{-\infty}^{\eta_i} \{-\partial^2 \ln[f(y_i;\mu,\sigma)]/\partial\mu^2\}_0 f(y_i;\mu_0,\sigma_0)\,dy_i$$

$$+ \{-\partial^2 \ln[R(\eta_i;\mu,\sigma)]/\partial\mu^2\}_0 R(\eta_i;\mu_0,\sigma_0);$$

$$E_0\{-\partial^2 \mathcal{L}_i/\partial\sigma^2\}_0 = \int_{-\infty}^{\infty} \{-\partial^2 \mathcal{L}_i/\partial\sigma^2\}_0 f(y_i;\mu_0,\sigma_0)\,dy_i \qquad (5.4.10)$$

$$= \int_{-\infty}^{\eta_i} \{-\partial^2 \ln[f(y_i;\mu,\sigma)]/\partial\sigma^2\}_0 f(y_i;\mu_0,\sigma_0)\,dy_i$$

$$+ \{-\partial^2 \ln[R(\eta_i;\mu,\sigma)]\partial\sigma^2\}_0 R(\eta_i;\mu_0,\sigma_0);$$

$$E_0\{-\partial^2 \mathcal{L}_i/\partial\mu\partial\sigma\}_0 = E_0\{-\partial^2 \mathcal{L}_i/\partial\sigma\partial\mu\}_0$$

$$= \int_{-\infty}^{\infty} \{-\partial^2 \mathcal{L}_i/\partial\mu\partial\sigma\}_0 f(y_i;\mu_0,\sigma_0)\,dy_i$$

$$+ \{-\partial^2 \ln[R(\eta_i;\mu,\sigma)]/\partial\mu\partial\sigma\}_0 R(\eta_i;\mu_0,\sigma_0).$$

For a discrete distribution, these integrals are replaced by sums over the outcomes.

Since $\mathcal{L} = \Sigma_i \mathcal{L}_i$, then, for the sample log likelihood,

$$E_0\{-\partial^2 \mathcal{L}/\partial\mu^2\}_0 = \sum_i E_0\{-\partial^2 \mathcal{L}_i/\partial\mu^2\}_0,$$

$$E_0\{-\partial^2 \mathcal{L}/\partial\sigma^2\}_0 = \sum_i E_0\{-\partial^2 \mathcal{L}_i/\partial\sigma^2\}_0, \qquad (5.4.11)$$

$$E_0\{-\partial^2 \mathcal{L}/\partial\mu\partial\sigma\}_0 = \sum_i E_0\{-\partial^2 \mathcal{L}_i/\partial\mu\partial\sigma\}_0.$$

The expectations of the second derivatives can also be expressed as

$$E_0\{-\partial^2\mathcal{L}_i/\partial\mu^2\}_0 = E_0\{(\partial\mathcal{L}_i/\partial\mu)_0^2\} = \int_{-\infty}^{\infty} (\partial\mathcal{L}_i/\partial\mu)_0^2 f(y_i; \mu_0, \sigma_0) \, dy_i,$$

$$E_0\{-\partial^2\mathcal{L}_i/\partial\sigma^2\}_0 = E_0\{(\partial\mathcal{L}_i/\partial\sigma)_0^2\} = \int_{-\infty}^{\infty} (\partial\mathcal{L}_i/\partial\sigma)_0^2 f(y_i; \mu_0, \sigma_0) \, dy_i,$$

$$E_0\{-\partial^2\mathcal{L}_i/\partial\mu\partial\sigma\}_0 = E_0\{(\partial\mathcal{L}_i/\partial\mu)_0(\partial\mathcal{L}_i/\partial\sigma)_0\} \qquad (5.4.12)$$

$$= \int_{-\infty}^{\infty} (\partial\mathcal{L}_i/\partial\mu)_0(\partial\mathcal{L}_i/\partial\sigma)_0 f(y_i; \mu_0, \sigma_0) \, dy_i.$$

(5.4.12) can be easier to evaluate than (5.4.10). The expectations of the first derivatives satisfy

$$E_0\{\partial\mathcal{L}_i/\partial\mu\}_0 = E_0\{\partial\mathcal{L}_i/\partial\sigma\}_0 = 0, \qquad (5.4.13)$$

which are similar to the likelihood equations (5.4.4). Relations (5.4.12) and (5.4.13) often simplify the calculation of theoretical results.

4. COVARIANCE MATRIX

As before, the true asymptotic (large-sample) covariance matrix of $\hat{\mu}$ and $\hat{\sigma}$ is the inverse of (5.4.6), namely,

$$\mathbf{\Sigma} = \begin{bmatrix} \text{Var}(\hat{\mu}) & \text{Cov}(\hat{\mu}, \hat{\sigma}) \\ \text{Cov}(\hat{\mu}, \hat{\sigma}) & \text{Var}(\hat{\sigma}) \end{bmatrix} = \mathbf{F}_0^{-1}. \qquad (5.4.14)$$

The square roots of the variances in (5.4.14) are the asymptotic standard errors of the ML estimates. The ML and local estimates of $\mathbf{\Sigma}$ are obtained as described in Section 5.3.

5. ESTIMATES OF FUNCTIONS OF THE PARAMETERS

Theory for estimating functions of the parameters is exactly the same as that in Section 5.3.

6. CONFIDENCE LIMITS

Theory for approximate confidence limits is exactly the same as that in Section 5.1.

5.5. ML Theory for Competing Failure Modes

This **advanced** section presents ML theory for fitting distributions to data with a mix of independent failure modes. The methods provide ML estimates and approximate confidence intervals. ML results are given without proof; for more details, see David and Moeschberger (1979), Moeschberger (1974), Moeschberger and David (1971), and Herman and Patell (1971). The descriptions cover the model, likelihood, ML estimates, Fisher and covariance matrices, estimation of a function, and approximate confidence intervals.

The model. The model employed here is the **series-system** model for **independent** competing failure modes, as described in Chapter 5. Suppose there are M independent failure modes. For the time to failure by the mth mode acting alone, denote the reliability function by $R_m(y; \mu_m, \sigma_m)$ and the probability density by $f_m(y; \mu_m, \sigma_m)$, where μ_m and σ_m are the distribution parameters, $m = 1, \ldots, M$. For concreteness, two parameters are used here, but any distribution may have any number. Also, the distributions for different modes may be different kinds, for example, exponential and lognormal. In addition, it is assumed that **no** distributions have the same value of some parameters. For example, distributions do not have a common standard deviation. David and Moeschberger (1979) discuss more complex models, including ones with common parameter values and dependent failure modes. The zero subscript does not appear on true parameter values in this section.

1. LIKELIHOOD

Suppose there are I statistically independent units in the sample. For test unit i, y_{mi} denotes the failure or running time for the mth mode. The likelihood l_i for sample unit i includes the failure times of failure modes that occurred and the running times of ones that did not occur, namely,

$$l_i = \prod_m {}' f_m(y_{mi}; \mu_m, \sigma_m) \prod_m {}'' R_m(y_{mi}; \mu_m, \sigma_m), \qquad (5.5.1)$$

where the first (second) product runs over modes that were (not) observed. This likelihood is the product of factors for the failure modes, because the modes are statistically independent. This likelihood extends that of Herman and Patell (1971) to data with more than one observed failure mode per test unit. This likelihood applies to both Type I and Type II censored data. The ML theory of Moeschberger and David (1971) and David and Moeschberger (1979) for dependent modes applies to such data. The likelihood l_0

for the sample of I statistically independent units is

$$l_0 = l_1 \times \cdots \times l_I. \tag{5.5.2}$$

This is a function of the model parameters μ_m, σ_m, $m = 1, \ldots, M$.

2. ML ESTIMATES

$\hat{\mu}_m$, $\hat{\sigma}_m$ are the parameter values that maximize l_0. Under certain conditions on the model and data that usually hold, these estimates are unique. Also, for asymptotic (large) sample sizes, they are approximately jointly normally distributed, with means equal to the true μ_m, σ_m and a covariance matrix given later. Moreover, their asymptotic variances are smaller than those for any other asymptotically normally distributed estimates. In particular, it is assumed that parameters of different modes do not have the same value.

The $\hat{\mu}_m$, $\hat{\sigma}_m$ can be iteratively calculated by numerically maximizing l_0 with respect to all the μ_m, σ_m. Also, for many models, l_0 can be maximized by the usual calculus method. Namely, the partial derivatives of l_0 with respect to each μ_m and σ_m are set equal to zero:

$$\partial l_0 / \partial \mu_m = 0, \qquad \partial l_0 / \partial \sigma_m = 0, \qquad m = 1, \ldots, M. \tag{5.5.3}$$

These are the likelihood equations. $\hat{\mu}_m$, $\hat{\sigma}_m$ are the solutions of these simultaneous nonlinear equations.

If M is not small, the iterative calculations are laborious and may not easily converge. The following splits the problem into simpler ones, which can be easily solved with existing computer programs.

The sample likelihood l_0 can be rewritten as

$$l_0 = l_{(1)} \times \cdots \times l_{(M)}, \tag{5.5.4}$$

where $l_{(1)}$ contains only the parameters μ_1, σ_1 for mode 1,..., and $l_{(M)}$ contains only the parameters μ_M, σ_M for mode M. For example,

$$l_{(1)} = \Pi_1 f_1(y_{1i}, \mu_1, \sigma_1) \, \Pi_1' R_1(y_{1i}'; \mu_1, \sigma_1), \tag{5.5.5}$$

where the first product Π_1 runs over units with mode 1 failure times y_{1i}, and the second product Π_1' runs over running times y_{1i}' on units without a mode 1 failure.

The ML values of the mode m parameters μ_m, σ_m that maximize $l_{(m)}$ are the values that maximize l_0. Moreover, $l_{(m)}$ is the likelihood for the multiply censored sample data on mode m. So the ML estimates $\hat{\mu}_m$, $\hat{\sigma}_m$ can be obtained for one mode at a time by maximizing one $l_{(m)}$ at a time. Thus data on each failure mode can be separately analyzed with a computer program for multiply censored data.

3. FISHER MATRIX

The Fisher information matrix is needed to obtain approximate confidence limits for the μ_m, σ_m and functions of them, for example, a percentile of the life distribution. Below, $\mathcal{L} = \ln(l_0)$ denotes the sample log likelihood.

First one needs the matrix of negative second partial derivatives

$$
\mathbf{F} = \begin{bmatrix}
-\partial^2\mathcal{L}/\partial\mu_1^2 & -\partial^2\mathcal{L}/\partial\mu_1\partial\sigma_1 & & \bigcirc \\
-\partial^2\mathcal{L}/\partial\mu_1\partial\sigma_1 & -\partial^2\mathcal{L}/\partial\sigma_1^2 & & \\
& & \ddots & \\
& \bigcirc & & -\partial^2\mathcal{L}/\partial\mu_M^2 & -\partial^2\mathcal{L}/\partial\mu_M\partial\sigma_M \\
& & & -\partial^2\mathcal{L}/\partial\mu_M\partial\sigma_M & -\partial^2\mathcal{L}/\partial\sigma_M^2
\end{bmatrix}
$$

$$(5.5.6)$$

For $m \neq m'$, these derivatives are zero, since, for example, $\partial\mathcal{L}/\partial\mu_{m'}$ does not contain μ_m or σ_m according to (5.5.4). The expectation of this matrix evaluated at the true μ_m, σ_m values is the true Fisher information matrix:

$$
\mathbf{F}_0 = \begin{bmatrix}
E_0\{-\partial^2\mathcal{L}/\partial\mu_1^2\}_0 & E_0\{-\partial^2\mathcal{L}/\partial\mu_1\partial\sigma_1\}_0 & & \bigcirc \\
E_0\{-\partial^2\mathcal{L}/\partial\mu_1\partial\sigma_1\}_0 & E_0\{-\partial^2\mathcal{L}/\partial\sigma_1^2\}_0 & & \\
& & \ddots & \\
& \bigcirc & & E_0\{-\partial^2\mathcal{L}/\partial\mu_M^2\}_0 & E_0\{-\partial^2\mathcal{L}/\partial\mu_M\partial\sigma_M\}_0 \\
& & & E_0\{-\partial^2\mathcal{L}/\partial\mu_M\partial\sigma_M\}_0 & E_0\{-\partial^2\mathcal{L}/\partial\sigma_M^2\}_0
\end{bmatrix}
$$

$$(5.5.7)$$

Such an expected value depends on the type of censoring (Type I, Type II, etc.) and on all of the models for the different failure modes. Moeschberger and David (1971) and David and Moeschberger (1979) give the complex formulas for (5.5.7).

4. COVARIANCE MATRIX

The inverse $\boldsymbol{\Sigma} = \mathbf{F}_0^{-1}$ is the true asymptotic covariance matrix of the $\hat{\mu}_m$, $\hat{\sigma}_m$. $\boldsymbol{\Sigma}$ is block diagonal, since $\mathbf{F}_0$ is block diagonal. Thus the $\hat{\mu}_m$, $\hat{\sigma}_m$ for different failure modes are asymptotically uncorrelated. Moreover, since the $\hat{\mu}_m$, $\hat{\sigma}_m$

are asymptotically jointly normally distributed, the $\hat{\mu}_m$, $\hat{\sigma}_m$ are asymptotically independent of $\hat{\mu}_{m'}$, $\hat{\sigma}_{m'}$, but $\hat{\mu}_m$ and $\hat{\sigma}_m$ for the same failure mode generally are correlated.

The ML estimate of Σ comes from using the $\hat{\mu}_m$, $\hat{\sigma}_m$ for the μ_m, σ_m in (5.5.7). Usually one estimates Σ with the simpler method next. The matrix (5.5.6) evaluated for $\mu_m = \hat{\mu}_m$ and $\sigma_m = \hat{\sigma}_m$ is called the **local Fisher information matrix**. Its inverse is another estimate of Σ. This local estimate is easier to calculate than the ML estimate, since it does not require the complicated expectations (5.5.7).

An estimate of Σ is used as explained below to obtain approximate confidence intervals. Below, the P model parameters are relabeled c_1, $c_2, \ldots, c_P$.

5. ESTIMATE OF A FUNCTION

Suppose that an estimate is desired for a function $h = h(c_1, \ldots, c_P)$ of $c_1, \ldots, c_P$. Distribution parameters, percentiles, and reliabilities are such functions. The ML estimate of the function is $\hat{h} = h(\hat{c}_1, \ldots, \hat{c}_P)$. Under certain conditions and for large samples, the cumulative distribution of $\hat{h}$ is approximately normal, with mean h and variance (5.5.8).

The following estimate of $\mathrm{Var}(\hat{h})$ is used to obtain approximate confidence intervals for h. Evaluate the column vector of partial derivatives $\hat{\mathbf{H}} = (\partial h/\partial c_1, \ldots, \partial h/\partial c_P)'$ at $c_p = \hat{c}_p$; the ' denotes transpose. By propagation of error (Wilks, 1962), the variance estimate is

$$\mathrm{var}(\hat{h}) = \hat{\mathbf{H}}'\hat{\Sigma}\hat{\mathbf{H}}, \qquad (5.5.8)$$

where $\hat{\Sigma}$ is an estimate of Σ. The estimate $s(\hat{h})$ of the standard error of $\hat{h}$ is

$$s(\hat{h}) = [\mathrm{var}(\hat{h})]^{1/2}. \qquad (5.5.9)$$

6. APPROXIMATE CONFIDENCE INTERVALS

As before, an approximate $100\gamma\%$ confidence interval for h has lower and upper limits

$$\underset{\sim}{h} = \hat{h} - K_\gamma s(\hat{h}), \qquad \tilde{h} = \hat{h} + K_\gamma s(\hat{h}), \qquad (5.5.10)$$

where K_γ is the $[100(1+\gamma)/2]$th standard normal percentile. This interval is appropriate if the range of h is from $-\infty$ to $+\infty$. The limits assume that the sample size is large enough that $\hat{h}$ is approximately normally distributed.

If $\hat{h}$ must be positive, approximate positive limits are

$$\underset{\sim}{h} = \hat{h}\exp\left[- K_\gamma s(\hat{h})/\hat{h}\right], \qquad \tilde{h} = \hat{h}\exp\left[K_\gamma s(\hat{h})/\hat{h}\right]. \quad (5.5.11)$$

The limits assume that the sample size is large enough that $\ln(\hat{h})$ is approximately normally distributed.

5.6. Motivation for ML Theory

The following advanced discussion motivates ML theory for complete data. It is a heuristic proof of theoretical results, and it will interest those who want motivation for ML theory. Moreover, the necessary regularity conditions on the sampled distribution are not stated. The key result is the following. Suppose the data are a complete sample of independent observations $y_1, \ldots, y_n$ from a one-parameter distribution with a probability density $f(y; \theta)$. Then the ML estimator $\hat{\theta}(y_1, \ldots, y_n)$ for θ_0 has an asymptotic cumulative distribution that is close to a normal one with mean θ_0, the true value, and variance $1/E_0\{-\partial^2\mathcal{L}/\partial\theta^2\}_0$. The subscript denotes that $\theta = \theta_0$. The heuristic proof extends to censored data and multiparameter distributions.

We first need to know the properties of the sample **score** defined as

$$\{\partial\mathcal{L}/\partial\theta\}_0 = \sum_i \{\partial\mathcal{L}_i/\partial\theta\}_0 = \sum_i (\partial f(y_i; \theta)/\partial\theta)_0/f(y_i; \theta_0), \quad (5.6.1)$$

where $\mathcal{L}_i = \ln[f(y_i; \theta)]$ is the log likelihood of observation i and $\mathcal{L} = \sum_i \mathcal{L}_i$ is the sample log likelihood. The score is a function of the random observations y_i. The mean score of observation i when $\theta = \theta_0$ is

$$E_0\{\partial\mathcal{L}_i/\partial\theta\}_0 = \int_{-\infty}^{\infty} \{(\partial f(y_i; \theta)/\partial\theta)_0/f(y_i; \theta_0)\}f(y_i; \theta_0)\,dy_i \quad (5.6.2)$$

$$= \int_{-\infty}^{\infty} (\partial f(y_i; \theta)/\partial\theta)_0\,dy_i = \left(\frac{\partial}{\partial\theta}\int_{-\infty}^{\infty} f(y_i; \theta)\,dy_i\right)_0 = 0,$$

since the last integral has the value 1. Thus the mean of the sample score is 0. The variance of the sample score for $\theta = \theta_0$ is

$$\mathrm{Var}\{\partial\mathcal{L}/\partial\theta\}_0 = \sum_i \mathrm{Var}\{\partial\mathcal{L}_i/\partial\theta\}_0 = nE_0\{(\partial\mathcal{L}_i/\partial\theta)_0^2\}, \quad (5.6.3)$$

since the $\{\partial\mathcal{L}_i/\partial\theta\}_0$ are statistically independent and $E_0\{\partial\mathcal{L}_i/\partial\theta\}_0 = 0$.

Another relation for the variance of the sample score is

$$\text{Var}\{\partial \mathcal{L}/\partial\theta\}_0 = nE_0\{-\partial^2\mathcal{L}_i/\partial\theta^2\}_0. \tag{5.6.4}$$

This comes from

$$E_0\{-\partial^2\mathcal{L}_i/\partial\theta^2\}_0 = \int_{-\infty}^{\infty}\{-\partial^2\ln[f(y_i;\theta)]/\partial\theta^2\}_0 f(y_i;\theta_0)\,dy_i$$

$$= \int_{-\infty}^{\infty}[-[f(y_i;\theta_0)]^{-1}(\partial^2 f(y_i;\theta)/\partial\theta^2)_0 \tag{5.6.5}$$

$$+ [f(y_i;\theta_0)]^{-2}(\partial f(y_i;\theta)/\partial\theta)_0^2] f(y_i;\theta_0)\,dy_i.$$

Then, because

$$0 = \frac{\partial^2}{\partial\theta^2}\{1\} = \left(\frac{\partial^2}{\partial\theta^2}\int_{-\infty}^{\infty} f(y_i;\theta)\,dy_i\right)_0 = \int_{-\infty}^{\infty}(\partial^2 f(y_i;\theta)/\partial\theta^2)_0\,dy_i,$$

$$\tag{5.6.6}$$

it follows that

$$E_0\{-\partial^2\mathcal{L}_i/\partial\theta^2\}_0 = \int_{-\infty}^{\infty}[f(y_i;\theta_0)]^{-1}(\partial f(y_i;\theta)/\partial\theta)_0^2\,dy_i \tag{5.6.7}$$

$$= \int_{-\infty}^{\infty}\{\partial\mathcal{L}_i/\partial\theta\}_0^2 f(y_i;\theta_0)\,dy_i = E_0\{(\partial\mathcal{L}_i/\partial\theta)_0^2\}.$$

For large sample size n, the sample score is approximately normally distributed with a mean of 0 and the variance derived above. This follows from the standard central limit theorem, since the score is the sum of independent identically distributed quantities $\{\partial\mathcal{L}_i/\partial\theta\}_0$.

The above properties of the sample score are the basis for the properties of the ML estimator, which are motivated below.

The ML estimate $\hat{\theta}(y_1,\ldots,y_n)$ for the observations $y_1,\ldots,y_n$ is the solution of

$$0 = \partial\mathcal{L}(y_1,\ldots,y_n;\theta)/\partial\theta. \tag{5.6.8}$$

Expand $\partial\mathcal{L}/\partial\theta$ in a power series in θ about θ_0 to get an approximate likelihood equation

$$0 \approx (\partial\mathcal{L}/\partial\theta)_0 + (\partial^2\mathcal{L}/\partial\theta^2)_0(\hat{\theta}-\theta_0) + \text{smaller terms}. \tag{5.6.9}$$

Thus,

$$\left(\hat{\theta} - \theta_0\right) \simeq \left\{\frac{1}{n}(\partial\mathcal{L}/\partial\theta)_0\right\} \Big/ \left\{-\frac{1}{n}(\partial^2\mathcal{L}/\partial\theta^2)_0\right\}. \qquad (5.6.10)$$

The numerator is $(1/n)$ times the sample score, and it has a mean of zero and a variance that decreases with increasing n. The denominator is the average $(1/n)\Sigma_i[-\partial^2\mathcal{L}_i/\partial\theta^2]_0$; its expectation is $(1/n)\Sigma_i$ $E_0\{-\partial^2\mathcal{L}_i(y_i;\theta)/\partial\theta^2\}_0 > 0$. For n large, the distribution of $(\hat{\theta} - \theta_0)$ is determined by the distribution of the numerator of (5.6.10) since the numerator distribution is centered on zero, and the denominator of (5.6.10) is close to its expected value. Thus the denominator's variation hardly affects the near-zero value of $(\hat{\theta} - \theta_0)$. So the denominator can be regarded as a constant equal to its expected value. Then

$$\left(\hat{\theta} - \theta_0\right) \simeq (\partial\mathcal{L}/\partial\theta)_0 \Big/ \left(nE\{-\partial^2\mathcal{L}_i/\partial\theta^2\}_0\right) \qquad (5.6.11)$$

is a multiple of the score $(\partial\mathcal{L}/\partial\theta)_0$. So $(\hat{\theta} - \theta_0)$ has an asymptotic normal distribution, since the score does. Asymptotically, $(\hat{\theta} - \theta_0)$ has a mean of zero like the score. Also,

$$\text{Var}\left(\hat{\theta} - \theta_0\right) \cong \text{Var}\{\partial\mathcal{L}/\partial\theta\}_0 \Big/ \left(n^2 E\{-\partial^2\mathcal{L}/\partial\theta^2\}_0\right)^2$$

$$= 1 \Big/ \left(nE_0\{-\partial^2\mathcal{L}_i/\partial\theta^2\}_0\right). \qquad (5.6.12)$$

So $\hat{\theta}$ has an asymptotic normal distribution with mean θ_0 and $\text{Var}(\hat{\theta}) = \text{Var}(\hat{\theta} - \theta_0)$ in (5.6.12).

This completes the motivation for the asymptotic normal distribution of ML estimators from a complete sample. The theory generalizes to multi-parameter distributions, censored data, dependent observations, and not identically distributed observations. For generalizations, see Hoadley (1971).

6. NUMERICAL METHODS

Often ML estimates cannot be explicitly expressed as functions of the data. Instead they must be iteratively calculated by numerically maximizing the log likelihood $\mathcal{L}$ or solving the likelihood equations. This section describes numerical methods for ML problems. The section (1) surveys some methods for calculating the ML estimates, (2) presents the Newton–Raphson method in detail, and (3) describes effective ways of handling computations. This section will especially aid those who wish to write ML programs, and it may aid those who encounter difficulties with existing programs.

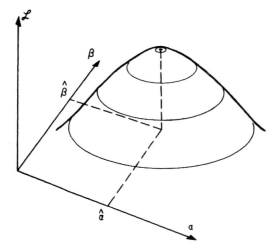

Figure 6.1. Sample log likelihood function.

Figure 6.1 depicts a sample log likelihood $\mathcal{L}(\alpha,\ \beta)$ that is a function of two parameters α and β. The maximum of $\mathcal{L}$ is at the ML estimates $(\hat{\alpha},\ \hat{\beta})$. The log likelihood differs from sample to sample with respect to its shape and location of its optimum. A numerical method iteratively calculates a series of trial estimates $(\hat{\alpha}_i,\ \hat{\beta}_i)$ that approach $(\hat{\alpha},\ \hat{\beta})$. The method finds $(\hat{\alpha}_i,\ \hat{\beta}_i)$ so that the $\mathcal{L}(\hat{\alpha}_i,\ \hat{\beta}_i)$ increase and approach $\mathcal{L}(\hat{\alpha},\ \hat{\beta})$. Such a method climbs up the likelihood function to its maximum. Alternatively, Figure 6.2

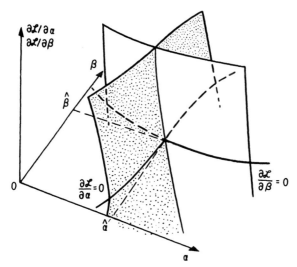

Figure 6.2. Depiction of likelihood equations.

depicts the likelihood partial derivatives $\partial \mathcal{L}/\partial \alpha$ and $\partial \mathcal{L}/\partial \beta$ as functions of α and β. The ML estimates $(\hat{\alpha}, \hat{\beta})$ are located at the point where the derivatives both equal zero. A numerical method iteratively calculates a series of trial estimates $(\hat{\alpha}_i, \hat{\beta}_i)$ that approach $(\hat{\alpha}, \hat{\beta})$. The method finds $(\hat{\alpha}_i, \hat{\beta}_i)$ so that both partial derivatives decrease and approach zero. These pictures of iterative methods extend to any number of parameters. For large samples and for (α, β) near $(\hat{\alpha}, \hat{\beta})$, $\mathcal{L}$ is approximately a quadratic function of α and β, and $\partial \mathcal{L}/\partial \alpha$ and $\partial \mathcal{L}/\partial \beta$ are approximately linear functions of α and β.

6.1.   Survey of Numerical Methods

This section surveys some methods for calculating ML estimates. It describes whether a method (1) maximizes the sample log likelihood or solves the likelihood equations, (2) requires derivatives of the log likelihood with respect to parameters, and (3) converges quickly and surely. A method that requires analytic expressions for the partial derivatives burdens the user with their calculation. Choice of the starting values for the estimates is described in Section 6.3. Jacoby and others (1972), Kennedy and Gentle (1980), and Wilde and Beightler (1967) describe these methods in detail. They apply to models with any number of parameters.

Direct search (trial and error) involves calculating $\mathcal{L}$ for new trial ML estimates near the ones that yield large $\mathcal{L}$ values. This method does not use derivatives of $\mathcal{L}$. It converges slowly but surely. The differences between the latest trial estimates indicate how accurately the ML estimates are known. One search scheme involves maximizing $\mathcal{L}$ with respect to a parameter while holding all others at their current trial estimates. This is done for each parameter in turn and may require any number of passes through all parameters. The method generally converges, even with crude starting values for the parameters.

The Powell method without derivatives directly maximizes $\mathcal{L}$. It usually converges rapidly and relatively surely, even with crude starting values for the parameters. It does not use partial derivatives. Nelson, Morgan, and Caporal (1978) use this method in STATPAC.

The Fletcher–Powell–Davidon method requires the first derivatives of $\mathcal{L}$. It usually converges more rapidly and accurately than the Powell method, even with crude starting values for the parameters.

The method of steepest ascent involves the following steps: (1) at the latest $(\hat{\alpha}_i, \hat{\beta}_i)$, determine the direction of steepest ascent (where $\mathcal{L}$ increases fastest). (2) Search along the straight line in that direction to find the $(\hat{\alpha}_{i+1}, \hat{\beta}_{i+1})$ that maximize $\mathcal{L}$; they are the new trial estimates. Repeat (1) and (2) until the trial estimates are "close enough" to the ML estimates. The search in a direction can be done by trial and error or other schemes. The method requires first partial derivatives of $\mathcal{L}$. The method usually converges surely and faster than direct search, even with crude starting values for the parameters.

The Newton–Raphson method is described in detail in Section 6.2. The method uses the first and second partial derivatives of $\mathcal{L}$ to approximate the likelihood equations with linear equations and solves them. The method may fail to converge if the starting values are not close to the ML estimates; a modified method given below may avoid such difficulties. When it converges, it is faster than most methods.

The secant method uses the first derivatives of $\mathcal{L}$ and approximates the likelihood equations with linear equations and solves them. Like the Newton–Raphson method, it may fail to converge if the starting values are not close to the ML estimates. It is slower than the Newton–Raphson method. It does not require the second partial derivatives of $\mathcal{L}$; instead it approximates them with perturbation calculations.

6.2. Newton–Raphson Method (Method of Scoring)

Statisticians often use the Newton–Raphson method for solving likelihood equations. Often recommended in its simplest form, the method may fail to converge. The method works well if the starting parameter values are "close" to the ML estimates. Improvements below help the method converge.

The following presents the method, an example, improvements, and the theory.

The method. Suppose the sample log likelihood is $\mathcal{L}(\alpha, \beta)$. It is a function of parameters α and β and the data. Although explained with two parameters, the method works with any number of parameters. The data values are constant throughout the calculation. Suppose that $\hat{\alpha}_i$ and $\hat{\beta}_i$ are the approximate ML estimates after the ith iteration. (1) Evaluate the first partial derivatives $\partial\mathcal{L}/\partial\alpha$ and $\partial\mathcal{L}/\partial\beta$ at $\alpha = \hat{\alpha}_i$ and $\beta = \hat{\beta}_i$. (2) Evaluate the second partial derivatives $\partial^2\mathcal{L}/\partial\alpha^2$, $\partial^2\mathcal{L}/\partial\beta^2$, and $\partial^2\mathcal{L}/\partial\alpha\partial\beta$ at $\alpha = \hat{\alpha}_i$ and $\beta = \hat{\beta}_i$. (3) Solve the linear equations for the adjustments a_i and b_i:

$$\partial\mathcal{L}/\partial\alpha = (-\partial^2\mathcal{L}/\partial\alpha^2)a_i + (-\partial^2\mathcal{L}/\partial\alpha\partial\beta)b_i,$$
$$\partial\mathcal{L}/\partial\beta = (-\partial^2\mathcal{L}/\partial\alpha\partial\beta)a_i + (-\partial^2\mathcal{L}/\partial\beta^2)b_i; \tag{6.1}$$

here the derivatives come from steps (1) and (2). Statisticians call the matrix of negative second partial derivatives the (local) Fisher information matrix; numerical analysts call the matrix of positive second partial derivates the

Hessian. (4) Calculate the new estimates

$$\hat{\alpha}_{i+1}=\hat{\alpha}_i+a_i, \qquad \hat{\beta}_{i+1}=\hat{\beta}_i+b_i. \qquad (6.2)$$

(5) Continue steps (1) through (4) until the estimates meet a convergence criterion. For example, stop when a_{i+1} and b_{i+1} are small, say, each a small fraction of the standard errors of $\hat{\alpha}_{i+1}$ and $\hat{\beta}_{i+1}$. Alternatively, stop when $\mathcal{L}(\hat{\alpha}_{i+1}, \hat{\beta}_{i+1})-\mathcal{L}(\hat{\alpha}_i, \hat{\beta}_i)$ is statistically small, say, less than 0.01.

Instead of using the second partial derivatives of the sample likelihood in (5.1), one can use their expected values, the terms of the true theoretical Fisher information matrix, where the true α_0 and β_0 are replaced by $\hat{\alpha}_i$ and $\hat{\beta}_i$. Then the iterative method is called the **method of scoring**. Also, it is not necessary to calculate these terms anew in each iteration; only the first derivatives must be calculated on each iteration.

Example. Suppose a complete sample of $n=10$ observations from a normal distribution has a mean $\bar{y}=4$ and standard deviation $s'=\sum_{i=1}^n(y_i-\bar{y})^2/n)=3$. Then the sample log likelihood from (2.1) is

$$\mathcal{L}(\mu,\sigma)=-10\ln(\sigma)-(5/\sigma^2)\left[9+(4-\mu)^2\right].$$

Of course, $\hat{\mu}=4$ and $\hat{\sigma}=3$ maximize this. Suppose that the current approximations are $\hat{\mu}_1=3$ and $\hat{\sigma}_1=2$.

1. The first partial derivatives at $\mu=\hat{\mu}_1=3$ and $\sigma=\hat{\sigma}_1=2$ are

$$\partial\mathcal{L}/\partial\mu=-(5/\sigma^2)(-2)(4-\mu)=2.5,$$

$$\partial\mathcal{L}/\partial\sigma=-(10/\sigma)+(10\sigma^3)\left[9+(4-\mu)^2\right]=7.5.$$

2. The second partial derivatives at $\mu=\hat{\mu}_1=3$ and $\sigma=\hat{\sigma}_1=2$ are

$$\partial^2\mathcal{L}/\partial\mu^2=-10/\sigma^2=-2.5, \qquad \partial^2\mathcal{L}/\partial\mu\,\partial\sigma=-(20/\sigma^3)(4-\mu)=-2.5,$$

$$\partial^2\mathcal{L}/\partial\sigma^2=(10/\sigma^2)-(30/\sigma^4)\left[9+(4-\mu)^2\right]=-16.25.$$

3. The solution of

$$2.5=2.5a_1+2.5b_1,$$

$$7.5=2.5a_1+16.25b_1$$

is $a_1=0.6363$ and $b_1=0.3636$, repeating decimals.

4. The new approximations are

$$\hat{\mu}_2 = 3 + 0.6363 = 3.6363, \qquad \hat{\sigma}_2 = 2 + 0.3636 = 2.3636.$$

These are closer to the correct ML estimates $\hat{\mu} = 4$ and $\hat{\sigma} = 3$ than the previous approximations $\hat{\mu}_1 = 3$ and $\hat{\sigma}_1 = 2$. Repeated iterations would bring the approximations closer to $\hat{\mu}$ and $\hat{\sigma}$.

Improvements. The method may fail to converge. Then the successive approximate estimates may march off to infinity. To avoid this, check whether $\mathcal{L}(\hat{\alpha}_{i+1}, \hat{\beta}_{i+1}) > \mathcal{L}(\hat{\alpha}_i, \hat{\beta}_i)$. If so, $(\hat{\alpha}_{i+1}, \hat{\beta}_{i+1})$ is better than $(\hat{\alpha}_i, \hat{\beta}_i)$. Otherwise, $(\hat{\alpha}_{i+1}, \hat{\beta}_{i+1})$ is poorer; then directly search along the line through the old and new values to find $(\hat{\alpha}'_{i+1}, \hat{\beta}'_{i+1})$ such that $\mathcal{L}(\hat{\alpha}'_{i+1}, \hat{\beta}'_{i+1}) > \mathcal{L}(\hat{\alpha}_i, \hat{\beta}_i)$. Continue the iteration with the new $(\hat{\alpha}'_{i+1}, \hat{\beta}'_{i+1})$.

In the example above, $\mathcal{L}(3, 2) \cong -19.4 < -16.8 \cong \mathcal{L}(3.6363, 2.3636)$. So the new values increase the likelihood.

In practice, perturbation calculations can approximate the needed partial derivatives. This helps when analytic calculation of the derivatives is difficult. One may save labor by using the same values of the **second** partial derivatives for a number of iterations, rather than recalculate them for each iteration. The method may then take more iterations, but each iteration is easier.

Theory. The following motivates the Newton–Raphson method. The method assumes that a quadratic function adequately approximates the log likelihood. The quadratic function is the Taylor series about the latest $(\hat{\alpha}_i, \hat{\beta}_i)$. That is,

$$\mathcal{L}(\alpha, \beta) \approx \mathcal{L}(\hat{\alpha}_i, \hat{\beta}_i) + (\partial \mathcal{L}/\partial \alpha)(\alpha - \hat{\alpha}_i) + (\partial \mathcal{L}/\partial \beta)(\beta - \hat{\beta}_i)$$
$$+ \tfrac{1}{2}(\partial^2 \mathcal{L}/\partial \alpha^2)(\alpha - \hat{\alpha}_i)^2 + \tfrac{1}{2}(\partial^2 \mathcal{L}/\partial \beta^2)(\beta - \hat{\beta}_i)^2$$
$$+ (\partial^2 \mathcal{L}/\partial \alpha \partial \beta)(\alpha - \hat{\alpha}_i)(\beta - \hat{\beta}_i) + \text{neglected terms,} \quad (6.3)$$

where the derivatives are evaluated at $\alpha = \hat{\alpha}_i$ and $\beta = \hat{\beta}_i$. The α and β values that maximize this quadratic are the new $\hat{\alpha}_{i+1} = \hat{\alpha}_i + a_i$ and $\hat{\beta}_{i+1} = \hat{\beta}_i + b_i$. The maximizing α and β values come from the usual calculus method; namely, set the partial derivatives of (6.3) equal to zero and solve the resulting linearized likelihood equations (6.1). The method works well if (6.3) approximates $\mathcal{L}(\alpha, \beta)$ well between $(\hat{\alpha}_i, \hat{\beta}_i)$ and $(\hat{\alpha}_{i+1}, \hat{\beta}_{i+1})$.

6.3.   Effective Computations

The following describes a number of ways of making numerical computations more effective. Ross (1970) describes a number of other methods that help maximum likelihood iterations converge more quickly and surely.

Checks to assure convergence. Methods like Newton–Raphson that solve the likelihood equations sometimes fail to converge when $\mathcal{L}(\hat{\alpha}_{i+1}, \hat{\beta}_{i+1}) < \mathcal{L}(\hat{\alpha}_i, \hat{\beta}_i)$, that is, when the log likelihood **decreases** for successive trial estimates. Then the trial estimates often shoot off to infinity. To avoid this, check that the latest trial estimates have increased $\mathcal{L}$. If not, discard them and find new ones by some other means, such as direct search along the line through $(\hat{\alpha}_i, \hat{\beta}_i)$ and $(\hat{\alpha}_{i+1}, \hat{\beta}_{i+1})$.

The log likelihood should have an optimum at the final values of the ML estimates. Equivalently, at those values, the local negative Fisher matrix should be positive definite. The Cholesky square root method automatically checks for this, as it inverts the matrix to obtain the "local" estimate of the covariance matrix of the ML estimates.

Enough data. A k-parameter distribution can be fitted only if the sample contains at least k distinct failure times. Otherwise, the estimates do not exist, and an iteration will fail to converge. For example, for an exponential distribution there must be at least one failure time. For an extreme value, Weibull, normal, or lognormal distribution, there must be at least two different failure times.

Starting values. The parameter values for the first iterative step determine whether an iteration converges. The starting values should be as close to the ML estimates as possible. This makes the convergence surer and faster. Often the sample data can be plotted, and graphical estimates of parameters can be used as starting values. Some iteration methods work with almost any reasonable starting values, and others are quite sensitive to the starting values.

Stopping criteria. Two criteria for stopping an iteration are often used. (1) Stop when an iteration increases the log likelihood less than some small amount, say, 0.001 or 0.0001. (2) Stop when an iteration changes each trial estimate by less than its prescribed stopping value. Each such stopping value should be a small fraction (say, 0.001) of the standard error of the corresponding estimate. The standard errors can be guessed before one starts the iteration, or they can be calculated during the iteration for the latest trial estimates.

Approximate derivatives of $\mathcal{L}$. Certain methods require the first or second derivatives of $\mathcal{L}$. Sometimes it is difficult to derive analytic expressions for such derivatives. Then it may be easier to approximate such derivatives by perturbation calculations. A perturbation should be small. In practice, this means one that changes $\mathcal{L}$ by about 0.1. A direct search for such perturbations is effective. Too small a perturbation produces appreciable round off in the calculation. Too large a one introduces appreciable nonquadratic behavior of $\mathcal{L}$.

Multiple ML estimates. The likelihood equations may have more than one set of solutions, perhaps at saddle points of the log likelihood. Also, the log likelihood may have local maximums. In either case, use the parameter estimates at the global maximum.

Number of iterations. Speed of convergence depends on a number of factors:

Fitted distribution.
Data.
Starting values.
Stopping criterion.

Of course, more iterations are required for larger numbers of parameters, poorer starting values, and more precise stopping criteria.

Constraints on parameters. Constraints may cause numerical difficulties. Some distribution parameters cannot take on all values from $-\infty$ to $+\infty$. For example, the standard deviation of a normal distribution must be positive. When parameters are constrained, computer maximization of a sample likelihood is difficult. An iterative method usually fails if it uses a forbidden value for a trial parameter estimate. To avoid this, express the parameter as a function of a new parameter that can take on all values from $-\infty$ to $+\infty$. For example $\sigma = \exp(\sigma')$ is always positive where $-\infty < \sigma' < \infty$. Then maximize the sample likelihood in terms of the new parameter, and the optimization is unconstrained. Nelson and others (1978) use such transformed parameters in the STATPAC program. To transform a parameter p where $0 < p < 1$, use $p = 1/(e^{-p'} + 1)$, where $-\infty < p' < \infty$.

Numerical overflow. An iteration scheme often fails when calculating a log likelihood value that is too large (overflow) or too small (underflow) for the computer. For example, the trial estimates of a normal $\hat{\mu}_i$ and $\hat{\sigma}_i$ may be off enough so that $z_j = (y_j - \hat{\mu}_i)/\hat{\sigma}_i$ for a censored value y_j may exceed 10 standard deviates. Then the calculation of $\mathcal{L}_j = \ln\{1 - \Phi[(y_j - \hat{\mu}_i)/\hat{\sigma}_i]\}$ may underflow the computer when { } is calculated. This can be avoided through the use of a transformed standardized deviate.

$$
z'_j = \begin{cases}
K_1 + \left\{ (z_j - K_1)/\left[1 + (z_j - K_1)/K'_1\right] \right\}, & \text{for } z_j > K_1 \\
z_j, & \text{for } -K_2 \leq z_j \leq K_1, \\
-K_2 - \left\{ (2 + K_2)/\left[-1 + (z_j + K_2)/K'_2\right] \right\}, & \text{for } z_j < -K_2,
\end{cases}
$$

where K_i and K'_i are constants around 5. Then the z'_j are in the range $-(K_2 + K'_2)$ to $K_1 + K'_1$. The scheme uses the z'_j in place of the z_j until the iteration is close to converging. Then the iteration switches to using the z_j and completes the iteration. The transformed standardized deviates help an iteration converge even when the starting values of the parameters are bad. This method applies to a distribution with standardized deviates:

Normal: $z = (y - \mu)/\sigma$
Lognormal: $z = [\log(t) - \mu]/\sigma$
Extreme value: $z = (y - \lambda)/\delta$
Weibull: $z = [\ln(t) - \ln(\alpha)]\beta$
Exponential: $z = y/\theta$

The choice of K_i and K'_i depends on the distribution. Nelson and others (1978) use such a z' in the STATPAC program.

There is a simpler method of avoiding such numerical problems. Use starting values of σ, δ, and θ that are several times larger than the ML estimate. Similarly, use a starting β value that is several times too small.

Another covariance matrix estimate. Local and ML estimates of the covariance matrix of the ML estimators of distribution parameters were given in previous sections. These estimates involve the second partial derivatives of the sample log likelihood with respect to the parameters. These derivatives (and their expectations) may be difficult to derive analytically or to approximate accurately by numerical perturbation. The following estimate involves only the first partial derivatives. For concreteness, assume that the distribution has two parameters μ and σ and that the log likelihood for sample unit i is $\mathcal{L}_i$, $i = 1, 2, \ldots, n$. The estimates of the terms of the Fisher information matrix are

$$
F^{**}_{\mu\mu} = \sum_i (\partial \mathcal{L}_i / \partial \mu)^2, \quad F^{**}_{\sigma\sigma} = \sum_i (\partial \mathcal{L}_i / \partial \sigma)^2, \quad F^{**}_{\mu\sigma} = \sum_i (\partial \mathcal{L}_i / \partial \mu)(\partial \mathcal{L}_i / \partial \sigma),
$$

where the first partial derivatives are evaluated at $\mu = \hat{\mu}$ and $\sigma = \hat{\sigma}$. These estimates are motivated by equations (5.4.11) and (5.4.12). The inverse of this estimate of the Fisher information matrix is the estimate of the covariance matrix, namely,

$$\hat{\Sigma}^{**} = \begin{pmatrix} F_{\mu\mu}^{**} & F_{\mu\sigma}^{**} \\ F_{\mu\sigma}^{**} & F_{\sigma\sigma}^{**} \end{pmatrix}^{-1}$$

This "first-derivative" estimate and the local and ML estimates of $\hat{\Sigma}$ have not been studied for small samples. For large samples (many failures), they are nearly equal. This first-derivative estimate extends to distributions with three or more parameters in the obvious way.

PROBLEMS

8.1. Insulating fluid (exponential). Analyze the 25-kV data on time to breakdown of an insulating fluid in Table 2.1 of Chapter 7. Assume that the distribution is exponential

(a) Make a Weibull plot of the data, graphically fit an exponential distribution to the plotted data, and estimate the distribution mean. Does the plot look satisfactory?

(b) Calculate the ML estimate of the mean.

(c) Calculate two-sided 95% confidence limits for the mean based on the normal approximation.

(d) Do (c), based on the chi-square approximation.

(e) Calculate the ML estimate of the 10th percentile and two-sided 95% confidence limits bases on the chi-square approximation.

(f) Using (d) and (e), plot the ML estimate of the cumulative distribution function and its 95% confidence limits on the Weibull plot from (a).

(g) Repeat (a) through (f) for data from other test voltages.

8.2. Class B insulation. Use the 220°C data on (log) time to failure of insulation specimens in Table 3.1 of Chapter 7.

(a) Make a lognormal plot of the data, and graphically estimate the lognormal parameters. Using the estimate of σ, describe whether the failure rate increases or decreases with age.

(b) Calculate the ML estimates of the parameters and the lognormal median. The referenced tables for singly censored samples may help reduce the effort, as would a computer program.

(c) Calculate the sample local Fisher information matrix for the parameter estimates.

(d)   Calculate the local estimate of the asymptotic covariance matrix of the parameter estimates.

(e)   Calculate two-sided approximate 95% confidence limits for the parameters and lognormal median.

(f)   Calculate the ML estimate for the 10th percentile of the lognormal distribution.

(g)   Calculate the local estimate of the asymptotic variance of the **log** of the ML estimator (f) of the 10th percentile.

(h)   Calculate two-sided approximate 95% confidence limits for the 10th (lognormal) percentile.

(i)   Using (b) and (f), plot the ML fitted distribution on the lognormal plot from (a). Also, plot the confidence limits from (e) and (h).

(j*)   Calculate limits for the parameters and lognormal median using exact tables referenced in Section 2.

(k)   Repeat (c) through (h) using the first-derivative estimate of Section 6.3 for the Fisher information matrix.

8.3.   Insulating fluid (Weibull).   Use the 26-kV data on time to break-down of an insulating fluid in Table 1.1 of Chapter 3. Use a computer program if you have one.

(a)   Make a Weibull plot of the data, and graphically estimate the Weibull parameters.

(b)   Iteratively solve (3.3) to get the ML estimate of the shape parameter accurate to two decimal places; use the estimate from (a) to start the iteration.

(c)   Use (3.4) to get the ML estimate of the scale parameter.

(d)   Use (3.11) to calculate the sample local Fisher information matrix for the corresponding extreme value parameter estimates.

(e)   Calculate the local estimate of the asymptotic covariance matrix for the ML estimates of the parameters of the corresponding extreme value distribution.

(f)   Calculate two-sided approximate 95% confidence limits for the extreme value scale parameter and corresponding limits for the Weibull shape parameter. Are the data consistent with a true shape parameter of 1?

(g)   Calculate the ML estimate for the 10th percentile of the Weibull distribution.

*Asterisk denotes laborious or difficult.

(h) Calculate the local estimate of the asymptotic variance of the ML estimator for the 10th percentile of the corresponding extreme value distribution.

(i) Calculate two-sided approximate 95% confidence limits for the extreme value 10th percentile and corresponding limits for the Weibull 10th percentile.

(j*) Calculate exact limits for the parameters and 10th percentile using tables referenced in Section 3.

8.4. Shave die. This problem uses the data on time to wear out of shave dies in Problem 4.1 of Chapter 4.

(a) Plot the product-limit estimate of the cumulative distribution function on Weibull paper and assess the validity of the data and the fit of the Weibull distribution.

Use the following STATPAC output. CENTER denotes the Weibull α and SPREAD denotes β.

* MAXIMUM LIKELIHOOD ESTIMATES FOR DIST. PARAMETERS
  WITH APPROXIMATE 95% CONFIDENCE LIMITS

PARAMETERS	ESTIMATE	LOWER LIMIT	UPPER LIMIT
CENTER	78.35229	60.83119	100.9200
SPREAD	2.755604	1.710175	4.440104

* COVARIANCE MATRIX

PARAMETERS	CENTER	SPREAD
CENTER	102.3808	
SPREAD	−3.605433	0.4498142

* MAXIMUM LIKELIHOOD ESTIMATES FOR DIST. PCTILES
  WITH APPROXIMATE 95% CONFIDENCE LIMITS

PCT.	ESTIMATE	LOWER LIMIT	UPPER LIMIT
0.1	6.389104	2.163791	18.86534
0.5	11.46595	5.094722	25.80474
1	14.75874	7.360155	29.59454
5	26.66470	17.24049	41.24048
10	34.62458	24.81156	48.31868
20	45.46268	35.46651	58.27625
50	68.59425	54.72485	85.97870
80	93.12218	68.64407	126.3290
90	106.0471	74.59757	150.7553
95	116.6737	79.08319	172.1322
99	136.3771	86.71072	214.4915

(b)   Does the ML estimate of the shape parameter suggest that the failure rate of the dies increases or decreases with die age?

(c)   Do the confidence limits for the shape parameter provide convincing evidence that the failure rate increases or decreases with die age?

(d)   On the Weibull plot from (a), plot the fitted distribution and confidence limits for the percentiles.

8.5.   Locomotive control (Weibull).   The following STATPAC output shows the ML fit of a Weibull distribution to the locomotive control data of Section 2. CENTER $= \alpha$ and SPREAD $= \beta$.

```
* MAXIMUM LIKELIHOOD ESTIMATES FOR DIST. PARAMETERS
  WITH APPROXIMATE 95% CONFIDENCE LIMITS

PARAMETERS     ESTIMATE      LOWER LIMIT      UPPER LIMIT

CENTER       183.4091        153.7765         218.7519
SPREAD         2.331076        1.720961         3.157489

* COVARIANCE MATRIX

PARAMETERS     CENTER           SPREAD

CENTER        271.9164
SPREAD         -3.691133        0.1302468


* MAXIMUM LIKELIHOOD ESTIMATES FOR   DIST.   PCTILES
  WITH APPROXIMATE 95% CONFIDENCE LIMITS

PCT.            ESTIMATE      LOWER LIMIT      UPPER LIMIT
0.1             9.474467        4.249243         21.12506
0.5            18.91372        10.41828         34.33667
1              25.49080        15.32750         42.39314
5              51.29256        37.62473         69.92547
10             69.84944        55.50577         87.89977
20             96.37678        81.89924        113.4136
50            156.7246        134.7152         182.3299
80            224.9476        180.5117         280.3222
90            262.3050        202.6526         339.5165
95            293.6528        220.2723         391.4791
99            353.1379        251.9352         494.9937


PERCENT(LIMIT 80.)

* MAXIMUM LIKELIHOOD ESTIMATES FOR   % WITHIN LIMITS
  WITH APPROXIMATE 95% CONFIDENCE LIMITS

                ESTIMATE      LOWER LIMIT      UPPER LIMIT
PCT       13.45950          8.643893         20.36003
```

(a)   Make a Weibull plot of the data, and assess how well the Weibull distribution fits the data (both on an absolute basis and in comparison with the lognormal fit).

(b)   Does the shape parameter estimate suggest that the failure rate increases or decreases with age? Do the confidence limits for the shape parameter provide convincing evidence that the failure rate increases (or decreases)?

(c)   What are the ML estimate and the two-sided approximate 95% confidence limits of the percentage failing on an 80-thousand mile warranty? How do they compare with those from the lognormal fit?

(d)   Plot the fitted Weibull distribution and the confidence limits for the percentiles on the Weibull plot from (a). Also, plot the fitted lognormal distribution (a curve) and the lognormal confidence limits for the percentiles on the same plot. How do the estimates and confidence limits compare (for practical purposes) in (1) the range of the data (particularly the middle), in (2) the lower tail beyond the data, and in (3) the upper tail beyond the data.

(e)   Would the key conclusions concerning the failure rate and the percentage failing or warranty depend on which distribution is used?

8.6.   Known Weibull shape.   For a **complete** sample of size n from a Weibull distribution with a known shape parameter value β_0 and an unknown scale parameter α_0, the scale parameter is to be estimated.

(a)   Give the sample likelihood function and the log likelihood.

(b)   Obtain and solve the likelihood equation for the maximum likelihood estimate $\hat{\alpha}$ of the scale parameter.

(c)   Derive the expression for the sample local Fisher information both in terms of the true value α_0 of the scale parameter and in terms of the sample estimate $\hat{\alpha}$.

(d)   Derive the expression for the asymptotic Fisher information both in terms of the true value α_0 of the scale parameter and in terms of the sample estimate $\hat{\alpha}$.

(e)   Derive the expressions for the asymptotic variance and standard error for the maximum likelihood estimate $\hat{\alpha}$ both in terms of the true value α_0 and the estimate $\hat{\alpha}$.

(f)   Give expressions for the approximate confidence limits for the true value of the scale parameter—this must be in terms of the estimate $\hat{\alpha}$.

(g)   Use the results from (b) through (f) and obtain the corresponding sample quantities for the 35-kV data of Table 2.1 of Chapter 7, assuming $\beta_0 = \frac{1}{2}$.

8.7. Poisson λ. Suppose Y_k has a Poisson distribution with mean $\lambda_0 t_k$, where the exposure t_k is known, $k = 1, \ldots, K$, and the Y_k are statistically independent.

(a) Write the likelihood function for $Y_1, \ldots, Y_K$.

(b) Set the derivative of the log likelihood with respect to λ equal to zero and solve for the ML estimate $\hat{\lambda}$.

(c) Calculate the true Fisher information for $\hat{\lambda}$.

(d) Calculate the true asymptotic variance and standard error of $\hat{\lambda}$.

(e) Give the expression for the large-sample two-sided approximate $100\gamma\%$ confidence limits for λ_0 (positive limits).

(f) Calculate such two-sided 95% confidence limits for the yearly failure rate of a power line that had $Y_1 = 2$, $Y_2 = 6$, $Y_3 = 4$, and $Y_4 = 4$ failures over four years.

8.8.* Binomial p. Suppose Y_k has a binomial distribution with probability p_0 and sample size n_k, $k = 1, \ldots, K$, and the Y_k are statistically independent.

(a) Write the likelihood function for $Y_1, \ldots, Y_K$.

(b) Derive the ML estimate $\hat{p}$ for p_0.

(c) Calculate the Fisher information for $\hat{p}$.

(d) Calculate the true asymptotic variance and standard error of $\hat{p}$.

(e) Give the expression for large-sample two-sided approximate $100\gamma\%$ confidence limits for p_0.

(f) Calculate such two-sided 95% confidence limits for the percentage of locomotive controls failing on warranty (Chapter 3). $Y = 15$ of $n = 96$ sample controls failed on warranty.

(g) Compare (f) with exact binomial limits and the limits from Problem 8.5c.

8.9.* Mixture. A distribution with a "bathtub" hazard function is a mixture of two Weibull distributions with parameters α_k, β_k and proportion p_k, $k = 1, 2$, $\beta_1 < 1 < \beta_2$, $p_1 + p_2 = 1$.

(a) Give the log likelihood function of the mixture distribution for a singly time censored sample.

(b) Comment on the theoretical and numerical labor to obtain the ML estimates and asymptotic covariance matrix.

8.10.* Trading stamps. A trading stamp company wished to estimate the proportion p_0 of stamps that will eventually be redeemed. Specially marked stamps were simultaneously released and their times to redemption were observed until a time y_0 after release. Suppose that the cumulative distribution of time to redemption of redeemed stamps is $F(y)$. The cumulative distribution of time to redemption for all stamps is $p_0 F(y)$; it is degenerate at infinity, since a proportion $1 - p_0$ of the stamps are not redeemed. Assume (wrongly) that stamps are statistically independent and that $F(y)$ is an exponential distribution with mean "failure" rate λ_0.

(a) Give the expression for the probability a stamp is not redeemed by time y_0.

(b) Give the likelihood for r observed stamps (redemption time y_i) in a sample of n stamps where $n - r$ times are singly censored at time y_0.

(c) Assume that y_0 is much greater than θ and simplify the likelihood so that it lacks y_0. Give the **log** likelihood.

(d) Calculate the likelihood equations for $\hat{p}$ and $\hat{\lambda}$.

(e) Calculate the local estimate of asymptotic covariance matrix of the ML estimates $\hat{p}$ and $\hat{\lambda}$. Leave in terms of $\hat{p}$ and $\hat{\lambda}$.

8.11.*   Trading stamps.   Repeat the preceding problem with a Weibull distribution.

8.12.*   Unidentified competing causes.   Consider a series system with two independent competing failure causes. Suppose that cause k has a Weibull distribution with parameters α_k and β_k, $k = 1, 2$. Also, suppose that the cause of any failure is not identified.

(a) Give the log likelihood for a sample multiply censored on the right.

(b) Comment on the theoretical and numerical labor to obtain the ML estimates and asymptotic covariance matrix.

(c) Under the assumption $\beta_1 = \beta_2 = 1$ (exponential distributions), can α_1 and α_2 be estimated separately? Explain.

(d) Under the assumption $\beta_1 = \beta_2$, can α_1 and α_2 be estimated separately? Explain.

8.13.*   Exponential left censored.   Suppose that a type of unit has an exponential life distribution with mean θ_0. Suppose that unit i is not observed until time τ_i (left censored) but a failure time T_i after τ_i is observed exactly, $i = 1, \ldots, n$.

(a) Write the sample log likelihood $\mathcal{L}$, distinguishing between left censored and observed units.

(b) Derive $\partial \mathcal{L} / \partial \theta$, and give the likelihood equation for θ.

(c) Use the insulating fluid data at 45 kV in Table 10.3 of Chapter 3, and iteratively calculate the ML estimate $\hat{\theta}$ accurate to two figures.

(d) Derive the formula for $\partial^2 \mathcal{L}/\partial\theta^2$ and evaluate it for the 45-kV data.

(e) Give the formula for the local estimate of $\text{Var}(\hat{\theta})$.

(f) Calculate the local estimate of $\text{Var}(\hat{\theta})$ for the 45-kV data.

(g) Calculate positive approximate 95% confidence limits for θ_0, using (f).

(h) Express the sample log likelihood in terms of indicator functions, and derive the formula for the true asymptotic variance of $\hat{\theta}$.

(i) Evaluate the ML estimate of $\text{Var}(\hat{\theta})$ for the 45-kV data.

(j) Calculate positive approximate 95% confidence limits for θ_0, using (i).

(k) On Weibull paper, plot the data, the fitted exponential distribution, and approximate 95% confidence limits for percentiles.

8.14. Newton–Raphson method. Continue the Newton–Raphson iteration through two more iterations on the example in Section 6.2. Check that the log likelihood increases with each iteration.

8.15.* Unidentified competing failure modes. A certain product contains many components which can fail, including a motor. The motor manufacturer reimburses the product manufacturer for each motor failure on warranty. In a particular production period, the method of motor manufacture was changed, resulting in a defective motor with a high failure rate. The following model and analyses were used to predict the motor manufacturer's liability. The cause of product failure was not identified.

(a) Prior to any motor problem, experience with all other failure modes indicated that (1) the product had a small proportion π that were found failed when installed and (2) the remaining proportion $(1 - \pi)$ followed a Weibull distribution for time t to failure with scale parameter α and shape parameter β near 1. Write this cumulative distribution function for the product, assuming no motor failures.

(b) The defective motor was assumed to have a Weibull life distribution with scale parameter α' and shape parameter β'. Assuming these motors are not failed on installation (not in the proportion π), write the combined cumulative distribution function for time t to failure due to both motors and all other causes, assuming motor failures are independent of all others.

(c)   Ordinarily the Weibull distribution is written in terms of the scale parameter. Rewrite the distributions from (a) and (b) in terms of small percentiles $t_p = \alpha - [\ln(1 - P)]^{1/\beta}$ and $t'_{p'} = \alpha' - [\ln(1 - P')]^{1/\beta'}$, eliminating α and α'. This reparametrization makes the ML calculations converge more easily, when P and P' are chosen near the corresponding sample fractions failed.

(d)   The accompanying figure shows maximum likelihood estimates and the local estimate of their covariance matrix for the coefficients $C1, \ldots, C5$ where

$$\pi = \sin^2(C1), \qquad t_{.01} = \exp(C2), \qquad \beta = \exp(C3),$$

$$t'_{.01} = \exp(C4), \qquad \beta' = \exp(C5).$$

Here $t_{.01}$ and $t'_{.01}$ are expressed in months and $C1$ in radians. Calculate two-sided (approximate) 95% confidence limits for π, β and β'. Is $\hat{\beta}$ near 1 as expected? Is $\hat{\beta}'$ significantly different from 1, and what does it indicate regarding future numbers of motor failures in service? Is $\hat{\pi}$ significantly different from zero, suggesting π in the model improves the fit?

(e)   Write the algebraic formula for the fraction failing from all causes on a warranty of t^* months. Calculate the numerical ML estimate of this fraction for $t^* = 60$ months.

★ MAXIMUM LIKELIHOOD ESTIMATES FOR MODEL COEFFICIENTS WITH APPROXIMATE 95% CONFIDENCE LIMITS

COEFFICIENTS		ESTIMATE	LOWER LIMIT	UPPER LIMIT	STANDARD ERROR
C	1	-0.3746865E-01	-0.4319699E-01	-0.3174031E-01	0.2922623E-02
C	2	3.234415	3.058929	3.409902	0.8953391E-01
C	3	-0.2541079	-0.4578389	-0.5037680E-01	0.1039444
C	4	3.358588	3.312264	3.404913	0.2363509E-01
C	5	1.384372	1.158305	1.610439	0.1153402

★ COVARIANCE MATRIX

COEFFICIENTS	C	1	C	2	C	3	C	4
C	1	0.8541726E-05						
C	2	0.1161638E-03	0.8016321E-02					
C	3	-0.2363367E-03	-0.6771157E-02	0.1080444E-01				
C	4	-0.4079892E-04	-0.1715020E-02	0.2035757E-02	0.5586175E-03			
C	5	-0.1697718E-03	-0.8399796E-02	0.8862306E-02	0.1986809E-02			

COEFFICIENTS	C	5
C	5	0.1330337E-01

(f)   Write the algebraic formula for the fraction failing from all other causes (except motor) on a warranty of t^* months. Calculate the numerical ML estimate of this fraction for $t^* = 60$ months.

(g)   The fraction from (e) minus the fraction from (f) is the increase in failures due to motors and is paid by the motor manufacturer. Write the algebraic formula for this difference, and calculate its numerical ML estimate for $t^* = 60$ months.

(h)   Numerically calculate two-sided approximate 95% confidence limits for the difference in (g).

(i)   Explain why one-sample prediction limits are preferable to the confidence limits in (h), and discuss how much the two types of limits differ in this application.

(j)   Criticize the model in (b).

9

Analyses of Inspection Data (Quantal-Response and Interval Data)

Introduction

For some products, a failure is found only on inspection, for example, a cracked component inside a turbine. Two types of such inspection data are treated here: (1) quantal-response data (there is exactly one inspection on each unit to determine whether it is failed or not) and (2) interval data (there is any number of inspections on a unit). This chapter describes graphical and maximum likelihood (ML) methods for estimating the life distribution of units that are inspected and found to be either failed or else running. Most analyses of life data assume that each failure time is known exactly. Such analyses are incorrect for inspection data, since only the interval in which a failure occurred is known.

Needed background for this chapter includes the basic distributions in Chapter 2, the basics of probability plotting from Chapter 3, and, for ML methods, acquaintance with Chapter 8.

Quantal-Response Data

Definition. Suppose each unit is inspected only **once**. If a unit is found failed, one knows only that its failure time was before its inspection time. Similarly, if a unit is found unfailed, one knows only that its failure time is beyond its inspection time. Such inspection data are called **quantal-response**

data, sensitivity data, probit data, binary data, logit data, and all-or-none response data. Such life data are often **wrongly** analyzed as multiply censored data; then the inspection time when a failure is found is wrongly treated as the failure time. The failure occurs before the inspection, and this must be properly taken into account as described here.

Other applications.   Quantal-response data arise in many other applications besides product life. For example, fan blades of jet engines are tested under bird impact (Nelson and Hendrickson, 1972). A bird is "fired" with a known velocity at a blade. The impact cracks the blade if the bird's velocity is above the blade's cracking velocity; otherwise, the bird does not crack the blade. Each blade (and bird) is impacted only once. The data are used to estimate the distribution of cracking velocities of such blades. Other applications include the distribution of

1.   Drop height that causes bombs to explode (Golub and Grubbs, 1956).

2.   Endurance limit of metal specimens in a fatigue test (Little and Jebe, 1975, Chap. 10).

3.   Insecticide dose that kills insects (Finney, 1968; Berkson, 1953).

4.   Penetrating velocity of shells fired at a given thickness of armor (Golub and Grubbs, 1956).

5.   Voltage that blows a fuse (Sheesley, 1975).

6.   Cyclamate dose that causes cancer in rats.

In such applications, each test unit is subjected to a value of some "stress" and is a failure or success. A unit is not retested at another stress. The relationship between the probability of failure and "stress" can be expressed in terms of a cumulative distribution function. For example, the probability $F(y)$ of bomb explosion as a function of drop height y is assumed to have the form

$$F(y) = \Phi\big[(y-\mu)/\sigma\big],$$

where $\Phi(\ )$ is the standard normal cumulative distribution function, and μ and σ are to be estimated from data.

Overview.   Analysis of quantal-response data involves fitting a distribution to the data. Then the fitted distribution provides information on product performance or life. Section 1 explains simple graphical analyses of such data. Section 2 explains ML methods, which require special computer programs. Section 5 briefly presents advanced ML theory for quantal-response data.

Interval Data

Definition. For some products, a failure is found only on inspection, for example, a cracked part inside a machine. This chapter presents methods for estimating the distribution of time to failure when each unit is inspected periodically. If a unit is found failed, one knows only that the failure occurred between that inspection and the previous one. Also, if a unit is unfailed on its latest inspection, one knows only that its failure time is beyond the inspection time. Such periodic inspection data are called **interval data**, grouped data, and coarse data.

If the intervals are small, then each failure time can be approximated by the midpoint of its interval, and these times can be analyzed like complete or censored data. The approximate results are useful for practical purposes if the interval widths are less than, say, one-fifth of the distribution standard deviation. This chapter describes exact methods that apply to any interval widths.

Other applications. Interval data arise in many applications. Indeed most data values are recorded to a finite number of significant figures. For example, a data value of 9.6 usually means that the true value is in the interval from 9.55 to 9.65. However, in most applications, the interval width is small enough to neglect.

Overview. Analysis of interval data involves fitting a distribution to the data and using the fitted distribution to obtain desired information. Section 3 describes simple graphical analyses of such data with probability plots. Section 4 explains maximum likelihood analyses, which require special computer programs. Section 5 briefly presents the underlying (advanced) maximum likelihood theory and surveys literature on analysis of interval data.

1. GRAPHICAL ANALYSES OF QUANTAL-RESPONSE DATA

Plots of quantal-response data serve several purposes. They estimate the life distribution; i.e., the relationship between the cumulative percentage failing and age. Also, they assess the validity of the assumed distribution and data. The following explains how to make and interpret a probability plot of such data. Ordinary probability plotting of Chapter 3 and hazard plotting of Chapter 4 must be modified as described here.

Turbine Wheel Data

The methods are illustrated with data on turbine wheels. Each wheel was inspected once to determine if it had started to crack or not. Some of the

Table 1.1. Wheel Crack Initiation Data (Hours)

```
3322+
4009+
1975-
1967-
1892-
2155+
2059+
4144+
1992+
1676+
4079+
2278-
1366+
etc.
```

```
+ unfailed on inspection
- found failed on inspection
```

data are shown in Table 1.1, which contains the wheel's age at inspection (in hours) and condition (" − " denotes cracked and " + " denotes not cracked).

The purpose of the analysis is to estimate the distribution of time to crack initiation. This information was needed to schedule regular inspections. Also, engineering wanted to know if the failure rate increases or decreases with wheel age. An increasing failure rate requires replacement of wheels by some age when the risk of cracking gets too high.

How to Make a Probability Plot

First, divide the range of the data into intervals that each contain inspections of at least 10 units. The plot will be crude unless the data set is moderately large, say, over 100 units. Figure 1.1 shows this for the entire set of wheel data. For the units inspected in an interval, calculate the percentage that are failed. For example, for the 30 wheels inspected between 2000 and 2400 hours, the cumulative percentage failed is $100(5/30) = 16.7\%$. On probability paper for an assumed distribution, plot each percentage failed against the midpoint of its interval. For example, 16.7% is plotted against 2200 hours in the Weibull probability plot in Figure 1.2. If such a percentage is 0 or 100%, the plotting position is off scale; the intervals might be chosen to avoid this.

If the plotted points follow a straight line reasonably well, the distribution adequately fits the data. Then draw a line through the plotted points to estimate the life distribution. In fitting the line, try to weight each point according to its number of units, or (better) try to weight each point according to its confidence interval described below. If the plotted points do not follow a straight line, plot the data on probability paper for another distribution or draw a curve through the points (nonparametric fit).

HOURS	FAILED /OBS'D	% FAILED	BINOM. 95% CONF. LIMITS	
4400+	21/36 =	58.4	40.8	74.5
4000-4400	21/40 =	52.5	36.1	68.5
3600-4000	22/34 =	64.8	46.5	80.3
3200-3600	6/13 =	46.2	15.9	74.9
2800-3200	9/42 =	21.4	10.3	36.8
2400-2800	9/39 =	23.1	11.1	39.3
2000-2400	5/30 =	16.7	5.6	34.7
1600-2000	7/73 =	9.59	4.9	18.8
1200-1600	2/33 =	6.06	0.7	20.2
800-1200	4/53 =	7.55	2.1	18.2
0- 800	0/39 =	0	0.0	9.0

Figure 1.1. Wheel data and plotting positions.

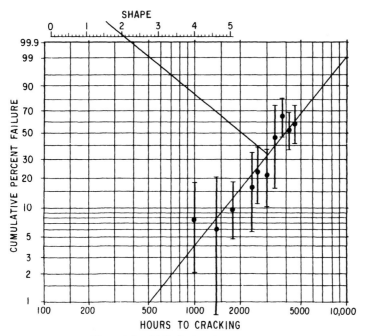

Figure 1.2. Weibull plot of wheel data.

409

The basis for the plotting method. The units inspected in an interval are all regarded as having the interval midpoint as their age. Then the fraction failed among those units estimates the population cumulative fraction failing by that age.

Such plots may be peculiar in that the observed cumulative fraction failing (an increasing function) may decrease from one interval to the next. When this happens one may wish to combine the data in two such intervals if that eliminates the decrease in the sample cumulative distribution. For example, in Figure 1.2 the points 7.55% at 1000 hours and 6.06% at 1400 hours could be replaced by the point $100(4+2)/(53+33)=6.98\%$ at 1200 hours. This simple subjective method gives an increasing cumulative distribution function; Peto's (1973) complex method does this objectively and does not group the data into intervals.

Confidence limits. Limits for the sample cumulative distribution are obtained as follows. Each unit in an interval is a binomial trial that is either a failure or a success. The value of the cumulative distribution function for the interval midpoint is (roughly) the binomial probability of failure. So binomial confidence limits apply. Section 3 of Chapter 6 and Hahn and Shapiro (1967) give binomial confidence limits. For the wheel example, the interval from 2000 to 2400 hours has five failures among 30 wheels. The binomial 95% confidence limits are 5.6 and 34.7%. Such two-sided 95% limits are given in Figure 1.1. Also, the bars in Figure 1.2 show these limits. These binomial limits involve no assumption about the form of the product life distribution; that is, they are nonparametric.

Peto (1973) gives approximate limits for the cumulative distribution. His limits have the advantage of using all data, not just that in a particular interval. However, his limits require a special computer program.

A sample of quantal-response data is less informative than one where the exact failure times are observed. In other words, for samples with the same number of units from the same population, confidence limits from exact failure times tend to be narrower than ones from quantal-response data.

How to Interpret a Probability Plot

The plot is interpreted like any other probability plot and provides

1. A check on the validity of the data and the fitted distribution.
2. Estimates of the distribution parameters.
3. The failure probability by any given age.
4. Distribution percentiles.
5. The behavior of the failure rate as a function of age.

These are explained below. Further discussions and practical details on how to use and interpret probability plots are given in Chapter 3.

Check validity of the data and distribution.   Departure of a probability plot from a straight line can indicate peculiar data or lack of fit of the assumed distribution. The plotting method uses a crude estimate of the cumulative fraction failing for each interval. So only extreme peculiarities in such a plot should be interpreted as an incorrect distribution or faulty data. Peto (1973) gives a more refined sample cumulative distribution.

Estimates of probabilities and percentiles.   Probabilities and percentiles are estimated from a plot as described in Chapter 3. For example, the estimate of the proportion of wheels failing by 1000 hours is 4.0% from Figure 1.2. Also, for example, the estimate of the 50th percentile, a nominal ·life, is 4000 hours from Figure 1.2. Such estimates can be obtained from a curve fitted to a plot, if the plot is not straight.

Parameter estimates.   Distribution parameters are estimated from a plot as described in Chapter 3. The estimate of the Weibull scale parameter for the wheel data is 4800 hours from Figure 1.2. The estimate of the Weibull shape parameter on the shape scale is 2.09.

Nature of the failure rate.   The following method assesses the nature of the failure rate, a basic question on the wheels. A Weibull failure rate increases (decreases) if the shape parameter is greater (less) than 1. The shape parameter estimate is 2.09 for the wheel data. This value indicates that the wheel failure rate increases with age, that is, a wear-out pattern. So wheels should be replaced at some age when they are too prone to failure. An appropriate replacement age can be determined from the relative costs of failure, of replacement before failure, and the remaining expected life of a wheel.

2.   MAXIMUM LIKELIHOOD FITTING FOR QUANTAL-RESPONSE DATA

This section presents fitting of a distribution to quantal-response data by maximum likelihood (ML). Analyses yield estimates and confidence limits for parameters, percentiles, probabilities, and other quantities. The section describes the ML fitting, other analyses, and available computer programs. Chapter 8 and Section 5 provide background for the ML method for those who wish to write their own programs or to acquire a deeper understanding of ML fitting.

Maximum Likelihood Fitting

The following presents the ML fit of the Weibull distribution to the wheel data. The method applies to other distributions. The ML analysis yields estimates of the Weibull parameters α and β for the wheels. ML calculations are laborious; they require special computer programs, which are referenced later. The example employs output from STATPAC of Nelson and others (1978).

STATPAC output from the Weibull fit to the wheel data appears in Figure 2.1. There the parameters α and β are denoted by "CENTER" and

```
* MAXIMUM LIKELIHOOD ESTIMATES FOR DIST. PARAMETERS
  WITH APPROXIMATE 95% CONFIDENCE LIMITS

PARAMETERS      ESTIMATE      LOWER LIMIT      UPPER LIMIT

CENTER  scale α̂   4809.557      4217.121         5485.219
SPREAD  shape β̂   2.091212      1.645579         2.657525

* COVARIANCE MATRIX

PARAMETERS       CENTER                SPREAD
                   α̂                     β̂
CENTER  α̂   104048.0
SPREAD  β̂   -58.57977      0.6537992E-01

* CORRELATION MATRIX

PARAMETERS      CENTER                SPREAD

CENTER   1.0000000
SPREAD  -0.7102457       1.000000


PCTILES

* MAXIMUM LIKELIHOOD ESTIMATES FOR DIST. PCTILES
  WITH APPROXIMATE 95% CONFIDENCE LIMITS

PCT.          ESTIMATE      LOWER LIMIT      UPPER LIMIT

0.1           176.8618      87.44939         357.6937
0.5           382.2012      226.8173         644.0328
1             533.0458      342.0775         830.6241
5             1162.162      892.7043         1512.955
10            1639.675      1357.619         1980.331
20            2347.489      2079.196         2650.402
50            4036.353      3631.013         4486.942
80            6038.581      5071.816         7189.627
90            7166.606      5806.924         8844.656
95            8127.646      6407.249         10309.98
99            9983.057      7515.869         13260.13
```

Figure 2.1. STATPAC output on Weibull fit to wheel data.

"SPREAD," respectively. The estimate of β is 2.09, and the approximate 95% confidence limits for the true β value are 1.65 and 2.66. The β estimate greater than 1 indicates that the wheels have an increasing failure rate. So wheels should be replaced at some age. The limits do not enclose 1, which corresponds to a constant failure rate. This is statistically significant evidence that the wheel failure rate increases with age.

STATPAC also gives the ML estimates of distribution percentiles. These percentile estimates plot as a straight line in Figure 2.2. The two-sided approximate 95% confidence limits plot as curves. Note that each confidence interval encloses the true distribution line with high probability at the corresponding age, but it does not necessarily enclose the data point. These limits are narrower than the binomial ones, since they use all of the data (not just the data in an interval) and employ a specified (Weibull) life distribution. Peto's (1973) confidence limits do not employ a specified distribution. The ML and Peto's intervals are approximate and tend to be too short for small samples.

The output shows seven-figure results. They are accurate and useful to about three figures.

The ML fitting is iterative and may fail to converge. This happens if the sample cumulative distribution does not increase with age. For example, if there were just two intervals, there is a chance that the observed fraction

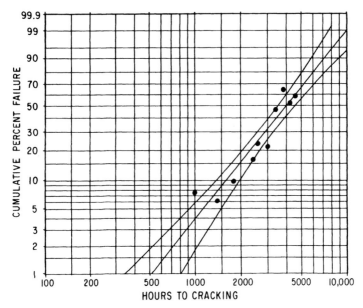

Figure 2.2. Plot of Weibull fit to the wheel data.

failed in the second interval is less than that in the first interval. This is more possible with small samples than with large ones.

Analytic Checks

The ML method can be used to check whether a distribution adequately fits the data. Then a likelihood ratio test (Chapter 12) is used to compare the chosen distribution with a more general one. The method for doing this is described by Prentice (1974), by Nelson and Hendrickson (1972), and in Chapter 12.

Other Analyses

Almost any analysis of complete data extends to quantal-response data. For example, two samples can be compared with a hypothesis test (Chapter 12) to assess whether they differ significantly. Nelson and Hendrickson (1972) give such an example. They compare the distributions of breaking velocity of two designs of jet engine fan blades in a bird impact test. Also, for example, a parametric fitted distribution like the Weibull distribution can be compared with Peto's (1973) nonparametric estimate to assess how well the parametric distribution fits the data.

Computer Programs

Some computer programs for fitting distributions to quantal-response data are listed here. The STATPAC program of Nelson and others (1978) fits by ML the normal, lognormal, Weibull, extreme value, and exponential distributions; it provides approximate confidence limits for parameters and functions of them. The CENSOR program of Meeker and Duke (1979) and SURVREG of Preston and Clarkson (1980) do the same and include the logistic and log-logistic distributions. The BMD03S program of Dixon (1974) and the program of Hahn and Miller (1968) do the same for the normal and lognormal distributions. The program of Sheesley (1975) fits the normal and lognormal distributions; it fits with iterative weighted least squares and does not provide confidence limits. Finney (1968) references programs for analyses. Some of these programs use minimum chi-square fitting, which applies to dosage–mortality data; such programs apply only to data with a small number of inspections (dosages). Peto's (1973) program does ML fitting without assuming a form for the distribution (nonparametric fitting). Theory for such programs appears in Section 5.

3. GRAPHICAL ANALYSES OF INTERVAL DATA

Plots of interval data are simple and serve several purposes. They estimate the life distribution, i.e., the relationship between the cumulative percentage failing and age. Also, they assess the validity of the assumed distribution and of the data. The following explains how to make and interpret a probability plot of interval data for the special case where all units have the same inspection intervals. The plotting methods of Chapter 3 must be modified for interval data as described here.

If inspection intervals differ, it may suffice to treat each failure as if it occurred at the middle of its interval. Such failure times may then be plotted as described in Chapters 3 and 4. This method is crude if the intervals are wide compared to the distribution width, say, if any interval contains more than 20% of the sample.

Part Data

The methods are illustrated with data on 167 identical parts in a machine. At certain ages the parts were inspected to determine which had cracked since the previous inspection. The data appear in Table 3.1; it shows the months in service at the start and end of each inspection period and the number of cracked parts found in each period. For example, between 19.92 and 29.64 months, 12 parts cracked. 73 parts survived the latest inspection at 63.48 months. The data are simple in that all parts were inspected at the same ages; this is not so for many sets of interval data.

The purpose of the analysis is to estimate the distribution of time to crack initiation. This information is needed to schedule manufacture of replacement parts. Also, engineering wanted to know if the failure rate increases or decreases with part age. An increasing failure rate implies that parts should be replaced by some age when the risk of cracking gets too high.

Table 3.1. Part Cracking Data

Inspection (Months)		Number		Cumulative Percentage		
Start	End	Cracked	Cumulative	Estimate	Lower	Upper
0	6.12	5	5	2.99	.98	6.85
6.12	19.92	16	21	12.6	7.95	18.6
19.92	29.64	12	33	19.8	14.0	26.6
29.64	35.40	18	51	30.5	23.7	38.1
35.40	39.72	18	69	41.3	33.8	49.2
39.72	45.24	2	71	42.5	34.9	50.4
45.24	52.32	6	77	46.1	38.4	54.0
52.32	63.48	17	94	56.3	48.4	63.9
63.48+	Survived	73	167			
	Total	167				

How to Make a Probability Plot

The following probability plotting method applies to data where units have a **common inspection schedule**. The data on 167 parts in Table 3.1 illustrate the method. Chapter 3 describes probability plotting in detail.

The steps.   The steps in making the plot follow.

1.   For each inspection time, calculate the cumulative number of failures by that time. For example, for the inspection at 29.64 months, the cumulative number of cracked parts is $33 = 5 + 16 + 12$. Table 3.1 shows the cumulative number of cracked parts for each inspection time.

2.   For each inspection time, calculate the sample cumulative percentage failed. For example, for the inspection at 29.64 months, this is $100(33/167)$ $= 19.8\%$. Table 3.1 shows this percentage for each inspection time.

3.   Choose a probability plotting paper. The Weibull paper in Figure 3.1 was chosen for the part data. Engineering experience may suggest a specific distribution; otherwise, try papers for different distributions. Label the data scale on the paper to cover the range of the data as in Figure 3.1.

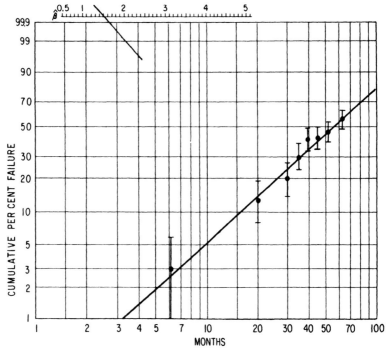

Figure 3.1.   Weibull plot of part data.

4. Plot each sample cumulative percentage against its inspection time. Figure 3.1 shows this for the part data. A cumulative percentage of 100% cannot be plotted on the paper; for this reason, some people use the sample size plus one (168) to calculate the sample cumulative percentages. Chapter 3 discusses alternate plotting positions.

5. Draw a straight line through the plotted points. This line is an estimate of the cumulative life distribution.

The method above applies only when the units have a common inspection schedule and all unfailed units have run beyond all failed units. Peto (1973) gives a complex general method that applies when units have different inspection schedules. His method employs the nonparametric maximum likelihood estimate; the calculations require a special computer program.

Confidence limits. The following method provides simple confidence limits for the cumulative distribution function at each inspection time. Each sample unit fails either before or else after a particular inspection time. Each unit can be regarded as a binomial trial, and its binomial probability of failure is the value of the cumulative distribution function at the inspection time. So confidence limits for a binomial probability apply; see Chapter 6 or Hahn and Shapiro (1967). For the part example, at 29.64 months there are 33 cracked parts out of 167. Two-sided binomial 95% confidence limits at that age are 14.0 and 26.6%. Table 3.1 gives such confidence limits. Also, the bars in Figure 3.1 show these binomial limits, which can be put on any plotting paper.

The limits above apply only when the units have a common inspection schedule and all unfailed units have run beyond all failed units. Peto (1973) gives approximate confidence limits for the general situation where units have different inspection schedules; their calculation requires a special computer program.

How to Interpret a Probability Plot

The plot is interpreted like any other probability plot, as described in Chapter 3.

Check validity of the data and distribution. Departure of a probability plot from a straight line can indicate peculiar data or lack of fit of the assumed distribution. Only extreme peculiarities should be interpreted as an incorrect distribution or faulty data, since the sample cumulative distribution function for interval data is usually coarse. The points in Figure 3.1 follow a reasonably straight line.

Estimates of probabilities and percentiles. Probabilities and percentiles are estimated from a plot as described in Chapter 3. For example, the estimate of the percentage of parts cracking by 12 months is 6.8% from Figure 3.1. Also, for example, the estimate of the 50th percentile, a nominal life, is 56 months from Figure 3.1. Such estimates can also be obtained from a curve fitted to a curved plot.

Parameter estimates. Distribution parameters are estimated from a plot as described in Chapter 3. The estimate of the Weibull scale parameter is 72 months in Figure 3.1. The Weibull shape parameter estimate on the shape scale is 1.49 for the part data.

Nature of the failure rate. For a Weibull plot, the following method assesses the nature of the failure rate, a basic question on the parts. A Weibull failure rate increases (decreases) if the shape parameter is greater (less) than 1. The shape parameter estimate of 1.49 indicates that the part failure rate increases with age, a wear-out pattern. So parts should be replaced at some age when they are too prone to cracking.

4. MAXIMUM LIKELIHOOD ANALYSIS FOR INTERVAL DATA

This section presents an example of a distribution fitted to interval data by maximum likelihood (ML). The analyses yield estimates and confidence limits for parameters, percentiles, probabilities, and other quantities. This section describes the ML model fitting, other analyses, and available computer programs. Chapter 8 and Section 5 provide background on the ML method for those who wish to write their own programs or to acquire a deeper understanding of ML fitting.

Model Fitting

The following presents the ML fit of the Weibull distribution to the part data. The method extends to other distributions. The fitting yields estimates of the Weibull parameters α and β for the parts. Laborious ML calculations require a computer program such as STATPAC of Nelson and others (1978).

STATPAC output from the Weibull fit to the part data appears in Figure 4.1. There the parameters α and β are denoted by "CENTER" and "SPREAD," respectively. The estimate of β is 1.49. This β estimate greater than 1 indicates that the parts have an increasing failure rate. So parts should be replaced at some age. In Figure 4.1, the approximate 95% confidence limits for the true β value are 1.22 and 1.80. This interval does not enclose 1, which corresponds to a constant failure rate. So there is statistically significant evidence that the part failure rate increases with age.

```
* MAXIMUM LIKELIHOOD ESTIMATES FOR DIST. PARAMETERS
  WITH APPROXIMATE 95% CONFIDENCE LIMITS

PARAMETERS    ESTIMATE      LOWER LIMIT       UPPER LIMIT

CENTER     71.68742  [â]    61.96245          82.93873
SPREAD      1.485506 [β̂]    1.224347           1.802371

* COVARIANCE MATRIX

PARAMETERS    CENTER           SPREAD

CENTER     28.43218
SPREAD     -0.2791416       0.2147405E-01


PCTILES

* MAXIMUM LIKELIHOOD ESTIMATES FOR  DIST.  PCTILES
  WITH APPROXIMATE 95% CONFIDENCE LIMITS

PCT.          ESTIMATE       LOWER LIMIT      UPPER LIMIT

0.1        0.6856063       0.2907563        1.616667
0.5        2.028559        1.057320         3.891964
1          3.240171        1.844604         5.691578
5          9.707168        6.764608         13.92972
10         15.75932        11.95060         20.78190
20         26.11732        21.43519         31.82217
50         56.01331        48.87925         64.18860
80         98.75752        82.68735         117.9509
90         125.6829        101.8173         155.1425
95         150.0414        118.2933         190.3103
99         200.4108        150.6475         266.6123


PERCENT(LIMIT 12.)

* MAXIMUM LIKELIHOOD ESTIMATES FOR  % WITHIN LIMITS
  WITH APPROXIMATE 95% CONFIDENCE LIMITS

              ESTIMATE      LOWER LIMIT       UPPER LIMIT

PCT     6.787357  [F̂(12)]   4.238944          10.69671
```

Figure 4.1.   STATPAC output on Weibull fit to part data.

STATPAC also gives the ML estimates of the distribution percentiles. These percentile estimates plot as a straight line on Weibull paper in Figure 4.2. The corresponding two-sided approximate 95% confidence limits plot as curves. These limits are narrower than the binomial ones, since (1) they use all of the data (not just the data for an inspection time) and (2) they employ a parametric (Weibull) life distribution. Peto (1973) gives nonparametric limits that apply to any mix of inspection schedules. The ML and Peto's limits are approximate and tend to be too short for small samples.

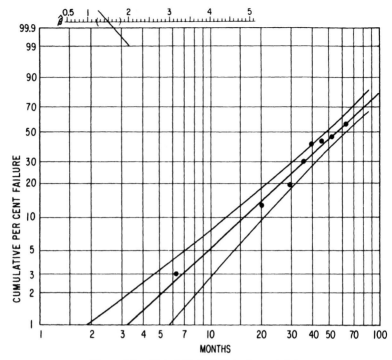

Figure 4.2.   Plot of Weibull fit to the part data.

The output shows seven-figure results. They are accurate and useful to about three figures.

Analytic Checks

The ML method can be used to check whether a distribution adequately fits the data. Then a likelihood ratio test (Chapter 12) is used to compare the chosen distribution with a more general one. The method for doing this is described by Prentice (1974) and by Nelson and others (1978). The minimum chi-square test of fit also applies (Cramer, 1945).

Other Analyses

Almost any analysis of complete data extends to interval data. For example, two samples can be compared with a hypothesis test (Chapter 12) to assess whether they differ significantly. Nelson and Hendrickson (1972) and Chapter 12 describe this.

Computer Programs

The STATPAC program of Nelson and others (1978) fits by ML the normal, lognormal, Weibull, extreme value, and exponential distributions to interval data; it provides approximate confidence limits for parameters and functions of them. The CENSOR program of Meeker and Duke (1979) and SURVREG of Preston and Clarkson (1980) do the same for the logistic and log-logistic distributions and for the STATPAC distributions except the exponential. The routine MLP FIT FREQUENCY of Ross and others (1976) fits by ML the STATPAC distributions plus a mixture of two normal distributions and the gamma, beta, and three-parameter lognormal distributions. Peto's (1973) computer program does ML fitting without assuming a form for the distribution (nonparametric fitting).

5. ML THEORY FOR INTERVAL AND QUANTAL-RESPONSE DATA

This **advanced** section presents maximum likelihood (ML) methods for fitting distributions to interval and quantal-response data. The section merely explains the computations. For detailed general theory and derivations of the ML method, refer to Rao (1973), Wilks (1962), and Hoadley (1971). The contents of the section are

1. Basic distributions.
2. Sample likelihood.
3. Maximum likelihood estimates.
4. Fisher information matrix.
5. Covariance matrix of ML estimates.
6. ML estimate and confidence limits for any quantity.
7. Literature on theory.

This section is for those who wish to understand the theory, apply it to other distributions, or write computer programs. Chapter 8 is needed background.

Basic Distributions

The ML method is given for distributions having a location parameter μ and a scale parameter σ. That is, the cumulative distribution function (cdf) for the population fraction below y has the form

$$F(y) = \Phi\big[(y - \mu)/\sigma\big],$$

and the probability density function (pdf) has the form

$$f(y)=(1/\sigma)\phi[(y-\mu)/\sigma],$$

where $\Phi(\ )$ and $\phi(\ )$ are the standard cdf and pdf of the distribution ($\mu=0, \sigma=1$). These include the normal, lognormal, smallest extreme value, Weibull, and exponential distributions. These distributions are given below in the general form. Of course, ML theory applies to other types of distributions.

Sample Likelihood

Suppose a sample contains I units with statistically independent (log) lifetimes y_i ($i=1,2,\dots,I$) from a cdf $F(y)=\Phi[(y-\mu_0)/\sigma_0]$. Here μ_0 and σ_0 are the true values of the location and scale parameters. The end points of the J_i inspection intervals for unit i are $\eta_{i0}, \eta_{i1}, \eta_{i2}, \cdots, \eta_{iJ_i}$, which are *independent* of y_i and are assumed to be *known*. The unit is found failed in interval j if $\eta_{i,j-1}<y_i\leqslant\eta_{ij}$, and the unit is unfailed if $y_i>\eta_{iJ_i-1}$. η_{i0} is the lower limit of the distribution, namely, $-\infty$ for the normal and extreme value distributions and 0 for the exponential distribution. $\eta_{iJ_i}=\infty$ for the distributions here; that is, units that fail in interval J_i are ones surviving the last inspection at η_{i,J_i-1}. For the lognormal and Weibull distributions, y_i and η_{ij} are the log times. For quantal-response data, unit i is inspected once at time η_{i1}; many of the formulas below simplify for this special case.

The sample likelihood is the probability model for the sample data. Suppose that y_i falls into period j_i. Then unit i contributes the probability $F(\eta_{ij_i})-F(\eta_{i,j_i-1})$ to the sample likelihood for independent y_i's; namely,

$$L=\prod_i\left[F(\eta_{ij_i})-F(\eta_{i,j_i-1})\right]=\prod_i\left\{\Phi[(\eta_{ij_i}-\mu)/\sigma]-\Phi[(\eta_{i,j_i-1}-\mu)/\sigma]\right\},$$
$$(5.1)$$

where μ and σ denote arbitrary values of the parameters.

The sample log likelihood is the base e log of the likelihood, namely,

$$\mathcal{L}=\sum_i\ln\left[F(\eta_{ij_i})-F(\eta_{i,j_i-1})\right].\qquad(5.2)$$

The sample (log) likelihood takes into account the interval form of the data. The rest of the ML method is the same as in Chapter 8, but explicit variance formulas can be obtained here.

Maximum Likelihood Estimates

The ML estimates of the parameters μ_0 and σ_0 are the values $\hat{\mu}$ and $\hat{\sigma}$ that maximize the likelihood L or, equivalently, $\mathcal{L}$. For most life distributions and choices of inspection times, the asymptotic (large-sample) sampling distribution of $\hat{\mu}$ and $\hat{\sigma}$ is approximately joint normal with means equal to the true μ_0 and σ_0 and a covariance matrix (5.10). Usually no other asymptotically normally distributed estimators for μ_0 and σ_0 have smaller asymptotic variances.

The $\hat{\mu}$ and $\hat{\sigma}$ can be obtained by numerically optimizing $\mathcal{L}$. The STATPAC program of Nelson and others (1978) does this. Alternatively, one can use the usual calculus method of setting the partial derivatives of $\mathcal{L}$ with respect to μ and σ equal to zero. This yields the **likelihood equations**

$$0 = \partial\mathcal{L}/\partial\mu$$
$$= \sum_i (-1/\sigma)\left[\phi(\zeta_{ij_i}) - \phi(\zeta_{i,j_i-1})\right] / \left[\Phi(\zeta_{ij_i}) - \Phi(\zeta_{i,j_i-1})\right], \qquad (5.3)$$

$$0 = \partial\mathcal{L}/\partial\sigma$$
$$= \sum_i (-1/\sigma)\left[\zeta_{ij_i}\phi(\zeta_{ij_i}) - \zeta_{i,j_i-1}\phi(\zeta_{i,j_i-1})\right] / \left[\Phi(\zeta_{ij_i}) - \Phi(\zeta_{i,j_i-1})\right],$$

where $\zeta_{ij_i} \equiv (\eta_{ij_i} - \mu)/\sigma$ is the standardized deviate. The values $\hat{\mu}$ and $\hat{\sigma}$ that satisfy these equations are the ML estimates.

Fisher Information Matrix

The Fisher information matrix is used to calculate the covariance matrix of $\hat{\mu}$ and $\hat{\sigma}$ and confidence limits. Its derivation follows.

It is convenient to work with the sample log likelihood in the form

$$\mathcal{L} = \sum_i \sum_j I_{ij} \ln\left[F(\eta_{ij}) - F(\eta_{i,j-1}) \right], \qquad (5.4)$$

where

$$I_{ij} = \begin{cases} 1 & \text{if } \eta_{i,j-1} < y_i \leq \eta_{ij} \quad (\text{unit } i \text{ fails in interval } j), \\ 0 & \text{otherwise}, \end{cases} \qquad (5.5)$$

is an indicator function and a random variable.

The Fisher information matrix requires the second partial derivatives of $\mathcal{L}$ with respect to μ and σ. Let $\Phi_{ij} = \Phi(\zeta_{ij})$ and $\phi_{ij} = \phi(\zeta_{ij})$. Then the first partial derivatives are

$$\partial\mathcal{L}/\partial\mu = \sum_i \sum_j (-1/\sigma) I_{ij} (\phi_{ij} - \phi_{i,j-1}) / (\Phi_{ij} - \Phi_{i,j-1}),$$
$$\qquad (5.6)$$
$$\partial\mathcal{L}/\partial\sigma = \sum_i \sum_j (-1/\sigma) I_{ij} (\zeta_{ij}\phi_{ij} - \zeta_{i,j-1}\phi_{i,j-1}) / (\Phi_{ij} - \Phi_{i,j-1}).$$

Let $\phi'_{ij} \equiv d\phi(\zeta)\,d\zeta$ evaluated at $\zeta = \zeta_{ij}$. The second partial derivatives are

$$\partial^2 \mathcal{L}/\partial\mu^2 = \sum_i \sum_j (1/\sigma^2) I_{ij}(\phi'_{ij} - \phi'_{i,j-1})/(\Phi_{ij} - \Phi_{i,j-1})$$
$$+ \sum_i \sum_j (-1/\sigma^2) I_{ij}(\phi_{ij} - \phi_{i,j-1})^2/(\Phi_{ij} - \Phi_{i,j-1})^2,$$

$$\partial^2 \mathcal{L}/\partial\sigma^2 = \sum_i \sum_j (-1/\sigma^2) I_{ij}(\zeta_{ij}\phi_{ij} - \zeta_{i,j-1}\phi_{i,j-1})^2/(\Phi_{ij} - \Phi_{i,j-1})^2$$
$$+ \sum_i \sum_j (1/\sigma^2) I_{ij}(\zeta_{ij}\phi_{ij} - \zeta_{i,j-1}\phi_{i,j-1})/(\Phi_{ij} - \Phi_{i,j-1})$$
$$+ \sum_i \sum_j (1/\sigma^2) I_{ij}(\zeta^2_{ij}\phi'_{ij} - \zeta^2_{i,j-1}\phi'_{i,j-1})/(\Phi_{ij} - \Phi_{i,j-1}),$$

$$\partial^2 \mathcal{L}/\partial\mu\,\partial\sigma = \sum_i \sum_j (1/\sigma^2) I_{ij}(\zeta_{ij}\phi'_{ij} - \zeta_{i,j-1}\phi'_{i,j-1})/(\Phi_{ij} - \Phi_{i,j-1})$$
$$+ \sum_i \sum_j (-1/\sigma^2) I_{ij}(\phi_{ij} - \phi_{i,j-1}) \tag{5.7}$$
$$\cdot (\zeta_{ij}\phi_{ij} - \zeta_{i,j-1}\phi_{i,j-1})/(\Phi_{ij} - \Phi_{i,j-1})^2 - (1/\sigma)\partial\mathcal{L}/\partial\mu.$$

The terms of the Fisher information matrix are the expectations of (5.7), evaluated at $\mu = \mu_0$ and $\sigma = \sigma_0$, namely,

$$E_0\{-\partial^2\mathcal{L}/\partial\mu^2\}_0 = -\partial^2\mathcal{L}/\partial\mu^2, \qquad E_0\{-\partial^2\mathcal{L}/\partial\sigma^2\}_0 = -\partial^2\mathcal{L}/\partial\sigma^2,$$
$$E_0\{-\partial^2\mathcal{L}/\partial\mu\,\partial\sigma\}_0 = -\partial^2\mathcal{L}/\partial\mu\,\partial\sigma. \tag{5.8}$$

Each right-hand side is evaluated at $I_{ij} = \Phi_{ij} - \Phi_{i,j-1}$, where $\mu = \mu_0$ and $\sigma = \sigma_0$, since $E_0 I_{ij} = \Phi_{ij} - \Phi_{i,j-1}$. The true Fisher information matrix is

$$\mathbf{F}_0 = \begin{bmatrix} E_0\{-\partial^2\mathcal{L}/\partial\mu^2\}_0 & E_0\{-\partial^2\mathcal{L}/\partial\mu\,\partial\sigma\}_0 \\ E_0\{-\partial^2\mathcal{L}/\partial\mu\,\partial\sigma\}_0 & E_0\{-\partial^2\mathcal{L}/\partial\sigma^2\}_0 \end{bmatrix} \tag{5.9}$$

Covariance Matrix of ML Estimates

The true asymptotic covariance matrix of $\hat{\mu}$ and $\hat{\sigma}$ is

$$\mathbf{\Sigma} = \mathbf{F}_0^{-1} = \begin{bmatrix} \mathrm{Var}(\hat{\mu}) & \mathrm{Cov}(\hat{\mu}, \hat{\sigma}) \\ \mathrm{Cov}(\hat{\mu}, \hat{\sigma}) & \mathrm{Var}(\hat{\sigma}) \end{bmatrix} \tag{5.10}$$

An estimate of Σ is needed for the confidence limits for parameters and other quantities.

To obtain the ML estimate of Σ, substitute the ML estimates $\hat{\mu}$ and $\hat{\sigma}$ in place of μ_0 and σ_0 in (5.9) to get $\hat{F}_0$. Then by (5.10), the ML estimate is

$$\hat{\Sigma} = \hat{F}_0^{-1} = \begin{bmatrix} \text{Vâr}(\hat{\mu}) & \text{Côv}(\hat{\mu}, \hat{\sigma}) \\ \text{Côv}(\hat{\mu}, \hat{\sigma}) & \text{Vâr}(\hat{\sigma}) \end{bmatrix}. \qquad (5.11)$$

To obtain the local estimate of Σ, substitute $\mu = \hat{\mu}$ and $\sigma = \hat{\sigma}$ in (5.7). Then use the observed negative second partial derivatives of $\mathcal{L}$ in place of the expected values in (5.9). The inverse (5.10) is the local estimate of Σ. The STATPAC program of Nelson and others (1978) calculates it.

The local and ML estimates usually differ. For interval data, the ML estimate is harder to calculate, but it should be more precise than the local one. However, the ML estimate requires the values of all inspection times, even those that are planned but not observed when a unit fails earlier. In practice, inspection times in the field may be haphazard and not known in advance. The local estimate requires only the inspection times for the interval when a unit fails and the running time of an unfailed unit; these are always known.

ML Estimate and Confidence Limits for Any Quantity

Section 5.3 of Chapter 8 presents the ML estimate of any function of the distribution parameters, the variance of that ML estimate, and approximate confidence limits for the true function value eg., percentiles and reliabilities.

Literature on Theory for Quantal-Response Data

The following briefly surveys theory on analysis of quantal-response data. Much work has been done on analysis of such bioassay data; such data are often taken at a small number of carefully prechosen dosage levels. However, such life data usually involve a large number of haphazardly chosen inspection times. So the bioassay work has limited application to life data. The survey covers the major life distributions–exponential, Weibull, extreme value, normal, lognormal, logistic, and nonparametric.

Normal and lognormal.   These distributions have been studied for analysis of quantal-response data by Easterling (1975), Finney (1968), Golub and Grubbs (1956), and Dixon and Massey (1969). In some experiments, the test units are run one after another. If the stress level can be adjusted for each test unit, the staircase method provides accurate estimates. Dixon and Massey (1969) describe this method for the normal and lognormal distributions. Methods for the normal distribution apply to the lognormal distribution, but one then works with the logs of lognormal data.

Weibull, extreme value, and exponential.   Dubey (1965) briefly indicates the use of the Weibull distribution for ML analysis of quantal-response data. Methods for the extreme value distribution apply to the Weibull distribution, but one then works with the (base e) logs of the Weibull data. The exponential distribution is a special case of the Weibull distribution.

Logistic.   Berkson (1953) describes ML and minimum chi-square methods for fitting a logistic distribution to such data. He also gives charts that simplify ML estimation for bioassay applications with equally spaced stress levels. The methods of Cox (1970) also apply. Meeker and Hahn (1977) give two optimum inspection times for estimating a low percentile.

Nonparametric.   Peto (1973) and Turnbull (1976) give the ML method for a nonparametric estimate of a distribution from quantal-response data.

Other distributions.   Papers on analysis of quantal-response data with the preceding and other distributions are referenced in the bibliographies of Buckland (1964), Mendenhall (1958), and Govindarajulu (1964).

Literature on Theory for Interval Data

The following briefly surveys theory on analysis of interval data. There has been much work, particularly for equal length intervals. However, interval life data often involve a large number of haphazardly chosen inspection times. So previous work has limited application to life data. The survey covers the major life distributions—exponential, Weibull, extreme value, normal, lognormal, and nonparametric. The ML method is most widely used for analysis of such data. Also, the minimum chi-square method applies (Rao, 1973).

Normal and lognormal.   These distributions are well studied for analysis of interval data particularly by Kulldorff (1961). Kulldorff gives optimum inspection times for a fixed number of equally spaced and of optimally spaced inspections. Methods for the normal distribution apply to the lognormal distribution, but one then works with the logs of lognormal data.

Suppose all units have the same inspection times and all intervals have width h. Suppose one assigns each observation the midvalue of its interval, and one calculates the mean m and variance v of the midvalues. m is an estimate of the normal mean μ. Also, $\sigma^{*2} = v - (h^2/12)$ is Sheppard's corrected estimate of the normal variance σ^2. This correction for the normal distribution applies to the lognormal distribution if the intervals between the **log** inspection times all have width h.

Weibull, extreme value, and exponential.   Dubey (1965) briefly indicates the use of the Weibull distribution for ML analysis of interval data.

Methods for the extreme value distribution apply to the Weibull distribution, but one then works with the (base e) logs of the Weibull data. Meeker (1980) gives optimum test plans with equally spaced inspection times for ML estimates of selected percentiles of a Weibull distribution. The exponential distribution is a special case of the Weibull distribution, and Kulldorff (1961) and Ehrenfeld (1962) give ML estimates and optimum inspection times. Nelson (1977) gives optimum demonstration test plans for the exponential distribution.

Nonparametric. Peto (1973) and Turnbull (1976) give the ML nonparametric estimate of a distribution from interval data. Kalbfleisch and Prentice (1980) present a variety of methods for interval data.

Other distributions. Papers on analysis of interval data with the preceding and other distributions are referenced in the bibliographies of Buckland (1964), Mendenhall (1958), and Govindarajulu (1965).

PROBLEMS

9.1. Exponential quantal-response. Suppose a type of unit has an exponential life distribution with mean θ_0. Suppose that unit i is inspected once at time η_i to determine whether it has failed or not, $i=1,\dots,I$.

(a) Write the sample log likelihood $\mathcal{L}$, distinguishing between failed and unfailed units.

(b) Derive $\partial\mathcal{L}/\partial\theta$, and give the likelihood equation.

(c) Use the wheel data of Section 1, and iteratively calculate the ML estimate $\hat{\theta}$ accurate to two figures. Treat each unit as if it were inspected at the middle of its time interval, and treat the last interval as if all inspections were at 4600 hours.

(d) Derive the formula for $\partial^2\mathcal{L}/\partial\theta^2$, and evaluate it for the wheel data.

(e) Give the formula for the local estimate of $\mathrm{Var}(\hat{\theta})$.

(f) Evaluate the local estimate of $\mathrm{Var}(\hat{\theta})$ for the wheel data, using (d) and (e).

(g) Calculate positive approximate 95% confidence limits for θ_0, using (f).

(h) Express the sample log likelihood in terms of indicator functions, and derive the formula for the true asymptotic variance of $\hat{\theta}$.

(i) Evaluate the ML estimate of $\mathrm{Var}(\hat{\theta})$ for the wheel data, using (c) and (h).

(j)   Calculate positive approximate 95% confidence limits for θ_0, using (i).

(k)   On Weibull paper, plot the data, the fitted exponential distribution, and approximate 95% confidence limits for percentiles.

9.2.*   **Optimum exponential quantal response.**   For Problem 9.1, suppose that all I units have a common inspection time $\eta_i = \eta, i = 1, \ldots, I$. Suppose Y of the I units are failed on inspection.

(a)   Derive an explicit formula for the ML estimate $\hat{\theta}$; use part (b) of Problem 9.1.

(b)   Derive the formula for the true asymptotic variance of $\hat{\theta}$; use part (h) of Problem 9.1.

(c)   Derive the optimum inspection time η^* that minimizes the true asymptotic variance. It is a multiple c of the true unknown θ_0, $\eta^* = c\theta_0$. Numerically find c.

(d)   Evaluate the minimum variance for $\eta = \eta^*$, and compare it with the variance of $\hat{\theta}$ for a complete observed sample.

(e)   In practice, one must guess a value θ' for θ_0 and use $\eta' = c\theta'$. Calculate and plot $(1/\theta_0^2)\text{Var}_{\eta'}(\hat{\theta})$ versus θ'/θ_0 from $\frac{1}{10}$ to 10 on log-log paper.

(f)   Explain how to use "exact" confidence limits for a binomial proportion to get exact limits for θ_0.

9.3.   Grouped exponential.   Suppose a type of unit has an exponential life distribution with mean θ_0. Suppose that each unit has the same inspection times $\eta_1, \eta_2, \ldots, \eta_J = \infty$, and $\eta_0 = 0$. For a sample of I units, suppose that Y_j fail in period j: $(\eta_{j-1}, \eta_j), j = 1, 2, 3, \ldots$.

(a)   Write the sample log likelihood $\mathcal{L}$ in terms of the Y_j.

(b)   Derive $\partial \mathcal{L}/\partial\theta$, and solve the likelihood equation for $\hat{\theta}$.

(c)   Use the group 10 data in Table 3.1 of Chapter 10; calculate the ML estimate $\hat{\theta}$ accurate to two figures.

(d)   Derive $\partial^2 \mathcal{L}/\partial\theta^2$, and evaluate it for the data.

(e)   Give the formula for the local estimate of $\text{Var}(\hat{\theta})$.

(f)   Evalute the local estimate of $\text{Var}(\hat{\theta})$ for the data in (c).

(g)   Calculate two-sided positive approximate 95% confidence limits for θ_0, using (f).

(h)   Derive the formula for the true asymptotic variance of $\hat{\theta}$.

---

*Asterisk denotes laborious or difficult.

(i)   Evalute the ML estimate of $\text{Var}(\hat{\theta})$ for the data, using (c) and (h).

(j)   Calculate positive approximate 95% confidence limits for θ_0, using (i).

(k)   On Weibull paper, plot the data, the fitted exponential distribution, and approximate 95% confidence limits for percentiles.

9.4.* Optimum grouped exponential.   For Problem 9.3, suppose that all I units have a common time τ_I between inspections; that is, $\eta_j = j\tau_I (j = 1,\ldots,J-1)$ and $\eta_J = \infty$.

(a)   Derive an explicit formula for the ML estimate $\hat{\theta}$; use part (b) of Problem 9.3.

(b)   Derive the formula for the true asymptotic variance of $\hat{\theta}$; use part (h) of Problem 9.3.

(c)   Derive the optimum time τ_I^* that minimizes the true asymptotic variance. It is a multiple c_I of the true unknown θ_0, $\tau_I^* = c_I \theta_0$. Numerically find c_I for $J = 2, 3$.

(d)   Evaluate the minimum variance for $\tau_I = \tau_I^*$ and compare it with the variance of $\hat{\theta}$ for a complete observed sample.

(e)   In practice, one must guess a value θ' for θ_0 and use $\tau_I' = c_I \theta'$. Calculate and plot $(n/\theta_0^2)\text{Var}_{\tau_I'}(\hat{\theta})$ versus θ'/θ_0 from 0.1 to 10 on log-log paper for $J = 2$.

(f)   Explain how to use "exact" confidence limits for the parameter p of a geometric distribution (Chapter 2) to get exact limits for θ_0 when $J = \infty$.

9.5.* Exponential fit to grouped data with progressive censoring.   For motivation, first see the distribution transformer data in Table 3.1 of Chapter 10. Each group there is assumed here to have the same exponential life distribution with common mean θ_0. Suppose group k with n_k units was observed through year k, and the yearly numbers of failures are $y_{k1}, y_{k2}, \cdots, y_{kk}$, and the number of survivors is $y_{k,k+1} = n_k - (y_{k1} + \cdots + y_{kk})$, $k = 1, \cdots, K$.

(a)   Calculate an actuarial estimate (Chapter 4) of the common distribution and plot it on Weibull probability paper.

(b)   Write the sample log likelihood $\mathcal{L}_k(\theta)$ for group k in terms of θ.

(c)   Write the sample log likelihood $\mathcal{L}(\theta)$ for the K independent groups.

(d)   Derive $\partial\mathcal{L}/\partial\theta$, and solve the likelihood equation for $\hat{\theta}$.

(e)   Calculate $\hat{\theta}$ for the transformer data.

(f) Derive $\partial^2 \mathcal{L}/\partial\theta^2$ and its expectation when the true mean is θ_0.

(g) Calculate the local estimate of the asymptotic variance of $\hat{\theta}$.

(h) Calculate the ML estimate of the asymptotic variance of $\hat{\theta}$.

(i) Calculate two-sided positive approximate 95% confidence limits for θ from (g).

(j) Do (i), using (h).

(k) Plot the fitted exponential distribution and 95% confidence limits for percentiles (parallel straight lines) on the Weibull plot from (a).

9.6.* Nonparametric fit to grouped data with progressive censoring. For motivation, first see the distribution transformer data in Table 3.1 of Chapter 10. Suppose each of the K groups there is assumed here to have the same proportion π_1 failing in year 1, the same proportion π_2 failing in year 2, etc. The following method yields ML estimates of the proportions $\pi_1, \pi_2, \cdots$ from the data on all groups.

(a) Write the separate multinomial log likelihoods for groups K through 1, denoting the number from group k failing in year m by y_{km} and the number surviving the last year by $y_{k,k+1}$.

(b) From the total log likelihood calculate the likelihood equations.

(c) Solve the equations to obtain the ML estimates of $\pi_1,\ldots,\pi_{K+1}$.

(d) Derive formulas for all second partial derivatives of the total log likelihood with respect to $\pi_1,\ldots,\pi_K$.

(e) Derive the expectations of minus one times those derivatives.

(f) Give the asymptotic covariance matrix.

(g) Calculate the "local" estimate of the covariance matrix.

(h) Calculate approximate 90% confidence limits for $\pi_1,\ldots,\pi_K$, and π_{K+1}.

9.7. Circuit breaker. Use the circuit breaker data from Problem 3.4. Use a computer program if available.

(a) Assume that cycles to failure has an exponential distribution. Write the sample log likelihood $\mathcal{L}(\theta)$ for the first sample of 18 specimens in terms of the observed numbers of failures in each interval.

(b) Derive $\partial\mathcal{L}/\partial\theta$, and iteratively solve the likelihood equation for $\hat{\theta}$.

(c) Derive $\partial^2 \mathcal{L}/\partial\theta^2$, and evaluate it for the data.

(d) Calculate the local estimate of $\text{Var}(\hat{\theta})$.

(e) Calculate two-sided positive approximate 95% confidence limits for the true mean.

(f)   Plot the fitted exponential distribution and parallel 95% confidence limits on the Weibull plot from Problem 3.4.

(g)   Assuming that cycles to failure has a Weibull distribution, write the sample log likelihood $\mathcal{L}(\alpha, \beta)$ for the first sample of 18 specimens in terms of the observed numbers of failures in each interval.

(h*)   Iteratively calculate the ML estimates $\hat{\alpha}$ and $\hat{\beta}$ by any means.

(i*)   Calculate the local estimate of the covariance matrix of the ML estimators.

(j*)   Calculate the ML estimate of the covariance matrix of the ML estimators.

(k*)   Calculate two-sided positive approximate 95% confidence limits for the Weibull shape parameter using (i) or (j). Is the shape parameter statistically significantly different from unity?

(l*)   Calculate the ML estimate of the fraction failing by 10,000 cycles. Calculate the binomial estimate of this fraction.

(m*) Calculate two-sided approximate 95% confidence limits for the fraction failing by 10,000 cycles. Obtain two-sided binomial 95% confidence limits for this fraction.

(n*)   Plot the fitted Weibull distribution on the Weibull plot from Problem 3.4.

(o*)   Repeat (a) through (n), using the pooled data on both samples of 18 circuit breakers.

9.8.   Vehicle motor. A sample of 43 large electric motors on vehicles were inspected to determine if a particular defect had occurred. Hours on each motor at inspection appear below where + indicates the defect had not occurred and − indicates it had.

7072+	1503−	2630−	1000+	4677+	5517+
3300−	800+	5700−	4000+	4786+	5948+
3329−	1100+	3300+	1400+	3038+	6563+
3200+	600+	3750−	1400+	1000+	913+
1228+	3397−	5200−	2000−	7199−	1914+
2328+	2981+	3108−	1203+	6000+	683+
2333+	3000−	4000+	2400−	6000+	7000+
					1171+

The results of ML fitting a Weibull distribution appear in the accompanying output.

(a)   Does the shape parameter estimate suggest that such motors get more or less prone to the defect as they age?

* MAXIMUM LIKELIHOOD ESTIMATES FOR DIST. PARAMETERS
WITH APPROXIMATE 95% CONFIDENCE LIMITS

PARAMETERS	ESTIMATE	LOWER LIMIT	UPPER LIMIT
CENTER $\hat{\alpha}$ 55628.70		88.97046	0.3478180E 08
SPREAD $\hat{\beta}$ 0.3735526		0.4301541E-01	3.243990

* COVARIANCE MATRIX

PARAMETERS	CENTER $\hat{\alpha}$	SPREAD $\hat{\beta}$
CENTER $\hat{\alpha}$ 0.3338934E 11		
SPREAD $\hat{\beta}$ -73141.19		0.1697079

* MAXIMUM LIKELIHOOD ESTIMATES FOR DIST. PCTILES
WITH APPROXIMATE 95% CONFIDENCE LIMITS

PCT.	ESTIMATE	LOWER LIMIT	UPPER LIMIT
0.1	0.5186608E-03	0.1144192E-17	0.2351083E 12
0.5	0.3875757E-01	0.9457868E-12	0.1588254E 10
1	0.2495389	0.3378272E-09	0.1843240E 09
5	19.59442	0.3152064E-03	1218063.
10	134.5887	0.1308441	138440.5
20	1003.367	57.30000	17569.73
50	20853.98	254.7561	1707077.
80	198868.3	21.42655	0.1845776E 10
90	518740.3	7.204192	0.3735207E 11
95	1049293.	3.218400	0.3421006E 12
99	3317470.	0.8580644	0.1282608E 14

(b)   Do the confidence limits for the shape parameter indicate that the information from (a) is conclusive or inconclusive?

(c)   Plot the fitted distribution and the confidence limits for the percentiles on Weibull paper. Does this small sample of quantal-response data yield accurate information for practical purposes?

(d)   Use the covariance matrix to calculate the positive 95% confidence limits for (1) α, (2) β, and (3) the 10th percentile. Do you think these limits are relatively crude or accurate for practical purposes?

(e)   Calculate a graphical estimate of the sample cumulative distribution function and plot it on the same Weibull paper.

(f)   It is possible that a fraction of motors have the defect when they start into service. Is the plot from (e) consistent with this possibility?

(g)   Assuming such defects are already present when some motors start service, estimate the fraction of motors with the defect.

(h*) Write the likelihood, assuming a fraction π of the motors start with the defect and the remaining $(1 - \pi)$ have a Weibull distribution for time to occurrence of the defect.

10

Comparisons (Hypothesis Tests) for Complete Data

INTRODUCTION

This chapter presents methods for comparing parameters of distributions with each other or with specified values, using complete samples. The methods include hypothesis tests and confidence intervals for one sample, two samples, and K samples. Useful background for this chapter is the corresponding material on estimation in Chapter 6. This introduction overviews the chapter and presents basic background on comparisons. Most statistical computer programs calculate the results described in this chapter.

Overview

For each of the following distributions, the corresponding section gives methods for complete data for hypothesis tests, confidence intervals, and pooled estimates: Poisson (Section 1), binomial (Section 2), multinomial (Section 3), exponential (Section 4 covers multiply censored data, too), normal and lognormal (Section 5), and Weibull, nonparametric, and others (Section 6). Sections 1 through 6 are simple (particularly the graphical comparisons), require a modest background, and depend on just Chapter 6. Section 7 is a general abstract presentation of hypothesis testing, including definitions, and may interest advanced readers. Chapter 11 presents comparisons for singly censored data and employs linear estimates; that chapter is more advanced and depends on Chapter 7. Chapter 12 presents comparisons for multiply censored and other data; it employs maximum likelihood estimates. Chapter 12 is still more advanced and depends on Chapter 8.

Comparisons with Hypothesis Tests and Confidence Intervals

The following paragraphs briefly review basic ideas of hypothesis testing, including the use of confidence intervals. A more formal presentation appears in Section 7. The ideas here include reasons for comparisons, parametric distributions, hypotheses, actions, tests, confidence limits, significance (statistical and practical), and performance and sample size. Standard statistical texts discuss hypothesis testing in more detail; without such previous background, readers may find this and the next two chapters difficult.

Reasons for comparisons. The following are some reasons for comparing life distributions. In reliability demonstration testing, a product must demonstrate that its reliability, mean life, failure rate, or whatever is better than a specified value. In verifying engineering theory, one may check that parameters of life distributions have specified theoretical values, for example, that a Weibull shape parameter is unity. In development work, one may compare two or more designs to select one. In analyzing sets of data collected over time, one may want to confirm that the life distribution is not changing; this is often done before pooling data to get a more precise pooled estimate of a common distribution parameter. There are two basic objectives in such comparisons. One is to **demonstrate** that a product **surpasses** a requirement or that a product surpasses others. The other is to assess whether a product parameter is **consistent** with a specified value or whether corresponding parameters of a number of products are **comparable** (equal). Here "parameter" means any distribution value, including percentiles and reliabilities. Chapters 10, 11, and 12 give examples of these reasons.

Parametric distributions. In what follows, the data are assumed to be independent random samples from parametric distributions, which are assumed to be correct models. In practice, one does not usually know whether a Weibull or some other distribution is correct; such assumptions must be assessed through plots or formal tests of fit (Section 7). In what follows, the assumed distribution is assumed to be adequate for the intended purposes. Section 7 briefly references nonparametric hypothesis tests. In most engineering work, parametric distributions are used for a number of reasons. If correct, they make the most efficient and informative use of the data; this is especially important for small samples. Engineering experience and theory indicate that certain distributions describe the life of certain products. Also, there is a well-developed exact statistical theory for the most commonly used parametric distributions and complete data.

Hypotheses. A **hypothesis** is a proposed statement about the value(s) of one or more distribution parameters. Some examples are

1. The mean of an exponential distribution exceeds a specified value (this is common in reliability demonstration tests of hardware).
2. Product reliability at a specified age exceeds a given value.
3. A Weibull shape parameter equals 1; that is, product life has an exponential distribution.
4. The mean of normal distribution 1 exceeds the mean of normal distribution 2 (common in comparing alternative designs, materials, methods of manufacture, manufacturing periods, etc.).
5. The means of a number of normal distributions are equal (a common hypothesis in analysis of variance to compare a number of designs, materials, vendors, production periods, etc.).
6. The means of a number of exponential distributions are equal.
7. The shape parameters of a number of Weibull distributions are equal.
8. The 10th percentiles of a number of Weibull distributions are equal (a common hypothesis in ball bearing life tests).

The **alternative** to a hypothesis is the statement that the hypothesis is not true. In contrast, the "hypothesis" above is also called the **null hypothesis**. Some examples of alternatives for some preceding examples are

1. The mean of an exponential distribution is below the specified value.
3. The Weibull shape parameter differs from 1 (greater or smaller).
4. The mean of normal distribution 1 does not exceed the mean of normal distribution 2.
5. Two or more of a number of normal means differ.
7. Two or more of the shape parameters of a number of Weibull distributions differ.

Such a hypothesis (or alternative) about a parameter may be **one sided**. That is, the parameter is above (below) a specified value. (1) and (2) are examples of this. (4) is a one-sided example concerning two parameters. Also, a hypothesis may be **two sided**. That is, a parameter has a specified value, or parameters of different populations are equal. (3) and (5) through (8) are examples of two-sided (or equality) hypotheses (and alternatives). In practice, one must decide whether a one- or two-sided hypothesis is appropriate. The choice is determined by the practical consequences of the true parameter value(s).

Actions. In engineering and other applied fields, if the hypothesis is true, the practitioner wants to take one course of action. If the alternative is true, the practitioner wants to take another, which may depend on the parameter values. Some examples from above are

1. In reliability demonstration, hardware with an exponential mean that "exceeds a specified value" (the hypothesis) is accepted by the customer. Otherwise, it is rejected by the customer (the alternative), and then the product must be redesigned or the contract renegotiated.

4. If the mean of the normal distribution of performance of new product 1 exceeds that of the standard product 2, then product 1 replaces product 2 (the hypothesis); otherwise, the standard product 2 is retained (the alternative).

5. If the means of a number of normal life distributions of groups of specimens are equal (the hypothesis), data from the groups may be pooled to estimate the common mean; otherwise, the means must be estimated separately (the alternative). In another situation, one may have field data on units made in different production periods. If the means are all equal, the cause of failure is not related to production period (the hypothesis); otherwise, product life is related to production period (the alternative), and periods with high and low means should be compared to determine causes. Also, then preventive replacement policies can be determined to minimize costly field failure (the most prone to failure are replaced first).

7. If the shape parameters of a number of Weibull distributions are equal (the hypothesis), then data from the populations can be pooled to estimate the common shape parameter. Otherwise, a separate estimate for each population is appropriate (the alternative). Such pooling is often considered for life test data collected on the same product under different conditions or in different time periods.

Hypothesis test. In real life, distribution parameters are not known, and one must take actions on the basis of sample data. One then wants convincing evidence in the data that a particular action is appropriate. For example, observed differences between a sample estimate for a parameter and a specified value or between sample estimates should be greater than normal random variations in such estimates; then it is convincing that the observed differences are due to real differences in the true parameter values. A **statistical test** involves a **test statistic**, which is some function of the sample data. Examples of such statistics are sample means, medians, and t statistics. For true parameter values under the hypothesis, the statistic has a known "null" distribution; for true parameter values under the alternative, the statistic tends to have larger (smaller) values. If the observed value of the statistic is unusual (in an extreme tail of its null distribution), this is evidence that the hypothesis is not true, and an alternative action is appropriate. If the observed statistic is beyond the upper (or lower) 5% point, it is said to be **statistically significant**. If beyond the (0.1%) 1% point, it is said to be **(very) highly statistically significant**. The exact percentage of the null distribution beyond the observed statistic is called the **significance level**.

Confidence intervals. Many comparisons can also be made as follows with confidence intervals. Such intervals are usually equivalent to but more informative than a corresponding hypothesis test. Thus intervals are often preferable. Such intervals can (1) indicate that the data are consistent with specified parameter values or (2) demonstrate that the data surpass specified parameter values.

1. A $100\gamma\%$ confidence interval for a parameter is **consistent** with a specified value of the parameter if the interval encloses the specified value. A $100\gamma\%$ confidence interval that does not enclose the specified value indicates that a corresponding test statistic is statistically significant at the $[100(1-\gamma)]\%$ level. For example, if the confidence interval for a Weibull shape parameter encloses the value 1, the data are consistent with an assumed exponential life distribution (for this way of assessing adequacy of the exponential distribution).

2. A confidence interval for a parameter **demonstrates** a specified parameter value (or **better**) if the interval encloses only "better" parameter values. For example, a specified mean θ_1 of an exponential life distribution is demonstrated with $100\gamma\%$ confidence if the $100\gamma\%$ lower confidence limit for the true θ is above θ_1.

3. A confidence interval for the difference (ratio) of corresponding parameters of two distributions is consistent with equality of the parameters if it encloses zero (one). Similarly, such an interval that does not enclose zero (one) **"demonstrates"** that one parameter exceeds the other. For example, one may wish that the mean life of a new design prove superior to that of the standard design before adopting the new one.

4. Simultaneous confidence intervals for corresponding parameters of K distributions are consistent with equality of those parameters if all such intervals enclose zero (one).

Significance. It is important to distinguish between practical and statistical significance in comparisons for equality. An observed difference is **statistically significant** if the data are convincing evidence that the observed differences are greater than would be observed by chance and are hence real; then observed differences are large compared to the uncertainty in the data. Observed differences are **practically significant** if they are large enough to be important in real life. Observed differences can be statistically significant but not practically significant (that is, convincing but so small that they have no practical value). This can happen particularly for large samples that reveal even small differences. Then the corresponding population parameters, although different, are equal for practical purposes. Observed differences can be practically significant but not statistically significant; this can happen when sample sizes are small. Then a larger sample is needed to resolve whether the observed important differences are real. In practice, one needs observed differences that are **both** statistically (convincing) and practically (important) significant. Confidence intervals are very informative in judging both practical and statistical significance, whereas hypothesis tests merely indicate statistical significance. A confidence interval for a difference should ideally be smaller than an important practical difference. If it is not, then one needs more data to discriminate adequately. Mace (1964) provides guidance on how to choose sample size in terms of the desired length of confidence intervals.

Performance and sample size.   The performance of a confidence interval is usually judged in terms of its "typical" length. That of a hypothesis test is judged in terms of its operating characteristic (OC) function, defined in Section 7. Such performance, of course, depends on the assumed life distribution(s), the parameter(s) compared, the sample statistic(s) used, and the sample size(s). Mace (1974) gives methods for choosing sample sizes for confidence intervals. Cohen (1977) gives methods for choosing sample sizes for hypothesis tests. In this book, it is assumed that the sample size has been determined. In practice, it is usually determined by nonstatistical considerations, such as limited budget or time, and by the number of units in service or available for a test. Sequential sampling plans have been developed to reduce sample size; they are presented in various military standards referenced below and are surveyed by Aroian (1976).

1.   COMPARISON OF POISSON SAMPLES

This section presents methods for comparing Poisson samples, namely,

1.   A sample with a given occurrence rate (demonstration and acceptance testing).

2.   Two samples for equal occurrence rates.

3. K samples for equal occurrence rates.

Also, this section shows how to pool samples to estimate a common occurrence rate.

Comparison of a Poisson Sample with a Specified λ_0

Below are two ways of comparing a sample occurrence rate $\hat{\lambda}$ with a given λ_0. Suppose Y occurrences are observed in a "length" t.

Confidence interval. Suppose a $100\gamma\%$ confidence interval for the true λ encloses λ_0. Then the observed $\hat{\lambda}$ is **consistent** with λ_0 at the $100\gamma\%$ confidence level. If the interval does not enclose λ_0, then $\hat{\lambda}$ differs from λ_0 at the $[100(1-\gamma)]\%$ significance level. The interval may be one or two sided.

Demonstration test. A reliability demonstration test for equipment failure rate is usually stated by the consumer as follows: a failure rate better than specified λ_0 is to be **demonstrated** with $100\gamma\%$ confidence. This means that the equipment passes the test if the observed upper $100\gamma\%$ confidence limit $\tilde{\lambda}$ for the true λ is below λ_0; otherwise, the equipment fails. Equivalently, the equipment passes the test if the observed number Y of failures in a total running time t is less than a specified acceptance number y_0.

One chooses y_0 and t so that the probability of passing the test is

$$P_{\lambda_0 t}\{Y \leqslant y_0\} = 1 - \gamma \qquad (1.1)$$

when $\lambda_0 t$ is the true Poisson mean. In other words, when the true failure rate λ equals λ_0, the probability of the equipment passing the test is a **low** value $1 - \gamma$. This means that the true failure rate λ must be below λ_0 if the equipment is to have a high probability of passing the test. The Poisson probability

$$P(\lambda) = P_{\lambda t}\{Y \leqslant y_0\} = \sum_{y=0}^{y_0} (\lambda t)^y \exp(-\lambda t)/y! \qquad (1.2)$$

of passing the test as a function of λ is called the **operating characteristic (OC) curve**. The OC curve is easy to obtain from a table of the Poisson distribution. Such a test for the number of defects in manufactured products is called an "acceptance sampling test for defects" (Grant and Leavenworth, 1980).

Capacitor demonstration example. A contract for a high-reliability capacitor specified that a rate of $\lambda_0 = 0.01$ failures per million hours be demonstrated with 60% confidence. (Many demonstration tests require higher confidence, say, 90 or 95%. The minimum confidence used in practice is 50%). The minimum total test time t results if $y_0 = 0$, no failures.

Then the required test time is the solution of (1.1), that is,

$$1-0.60= P_{10^{-8}t}\{Y\leqslant 0\}=\exp(-10^{-8}t),$$

namely,

$$t=-10^8\ln(0.40)=9.163\times 10^7 \text{ hours.}$$

This time was accumulated by running 9163 capacitors each for 10,000 hours. The capacitors are assumed to have an exponential life distribution. So a different number of capacitors could equivalently be run a different duration to accumulate this time. The relationship between the Poisson and exponential distributions is described in Chapter 2.

The corresponding OC function (1.2) is

$$P(\lambda)=\exp(-\lambda 9.163\times 10^7),$$

Figure 1.1 shows this function. To have a 90% probability of passing the test, such capacitors need a true λ such that the probability of no failure satisfies

$$0.90=\exp(-\lambda 9.163\times 10^7),$$

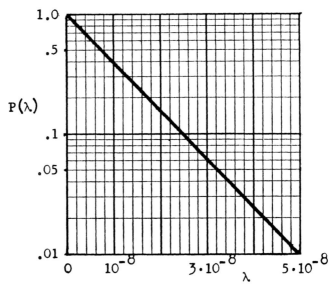

Figure 1.1.   OC curve of the demonstration plan.

namely, a design failure rate of

$$\lambda = -\ln(0.90)/9.163 \times 10^7 = 0.11 \times 10^{-8}.$$

So λ must be one-ninth of the specified $\lambda_0 = 10^{-8}$.

The probability of passing a test when $\lambda < \lambda_0$ can be increased if one uses a test that allows more failures y_0' in a greater total test time t'. Then y_0' and t' are still chosen to meet the demonstration requirement $P_{\lambda_0 t'}\{Y \leq y_0'\} = 1 - \gamma$. Of course, such a test requires more test time and cost. For example, the capacitor test with $y_0' = 1$ requires $t_0' = 20.2 \times 10^7$ hours; this is more than twice the time for $y_0 = 0$.

Poisson demonstration. Standard plans and their OC curves are given by Schilling (1981) and the references below. The references include two-stage, multistage, sequential, and other sampling plans; such plans result in a smaller average "length" of observation than the above single-stage plans with comparable OC functions. The following references may be bought from the U.S. Government Printing Office, Washington, D.C. 20402.

1. MIL-STD-105D, "Sampling Procedures and Tables for Inspection by Attributes," 29 April 1963. To use for Poisson data, treat the binomial probability as the number of occurrences per 100 sample units. Undergoing revision.

2. MIL-STD-690B, "Failure Rate Sampling Plans and Procedures," 17 April 1968.

3. MIL-STD-781C, "Failure Tests: Exponential Distribution," 21 August 1977. Consult material on sequential and fixed-length tests.

4. MIL-HDBK-108, "Sampling Procedures and Tables for Life and Reliability Testing," 29 April 1960. Consult material on testing with replacement.

Comparison of Two Poisson Occurrence Rates

This section gives methods for comparing two Poisson occurrence rates for consistency. Suppose Y is a Poisson count in a "length" t where the occurrence rate is λ, and X is an independent Poisson count in a length s where the occurrence rate is v.

Suppose a $100\gamma\%$ confidence interval for the relative failure rate

$$\rho = \lambda/v \tag{1.3}$$

does not enclose the value 1. Then $\hat{\lambda} = Y/t$ and $\hat{v} = X/s$ differ at the $[100(1-\gamma)]\%$ significance level. That is, there is convincing evidence that λ and v differ when $1 - \gamma$ is small, say, below 0.10 or 0.05. Otherwise, if the interval encloses 1, the data are consistent with the hypothesis $\lambda = v$.

Confidence limits.   Limits of a two-sided $100\gamma\%$ confidence interval for ρ are

$$\underline{\rho}=\{(Y/t)/[(X+1)/s]\}/F[(1+\gamma)/2;2X+2,2Y],$$

$$\tilde{\rho}=\{[(Y+1)/t]/(X/s)\}\cdot F[(1+\gamma)/2;2Y+2,2X],$$

(1.4)

where $F(\delta;\,a,\,b)$ is the 100δth percentile of the F distribution with a degrees of freedom in the numerator and b in the denominator. Nelson (1970) gives simple charts for these limits.

If Y and X are large, approximate $100\gamma\%$ limits are

$$\underline{\rho}\cong\hat{\rho}/\phi\quad\text{and}\quad\tilde{\rho}\cong\hat{\rho}\cdot\phi,\tag{1.5}$$

where $\hat{\rho}=\hat{\lambda}/\hat{\nu}=(Y/t)/(X/s)$ estimates ρ,

$$\phi=\exp\left[K_{\gamma}(Y^{-1}+X^{-1})^{1/2}\right],\tag{1.6}$$

and K_{γ} is the $[100(1+\gamma)/2]$th standard normal percentile. These limits are usually accurate enough in practice if X and Y are above 10.

A one-sided $100\gamma\%$ confidence limit comes from the corresponding two-sided limit above. Then γ replaces $(1+\gamma)/2$.

Tree and bare wire example.   The failure rates of two types of power line wire in a region were to be compared. The wires were subjected to the same weather and operation. It was assumed that the number of line failures for a given exposure has a Poisson distribution. The standard bare wire had $X=69$ failures in $s=1079.6$ 1000-ft·years of exposure, and a polyethylene-covered tree wire had $Y=12$ failures in $t=467.9$ 1000-ft·years. Then $\hat{\lambda}=69/1079.6=0.0639$ per 1000 ft per year for the bare wire, and $\hat{\nu}=12/467.9=0.0256$ for the tree wire. Figure 1.2 shows these estimates; each

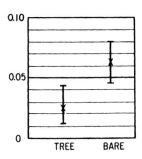

Figure 1.2.   Power line comparison.

with its 95% confidence interval. The intervals do not overlap. This implies that the two rates differ significantly at least at the 5% significant level. Also $\hat{\rho} = 0.0256/0.0639 = 0.401$; that is, the observed failure rate for tree wire is 40% of that for bare wire.

Two-sided 95% confidence limits for ρ are

$$\underset{\sim}{\rho} = \{(12/467.9)/[(69+1)/1079.6]\}/F[(1+0.95)/2; 2 \cdot 12 + 2, 2 \cdot 69]$$

$$= 0.20,$$

$$\tilde{\rho} = \{[(12+1)/467.9]/(69/1079.6)\} \cdot F[(1+0.95)/2; 2 \cdot 69 + 2, 2 \cdot 12]$$

$$= 0.75.$$

Calculate approximate 95% limits as follows:

$$\phi = \exp\left[1.960(12^{-1}+69^{-1})^{1/2}\right] = 1.846,$$

$$\underset{\sim}{\rho} \cong 0.401/1.846 = 0.22, \qquad \tilde{\rho} \cong 0.401 \cdot 1.846 = 0.74.$$

The limits do not enclose 1. So the failure rate of tree wire is statistically significantly below that of bare wire. Convinced of a real difference, engineering could use the confidence interval to help decide if tree wire is economically enough better than bare wire.

Comparison of K Poisson Occurrence Rates

The following explains how to compare K Poisson occurrence rates for equality.

For $k = 1, \ldots, K$, suppose that Y_k is a Poisson count in an observed "length" t_k where the occurrence rate is λ_k. Also, suppose $Y_1, \ldots, Y_K$ are statistically independent.

Test. The following tests the equality hypothesis $\lambda_1 = \lambda_2 = \cdots = \lambda_K$ against the alternative $\lambda_k \neq \lambda_{k'}$ for some k and k'. First calculate $Y = Y_1 + \cdots + Y_K$, $t = t_1 + \cdots + t_K$, and $\hat{\lambda} = Y/t$. The test statistic is

$$Q = \sum_{k=1}^{K} \left(Y_k - \hat{\lambda}t_k\right)^2 / \left(\hat{\lambda}t_k\right). \tag{1.7}$$

Here $E_k = \hat{\lambda}t_k$ estimates the expected number of occurrences. If the equality hypothesis is true, the distribution of Q is approximately chi square with $K - 1$ degrees of freedom. If the alternative is true, Q tends to have larger values. So the test is

1. If $Q \leqslant \chi^2(1 - \alpha; K - 1)$, accept equality.
2. If $Q > \chi^2(1 - \alpha; K - 1)$, reject equality at the $100\alpha\%$ significance level.

Here $\chi^2(1 - \alpha; K - 1)$ is the $100(1 - \alpha)$th chi-square percentile with $K - 1$ degrees of freedom. Q is called the chi-square statistic. The chi-square approximation is more precise the larger the Y_k are. It is usually satisfactory if all $Y_k \geqslant 5$.

If there is a statistically significant difference, examine the $\hat{\lambda}_k = Y_k / t_k$ to see how they differ. A figure like Figure 1.3 can reveal the differences. Also, the $\hat{\lambda}_k$ that give the largest terms $(Y_k - \hat{\lambda} t_k)^2 / (\hat{\lambda} t_k)$ of Q are ones that differ most from the pooled $\hat{\lambda}$.

Power lines example. Table 1.1 shows outage data on seven power transmission lines from Albrecht, Nelson, and Ringlee (1968). Figure 1.3 depicts the estimate and confidence limits for the outage rate of each line. For line k, Y_k is its total number of outages, L_k is its length in miles, N_k is the number of years of observation, and $t_k = N_k L_k$ is the total exposure. Power line engineers wanted an estimate of the outage rate of such lines to aid maintenance planning. The pooled estimate (1.8) of the common outage rate is $\hat{\lambda} = 0.414$. $Q = 34.30$ is calculated in Table 1.1. Since $Q = 34.30 > 22.5 = \chi^2(0.999, 7 - 1)$, the seven outage rates differ very highly significantly (0.1% level). Figure 1.3 reveals that the outage rates of the three shortest lines differ most from the pooled rate. Also, the three shortest lines make the greatest contributions to Q.

Table 1.1.   Power Line Outage Data and Calculations

Line k	Length L_k mi.	Years N_k	Exposure $t_k = N_k L_k$	Outages Y_k	Expected $E_k = \hat{\lambda} t_k$	Q Term $(Y_k - E_k)^2 / E_k$	Outage Rate $\hat{\lambda}_k$
1	10 x	5 =	50	10	20.72	5.54	0.20
2	13	1	13	13	5.39	10.76	1.00
3	17	5	85	17	35.22	9.43	0.20
4	24	9	216	102	89.51	1.74	0.47
5	42	6	252	124	104.42	3.67	0.49
6	61	2	122	53	50.56	0.12	0.43
7	69	2	138	44	57.18	3.04	0.32
			t = 876	363 = Y		34.30 = Q	

$$\hat{\lambda} = 363/876 = 0.414$$

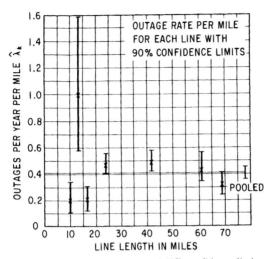

Figure 1.3. Line outage rates and 90% confidence limits.

Because the data contain many outages, the test detects small differences in the true outage rates. Such differences may be unimportant from an engineering point of view. An engineer can judge this from Figure 1.3. The high significance of Q indicates that confidence and prediction intervals calculated from the pooled $\hat{\lambda}$ (described below) are too narrow. Also, there may be variables in addition to line length that significantly affect outage rate, for example, the yearly number of thunderstorms at each line.

Chapter 12 gives a likelihood ratio test for equality of Poisson occurrence rates. General theory for that test appears in Cox and Lewis (1966). The statistical performance (OC curve) of that test is similar to that of the test above. Also, quality control charts (Grant and Leavenworth, 1980) are a means of comparing K Poisson samples.

Other uses of the test.   The test is often used on counts from a Poisson process for a number of time periods. The test checks for a constant occurrence rate over time. For example, suppose that monthly failure data are collected on a stabile population of equipment. The test would be used to reveal seasonal and other effects on the failure rate. Also, the effect of seasonal variations can be eliminated if one can use an entire year as an observation period.

Nonconstant failure rate can also be assessed from a plot of the cumulative number of occurrences versus time. Departures from a straight plot may indicate a nonconstant failure rate.

The above chi-square test is also used to test the assumption of a Poisson distribution when there are K observed counts from the distribution.

Pooled Estimate of a Poisson Occurrence Rate

Often one wishes to combine Poisson data to get more informative analyses. The following describes a pooled estimate of a common Poisson occurrence rate, its variance, and confidence and prediction limits.

Estimate.   For $k = 1, \ldots, K$, suppose that Y_k is a Poisson count in an observed "length" t_k where the occurrence rate is λ. The pooled estimate of the common λ is

$$\hat{\lambda} = (Y_1 + \cdots + Y_K)/(t_1 + \cdots + t_K). \tag{1.8}$$

This is the total number of occurrences $Y = Y_1 + \cdots + Y_K$ divided by the total length of observation $t = t_1 + \cdots + t_K$. This estimate is unbiased, and its variance is

$$\mathrm{Var}(\hat{\lambda}) = \lambda/t. \tag{1.9}$$

Limits.   Section 2 of Chapter 6 gives confidence limits for the common λ and prediction limits from Y occurrences in a length of observation t. Before using a pooled estimate, one should check that the failure rates do not differ significantly. The previous section presents such a check.

Power lines example.   Table 1.1 shows data on outages of seven power transmission lines. For the seven lines, the total number of outages is $Y = 363$, and the total exposure is $t = 876$ mile·years. The estimate of a (wrongly assumed) common Poisson outage rate is $\hat{\lambda} = 363/876 = 0.414$ outages per mile per year. Two-sided 90% confidence limits for the true common failure rate are $\lambda = 0.378$ and $\tilde{\lambda} = 0.450$ from Section 2 of Chapter 6. Predictions and prediction limits for the numbers of outages can be calculated for future years, assuming a common outage rate, which is not a valid assumption for the power lines.

Other Methods for Poisson Data

For other methods for Poisson data, consult Haight (1967) and Cox and Lewis (1966). Such methods include, for example, the following.

1.   Fitting regression equations that express the Poisson mean as a function of numerical variables.

2. Doing analyses of variance where the Poisson mean is a function of qualitative variables.

3. Fitting a Poisson distribution to censored and interval data.

4. Predicting reliability growth with a Duane plot (see Section 7 of Chapter 13).

2. COMPARISON OF BINOMIAL SAMPLES

This section presents methods for comparing binomial samples, namely,

1. A sample with a specified binomial proportion (acceptance sampling).

2. Two samples for equal binomial proportions.

3. K samples for equal binomial proportions.

Also, this section presents methods for

4. Pooling samples to estimate a common binomial proportion.

Comparison of a Binomial Sample with a Given p_0

The following describes three equivalent ways to compare a sample binomial proportion $\hat{p}$ with a given p_0 to assess whether they are consistent with each other. Suppose there are Y category units in a sample of n units.

Confidence interval comparison. Suppose that a $100\gamma\%$ confidence interval for the true p encloses p_0. Then $\hat{p} = Y/n$ is **consistent** with p_0 at the $100\gamma\%$ confidence level. If the interval does not enclose p_0, then $\hat{p}$ differs significantly from p_0 at the $[100(1-\gamma)]\%$ significance level. Such an interval may be one or two sided, as described in Section 3 of Chapter 6.

Rat survival. An experiment employed 17 pairs of rats that were poisoned (Sampford and Taylor, 1959). One rat of each pair was randomly chosen and treated to prolong life, and the other rat served as a control. In seven pairs the treated rat survived the control rat. If the treatment has no effect, the true proportion of pairs where the treated rat outlives the control rat is $p_0 = \frac{1}{2}$. Two-sided 90% confidence limits based on 7 out of 17 are $p = 0.21$ and $\check{p} = 0.64$, as described in Chapter 6. These enclose $p_0 = \frac{1}{2}$. So the treatment and control do not differ significantly at even the 10% level. Of course, 17 is a small sample, as the wide confidence interval shows.

Hypothesis test. A hypothesis test can be used to check if a binomial sample is **consistent** with a specified population proportion p_0 in a category. Suppose a sample of n units contains Y category units. For a two-sided test, the significance level is the probability of the observed number of category units being $|Y - np_0|$ or more away from np_0, namely,

$$P = \sum_{y=0}^{L} \binom{n}{y} p_0^y (1 - p_0)^{n-y} + \sum_{y=U}^{n} \binom{n}{y} p_0^y (1 - p_0)^n, \qquad (2.1)$$

where L is $np_0 - |Y - np_0|$ rounded down to the nearest integer, and U is $np_0 + |Y - np_0|$ rounded up to the nearest integer. If P is small, there is significant evidence that the true proportion differs from the specified one. Binomial tables (referenced in Chapter 2) readily provide the above sums. When $p_0 = \frac{1}{2}$, the test is called a "sign test." This is a two-sided test.

Rat survival. The test can be used on the above data on rat survival. Then $17(\frac{1}{2}) - |7 - 17(\frac{1}{2})| = 7$ is L, and $17(\frac{1}{2}) + |7 - 17(\frac{1}{2})| = 10$ is U. The two-sided significance level is

$$P = \sum_{y=0}^{7} \binom{17}{y} (\tfrac{1}{2})^y (1 - \tfrac{1}{2})^{17-y} + \sum_{y=10}^{17} \binom{17}{y} (\tfrac{1}{2})^y (1 - \tfrac{1}{2})^{17-y} = 0.629.$$

This probability is large; so there is no convincing evidence that the treatment has an effect.

Acceptance sampling. An acceptance sampling plan is used to decide whether a population, such as a manufactured lot or shipment, is acceptable. When units are categorized as good or bad, the plan usually states that the lot must have an acceptable quality level (AQL) of p_0, the fraction defective. Binomial data are usually called **attribute data** in acceptance sampling.

An **acceptance sampling plan** specifies the number n of sample units and the **acceptance number** y_0. If a consumer finds more than y_0 defectives in a sample, the shipment fails.

The n and y_0 are chosen so the plan accepts most "good" shipments and rejects most "poor" ones. A plan had $n = 20$ and $y_0 = 1$. If a shipment has a proportion defective of $p = 0.01$, the chance of it passing inspection is $F(1) = 0.983$, from a binomial table. If a shipment has a proportion defective of $p = 0.10$, the chance of it passing inspection is $F(1) = 0.392$. When the true proportion defective is p, the probability of a lot passing the test with n and y_0 is

$$P_p\{Y \le y_0\} = \sum_{y=0}^{y_0} \binom{n}{y} p^y (1 - p)^{n-y}. \qquad (2.2)$$

This function of p is called the **operating characteristic (OC) curve.** The OC

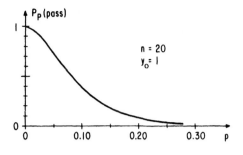

Figure 2.1. OC curve of acceptance sampling plan.

curve for the sampling plan above appears in Figure 2.1. This is a one-sided test.

In practice, n and y_0 are chosen so that the plan accepts lots with a low proportion defective p_0 with high probability $1 - \alpha$ and accepts lots with a high proportion defective p_1 with a low probability β. p_0 is called the **acceptable quality level** (AQL). The probability of such an acceptable lot failing is α and is called the **producer's risk**. The high proportion p_1 is called the **lot tolerance percent defective (LTPD)**; it is an unacceptable quality level. The probability of such an unsatisfactory lot passing is β and is called the **consumer's risk**. For the plan above, if the AQL is 1% defective, the producer's risk is $1 - 0.980 = 0.020$, or 2%. Also, if the LTPD is 10%, the consumer's risk is 0.392, or 39%. Acceptance sampling plans are treated in detail in most books on statistical quality control; see, for example, Grant and Leavenworth (1980) and Schilling (1981).

Binomial plans. Standard binomial plans and their OC curves appear in the references below. The references include two-stage, multistage, and other sampling plans. Such plans result in a smaller average sample size than the above single-stage plans with comparable OC functions.

1. MIL-STD-105D, "Sampling Procedures and Tables for Inspection by Attributes," 29 April 1963, U.S. Government Printing Office, Washington, D.C. 20402.

2. Dodge, H. F., and Romig, H. G. (1959), "Single Sampling and Double Sampling Inspection Tables," 2nd ed., Wiley, New York.

Comparison of Two Binomial Proportions

This section gives the Fisher exact test for equality of two binomial proportions. An approximate K-sample test is given later; it also applies to the two-sample problem. Suppose Y is the number of category units in a sample of n units where the population proportion is p_Y, and X is the number in another sample of m units where the population proportion is p_X.

Test.   Fisher's one-sided test assesses whether p_Y is significantly larger than p_X. Calculate the hypergeometric tail probability

$$P = \sum_{y=Y}^{M} \binom{X+Y}{y}\binom{m+n-X-Y}{n-y} \bigg/ \binom{m+n}{n}, \qquad (2.3)$$

where $M = \min(X+Y, n)$. Lieberman and Owen (1961) tabulate such probabilities. Then P is the significance level of the test. If P is small (say, less than 0.05), then $\hat{p}_Y = Y/n$ is statistically significantly larger than $\hat{p}_X = X/m$. That is, there is evidence that $p_Y > p_X$. Otherwise, Y/n is not significantly greater than X/m. This is a one-sided hypothesis test. A two-sided test employs the hypergeometric probabilities in both tails [see (2.1)].

Printed circuit example.   A printed circuit had been manufactured in two lots. The newer lot had some design improvements, and the engineers wanted to know if the newer lot was really better. The older lot had $Y=4$ failed circuits among $n=119$, and the newer lot had $X=0$ failed among $m=254$ circuits. If the probability of a failure is the same for all circuits, the probability of all four failures falling into the older lot is

$$P = \binom{0+4}{4}\binom{254+119-0-4}{119-4} \bigg/ \binom{254+119}{119} = 0.010.$$

This small probability indicates that the observed results are highly unlikely if circuits from both lots have the same probability of failing. This is convincing evidence that the newer design is better.

Sample size.   Cohen (1977) explains how to choose the two sample sizes to compare two binomial proportions.

Comparison of K Binomial Proportions

Test.   The following hypothesis test compares K binomial proportions for equality. For $k=1,\ldots,K$, suppose there are Y_k category units in a sample of n_k units where the population proportion is p_k.

The following is an approximate test of the hypothesis of equality (or homogeneity) $p_1 = \cdots = p_K$ against the alternative $p_k \neq p_{k'}$ for some k and k'. First calculate $Y = Y_1 + \cdots + Y_K$, $n = n_1 + \cdots + n_K$, and $\hat{p} = Y/n$. The test statistic is

$$Q = \sum_{k=1}^{K} (Y_k - n_k \hat{p})^2 / [n_k \hat{p}(1-\hat{p})], \qquad (2.4)$$

where $n_k \hat{p}$ estimates the expected number of category units. If the equality hypothesis is true, the distribution of Q is approximately chi square with $K-1$ degrees of freedom. If the alternative is true, the distribution of Q shifts toward larger values. So the test is

1. If $Q \leqslant \chi^2(1-\alpha; K-1)$, accept the equality hypothesis.
2. If $Q > \chi^2(1-\alpha; K-1)$, reject the equality hypothesis at the $100\alpha\%$ significance level.

Here $\chi^2(1-\alpha; K-1)$ is the $[100(1-\alpha)]$th chi-square percentile with $K-1$ degrees of freedom. Q is called the chi-square statistic. The chi-square approximation is more precise the larger Y and $n-Y$ are. It is usually satisfactory if all $Y_k \geqslant 5$ and $n_k - Y_k \geqslant 5$. There are many computer programs for contingency table analysis that do the calculations above.

If Q is statistically significant, examine the $\hat{p}_k = Y_k / n_k$ to see how they differ. A figure like Figure 2.2 can reveal the nature of the differences. Also, the $\hat{p}_k$ that give the largest terms of $(Y_k - n_k \hat{p})^2 / [n_k \hat{p}(1 - \hat{p})]$ of Q are the ones that differ most from the pooled $\hat{p}$.

Capacitor failure example. Table 2.1 shows the numbers of capacitor of the same type and how many of them failed on nine different circuit boards. Engineering wanted to know if there were real differences among such boards. Figure 2.2 shows the $\hat{p}_k$ with 90% confidence limits. The pooled estimate of the common binomial proportion is $\hat{p} = 0.03936$. $Q = 56.8$ is calculated in Table 2.1; it has $9 - 1 = 8$ degrees of freedom. Since $Q = 56.8 > 26.1 = \chi^2(0.999; 8)$, the nine circuit boards differ very highly significantly (0.1% level). Figure 2.2 reveals that board 4 has an unusually high proportion of capacitors failing, and board 7 has an unusually low proportion. Also, those boards make the greatest contribution to Q. This indicates that these two boards should be physically contrasted to determine and eliminate causes of failure.

Chapter 12 gives a likelihood ratio test for equality of binomial proportions. General theory for that appears in Cox (1970). The statistical performance of that test is similar to that of the test above. Also, binomial plotting

Table 2.1. Capacitor Failure Data and Calculations

Board k:	1	2	3	4	5	6	7	8	9	Pooled
Sampled n_k:	84 +	72 +	72 +	119 +	538 +	51 +	517 +	462 +	143 =	2058 = n
Failed Y_k:	2 +	3 +	5 +	19 +	21 +	2 +	9 +	18 +	2 =	81 = Y
$E_k = n_k\hat{p}$:	3.309	2.837	2.837	4.688	21.196	2.009	20.368	18.123	5.634	.03936 = $\hat{p}$
$(Y_k - E_k)^2 / [E_k(1-\hat{p})]$:	0.54 +	0.01 +	1.72 +	45.48 +	0.00 +	0.00 +	6.61 +	0.00 +	2.44 =	56.8 = Q
$\hat{p}_k$:	.0238	.0417	.0694	.1597	.0390	.0392	.0174	.0390	.0140	

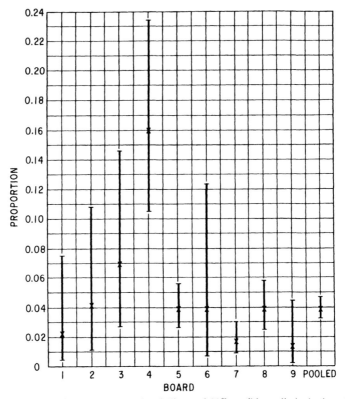

Figure 2.2.   Capacitor proportion failing and 90% confidence limits by board.

paper (King, 1971) and quality control charts (Grant and Leavenworth, 1980) are means of comparing binomial proportions.

Pooled Estimate of a Binomial Proportion

The following describes a pooled estimate of a common binomial proportion, its variance, and confidence and prediction limits.

Estimate.   For $k = 1, \ldots, K$, suppose that there are Y_k category units in a sample of n_k units. The pooled estimate of the common proportion p is

$$\hat{p} = (Y_1 + \cdots + Y_K)/(n_1 + \cdots + n_K). \tag{2.5}$$

This is the total number of category units $Y = Y_1 + \cdots + Y_K$ divided by the total number $n = n_1 + \cdots + n_K$ of sample units. So $\hat{p}$ is the pooled sample proportion. It is unbiased, and its variance is

$$\text{Var}(\hat{p}) = p(1-p)/n. \tag{2.6}$$

Use the Y category units in n sample units to calculate confidence limits and prediction limits as described in Section 3 of Chapter 6.

Before using the pooled estimate, one should check that the sample proportions do not differ significantly. The previous subsection presents such a check.

Capacitor example.   Table 2.1 shows data on identical capacitors used in $K = 9$ circuit boards. For the nine circuit boards, there are $Y = 81$ failures among $n = 2058$ capacitors. The estimate of a common proportion failing is $\hat{p} = \frac{81}{2058} = 0.039$. Two-sided 90% confidence limits for the common p are $p = 0.032$ and $\tilde{p} = 0.046$. Figure 2.2 depicts the pooled and individual estimates and their 90% confidence limits. Of course, the previous subsection indicated that pooling the data is not appropriate.

Other Methods for Binomial Data

For other methods for binomial data, consult Cox (1970), Finney (1968), Fleiss (1973), Maxwell (1961), Plackett (1974), Patil and Joshi (1968), and Johnson and Kotz (1969). These references include, for example,

1. Fitting regression equations that express a binomial proportion as a function of numerical variables.
2. Doing analyses of variance where the binomial proportion is a function of qualitative variables.
3. Fitting a binomial distribution to censored and interval data.

3. COMPARISON OF MULTINOMIAL SAMPLES

This section presents statistical hypothesis tests for comparing

1. A multinomial sample with specified multinomial proportions.
2. K multinomial samples for equality of corresponding multinomial proportions.

Also, this section presents methods for

3. Pooling samples to estimate common multinomial proportions. To compare samples with respect to the proportion in **one** particular category, use the binomial methods of Section 2.

Comparison of a Multinomial Sample with Given Proportions

Test.   The following hypothesis test compares the observed sample numbers $Y_1,\ldots,Y_M$ in M multinomial categories with specified proportions $\pi_1,\ldots,\pi_M$. The test statistic is

$$Q = \sum_{m=1}^{M} (Y_m - n\pi_m)^2 / (n\pi_m) \tag{3.1}$$

where $n = Y_1 + \cdots + Y_M$. Here $E_m = n\pi_m$ is the expected number of observations in category m, and Y_m is the observed number. If the π_m's are the true proportions, the distribution of Q is approximately chi square with $M-1$ degrees of freedom. The approximation is more exact the larger the E_m are, and it is usually satisfactory if they all exceed 5. If the π_m's are not the true proportions, the distribution of Q shifts toward larger values. So the test is

1. If $Q \leqslant \chi^2(1-\alpha; M-1)$, accept the specified proportions.
2. If $Q > \chi^2(1-\alpha; M-1)$, reject the specified proportions

at the $100\alpha\%$ significance level; here $\chi^2(1-\alpha; M-1)$ is the $[100(1-\alpha)]$th percentile of the chi-square distribution with $M-1$ degrees of freedom. This is the **chi-square test of fit** for the specified (multinomial) distribution.

If Q is statistically significant, locate the largest chi-square contributions $(Y_m - n\pi_m)^2/(n\pi_m)$; these correspond to the hypothetical π_m that differ most from the corresponding observed proportions Y_m/n.

Transformer example.   Table 3.1 shows data on distribution transformers. Each of the $K=10$ lines of data corresponds to a group of transformers installed in a particular year. A simple model for such transformer life has the geometric probability $\pi_m = 0.01166(0.98834)^{m-1}$ for failure in year $m=1,2,\ldots,M-1$ and survival probability $\pi_M = 0.98834^{M-1}$ for survival beyond year $M-1$. The following tests whether this model fits the data on the $n=652$ units of group 8. Table 3.2 shows the calculation of $Q = 25.83 > 20.09 = \chi^2(0.99; 9-1)$. So the simple model is rejected at the 1% significance level (highly statistically significant rejection). The chi-square contributions show that years 2 and 6 have unusually high numbers of failures. The data are from Albrecht and Campbell (1972).

Comparison of K Multinomial Samples

Test.   The following hypothesis test compares K multinomial samples for equality of corresponding category proportions. Suppose population k has M true category proportions $\pi_{k1},\ldots,\pi_{kM}$, where $k=1,\ldots,K$. The equality

Table 3.1. Yearly Numbers of Failed Distribution Transformers

Installed		Year in Service									
Group	Number	1	2	3	4	5	6	7	8	9	10
10	422	9	5	4	6	2	3	5	7	5	3
9	541	10	16	24	21	19	5	21	17	17	
8	652	6	17	10	5	8	16	8	8		
7	1139	1	18	16	9	16	11	5			
6	1608	0	15	12	15	19	12				
5	1680	9	31	19	15	12					
4	1404	5	81	33	22						
3	2427	65	63	41							
2	3341	3	24								
1	2265	22									

hypothesis is $\pi_{1m} = \cdots = \pi_{Km}$ for all $m = 1, \ldots, M$. The alternative is some $\pi_{km} \neq \pi_{k'm}$. Also, suppose the sample from population k contains $Y_{k1}, \ldots, Y_{kM}$ units in the M categories and $n_k = Y_{k1} + \cdots + Y_{kM}$ units. Calculate the total number of observations $n = n_1 + \cdots + n_K$ and the observed pooled proportions $p_m = (Y_{1m} + \cdots + Y_{Km})/n$. The test statistic is

$$Q = \sum_{m=1}^{M} \sum_{k=1}^{K} (Y_{km} - n_k p_m)^2 / (n_k p_m), \qquad (3.2)$$

where $n_k p_m$ estimates the expectation of Y_{km}. If the equality hypothesis is

Table 3.2. Test Statistic Calculation

Year m	π_m	$E_m = n\pi_m$	Y_m	$(Y_m - E_m)^2 / E_m$
1	0.01166	7.60	6	0.34
2	0.01152	7.51	17	11.99
3	0.01139	7.43	10	0.89
4	0.01126	7.34	5	0.75
5	0.01113	7.25	8	0.08
6	0.01100	7.17	16	10.87
7	0.01087	7.09	8	0.12
8	0.01074	7.00	8	0.14
>8	0.91044	593.61	574	0.65
		n = 652		25.83 = Q

true, the distribution of Q is approximately chi square with $\nu = (M-1)(K-1)$ degrees of freedom. The approximation is more exact the larger the $n_k p_m$ are, and it is usually satisfactory if they all exceed 5. If the alternative is true, Q tends to have larger values. So the test is

1. If $Q \leq \chi^2(1-\alpha; \nu)$, accept the equality hypothesis.
2. If $Q > \chi^2(1-\alpha; \nu)$, reject the equality hypothesis

at the $100\alpha\%$ significance level; here $\chi^2(1-\alpha; \nu)$ is the $[100(1-\alpha)]$th percentile of the chi-square distribution with $\nu = (M-1)(K-1)$ degrees of freedom. Q is called the chi-square statistic for a two-way **contingency table analysis**. There are many computer programs for the contingency table calculations above.

If Q is statistically significant, locate the largest chi-square contributions $(Y_{km} - n_k p_m)^2/(n_k p_m)$; each corresponds to a p_{km} that differs much from its corresponding proportions in the other populations. Plots of the p_{km} and corresponding confidence limits can help one spot differences.

This test can also be used to compare K distributions for equality. One divides the range of the data into M intervals. Then Y_{km} is the number of observations of sample k falling into interval m, and π_{km} is the proportion of distribution k in interval m. This test makes no assumption about the form of the distributions. The following example can be regarded as such a test.

Transformer example. For the top $K = 5$ groups of Table 3.1, Table 3.3 shows the numbers of failures in the first $M-1 = 6$ years in service and the numbers of survivors. Figure 3.1 shows computer output for a contingency table test for equality of the life distributions of the five groups. The test statistic is $Q = 174.247 > 51.18 = \chi^2[0.999; (5-1)(7-1)]$. This indicates that the five life distributions differ very highly significantly. Examination of the (large) chi-square contributions indicates that group 2 has higher proportions failing than the other groups, attributed to poorer quality that year. In a preventive replacement program, group 2 would be replaced first.

Table 3.3. Contingency Table Data on Distribution Transformers

Group	Installed Number	1	2	3	4	5	6	7+
10	422	9	5	4	6	2	3	393
9	541	10	16	24	21	19	5	446
8	652	6	17	10	5	8	16	590
7	1139	1	18	16	9	16	11	1068
6	1608	0	15	12	15	19	12	1535

* CONTINGENCY TABLE

GROUP	YEAR	----NUMBER---- OBSERVED	EXPECTED.	CHI-SQ CONTRIB
GROUPX	YEAR1	9	2.515	16.718
GROUPX	YEAR2	5	6.869	0.508
GROUPX	YEAR3	4	6.385	0.891
GROUPX	YEAR4	6	5.418	0.063
GROUPX	YEAR5	2	6.192	2.838
GROUPX	YEAR6	3	4.547	0.526
GROUPX	YEAR7	393	390.074	0.022
GROUP9	YEAR1	10	3.225	14.236
GROUP9	YEAR2	16	8.806	5.877
GROUP9	YEAR3	24	8.186	30.552
GROUP9	YEAR4	21	6.945	28.440
GROUP9	YEAR5	19	7.938	15.417
GROUP9	YEAR6	5	5.829	0.118
GROUP9	YEAR7	446	500.072	5.847
GROUP8	YEAR1	6	3.886	1.150
GROUP8	YEAR2	17	10.613	3.844
GROUP8	YEAR3	10	9.865	0.002
GROUP8	YEAR4	5	8.370	1.357
GROUP8	YEAR5	8	9.566	0.256
GROUP8	YEAR6	16	7.025	11.465
GROUP8	YEAR7	590	602.674	0.267
GROUP7	YEAR1	1	6.789	4.936
GROUP7	YEAR2	18	18.539	0.016
GROUP7	YEAR3	16	17.234	0.088
GROUP7	YEAR4	9	14.623	2.162
GROUP7	YEAR5	16	16.712	0.030
GROUP7	YEAR6	11	12.273	0.132
GROUP7	YEAR7	1068	1052.831	0.219
GROUP6	YEAR1	0	9.585	9.585
GROUP6	YEAR2	15	26.173	4.770
GROUP6	YEAR3	12	24.330	6.249
GROUP6	YEAR4	15	20.644	1.543
GROUP6	YEAR5	19	23.593	0.894
GROUP6	YEAR6	12	17.326	1.637
GROUP6	YEAR7	1535	1486.349	1.592
TOTAL		4362	4362.000	174.247

* THE CHI-SQUARE STATISTIC TO TEST THAT EACH GROUP
HAS THE SAME PROPORTION IN EACH YEAR   CATEGORY
HAS THE VALUE 174.247; THIS STATISTIC HAS 24
DEGREES OF FREEDOM.

THE PROBABILITY OF EXCEEDING THIS CHI-SQUARE
VALUE BY CHANCE ALONE IF THERE ARE NO
TRUE EFFECTS IS 0.   PER CENT

Figure 3.1. Contingency table calculations for transformer data.

Sample size.   Cohen (1977) explains how to choose sample sizes when comparing multinomial samples.

Pooled Estimate of Multinomial Proportions

The following describes a pooled estimate of a common multinomial proportion, its variance, and confidence limits.

Estimate.   For $k = 1, \ldots, K$, suppose there are Y_{km} observations in category m in a sample of n_k units. The pooled estimate of the common proportion π_m in category m is

$$p_m = Y_m / n, \tag{3.3}$$

where $Y_m = Y_{1m} + \cdots + Y_{Km}$ is the total number of units in the category, and $n = n_1 + \cdots + n_K$ is the total number of sample units. So p_m is the pooled sample proportion. It is an unbiased estimator for π_m, and

$$\mathrm{Var}(p_m) = \pi_m(1 - \pi_m)/n. \tag{3.4}$$

One uses the observed total number Y_m of category units in n sample units to calculate (1) binomial confidence limits for π_m from Section 3 of Chapter 6 and (2) binomial prediction limits (3.16) of Chapter 6 for the number of category units in a future sample.

Before using p_m, one should check that the sample proportions Y_{km}/n_k do not differ significantly.

Transformer example.   For the data in Table 3.1, the pooled estimate of the proportion failing in year 7 is $p_7 = (5+21+8+5)/(422+541+652+1139) = 0.014$. The four proportions should first be checked for equality.

Other Methods for Multinomial Data

For other methods for multinomial data, consult Finney (1968), Fleiss (1973), Maxwell (1961), Plackett (1974), Patil and Joshi (1968), and Johnson and Kotz (1969). Such methods

1. Fit equations that express a multinomial proportion as a function of numerical or qualitative variables.

2. Analyze three- and higher-way contingency tables.

3. Analyze contingency tables with missing data.

Bishop, Feinberg, and Holland (1975) give theory and analyses for such contingency table data by means of log-linear models. Olivier and Neff (1976) document the LOGLIN program for such analyses.

4. COMPARISON OF EXPONENTIAL SAMPLES

This section presents methods for comparing samples from exponential distributions. The methods include graphical and analytic comparison of

1. A sample with a specified mean (demonstration testing).
2. Two samples.
3. K samples.

The section also explains how to pool samples to estimate a common exponential mean. These methods are not robust; that is, they are valid only if the distributions are exponential, which needs checking with data plots.

For the exponential distribution, one compares only the means. If the means of exponential distributions are equal, then percentiles, reliabilities, failure rates, and other quantities are equal.

The analytic methods below are exact for Type II censored data, that is, failure censored data and therefore complete and singly censored data. The examples involve complete data. The methods are approximate for Type I censored data, that is, time censored data; Chapter 12 gives other methods for such data. The methods for exponential data are closely related to those for Poisson data; many formulas are similar, differing only in the degrees of freedom of chi-square statistics.

The analytic methods below are valid only for samples from an exponential distribution. So this should be checked, say, with probability or hazard plots. The exponential distribution is often wrongly assumed to apply to products.

Comparison of an Exponential Mean with a Specified Value

The following describes three ways to compare a sample mean with a specified value θ_0: (1) plots, (2) confidence intervals, and (3) demonstration tests.

Plots. A sample from an exponential distribution can be graphically compared with a specified mean. Plot a complete or singly censored sample on exponential or Weibull probability paper; plot a multiply censored sample on exponential or Weibull hazard paper. Weibull paper displays the lower tail of the sample better. Fit a straight line for an exponential distribution to the data, and compare the graphical estimate of the mean with the specified mean. This method can be used to compare reliabilities or percentiles. This method is subjective, but it reveals peculiarities in the data, and it allows one to assess whether the exponential distribution adequately fits the data.

Confidence interval. From the sample, calculate a $100\gamma\%$ confidence interval for the true mean θ from (1.8) of Chapter 8. If the interval encloses the specified θ_0, then the sample is **consistent** with θ_0 at the $100\gamma\%$ confidence level. If the interval does not enclose θ_0, then the sample differs from θ_0 at the $100(1-\gamma)\%$ significance level. Such an interval may be one or two sided. The interval length also indicates how accurate the estimate of θ is.

Demonstration test. A reliability demonstration test for equipment usually requires that a specified mean θ_0 (or failure rate λ_0) be **demonstrated** with $100\gamma\%$ confidence. This means that the equipment passes the test if the observed lower $100\gamma\%$ confidence limit $\underline{\theta}$ for the true θ is above θ_0. Otherwise, the equipment fails. Equivalently, this means that the total running time T summed over all tests units before the r_0th failure occurs (Type II censoring) must exceed T_0. The values of r_0 and T_0 are chosen so that the probability of passing the test when $\theta = \theta_0$ is

$$P_{\theta_0}\{T \geqslant T_0\} = 1 - \gamma. \tag{4.1}$$

This **low** probability $1 - \gamma$ is called the **consumer's risk**. So the true mean θ must be above θ_0 if the equipment is to have a high probability of passing the test. $100\gamma\%$ is the **confidence level** of the demonstration test.

The probability

$$P(\theta) = P_\theta\{T > T_0\} \tag{4.2}$$

of passing the test as a function of θ is called the **operating characteristic (OC) function** of the test with r_0 and T_0. The distribution of $2T/\theta$ is chi square with $2r_0$ degrees of freedom. So

$$P(\theta) = 1 - F_{2r_0}(2T_0/\theta) \tag{4.3}$$

in terms of the chi-square cumulative distribution function. The true value θ must be above θ_0 if the equipment is to pass with high probability. If $\theta = \theta'$ is the producer's true mean, $1 - P(\theta') = F_{2r_0}(2T_0/\theta')$ is called the **producer's risk** and is the probability the equipment fails the test.

Any number n of units may be run in such a test. Of course, the larger n is, the sooner the r_0 failures and total time T_0 are accumulated.

Capacitor demonstration example. A demonstration test of a high-reliability capacitor specified that a mean life of $\theta_0 = 10^8$ hours be demonstrated with 60% confidence. The minimum test time T results if $r_0 = 1$.

Then the required test time is the solution of (4.1), that is, of

$$1 - 0.60 = P_{10^8}\{T > T_0\} = \exp(-T_0 / 10^8).$$

Here $T_0 = 9.163 \times 10^7$ hours. This time was accumulated by running 9163 capacitors each for 10,000 hours. (Another number of capacitors could be run a different duration to accumulate the same time.) The corresponding OC function is

$$P(\theta) = P_\theta\{T > 9.163 \times 10^7\} = \exp(-9.163 \times 10^7 / \theta).$$

To pass the test with 90% probability, such capacitors need a true mean θ' that satisfies

$$0.90 = \exp(-9.163 \times 10^7 / \theta').$$

That is, $\theta' = 8.70 \times 10^8$ hours, which is 8.70 times θ_0. The probability of passing a demonstration requirement when $\theta > \theta_0$ increases if one uses a test that allows more failures r_0' in a greater accumulated time T_0'. Then r_0' and T_0' are still chosen to meet the demonstration requirement $P_{\theta_0}\{T > T_0'\} = 1 - \gamma$. Of course, such a test requires more test time and expense. For example, the capacitor test with $r_0' = 2$ requires $T_0' = 20.2 \times 10^7$ hours; this is over twice the time for $r_0 = 1$. This example is another way of viewing the Poisson demonstration test in Section 1.

Standard exponential demonstration plans. Such plans and their OC curves are given by Schilling (1981) and the following references, which may be bought from the U.S. Government Printing Office, Washington, DC 20402.

MIL-HDBK-108, "Sampling Procedures and Tables for Life and Reliability Testing," 29 April 1960.

MIL-STD-690B, "Failure Rate Sampling Plans and Procedures," 17 April 1968.

MIL-STD-781C, "Reliability Tests: Exponential Distribution," 12 August 1979.

Comparison of Two Exponential Means

The following describes two ways to compare the means of two exponential samples: (1) plots and (2) confidence intervals.

Plots. Samples from two exponential distributions can be compared graphically. Plot complete or singly censored data on exponential or Weibull probability paper; plot multiply censored data on exponential or Weibull hazard paper. Plot both samples on the same sheet of paper for easy comparison. Subjectively compare the plotted samples to assess whether the two distributions are equal. This comparison may be aided by fitting a straight line for each exponential distribution. Also, plotted confidence limits for the percentiles of each distribution help. The plots also help one to assess how well the exponential distribution fits the data, and they reveal peculiar observations.

Insulating fluid example. Table 4.1 shows 60 times to breakdown in minutes of an insulating fluid subjected to high voltage stress. The times in their observed order are divided into six groups. (The use of six groups is arbitrary.) If the experiment was under control, the mean time should be the same for each group. For comparison, the first two groups are plotted on Weibull probability paper in Figure 4.1. The figure also shows the fitted exponential distributions and their 90% confidence limits. The two groups appear to be similar.

Confidence interval. For $k = 1, 2$, suppose that sample k has Type II (failure) censoring, r_k failures, and a total running time T_k summed over all n_k sample units. Then the estimate of the exponential mean θ_k of population k is $\hat{\theta}_k = T_k / r_k$. A two-sided $100\gamma\%$ confidence interval for the ratio

$$\rho = \theta_1 / \theta_2 \tag{4.4}$$

has limits

$$\underline{\rho} = \left(\hat{\theta}_1 / \hat{\theta}_2\right) / F\left[(1+\gamma)/2; 2r_1, 2r_2\right],$$
$$\tilde{\rho} = \left(\hat{\theta}_1 / \hat{\theta}_2\right) \cdot F\left[(1+\gamma)/2; 2r_2, 2r_1\right], \tag{4.5}$$

Table 4.1. Times to Insulating Fluid Breakdown

Group					
1	2	3	4	5	6
1.89	1.30	1.99	1.17	8.11	2.12
4.03	2.75	.64	3.87	3.17	3.97
1.54	.00	2.15	2.80	5.55	1.56
.31	2.17	1.08	.70	.80	1.34
.66	.66	2.57	3.82	.20	1.49
1.70	.55	.93	.02	1.13	8.71
2.17	.18	4.75	.50	6.63	2.10
1.82	10.60	.82	3.72	1.08	7.21
9.99	1.63	2.06	.06	2.44	3.83
2.24	.71	.49	3.57	.78	5.13
T_k: 26.35	20.55	17.48	20.23	29.89	37.46
$\hat{\theta}_k$: 2.635	2.055	1.748	2.023	2.989	3.746

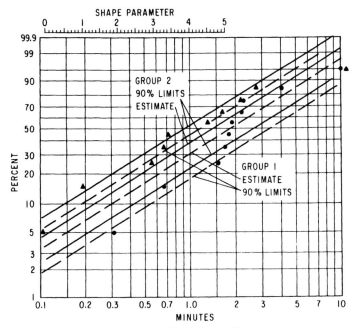

Figure 4.1.   Weibull plot of insulating fluid groups 1 and 2.

where $F[(1+\gamma)/2; a, b]$ is the $[100(1+\gamma)/2]$th percentile of the F-distribution with a degrees of freedom in the numerator and b in the denominator. If these limits do not enclose 1, $\hat{\theta}_1$ and $\hat{\theta}_2$ differ statistically significantly at the $[100(1-\gamma)]\%$ significance level. That is, there is convincing evidence that θ_1 and θ_2 differ. The limits are approximate for time censored data.

A one-sided $100\gamma\%$ confidence limit comes from the corresponding two-sided limit above. Then γ replaces $(1+\gamma)/2$.

Insulating fluid example. For the data in Table 4.1, the mean time should be the same for each group. Each pair of group means can be compared with the confidence interval (4.5). For example, two-sided 95% confidence limits for θ_1/θ_2 $(r_1 = r_2 = 10)$ are

$$\underset{\sim}{\rho}_{12} = (2.635/2.055)/F[(1+0.95)/2; 2\cdot 10, 2\cdot 10]$$

$$= (2.635/2.055)/2.46 = 0.521,$$

$$\tilde{\rho}_{12} = (2.635/2.055)\cdot 2.46 = 3.15.$$

These limits enclose 1. So the two groups do not differ significantly at the 5% level. This is consistent with the hypothesis that the two means are equal. To be statistically significant at the 5% level, the ratio of two such sample means would have to fall outside the range $F(0.975; 20, 20) = 2.46$ to $1/F(0.975; 20, 20) = 0.407$.

Comparison of K Exponential Means

The following describes three methods for comparing the means of K exponential samples: (1) plots, (2) test of homogeneity, and (3) simultaneous comparison of all pairs of means.

Plots. Samples from K exponential distributions can be compared graphically. If the data are complete or singly censored, plot each sample on a separate exponential or Weibull probability paper. If the data are multiply censored, plot each sample on a separate exponential or Weibull hazard paper. Subjectively compare the plots to assess whether the distributions differ. This comparison may be aided by fitting a straight line to each sample. Also, plotted confidence limits for the percentiles of each distribution help. Stack the plots and hold them up to the light to compare them. The plots also provide a check on the validity of the data and exponential distribution.

Another plot is useful for comparing complete and singly censored samples. Tabulate samples side by side, as shown in Figure 4.2. Plot the

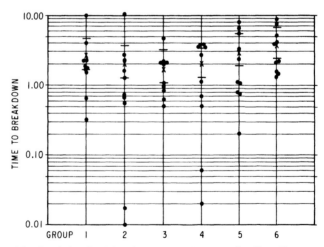

Figure 4.2.   Insulating fluid data by group: estimates and 90% confidence limits.

estimates and confidence limits for each mean to aid the comparison. For multiply censored samples, plot and compare just the estimates and confidence limits for the means.

Suppose each sample estimate of the mean is based on the same number r of failures. Then the estimates should plot as a straight line on probability paper for a chi-square distribution with $2r$ degrees of freedom. Departures from a straight line indicate unequal population means. The numbers of failures need to be equal, not the sample sizes.

Insulating fluid example.   Table 4.1 shows 60 times to breakdown of an insulating fluid in their observed order. The following checks whether the mean time to breakdown remained stabile over the experiment. The times are divided into six groups for comparison. The six groups are plotted side by side in Figure 4.2. The figure also shows the estimates and 90% confidence limits for the means. The means of the six groups appear to be comparable. The data were also divided into 12 groups of five times to breakdown. The 12 means are plotted on chi-square probability paper for 10 degrees of freedom in Figure 4.3. The slight curvature of the plot may indicate that the means differ significantly. A formal test of equality is needed.

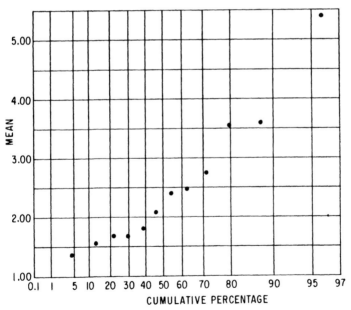

Figure 4.3.   Chi-square plot of 12 means.

Test of homogeneity. For $k = 1, \ldots, K$, suppose that sample k has Type II (failure) censoring, r_k failures, and a total running time T_k summed over all n_k sample units. Then the estimate of the mean of exponential population k is $\hat{\theta}_k = T_k / r_k$.

The following tests the homogeneity (equality) hypothesis $\theta_1 = \theta_2 = \cdots = \theta_K$ against the alternative that some $\theta_k \neq \theta_{k'}$. First calculate $r = r_1 + \cdots + r_K$, $T = T_1 + \cdots + T_K$, and $\hat{\theta} = T/r$. The test statistic is

$$Q = 2\left[r\ln(\hat{\theta}) - r_1\ln(\hat{\theta}_1) - \cdots - r_K\ln(\hat{\theta}_K)\right]/C, \qquad (4.6)$$

where

$$C = 1 + \left[6(K-1)\right]^{-1}\left(\sum_{k=1}^{K}(1/r_k) - (K/r)\right). \qquad (4.7)$$

If the equality hypothesis is true, the distribution of Q is approximately chi square with $K-1$ degrees of freedom. If the alternative is true, Q tends to have larger values. So the test is

1. If $Q \leqslant \chi^2(1-\alpha; K-1)$, accept the equality hypothesis.
2. If $Q > \chi^2(1-\alpha; K-1)$, reject the equality hypothesis at the $100\alpha\%$ significance level.

Here $\chi^2(1-\alpha; K-1)$ is the $[100(1-\alpha)]$th percentile of the chi-square distribution with $K-1$ degrees of freedom. The chi-square approximation is more precise the larger the r_k are. It is usually satisfactory if all $r_k \geqslant 5$. This test is **Bartlett's test,** and it is also used for time censored data.

If Q is statistically significant, examine the $\hat{\theta}_k$ and their confidence limits to see how they differ with a figure like Figure 4.2. Note that Bartlett's test is sensitive to departures from an exponential distribution and can then give misleading results.

Insulating fluid example. Table 4.1 shows 60 times to breakdown of an insulating fluid. The test above checks whether the mean time to breakdown remained constant over the experiment. For this purpose, the 60 times were divided into $K = 6$ groups with $r_k = n_k = 10$ observations in the order that they were observed. Table 4.1 shows the six sample $\hat{\theta}_k$. The calculation of Q is

$$r = 10 + \cdots + 10 = 60, \qquad T = 26.35 + \cdots + 37.46 = 151.96,$$

$$\hat{\theta} = 151.96/60 = 2.5326$$

$$C = 1 + \left[6(6-1)\right]^{-1}\left(\tfrac{1}{10} + \cdots + \tfrac{1}{10} - \tfrac{6}{60}\right) = 1.017,$$

$$Q = 2\left[60\ln(2.5326) - 10\ln(2.635) - \cdots - 10\ln(3.746)\right]/1.017 = 4.10.$$

Since $Q = 4.10 < 9.24 = \chi^2(0.90; 6 - 1)$, the six groups do not differ significantly at the 10% level. This is consistent with a constant true mean over time. Here all $r_k = n_k = 10$; in general, the r_k and n_k may all differ.

All pairwise comparisons. When there are K means, there are $K(K-1)/2$ pairs of means to compare. Simultaneous $100\gamma\%$ confidence intervals enclose **all** ratios $\theta_k / \theta_{k'}$, with probability γ. Then each interval encloses its true ratio with a probability higher than γ. If all intervals enclose 1, there are no significant differences among the means at the overall $[100(1-\gamma)]\%$ significance level. Approximate and exact simultaneous confidence limits follow.

Figure 4.4 displays all pairs of the six sample means of the insulating fluid on log-log paper. The means are marked on both scales. Perpendicular lines through the means intersect at the dots. The vertical distance of a dot from the 45° line is the log of the ratio of the corresponding two means. So the dots furthest from the 45° line correspond to the largest ratios.

Approximate simultaneous limits. Simultaneous two-sided approximate $100\gamma\%$ confidence limits for all ratios of pairs of means $(k, k' = 1, \ldots, K)$ are

$$\left(\hat{\theta}_k / \hat{\theta}_{k'}\right) / F(\gamma'; 2r_k, 2r_{k'}) \leqslant \theta_k / \theta_{k'} \leqslant \left(\hat{\theta}_k / \hat{\theta}_{k'}\right) F(\gamma'; 2r_{k'}, 2r_k), \quad (4.8)$$

where $\gamma' = 1 - (1 - \gamma)K^{-1}(K-1)^{-1}$ and $F(\gamma'; a, b)$ is the $100\gamma'$th F percentile with a degrees of freedom in the numerator and b in the denominator. These are the limits (4.5) for a single ratio, but γ' is higher. One-sided

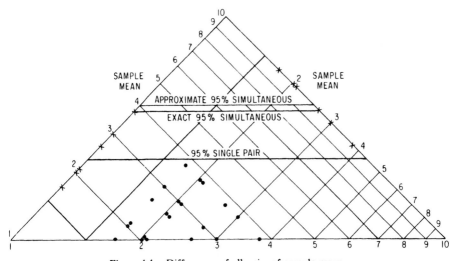

Figure 4.4. Differences of all pairs of sample means.

simultaneous limits are given by (4.8); then $\gamma' = 1 - 2(1 - \gamma)K^{-1}(1 - K)^{-1}$. These intervals can be discouragingly wide. These limits enclose all the true ratios with probability γ or higher. The excess probability is small when K is small and γ is large.

Insulating fluid example. For the $K = 6$ groups of data in Table 4.1, the simultaneous approximate 95% confidence limits are calculated as follows. All $r_k = 10$, $\gamma' = 1 - (1 - 0.95)6^{-1}(6 - 1)^{-1} = 0.99833$ (two-sided), and $F(0.99833; 20, 20) = 3.97$. For example, the limits for θ_1/θ_2 are (2.635 /2.055)/3.97 = 0.32 and (2.635/2.055)3.97 = 5.1. All $6(6 - 1)/2 = 15$ intervals enclose 1. So the six means do not differ significantly at the 5% significance level. Figure 4.4 shows a common line for the simultaneous 95% limit, $F(0.99833; 20, 20) = 3.97$, for all such ratios; all ratios are below it and are therefore not significant.

Exact simultaneous limits. If all $r_k = r$, then simultaneous two-sided exact $100\gamma\%$ confidence limits for all ratios of pairs of means $(k, k' = 1, \ldots, K)$ are

$$\left(\hat{\theta}_k/\hat{\theta}_{k'}\right)/F_{max}(\gamma; 2r, K) \leqslant \theta_k/\theta_{k'} \leqslant \left(\hat{\theta}_k/\hat{\theta}_{k'}\right)F_{max}(\gamma; 2r, K), \quad (4.9)$$

where $F_{max}(\gamma; 2r, K)$ is the 100γth percentile of the maximum F statistic among K chi-square statistics, each with $2r$ degrees of freedom. It is tabulated by Pearson and Hartley (1954) and by Owen (1962). If all $K(K - 1)/2$ intervals enclose 1, then there are no significant differences among the K means at the $[100(1 - \gamma)]\%$ significance level. Equivalently, calculate just the largest ratio. If it is less than $F_{max}(\gamma; 2r, K)$, then the K means do not differ significantly at the $[100(1 - \gamma)]\%$ level. This check for equality performs much like the above test of homogeneity. If the r_k differ slightly, use some average r to get approximate limits. Such intervals can be quite wide.

Insulating fluid example. For the $K = 6$ groups of data in Table 4.1, the simultaneous exact 95% confidence limits are calculated as follows. All $r_k = 10$, and $F_{max}(0.95; 20, 6) = 3.76$. Then, for example, the limits for θ_1/θ_2 are (2.635/2.055)/3.76 = 0.341 and (2.635/2.055)3.76 = 4.82. All 15 such intervals enclose 1, since the greatest ratio 3.746/1.748 = 2.14 is less than 3.76. Figure 4.4 shows lines for ratios of 3.76 and 3.97. They are the exact and approximate simultaneous upper 95% limits for all ratios. The other line (ratio of 2.46) is the upper 95% limit for a single ratio. So there are no significant differences among the six means at the 5% significance level, and the experiment appears stabile over time.

Pooled Estimate of an Exponential Mean

The following describes a pooled estimate of a common exponential mean from K samples, its variance, and confidence limits. For $k = 1, \ldots, K$, suppose that sample k has Type II (failure) censoring, r_k failures, and a total running time T_k summed over all n_k sample units.

Estimate. The pooled estimate of the common mean θ is

$$\hat{\theta} = T/r, \tag{4.10}$$

where $T = T_1 + \cdots + T_K$ and $r = r_1 + \cdots + r_K$. This is the total running time divided by the total number of failures, the usual estimate for an exponential mean. Then $2r\hat{\theta}/\theta$ has a chi-square distribution with $2r$ degrees of freedom. $\hat{\theta}$ is the maximum likelihood estimator and also the best linear unbiased estimator for θ. The variance of this unbiased estimator is

$$\text{Var}(\hat{\theta}) = \theta^2/r. \tag{4.11}$$

Calculate confidence limits for θ from (1.10) of Chapter 7 and prediction limits from Section 4 of Chapter 6 from a total running time T with r failures. Before using $\hat{\theta}$, check that the K means do not differ significantly. The previous subsection presents such checks. Also, combine the samples and plot the pooled sample on probability or hazard paper. The estimate is also used with time censored data.

Insulating fluid example. For the data in Table 4.2, the total running time for the six groups is $T = 26.35 + \cdots + 37.46 = 151.96$ minutes, and the total number of failures is $r = 10 + \cdots + 10 = 60$. The pooled estimate of the common exponential mean is $\hat{\theta} = 151.96/60 = 2.533$ minutes. The 95% confidence limits for θ Are $\underline{\theta} = 2 \cdot 60 \cdot 2.533/\chi^2(0.975, 2 \cdot 60) = 2 \cdot 60 \cdot 2.533/152.2 = 2.00$ and $\tilde{\theta} = 2 \cdot 60 \cdot 2.533/\chi^2(0.025, 2 \cdot 60) = 2 \cdot 60 \cdot 2.533/91.57 = 3.32$. The six means passed the previous tests for equality. So the pooled estimate and confidence limits appear satisfactory.

5. COMPARISON OF NORMAL AND LOGNORMAL SAMPLES

This section presents methods for comparing independent samples from (log) normal distributions. The methods include graphical and analytic comparison of

1. A sample with a specified distribution (demonstration testing).
2. Two samples.
3. K samples.

The methods compare means, standard deviations (or, equivalently, variances), percentiles, reliabilities, and other quantities. This section also explains how to pool samples to estimate a common mean or standard deviation. Methods for comparing means are robust to nonnormal distributions. Methods for comparing other quantities are sensitive to nonnormal distributions.

Needed background for this section includes (1) the basic properties of the normal and lognormal distributions (Chapter 2) and (2) the basic statistical methods for a single (log) normal distribution (Chapter 6).

The analytic methods below apply only to complete data, which consist entirely of failure times. Chapters 11 and 12 give analytic methods for singly and multiply censored data, which contain running times on unfailed units. (The graphical methods apply to complete and singly and multiply censored data.) The methods in this section assume that each sample is from a (log) normal distribution. So this should first be checked with, say, probability (Chapter 3) or hazard (Chapter 4) plots. In practice, combined graphical and analytic comparisons are most useful.

The analytic methods apply to both normal and lognormal data. However, one works with the logs of lognormal data. These methods appear in books on analysis of variance and experimental design; for example, Bartree (1968), Mendenhall (1968), Scheffé (1959), and Box, Hunter, and Hunter (1978).

Motor insulations example. The following example illustrates many of the methods. Table 5.1 shows life test data on specimens of three types of motor insulation tested at three temperatures (200, 225, 250°C). The test purpose was to find the longest-lived insulation at 200°C. Also, in use, the motors sometimes run up to 250°C; so the chosen insulation must compare well at 225 and 250°C. Engineering theory for such electrical insulation says that life has a lognormal distribution, and the log standard deviation of an insulation is the same at all temperatures. Periodic inspection of specimens revealed failures; Table 5.1 shows the days between inspections at each temperature. A tabulated failure time is the midpoint of the inspection period when the failure occurred. Such rounding slightly affects the results of the analyses below.

5.1. Comparison of a (Log) Normal Sample with a Specified Distribution

Often one wants to compare a sample standard deviation, mean, percentile, reliability, etc., with a specified value to assess whether a product is satisfactory. This section describes three ways to compare a complete sample with a specified (log) normal distribution: (1) plots, (2) confidence

Table 5.1. Failure Times (Hours) of Specimens of Three Motor Insulations

	Insulation		
Temp.	1	2	3
200°	1176	2520	3528
14 days	1512	2856	3528
	1512	3192	3528
	1512	3192	
	3528	3528	
225°	624	816	720
4 days	624	912	1296
	624	1296	1488
	816	1392	
	1296	1488	
250°	204	300	252
1 day	228	324	300
	252	372	324
	300	372	
	324	444	

intervals, and (3) demonstration (hypothesis) tests. In practical data analysis, a combination of plots and confidence limits is often most useful. Hypothesis tests are primarily useful for demonstration tests and are often specified in contracts.

Plots. Plot the sample on (log) normal probability or hazard paper, and draw a straight line through the data. Use the line to estimate the quantity of interest, and compare the estimate with the specified value. If a line fitted to the specified value passes "reasonably" through the data, then the data are consistent with that value. This method applies to standard deviations, means, reliabilities, percentiles, and other quantities. Chapter 3 describes such plots, and Figure 5.1 shows such plots.

Confidence intervals. From the sample, calculate a $100\gamma\%$ confidence interval for the quantity of interest θ, as described in Chapters 6, 7, and 8. If the interval encloses the specified value θ_0, then the sample is consistent with θ_0 at the $100\gamma\%$ confidence level. Otherwise, the sample differs significantly from θ_0 at the $[100(1-\gamma)]\%$ significance level. Such an interval may be one or two sided. The interval length indicates how accurate the estimate of θ is. Chapter 6 gives confidence intervals for a mean (5.11), standard deviation (5.6), percentile (5.16), and reliability (5.13).

Demonstration tests. A reliability demonstration test of equipment usually requires that the producer **demonstrate** better than a specified value θ_0

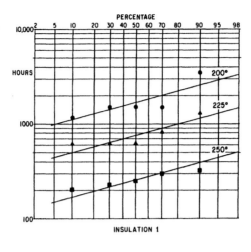

INSULATION 1

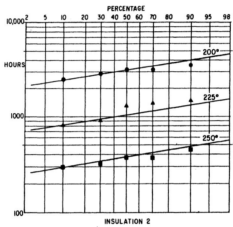

INSULATION 2

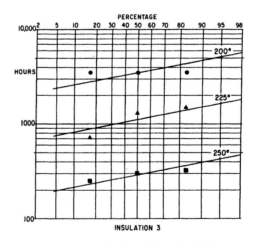

INSULATION 3

Figure 5.1. Lognormal plots of three insulation systems.

of a distribution parameter with $100\gamma\%$ confidence. This means that the equipment passes the test if the observed lower $100\gamma\%$ confidence limit θ for the true θ is above θ_0. Otherwise, the equipment fails the test. For a (log) normal life distribution, the parameter θ is usually the mean, a percentile, or reliability. An upper limit would be used for a fraction failing.

If the true $\theta = \theta_0$, then the probability of passing the test is

$$P_{\theta_0}\{\hat{\theta} \geqslant \theta_0\} = 1 - \gamma. \tag{5.1}$$

This low probability $1 - \gamma$ is called the **consumer's risk**. So the true value θ must be above θ_0 if the equipment is to pass with high probability. The probability

$$P(\theta) = P_\theta\{\hat{\theta} \geqslant \theta_0\} \tag{5.2}$$

of passing the test as a function of θ is the test's **operating characteristic (OC) curve**. $P(\theta)$ depends on the sample size n, θ_0, and γ. If $\theta = \theta'$ is the actual producer's value, then $1 - P(\theta') = P_{\theta'}\{\hat{\theta} < \theta_0\}$ is called the **producer's risk**; it is the probability of the equipment failing the test.

Standard (log) normal demonstration plans. Such plans and their OC curves are presented by Schilling (1981) and the references below. Demonstration tests for a reliability are related to those for a distribution percentile. So one need only consider tests for reliability. In particular, a test that the reliability $R(y_0)$ at a given age y_0 exceeds R_0 is equivalent to the test that the percentile y_{1-R_0} exceeds y_0. References for such reliability tests are

Bowker, A. H., and Goode, H. P. (1962), *Sampling Inspection by Variables*, McGraw-Hill, New York.

Duncan, A. J. (1965), *Quality Control and Industrial Statistics*, Chapters 12 and 13, Richard D. Irwin, Inc., Homewood, IL.

MIL-STD-414, "Sampling Procedures and Tables for Inspection by Variables for Percent Defective," 11 June 1957, U.S. Government Printing Office, Washington, DC 20402.

MIL-STD-471A, "Maintainability Demonstration," 27 March 1973, U. S. Government Printing Office, Washington, DC 20402.

For tests for the mean and standard deviation, see

Duncan, A. J. (1965), *Quality Control and Industrial Statistics*, Chapter 15. Richard D. Irwin, Inc., Homewood, IL.

Natrella (1963), Cohen (1977), and Owen (1962).

5.2. Pooled Estimates

This section shows how to combine data from a number of complete samples to estimate a **common** true standard deviation or mean. Such a pooled estimate is more precise than the individual sample estimates. Also, later analyses require a pooled estimate of the common true standard deviation. Before using a pooled estimate, first check that the samples do not differ significantly as described in Sections 5.3 and 5.4.

Pooled estimate of a common variance (standard deviation). Many analyses involve a pooled estimate of a common variance σ^2 from K **independent** samples. The following describes such an estimate, its variance, and confidence limits. For $k = 1, \ldots, K$, suppose that sample k has a variance S_k^2 from (2.3) of Chapter 6, with ν_k degrees of freedom.

The pooled estimate of a common variance σ^2 is

$$S^2 = \left(\nu_1 S_1^2 + \cdots + \nu_K S_K^2\right)/\nu, \tag{5.3}$$

where $\nu S^2/\sigma^2$ has a chi-square distribution with

$$\nu = \nu_1 + \cdots + \nu_K \tag{5.4}$$

degrees of freedom. S^2 is an unbiased estimator for σ^2, and its variance is

$$\mathrm{Var}(S^2) = 2\sigma^4/\nu. \tag{5.5}$$

The pooled estimate of a common standard deviation σ is

$$S = \left[\left(\nu_1 S_1^2 + \cdots + \nu_K S_K^2\right)/\nu\right]^{1/2}. \tag{5.6}$$

S is a biased estimator for σ, and its variance for large ν is approximately

$$\mathrm{Var}(S) \approx \sigma^2/(2\nu).$$

When the distribution means $\mu_1, \ldots, \mu_K$ differ, it is wrong to pool all samples and treat the pooled sample as a single sample to estimate σ^2. Such a wrongly pooled estimator tends to overestimate σ^2. Use (5.3). To calculate confidence limits for σ^2 or σ, use S^2 and its degrees of freedom ν in (5.6) of Chapter 6.

Motor insulations example. Table 5.2 shows log failure times of specimens of the three motor insulations at three temperatures. Theory for such an insulation says that life has a log normal distribution with the same standard deviation of log life at any temperature. For insulation 1, the sample standard deviations for the three temperatures are $S_{200} = 0.1832$, $S_{225} = 0.1384$, and $S_{250} = 0.0829$; these each have four degrees of freedom.

Table 5.2. Log Lives of Specimens of Three Insulations

	I n s u l a t i o n		
Temp.	1	2	3
200°	3.0704	3.4014	3.5475
	3.1796	3.4558	3.5475
	3.1796	3.5041	3.5475
	3.1796	3.5041	
	3.5475	3.5475	
225°	2.7952	2.9117	2.8573
	2.7952	2.9600	3.1126
	2.7952	3.1126	3.1726
	2.9117	3.1436	
	3.1126	3.1726	
250°	2.3096	2.4771	2.4014
	2.3579	2.5105	2.4771
	2.4014	2.5705	2.5105
	2.4771	2.5705	
$\overline{Y}_k$	2.5105	2.6474		$\overset{=}{Y}$ Pooled
200°	3.23134	3.48258	3.54750	3.40093
225°	2.88198	3.06010	3.04750	2.98868
250°	2.41130	2.55520	2.46300	2.95606
S_{temp}				
200°	0.1832	0.0558	0	
225°	0.1384	0.1166	0.1674	
250°	0.0829	0.0653	0.0559	
S_k	0.1408	0.0836	0.1019	0.1132
Degrees of freedom	12	12	6	30

So $\nu = 4+4+4 = 12$. The pooled estimate is $S_1^2 = [4(0.1832)^2 + 4(0.1384)^2 + 4(0.0829)^2]/12 = (0.1408)^2$, it has $\nu = 12$ degrees of freedom. The 95% confidence limits for σ^2 are $(0.101)^2$ and $(0.232)^2$. The three variances pass the later tests for equality. So the pooled estimate and confidence limits appear satisfactory.

Pooled estimate of a common mean. Many analyses involve a pooled estimate of a common mean from K **independent** samples. The following describes such an estimate, its variance, and confidence limits. For $k = 1, \ldots, K$, suppose that sample k has a mean $\overline{Y}_k$ of n_k observations from a distribution with variance σ_k^2. That is, the populations may have different variances. The pooled estimate of the common mean μ is

$$\hat{\mu} = \left[(n_1/\sigma_1^2)\bar{Y}_1 + \cdots + (n_K/\sigma_K^2)\bar{Y}_K\right]/V; \tag{5.7}$$

$$V = 1/\left[(n_1/\sigma_1^2) + \cdots + (n_K/\sigma_K^2)\right] \tag{5.8}$$

is the variance of $\hat{\mu}$. $\hat{\mu}$ is an unbiased estimator for μ. To use $\hat{\mu}$, one must know the σ_k^2 or use estimates of them, preferably from large samples. When the σ_k^2 differ, it is less accurate to pool all samples and to use the average of the pooled sample to estimate μ.

When the σ_k^2 have the same value σ^2, the pooled estimate is

$$\bar{\bar{Y}} = (n_1\bar{Y}_1 + \cdots + n_K\bar{Y}_K)/(n_1 + \cdots + n_K); \tag{5.9}$$

$\bar{\bar{Y}}$ does not involve σ^2, which is usually unknown. $\bar{\bar{Y}}$ is the average of all $n = n_1 + \cdots + n_K$ observations. $\bar{\bar{Y}}$ is an unbiased estimator for μ, and

$$\mathrm{Var}(\bar{\bar{Y}}) = \sigma^2/n. \tag{5.10}$$

Confidence limits for a common mean. Suppose that S^2 (5.3) is a pooled estimate of the common σ^2 and has ν degrees of freedom (5.4). Then two-sided $100\gamma\%$ confidence limits for μ are

$$\underline{\mu} = \bar{\bar{Y}} - t[(1+\gamma)/2; \nu](S/\sqrt{n}), \qquad \tilde{\mu} = \bar{\bar{Y}} + t[(1+\gamma)/2; \nu](S/\sqrt{n}), \tag{5.11}$$

where $t[(1+\gamma)/2; \nu]$ is the $[100(1+\gamma)/2]$th t percentile with ν degrees of freedom.

Before using $\bar{\bar{Y}}$ or the confidence limits, check that the $\bar{Y}_k$ do not differ significantly. Sections 5.3 and 5.4 present such checks.

Motor insulations example. A main analysis of the data in Table 5.2 is to compare the mean log life of the three insulations at $200°$. The three sample log means at $200°$ are $\bar{Y}_1 = 3.23134$, $\bar{Y}_2 = 3.48258$, and $\bar{Y}_3 = 3.54750$. Suppose that the true log standard deviations of the three insulations are equal. Then, by (5.9), the pooled estimate of the common log mean is

$$\bar{\bar{Y}} = [5(3.23134) + 5(3.48258) + 3(3.54750)]/13 = 3.40093.$$

The pooled estimate of the common log standard deviation is $S = 0.1132$, which has 30 degrees of freedom. Two-sided 95% confidence limits for μ are

$$3.40093 \pm 2.042(0.1132/\sqrt{13}) = 3.401 \pm 0.064.$$

The three means do not pass the test for equality. So the pooled estimate and confidence limits are not appropriate.

5.3. Comparison of Two (Log) Normal Samples

This section treats the comparison of two complete (log) normal samples by means of plots and confidence intervals. Such samples are compared with respect to standard deviations (variances) and means, and the methods extend to percentiles, reliabilities, and other quantities.

GRAPHICAL COMPARISONS

Plots. Samples from two (log) normal distributions can be compared graphically. Estimates of (log) standard deviations, percentiles, reliabilities, etc., are obtained graphically as described in Chapter 3 and compared visually. Section 5.4, on the comparison of K samples, presents such plots. The following example illustrates a comparison of two samples.

Motor insulations example. Table 5.1 shows times to failure of specimens of three types of insulations at three temperatures. The data are plotted on lognormal probability paper in Figure 5.1. A key question is "How do the insulations compare with respect to median life at 200°C?" The graphical estimate of this is 1700 hours for insulation 1 and 3000 hours for insulation 2. These medians differ appreciably compared to the scatter in the data. So the difference appears to be convincing. The medians of insulations 2 and 3 and of 3 and 1 can similarly be compared pairwise. Also, log standard deviations, percentiles, etc., can similarly be compared.

COMPARISON OF TWO (LOG) NORMAL VARIANCES

Confidence interval. For $k = 1, 2$, suppose that the variance S_k^2 of sample k has ν_k degrees of freedom. A two-sided $100\gamma\%$ confidence interval for the ratio $\rho = \sigma_1^2/\sigma_2^2$ has limits

$$\underset{\sim}{\rho} = \left(S_1^2/S_2^2\right)/F\left[(1+\gamma)/2; \nu_1, \nu_2\right], \qquad \tilde{\rho} = \left(S_1^2/S_2^2\right) \cdot F\left[(1+\gamma)/2; \nu_2, \nu_1\right],$$

$$(5.12)$$

where $F[(1+\gamma)/2; a, b]$ is the $[100(1+\gamma)/2]$th F percentile with a degrees of freedom in the numerator and b in the denominator. If these limits do not enclose 1, S_1^2 and S_2^2 differ statistically significantly at the $100(1-\gamma)\%$ significance level, convincing evidence that σ_1 and σ_2 differ. ν_1 and ν_2 are reversed in $\underset{\sim}{\rho}$ and $\tilde{\rho}$.

A one-sided $100\gamma\%$ confidence limit comes from the corresponding two-sided limit above. Then γ replaces $(1+\gamma)/2$.

This confidence interval applies only to variances of (log) normal distributions. It may be quite inaccurate for variances of other distributions. Robust methods for comparing variances of nonnormal distributions appear in the literature, for example, in Scheffé (1959).

Motor insulations example.   Table 5.2 shows log failure times of specimens of three motor insulations at three temperatures. Later analyses assume that the insulations have the same true standard deviation of log life. The following compares σ_1 and σ_3. $S_1 = 0.1408$ and $S_3 = 0.1019$ respectively have 12 and 6 degrees of freedom. Two-sided 95% confidence limits for σ_1^2/σ_3^2 are

$$\underline{\rho} = \left[(0.1408)^2/(0.1019)^2\right]/F(0.975; 12,6) = 0.36,$$

$$\tilde{\rho} = \left[(0.1408)^2/(0.1019)^2\right] \cdot F(0.975; 6,12) = 7.1.$$

These limits enclose 1; so the standard deviations do not differ statistically significantly.

COMPARISON OF TWO (LOG) NORMAL MEANS

The following confidence interval compares two (log) means μ_1 and μ_2. It is exact when the population (log) standard deviations are equal, this assumption can be checked with the confidence interval (5.12). For $k = 1,2$, suppose that sample k has n_k observations, a mean $\bar{Y}_k$, and a variance S_k^2 with ν_k degrees of freedom.

Confidence interval.   A two-sided $100\gamma\%$ confidence interval for $\Delta = \mu_1 - \mu_2$ is

$$\underline{\Delta} = (\bar{Y}_1 - \bar{Y}_2) - t[(1+\gamma)/2; \nu]S[(1/n_1) + (1/n_2)]^{1/2},$$

$$\tilde{\Delta} = (\bar{Y}_1 - \bar{Y}_2) + t[(1+\gamma)/2; \nu]S[(1/n_1) + (1/n_2)]^{1/2},$$

(5.13)

where $t[(1+\gamma)/2; \nu]$ is the $[100(1+\gamma)/2]$th t percentile with $\nu = \nu_1 + \nu_2$ degrees of freedom, and $S^2 = (\nu_1 S_1^2 + \nu_2 S_2^2)/\nu$ is the pooled variance. If these limits do not enclose 0, the means differ statistically significantly at the $[100(1-\gamma)]\%$ level.

A one-sided $100\gamma\%$ confidence limit comes from the corresponding two-sided limit above. Then γ replaces $(1+\gamma)/2$.

(5.13) is often a good approximate interval for a difference of means even if the parent distributions are not (log) normal. The approximation is better if the distributions are "close" to normal, if the sample sizes are large, if $n_1 \simeq n_2$, and if γ is not close to 1.

Sometimes two distributions are normal, but their variances differ. In this case the comparison of two means is called the Behrens–Fisher problem. (5.13) is a good approximation if the population variances are close to equal, if the sample sizes are large, if $n_1 \simeq n_2$, and if γ is not close to 1. Natrella (1963, Sec. 3-3.1.2) gives an improved approximate interval when variances are unequal.

Motor insulations example. For the data in Table 5.2, a key question is, "How do the log means of the three insulations compare at 200°C?" The following compares μ_1 and μ_2. The comparison employs the two variance estimates based on data from all three temperatures. The pooled number of degrees of freedom is $\nu = 12 + 12 = 24$, and the pooled variance is $S^2 = [12(0.1408)^2 + 12(0.0836)^2]/24 = (0.1158)^2$. Two-sided 95% confidence limits for $\mu_1 - \mu_2$ are

$$\underset{\sim}{\Delta} = (3.23134 - 3.48258) - t(0.975; 24)0.1158\left(\tfrac{1}{5} + \tfrac{1}{5}\right)^{1/2} = -0.408,$$

$$\tilde{\Delta} = -0.25124 + (2.042)0.1158(0.4)^{1/2} = -0.102.$$

These limits do not enclose 0; so the two log means differ significantly at the 5% level. Separate two-sided 95% limits for $\mu_2 - \mu_3$ are -0.205 and 0.075; such limits for $\mu_3 - \mu_1$ are 0.118 and 0.514. The intervals for the three differences indicate that μ_1 is significantly below μ_2 and μ_3, which are comparable. Section 5.4 describes simultaneous confidence limits for all three differences, in contrast to the separate limits given here.

Sample size. Cohen (1977) and Mace (1964) explain how to choose the sizes of the two samples when comparing two means.

COMPARISON OF PERCENTILES, RELIABILITIES, AND OTHER QUANTITIES

If (log) normal distributions have the same true (log) standard deviation, then percentiles, reliabilities, and other quantities are equal when the (log) means are equal. Then one need only compare means with the exact methods above.

If the true (log) standard deviations differ, then such quantities cannot be compared with exact methods. Maximum likelihood methods provide approximate comparisons and also apply to censored data; Chapter 12 describes such methods.

5.4. Comparison of K (Log) Normal Samples

This section describes how to compare means and standard deviations of K complete (log) normal samples. Methods include plots, hypothesis tests, and simultaneous confidence intervals. The methods extend to percentiles, reliabilities, and other quantities.

Plots. Samples from K (log) normal distributions can be compared graphically. If the samples are complete or singly censored, plot each on a separate (log) normal probability paper (Chapter 3). Stack the plots and hold them up to the light to compare them and to assess whether the distributions differ. A straight line fitted to each plot aids the comparison. Also, plotted confidence limits for the parameters and percentiles of each distribution help. The plots also provide a check on the validity of (1) the data and (2) the assumed (log) normal distribution. If a sample tail contains data points that do not fall in line with the straight line determined by the bulk of the sample, then these outlier points are suspect and should be considered for removal from the data. Of course, reasons for such data should be sought. Sometimes the plotted points do not roughly follow a straight line but clearly follow a curve. Then the (log) normal distribution does not fit the data well, and the confidence intervals and hypothesis tests are inaccurate. Another distribution or nonparametric methods should be used.

Motor insulations example. Figure 5.1 shows lognormal probability plots of the life data on the three insulations. The key question is, "How do they compare with respect to median life at 200°C?" The graphical median estimates are 1700 hours for insulation 1, 3000 hours for insulation 2, and 3500 hours for insulation 3. Compared to the data scatter, the median 1 is appreciably lower than medians 2 and 3, which are comparable.

Figure 5.2 shows the data (sample medians circled) and estimates of the log means at 200°C. The 95% confidence intervals employ three estimates of the log standard deviation at 200°C: (1) just the 200°C data on an insulation, (2) all data on an insulation, and (3) all data on all insulations. The intervals using (1) are usually widest and involve the fewest assumptions; the intervals using (3) are usually narrowest and involve the most assumptions. The assumptions of the intervals using (2) seem most appropriate, since the data and experience with insulation data suggest (2). Figure 5.2 yields the same conclusions as Figure 5.1. For insulation 3 the interval based on (1) has zero length and is not valid; the three specimen lives are equal owing to the periodic inspection scheme.

A common assumption for an insulation is that the log standard deviation is the same at all test temperatures. Experience with many insulations

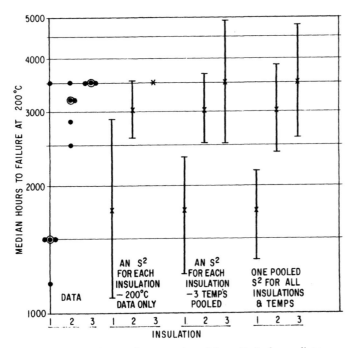

Figure 5.2.   Data, estimates, and confidence limits for medians.

indicates that this is usually so. This means that probability plots for the three temperatures should be parallel. The samples are small; so only extremely different slopes would convince one that the true standard deviations differ. For each insulation, the slopes at the three temperatures are similar. So the data appear consistent with the assumption. The fitted lines are parallel to reflect this.

Figure 5.3 depicts estimates and 95% confidence limits for the pooled log standard deviations of the three insulations. For each pair, the 95% confidence limits for one standard deviation encloses the estimate for the other. This indicates that the standard deviations do not differ significantly.

COMPARISON OF K (LOG) NORMAL VARIANCES

The following describes two methods for comparing (log) variances of K independent (log) normal samples: (1) test of homogeneity and (2) simultaneous comparison of all pairs of variances. Both methods test for equality of all variances. However, they differ in their sensitivity to different departures from equality. For $k = 1, \ldots, K$, suppose that the variance S_k^2 of sample k has ν_k degrees of freedom.

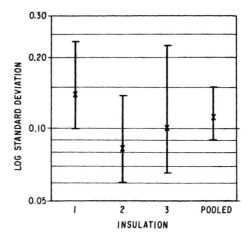

Figure 5.3. Display of estimates and 95% confidence limits for the log standard deviations.

Test of homogeneity. The following tests the homogeneity (equality) hypothesis $\sigma_1 = \sigma_2 = \cdots = \sigma_K$ against the alternative that some $\sigma_k \neq \sigma_{k'}$. The test statistic is

$$Q = C^{-1}\left[\nu\ln(S^2) - \nu_1\ln(S_1^2) - \cdots - \nu_K\ln(S_K^2)\right], \qquad (5.14)$$

where $\qquad \nu = \nu_1 + \cdots + \nu_K, \qquad S^2 = (\nu_1 S_1^2 + \cdots + \nu_K S_K^2)/\nu,$

$$C = 1 + \left[3(K-1)\right]^{-1}\left[(1/\nu_1) + \cdots + (1/\nu_K) - (1/\nu)\right]. \quad (5.15)$$

If the equality hypothesis is true, the distribution of Q is approximately chi square with $K-1$ degrees of freedom. If the alternative is true, Q tends to have larger values. So the test is

1. If $Q \leqslant \chi^2(1-\alpha, K-1)$, accept the equality hypothesis.
2. If $Q > \chi^2(1-\alpha, K-1)$, reject the equality hypothesis at the $100\alpha\%$ significance level.

Here $\chi^2(1-\alpha, K-1)$ is the $100(1-\alpha)$th percentile of the chi-square distribution with $K-1$ degrees of freedom. The chi-square approximation is more precise the larger the ν_k are; it usually suffices if all $\nu_k \geqslant 10$. This test is **Bartlett's test**.

If Q is statistically significant, examine the S_k to see how they differ; a figure like Figure 5.3 helps.

Bartlett's test applies only to variances of (log) normal distributions. It should not be used to compare variances of other distributions, since it can then be quite misleading. Before using the test, it is important to confirm normality with a probability plot of the data. Robust methods for comparing K variances of nonnormal distributions appear in the literature, for example, in Scheffé (1959).

Motor insulations example. Table 5.2 shows log times to failure of insulation specimens. The pooled sample standard deviations of the three insulations are $S_1 = 0.1408$, $S_2 = 0.0836$, and $S_3 = 0.1019$, which respectively have $\nu_1 = \nu_2 = 12$ and $\nu_3 = 6$ degrees of freedom. The calculation of Q is

$$\nu = 12 + 12 + 6 = 30, \qquad C = 1 + \left[3(3-1)\right]^{-1}\left(\tfrac{1}{12} + \tfrac{1}{12} + \tfrac{1}{6} - \tfrac{1}{30}\right) = 1.05,$$

$$S^2 = \left[12(0.1408)^2 + 12(0.0836)^2 + 6(0.1019)^2\right]/30 = (0.1132)^2,$$

$$Q = (1.05)^{-1}\left[30\ln(0.1132^2) - 12\ln(0.1408^2)\right.$$

$$\left. - 12\ln(0.0836^2) - 6\ln(0.1019^2)\right] = 3.13.$$

Since $Q = 3.13 < 4.60 = \chi^2(0.90, 2)$, the three σ_k do not differ significantly at the 10% level. The data are consistent with the hypothesis that the three insulations have equal true (log) standard deviations.

All pairwise comparisons. When there are K variances S_k^2, there are $K(K-1)/2$ pairs of variances to compare. The previous **separate** intervals (5.12) for each pair of variances each separately enclose the corresponding true ratio of two variances with probability γ. The probability that **all** such intervals enclose their true ratios is less then γ. **Simultaneous** $100\gamma\%$ confidence intervals enclose all ratios $\sigma_k^2/\sigma_{k'}^2$ with probability γ. If such limits for a ratio do not enclose 1, there is a "wholly significant" difference between the two variances. The simultaneous limits are wider than (5.12) for a single ratio. So a significant simultaneous interval is more convincing than a significant interval for a single ratio. If all intervals enclose 1, then there are no wholly significant differences among the variances at the $100(1-\gamma)\%$ significance level. Approximate and exact simultaneous confidence limits are presented below. These limits apply only to (log) normal variances. Miller (1966) discusses such intervals in detail.

Figure 5.4 displays all pairs of the three variances of motor insulation 1 on log-log paper. The variances are marked on both scales as X's. Perpendicular lines through the X's intersect at the dots. The distance of a dot from the 45° line is the log of the ratio of the corresponding two variances. So the dot farthest from the 45° line corresponds to the largest ratio.

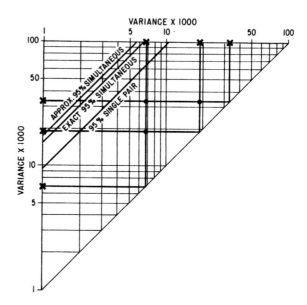

Figure 5.4.   Simultaneous comparison of three variances for insulation 1.

Approximate simultaneous limits.   Simultaneous two-sided approximate $100\gamma\%$ confidence limits for all ratios of pairs of variances $(k, k'=1,\ldots, K)$ are

$$\left(S_k^2/S_{k'}^2\right)/F(\gamma'; \nu_k, \nu_{k'})\leqslant \sigma_k^2/\sigma_{k'}^2\leqslant \left(S_k^2/S_{k'}^2\right)F(\gamma'; \nu_{k'}, \nu_k), \quad (5.16)$$

where $\gamma'=1-(1-\gamma)K^{-1}(K-1)^{-1}$ and $F(\gamma'; a, b)$ is the $100\gamma'$th percentile of the F-distribution with a degrees of freedom in the numerator and b in the denominator. These limits are the same as (5.12), but γ' in (5.16) is higher. One-sided simultaneous limits are given by (5.16); then $\gamma'=1-2(1-\gamma)K^{-1}(K-1)^{-1}$. These limits all enclose the true ratios with a probability of at least γ. The exact probability is close to γ when K is small and γ is large.

Motor insulations example.   The following compares the variances of insulation 1 at the $K=3$ temperatures in Table 5.2. The approximate 95% confidence limits (3.16) are calculated as follows. All $\nu_k=4$, $\gamma'=1-(1-0.95)3^{-1}(3-1)^{-1}=0.99167$ (two-sided), and $F(0.99167; 4,4)=17.6$. Then, for example, the limits for $\sigma_{200}^2/\sigma_{225}^2$ are $(0.1832^2/0.1384^2)/17.6=0.10$ and $(0.1832^2/0.1384^2)17.6=31$. All $3(3-1)/2=3$ intervals enclose 1. So the three variances do not differ significantly at the 5% level. Figure 5.4 shows this too.

Exact simultaneous limits. If all $\nu_k = \nu$, then simultaneous two-sided exact $100\gamma\%$ confidence limits for all ratios of pairs of variances ($k, k' = 1, \ldots, K$) are

$$\left(S_k^2/S_{k'}^2\right)/F_{\max}(\gamma; \nu, K) \leqslant \sigma_k^2/\sigma_{k'}^2 \leqslant \left(S_k^2/S_{k'}^2\right) \cdot F_{\max}(\gamma; \nu, K), \quad (5.17)$$

where $F_{\max}(\gamma; \nu, K)$ is the 100γth percentile of the maximum F statistic among K chi-square statistics, each with ν degrees of freedom. Pearson and Hartley (1954) and Owen (1962) tabulate these percentiles. If all $K(K-1)/2$ intervals enclose 1, there are no wholly significant differences among the K variances at the $[100(1-\gamma)]\%$ significance level. Equivalently, calculate just the largest ratio. If it is less then $F_{\max}(\gamma; \nu, K)$, then the K variances do not differ significantly at the $100(1-\gamma)\%$ significance level. This test of equality is about as sensitive in detecting inequality as is Bartlett's test of homogeneity. Moreover, the simultaneous limits (5.17) identify which variances differ significantly and are quicker to use. If the ν_k differ slightly, one can use some average ν to get approximate limits.

Motor insulations example. The following compares the variances of insulation 1 at the $K = 3$ temperatures in Table 5.2. The 95% confidence limits (5.17) are calculated as follows. All $\nu_k = 4$, and $F_{\max}(0.95; 4, 3) = 15.5$. Then, for example, the limits for $\sigma_{200}^2/\sigma_{225}^2$ are $(0.1832^2/0.1384^2)/15.5 = 0.11$ and $(0.1832^2/0.1384^2)15.5 = 27$. All three such intervals enclose 1; equivalently, the greatest ratio $0.1832^2/0.0829^2 = 4.88$ is less than 15.5. So there are no wholly significant differences among the three variances at the 5% significance level. Figure 5.4 shows lines for ratios of 15.5 and 17.6; these are the exact and approximate simultaneous 95% limits for all ratios. The other line (a ratio of 9.60) is the upper 95% limit for a single ratio.

COMPARISON OF K (LOG) NORMAL MEANS

The following describes two methods for comparing the means of K (log) normal samples: (1) an analysis of variance test of homogeneity and (2) simultaneous comparison of all pairs of means. Both methods test for equality of all means. However, they differ in their sensitivity to detecting different departures from equality (Scheffé, 1959; and Miller, 1966). The tests assume that the K population variances are equal; this assumption can be assessed with Bartlett's test above. For $k = 1, \ldots, K$, suppose that sample k has a mean $\overline{Y}_k$ of n_k observations and a variance estimate S_k^2 with ν_k degrees of freedom.

Test of homogeneity. The following tests the homogeneity (equality) hypothesis $\mu_1 = \mu_2 = \cdots = \mu_K$ against the alternative some $\mu_k \neq \mu_{k'}$. First calculate the total number of observations $n = n_1 + \cdots + n_K$, the total number of degrees of freedom $\nu = \nu_1 + \cdots + \nu_K$, the pooled estimate of the common variance $S^2 = (\nu_1 S_1^2 + \cdots + \nu_K S_K^2)/\nu$, and the pooled mean $\bar{\bar{Y}} = (n_1 \bar{Y}_1 + \cdots + n_K \bar{Y}_K)/n$. The test statistic is

$$F = \sum_{k=1}^{K} n_k \left(\bar{Y}_k - \bar{\bar{Y}} \right)^2 \Big/ \left[(K-1)S^2 \right] = \left(\sum_{k=1}^{K} n_k \bar{Y}_k^2 - n \bar{\bar{Y}}^2 \right) \Big/ \left[(K-1)S^2 \right].$$

$$(5.18)$$

The two formulas are equivalent. If the equality hypothesis is true, F has an F-distribution with $K-1$ degrees of freedom in the numerator and ν in the denominator. If the alternative is true, F tends to have larger values. So the test is

1. If $F \leqslant F(1-\alpha; K-1, \nu)$, accept the equality hypothesis.
2. If $F > F(1-\alpha; K-1, \nu)$, reject the equality hypothesis at the $100\alpha\%$ significance level.

Here $F(1-\alpha; K-1, \nu)$ is the $[100(1-\alpha)]$th percentile of the F-distribution with $K-1$ degrees of freedom in the numerator and ν in the denominator. This is the F-test for a one-way analysis of variance to compare K means. The F-test is usually a good approximate test for comparing means even when the distributions are not normal and the true variances differ, particularly when the n_k are large and (nearly) equal (Scheffé, 1959).

If F is statistically significant, examine the $\bar{Y}_k$ to see how they differ. A display of the sample means and confidence limits as in Figure 5.2 helps. Also, confidence intervals (5.13) or (5.19) for the differences of pairs of means help one to see how the means differ.

Standard computer routines for one-way analysis of variance calculate the F statistic (5.18) and the pooled variance (5.3) and mean (5.9). Many calculate the sample means and corresponding confidence limits and the pairwise comparisons (5.19) below.

Sample size.   Cohen (1977) and Mace (1964) explain how to choose the sample size to compare a number of means.

Motor insulations example.   The main information sought from the data in Table 5.2 is a comparison of the log means of the $K = 3$ insulations at

200°C. The calculations are

$$n = 5 + 5 + 3 = 13, \qquad v = 12 + 12 + 6 = 30,$$

$$S^2 = \left[12(0.1408)^2 + 12(0.0836)^2 + 6(0.1019)^2\right]/30 = (0.1132)^2,$$

$$\overline{\overline{Y}} = (5 \cdot 3.23134 + 5 \cdot 3.48258 + 3 \cdot 3.54750)/13 = 3.40093,$$

$$F = \left[5(3.23134)^2 + 5(3.48258)^2 + 3(3.54750)^2\right.$$

$$\left. - 13(3.40093)^2\right]/\left[(3-1)(0.1132)^2\right] = 9.43.$$

This F statistic has $K - 1 = 2$ degrees of freedom in the numerator and $v = 30$ in the denominator. Since $F = 9.43 > 8.77 = F(0.999; 2, 30)$, the three means differ very highly significantly (0.1% level). The sample means, their confidence limits, and the confidence limits for their differences show that insulation 1 has a significantly lower mean than do insulations 2 and 3, and insulations 2 and 3 have comparable means. Thus insulation 1 is statistically significantly poorer than the others.

All pairwise comparisons.   When there are K means $\overline{Y}_k$, there are $K(K - 1)/2$ pairs of means to compare. Simultaneous $100\gamma\%$ confidence intervals enclose all differences $\mu_k - \mu_{k'}$ with probability γ. If such limits for a difference do not enclose 0, there is a **wholly significant** difference between the two means. The simultaneous limits are wider than (5.13) for a single difference. So a significant simultaneous interval is more convincing than one for a single difference. If all intervals enclose 0, then there are no wholly significant differences among the means at the $100(1 - \gamma)\%$ significance level. Approximate and exact simultaneous confidence intervals are presented below. Miller (1966) discusses such intervals in detail. These limits are usually good approximations for comparing (log) means of distributions that are not (log) normal. The approximations are better under the same conditions as the above analysis of variance test.

Figure 5.5 displays all pairs of the three motor insulation log means at 200°C. The log means are marked on both scales as X's. Perpendicular lines through the means intersect at the dots. The vertical distance of a dot from the 45° line is the difference between the corresponding two means. So the dot farthest from the 45° line corresponds to the largest difference.

Approximate simultaneous confidence limits.   Simultaneous two-sided approximate $100\gamma\%$ confidence limits for all differences of pairs of means

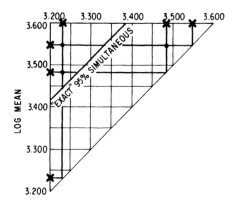

Figure 5.5. Simultaneous comparison of three log means.

$(k, k' = 1, \ldots, K)$ are

$$(\bar{Y}_k - \bar{Y}_{k'}) - t(\gamma'; \nu)S[(1/n_k) + (1/n_{k'})]^{1/2} \tag{5.19}$$

$$\leqslant \mu_k - \mu_{k'} \leqslant (\bar{Y}_k - \bar{Y}_{k'}) + t(\gamma'; \nu)S[(1/n_k) + (1/n_{k'})]^{1/2},$$

where $\gamma' = 1 - (1 - \gamma)K^{-1}(K - 1)^{-1}$, S^2 is the pooled estimate (5.3) of the common σ^2, and ν from (5.4) is its number of degrees of freedom. These are the same as the limits (5.13) for a single difference, but γ' is higher. The pooled estimate S^2 could also come from two samples k and k' or from all K samples. The latter requires the stronger assumption that all K true variances are equal. One-sided simultaneous limits are given by (5.19); then $\gamma' = 1 - 2(1 - \gamma)K^{-1}(K - 1)^{-1}$. If all such simultaneous limits enclose 0, then the means do not differ wholly statistically significantly.

The limits (5.19) enclose true differences with probability γ or higher. The excess probability is small when K is small and γ is large. The $100\gamma'$th percentile may be unusual and difficult to interpolate in a standard table of t percentiles.

Motor insulations example. The means of the $K = 3$ insulations at 200°C are simultaneously compared in pairs as follows. Calculations of simultaneous two-sided approximate 95% confidence limits involve

$$\gamma' = 1 - (1 - 0.95)3^{-1}(3 - 1)^{-1} = 0.99167,$$

$$S^2 = (0.1132)^2 \quad \text{with } \nu = 30 \text{ degrees of freedom,}$$

$$t(0.99167; 30) \cong 2.545 \quad \text{by interpolation.}$$

Then, for example, the limits for $\mu_1 - \mu_2$ are

$$(3.23134 - 3.48258) - 2.545(0.1132)(\tfrac{1}{5} + \tfrac{1}{5})^{1/2} = -0.4334,$$

$$(3.23134 - 3.48258) + 2.545(0.1132)(\tfrac{1}{5} + \tfrac{1}{5})^{1/2} = -0.0690.$$

These limits do not enclose 0; so the difference is wholly significant. The limits for $\mu_2 - \mu_3$ are -0.2753 and 0.1455, and the limits for $\mu_3 - \mu_1$ are 0.1058 and 0.5266. The three intervals indicate that μ_1 is significantly below μ_2 and μ_3, which are comparable. This is stronger evidence than the individual (nonsimultaneous) limits for the three differences.

Exact simultaneous limits. If all $n_k = n_0$, Tukey's exact simultaneous two-sided $100\gamma\%$ confidence limits for all differences of pairs of means $(k, k' = 1, \ldots, K)$ are

$$\left(\overline{Y}_k - \overline{Y}_{k'}\right) - q(\gamma; \nu, K)(S/\sqrt{n_0}) \leqslant \mu_k - \mu_{k'}$$

$$\leqslant \left(\overline{Y}_k - \overline{Y}_{k'}\right) + q(\gamma; \nu, K)(S/\sqrt{n_0}), \quad (5.20)$$

where S^2 is a pooled variance estimate with ν degrees of freedom and $q(\gamma; \nu, K)$ is the 100γth percentile of the Studentized range for K normal deviates and ν degrees of freedom. Pearson and Hartley (1954) and Owen (1962) tabulate $q(\gamma; \nu, K)$. If all $K(K-1)/2$ intervals enclose 0, then there are no (wholly) significant differences among the K means at the $100(1-\gamma)\%$ significance level. Equivalently, calculate the largest difference. If it is less than $q(\gamma; \nu, K)S/\sqrt{n_0}$, then the K means do not differ wholly significantly at the $100(1-\gamma)\%$ level.

This test is about as sensitive to departures from equality as is the above analysis of variance test. Moreover, the simultaneous limits identify which means differ, and they are quicker to use. If the n_k differ slightly, one can use some average n_0 to get approximate limits. Feder (1972) gives a plot on special paper like Figure 5.5; it lets one graphically determine the wholly significant differences.

Motor insulations example. The limits (5.20) are used to compare the 200°C log means of the three insulations in Table 5.2. Such "exact" 95% limits are calculated as follows, with the average sample size $n_0 = (5 + 5 + 3)/3 = 4.33$. This is crude but serves to illustrate the limits. Use the pooled estimate $S^2 = (0.1132)^2$, which has $\nu = 30$ degrees of freedom. Then $q(0.95; 30, 4.33) = 3.93$ by linear interpolation in n_0. Then, for example, the limits for $\mu_1 - \mu_2$ are $(3.23134 - 3.48258) \pm 3.93(0.1132/\sqrt{4.33}) = -0.251 \pm 0.214$, which do not enclose 0. The limits for $\mu_2 - \mu_3$ are 0.065 ± 0.214, which enclose 0. The limits for $\mu_3 - \mu_1$ are 0.316 ± 0.214, which do not enclose 0. These intervals, too, indicate that μ_1 is significantly below μ_2 and μ_3, which are comparable. Figure 5.5 depicts the simultaneous limit for a wholly significant difference.

COMPARISON OF PERCENTILES, RELIABILITIES, AND OTHER QUANTITIES

If distributions have the same true (log) standard deviation, then percentiles, reliabilities, and other such quantities are equal when the (log) means are equal. In that case, one need only compare means with the exact methods above.

If the true (log) standard deviations differ, then such quantities cannot be compared with exact methods, but separate estimates and confidence limits can be compared. Maximum likelihood methods provide approximate comparisons and apply to complete and censored data; Chapter 12 describes such methods.

6. NONPARAMETRIC, WEIBULL, AND OTHER DISTRIBUTIONS

This section briefly surveys and references hypothesis tests for nonparametric, Weibull, and other distributions. Graphical comparisons (probability and hazard plots) apply in the obvious way to complete, singly censored, and multiply censored data from such distributions.

Nonparametric

To use nonparametric methods, one does not assume a parametric form for a distribution. So they also apply to data from parametric distributions. This is useful, because parametric methods have not been developed for some parametric distributions. Nonparametric comparisons are less sensitive than parametric ones. They are commonly used in biomedical applications where, in general, one cannot assume a simple parametric form for a life distribution. Parametric distributions are usually used in engineering applications, because they are often adequate and yield more information from small samples.

Most books on nonparametric methods present methods for complete data only. For example, see Gibbons (1976), Hollander and Wolfe (1973), and Lehmann (1975). Methods include the sign test for paired comparisons, the Wilcoxon test for completely randomized comparisons, and other 2- and K-sample comparisons. Some of these methods extend to singly censored samples, provided that the needed order statistics are observed.

Nonparametric methods for singly and multiply censored data are not as well developed as those for complete data. Gross and Clark (1975), Lawless

(1982), and Kalbfleisch and Prentice (1980) present and reference methods for such data. Recent literature contains such methods; consult the journals and indices listed in Chapter 13.

Weibull and Extreme Value

Comparisons of Weibull and the related extreme value distributions generally employ linear or maximum likelihood estimates. Such comparisons are presented in Chapters 11 and 12.

Government standards based on a Weibull life distribution for reliability demonstration are described by Schilling (1981) and include the following.

TR-3, "Sampling Procedures and Tables for Life and Reliability Testing Based on the Weibull Distribution (Mean Life Criterion)," AD 613 183.

TR-4, "Sampling Procedures and Tables for Life and Reliability Testing Based on the Weibull Distribution (Hazard Rate Criterion)," AD 401 437.

TR-6, "Sampling Procedures and Tables for Life and Reliability Testing Based on the Weibull Distribution (Reliable Life Criterion)," AD 613-184.

These are distributed by National Technical Information Service, 5285 Port Royal Road, Springfield, Va. 22151.

Other Distributions

Much work has been done on methods for comparing complete samples from other distributions. Many references may be found in the journals, books, and indices listed in Chapter 13.

There are few methods for comparing singly and multiply censored samples from other distributions. Methods of Chapters 11 and 12 can be extended to many other distributions. Some references may be found in the journals, books, and indices listed in Chapter 13.

7.   THEORETICAL CONCEPTS FOR HYPOTHESIS TESTS

The following paragraphs briefly present general concepts needed for comparisons with hypothesis tests. These concepts include a probability model (or distribution), sample and parameter spaces, the null hypothesis and its alternative, a test and test statistic, and the operating characteristic (or power) function and size of a test. Many statistics texts provide further explanation. This advanced introduction is brief and abstract; some readers may wish to come back to this after or while reading the chapter. Lehmann (1959) presents advanced theory of hypothesis testing in detail. The Introduction of this chapter is necessary background for this section. This section is helpful background for Chapter 12.

Model. We assume that a sample of n observations $\mathbf{Y} = (Y_1, \ldots, Y_n)$ comes from a parametric cumulative distribution $F(y_1, \ldots, y_n; \boldsymbol{\theta})$ where $\boldsymbol{\theta} = (\theta_1, \ldots, \theta_r)$ consists of r parameters. The set of possible $\mathbf{Y}$ values is called the **sample space** and is denoted by $\mathcal{Y}$; it is a subset of n-dimensional Euclidean space. The set of possible $\boldsymbol{\theta}$ values is called the **parameter space** and is denoted by Ω; it is a subset of r-dimensional Euclidean space. All of the preceding specifies a **general model** for a problem.

For example, suppose that $Y_1, \ldots, Y_n$ are independent observations from a normal distribution with mean μ and standard deviation σ. The sample space is n-dimensional Euclidean space. The parameter space Ω consists of the values (μ, σ) that satisfy $-\infty < \mu < \infty$ and $\sigma > 0$; that is, Ω is the upper half-plane of two-dimensional space.

Hypotheses. A **hypothesis** is a statement that the true model is a special hypothetical case of the general model. The special case corresponds to certain parameter values $\boldsymbol{\theta}$ that are a subset ω of Ω. ω is called the **subspace of the null hypothesis**. The rest of the sample space $\Omega - \omega$ is called the subspace of the **alternative**. The problem is to decide whether $\boldsymbol{\theta}$ is in ω (accept the hypothesis) or in $\Omega - \omega$ (reject the hypothesis and accept the alternative).

For the preceding example, suppose that the hypothesis is that $\mu = \mu_0$, a specified value. The corresponding subspace ω consists of the values (μ_0, σ), where $\sigma > 0$. ω is a line in the half-plane Ω. The alternative is $\mu \neq \mu_0$, the rest of the half-plane.

Test. To decide whether the null hypothetical model holds, we use a statistical test. A **test** is specified by a **critical (rejection) region**, which is a subset $\mathcal{C}$ of the sample space $\mathcal{Y}$. Then, if the observed sample $y = (y_1, \ldots, y_n)$ is in $\mathcal{C}$, we reject the hypothesis; otherwise, if y is in the **acceptance region** $\mathcal{Y} - \mathcal{C}$, we accept the hypothesis. The critical region is usually specified by means of a **test statistic** $T = t(Y_1, \ldots, Y_n)$, which is a numerical function of the observations. Then, for T values in a critical set $\mathcal{T}$, we reject the null hypothesis. This is equivalent to specifying a critical region $\mathcal{C}$ of y values in $\mathcal{Y}$.

For the preceding example, suppose that the test statistic is the absolute t statistic $|T| = |n^{1/2}(\overline{Y} - \mu_0)/S|$, where $\overline{Y}$ is the sample average and S is the sample standard deviation. The critical set $\mathcal{T}$ usually consists of all T values such that $|T| > t'$, where t' is a specified value.

Operating characteristic (OC) function. The **OC function** of a test is the probability $P(\boldsymbol{\theta}) = P_{\boldsymbol{\theta}}\{\text{accept hypothesis}\} = P_{\boldsymbol{\theta}}\{\mathbf{Y} \text{ is in } \mathcal{Y} - \mathcal{C}\}$, which is a function of $\boldsymbol{\theta}$. A good test is one that accepts the hypothesis with high probability when $\boldsymbol{\theta}$ is in ω and accepts the hypothesis with low probability when θ is in $\Omega - \omega$, the alternative. If, for θ in ω, min $P(\theta) = 1 - \alpha$, then α is called the **size** or **level** of the test, and the test is called a "level-α test"; α is the (maximum) probability of rejecting the null hypothesis when it is true. $P_\theta\{$reject hypothesis$\} = 1 - P(\theta)$ as a function of θ is called the **power function** of the test. The OC or power functions of tests with different sample sizes can be examined to help one decide on a suitable sample size.

For the example, use $t' = t(1 - 0.5\alpha, n - 1)$, the $100(1 - 0.5\alpha)$th percentile of the t-distribution with $n - 1$ degrees of freedom. Then the OC function is $P(\mu, \sigma) = P_{\mu\sigma}(|T| > t')$. It has the value $1 - \alpha$ at $\mu = \mu_0$ and decreases as $|\mu - \mu_0|/\sigma$ increases. This and other OC curves are displayed by Natrella (1963), Cohen (1977), Grant and Leavenworth (1980), Owen (1962), and many statistics texts.

Statistical significance. Usually the test statistic T is chosen so that large values indicate that the hypothesis is false. If t'' is an observed value of T, then the value $p = P\{T > t''\}$, assuming that the hypothesis is true, is called the **significance level** of the test statistic. If p is small, say, 0.05 or less, then there is convincing evidence that the hypothesis is false, and the T value is called **statistically significant**. The smaller p is, the more convincing the evidence. In practice, one needs to first determine if there is a real (that is, statistically significant) difference from the hypothesis and then determine whether the difference is large enough to have practical importance. An observed important difference that is not statistically significant should not be regarded as real; more data are needed to determine if the difference is real and not just the result of random sampling variation.

Confidence limits. This chapter also presents individual and simultaneous confidence limits, which are defined in Chapter 6. Such intervals are often more informative than hypothesis tests, since the widths of such intervals indicate the precision of the sample estimates. Thus confidence intervals are generally preferred in place of hypothesis tests.

Tests of fit. Often one must decide if a distribution adequately fits a set of data. Probability and hazard plots are most versatile. Also, formal tests of fit may be useful. Mann, Schafer, and Singpurwalla (1974, Sec. 7.1) survey such tests for the exponential, Weibull (extreme value), and normal (lognormal) distributions and complete or censored data. Hahn and Shapiro (1967, Chap. 8) do so for complete data. Most such tests merely indicate whether or not a proposed distribution adequately fits a set of data; such tests do not suggest alternative distributions. In contrast, Bartholomew (1957) tests for an exponential distribution with a Weibull alternative. Similarly, Farewell and Prentice (1979) propose fitting a three-parameter log-gamma distribution that includes the normal (lognormal) and extreme value (Weibull) distributions as special cases; their method lets one assess, for example, whether a Weibull or lognormal distribution fits a set of data better. Michael and Schucany (1979), Lawless (1982), and Shapiro (1980) survey recent tests.

Of course, for all such tests, the sensitivity of a test to lack of fit depends on the sample size. Small samples seldom indicate lack of fit to most distributions. Thus large sample sizes are generally needed.

Choice of a distribution depends on many factors, and a test of fit is merely a possible aid in making a choice. Theory or experience may indicate a distribution. In some applications, if such a distribution does not adequately fit the data, the data are suspect first, rather than the distribution. Then the data may have outliers (which a plot would show) or other difficulties. Hawkins (1980) describes tests for outliers.

PROBLEMS

10.1. Poisson demonstration test.   For the capacitor example of Section 1, consider a demonstration test with acceptance number $y_0 = 2$.

(a) Find the required total test time to demonstrate a failure rate of 10^{-8} per hour with 60% confidence.

(b) How many capacitors must run 10,000 hours each in such a test?

(c) Calculate and plot the OC function for this test.

(d) What failure rate must the capacitors have to pass the test with 90% probability?

(e) Compare the tests for $y_0 = 0$, 1, and 2 with respect to test costs and reliability that must be designed into the capacitor to pass the test.

10.2. Binomial acceptance sampling.   For binomial acceptance sampling, calculate (using binomial tables) and plot the OC curve for

(a) $n = 20$, $y_0 = 0$.

(b) $n = 20$, $y_0 = 1$.

(c) $n = 20$, $y_0 = 2$.

(d) As a customer, which of these three plans would you prefer to use? Why?

(e) For plan (a), if the first sample unit is defective, is there a need to examine the remaining 19 units to arrive at a decision on the lot? Comment on the cost savings in practice.

10.3. Power line outages.   Use the data on the four longest power lines in Table 1.1

(a) Calculate a pooled estimate $\hat{\lambda}$ of their assumed common true failure rate.

(b) Calculate the test statistic (1.7). How many degrees of freedom does it have? Look up the 95 and 99% points of the corresponding chi-square distribution.

(c) Is the test statistic statistically significant? Which lines contribute significantly to the test statistic?

(d) Is the chi-square approximation adequate? Why?

10.4. Circuit board capacitors.   Use the data of Table 2.1, but omit the data on board 4.

(a) Calculate the pooled estimate $\hat{p}$ of the assumed common value of the binomial proportion.

(b) Calculate two-sided approximate 90% confidence limits for the common proportion.

(c) Calculate the test statistic (2.4). How many degrees of freedom does it have? Look up the 99 and 99.9% points of the corresponding chi-square distribution.

(d) Is the test statistic statistically significant? Which boards contribute significantly to the test statistic?

(e) Is the chi-square approximation adequate? Why?

10.5. Appliance component (renewal data).   The data in the accompanying table show the number of appliances that entered each month of service and the number of failures of a particular component. On failing, a component was replaced (renewed) with a new one, and an appliance can have any number of such failures over the months it is in service. Regard each month of service of an appliance as a binomial trial in which the component either fails or survives with the probability for that month.

(a) For each month, calculate the estimate of the binomial proportion failing (expressed as a percentage). Month 1 has a high percentage, typical of many appliance components.

(b) Test for equality of the binomial proportions of all months. Calculate the pooled estimate of the proportion.

(c) Repeat (b) for months 2 through 12.

(d) Repeat (b) for months 13 through 24.

(e)   Repeat (b) for months 25 through 29.

(f)   Cumulate the percentage failing for each month to estimate the cumulative percentage failing through each month. Such cumulative percentages could exceed 100%, as an appliance can have more than one failure. Plot the sample cumulative percentage failing (called the "renewal function") on log-log paper.

(g)   Estimate the cumulative percentage failing on the 12-month warranty and during the 15-year life of the appliance.

Month	Failed	Number Entering
1	83	22914
2	35	22914
3	23	22914
4	15	22914
5	22	22914
6	16	22914
7	13	22911
8	12	22875
9	15	22851
10	15	22822
11	24	22785
12	12	22745
13	7	2704
14	11	2690
15	15	2673
16	6	2660
17	8	2632
18	9	2610
19	9	2583
20	3	2519
21	6	2425
22	6	2306
23	6	2188
24	5	2050
25	7	862
26	5	845
27	5	817
28	6	718
29	3	590

(h) Criticize the preceding analyses.

(i) Comment on the accuracy of the 15-year estimate in view of the fact that data from months 1 through 12 come from 65% of the population, data from months 13 through 24 come from about 8% of the population that is on service contract (unlimited repair service, owner choses to pay a single premium for), and data from months 25 through 29 come from 2% of the population, whose owners elect a second year of service contract.

(j) Devise a better 15-year estimate of the cumulative percentage failing.

10.6. Insulating fluid breakdown. Use the data in Table 4.1. Combine groups 1 and 2, 3 and 4, and 5 and 6 to get three new groups, each with 20 times to breakdown. Assume that such times have an exponential distribution, and check for stability of the data over time as follows.

(a) Calculate the estimate of the mean for each of the three groups. How many degrees of freedom does each estimate have?

(b) Calculate the pooled estimate of the assumed common mean.

(c) Calculate Bartlett's test statistic (4.6) for the three groups. How many degrees of freedom does it have?

(d) Look up 90 and 95% points of the corresponding chi-square distribution. Do the three sample means differ statistically significantly? If so, why?

(e) Calculate simultaneous approximate 90% confidence limits for all ratios of pairs of the three means. Are any ratios wholly statistically significant?

10.7. Motor insulation 2. Use the log data and summary statistics on motor insulation 2 in Table 5.2. The model for such data assumes that time to breakdown has a lognormal distribution at each temperature, that the (log) standard deviation is the same at each temperature, and that the (log) mean decreases with temperature. The following analyses assess whether the data are consistent with these assumptions.

(a) Plot the data (Table 5.1) from the three temperatures on the same lognormal paper, and compare the distributions.

(b) Calculate a pooled standard deviation from those from the three test temperatures. How many degrees of freedom does it have?

(c) Calculate a separate two-sided 90% confidence interval for each ratio of pairs of (log) standard deviations for the three temperatures. Do any pairs differ statistically significantly?

(d) Calculate Bartlett's test statistic (5.14) for the (log) standard deviations of the three temperatures. How many degrees of freedom does the statistic have? Look up the 90 and 95% points of the corresponding chi-square distribution. Do the (log) standard deviations differ statistically significantly? If so, how?

(e) How good is the chi-square approximation for (d)?

(f) Calculate simultaneous two-sided approximate 90% confidence limits for all ratios of pairs of the three standard deviations. Do any pairs differ wholly statistically significantly?

(g) Would you expect the conclusions from (d) and (f) to usually be the same? Why?

(h) Calculate two-sided 95% confidence limits for the (log) mean at 200°C, using the pooled standard deviation. Calculate the corresponding estimate and confidence limits for the lognormal median at 200°C.

(i) Calculate a separate two-sided 90% confidence interval for the difference of each pair of (log) means for the three temperatures, using the pooled (log) standard deviation. Do any pairs differ statistically significantly?

(j) Calculate the F statistic (5.18) for a one-way analysis of variance. How many degrees of freedom does it have in the numerator and in the denominator? Look up the 90 and 95% points of the corresponding F-distribution. Do the (log) means differ statistically significantly? If so, how?

(k) Calculate simultaneous two-sided approximate 90% confidence limits for all differences of pairs of the three (log) means. Do any pairs differ wholly statistically significantly?

(l) Would you expect the conclusions from (i) and (j) to usually be the same? Why?

(m) In view of the plots from (a), do you think that the analytic comparisons (b) through (k) are necessary? Helpful? What further information do the plots yield that the analytic methods do not?

10.8. Circuit breaker. Use the circuit breaker data from Problem 3.4.

(a) The old design has a known proportion $p_0 = 0.50$ failing by 10,000 cycles. What is the binomial estimate of the proportion of the new design failing by 10,000 cycles, based on the first sample of 18 circuit breakers?

(b)   For a two-sided test for equality of these two proportions, calculate the significance level. Does this proportion for the new design differ statistically significantly from that for the old one?

(c)   Use Fisher's test to compare the two samples of the new breaker with respect to the proportion failed by 15,000 cycles.

(d)   For the sample sizes in (c), how big a difference in the proportions do you think the test will detect? A formal answer would involve calculating the OC curve of the test.

10.9.   Fatigue specimens.   Three labs tested fatigue specimens at two strain ranges. The failure of a specimen occurs either within or outside the length of the strain gage. According to specimen geometry and engineering theory, the expected fraction of failures within the gage length is $p_0 = 4/9$ for any strain range and lab. The fractions y/n of specimens failing within the gage are tabulated below.

Strain	Lab 1	Lab 2	Lab 3
1	218/352	76/173	74/132
2	200/336	40/76	69/107

(a)   Which of the six observed fractions differ significantly from the theoretical $p_0 = 4/9$?

(b)   For Lab 1, do the observed fractions at Strains 1 and 2 differ significantly? For Lab 2? For Lab 3?

(c)   For Strain 1, do the fractions for the three labs differ significantly? Simultaneously compare (1) all three labs and (2) each pair of labs.

(d)   Do (c) for Strain 2.

(e)   Pool data from each lab and compare the three labs as in (c).

(f)   State your overall conclusions.

(g)   Comment on the adequacy for practical purposes of any approximate tests or confidence intervals you used.

10.10.   Metal fatigue.   The following plot shows fatigue life data (log base 10 cycles to failure) on specimens from 27 blends of a metal alloy, tested at the same stress, temperature, etc. The basic question is: are the blends consistent? The following analyses assess this. Visually examine the plot.

```
NO.
IN            3.55        4.05        4.55
ROW   BLEND↓        3.80        4.30        4.80
             <.+....+....+....+....+....+>
  6   13+      2 11 1 1
      +
  6   15+        1  1 2      11
      +
  4   59+         2      11
      +
 14   07+       12211  11 2 1   11
      +
 22   10+       12253212   2  2
      +
 22   11+      1 251131       113 21
      +
  8   12+         11311      1
      +
 34   16+ 1 1335253 111   21112          1
      +
 20   17+       223342 11  1    1
      +
 46   20+ 2 328484211  2 231111
      +
 14   23+        1124  1    3  1    1
      +
 16   24+     13122 2 21       2
      +
 12   25+        1 1221 1      12    1
      +
 10   27+      111132  1
      +
 12   28+         2122 1       12 1
      +
 13   30+        13 41 11 1        1
      +
 10   32+     1 11122    2
      +
 10   33+     1 1511             1
      +
 24   34+     1 332151  121 121
      +
 10   35+      12  2121     1
      +
 24   36+      313422 11112   111
      +
  9   37+         1 11  2       1 1    1 1
      +
 10   38+         2232  1
      +
 10   39+     2 134
      +
  8   40+      111   2 2          1
      +
  8   47+      111111  1           1
      +
  9   50+        112111     2
```

(a) Assess whether the blends are similar or different with respect to typical life. Mark the sample median of each blend as an aid.

(b) Identify any blends with significantly longer or shorter life.

(c) How do the blends compare with respect to scatter in (log) fatigue life? Take into account the differing sample sizes.

(d) How do the blends compare in the lower tail, say, at roughly the 10 or 1% point. Early failures are important in fatigue work, and components are retired before they fail. Design life is cycles to usually 0.1% failure divided by 3. Estimate it.

(e) The blends are plotted in order of manufacture. Are there any time trends in fatigue life?

The following computer output shows a chi-square contingency table analysis to compare the 27 blends with respect to the (binomial) proportion failing by 7000 cycles, denoted by LE7K. Those surviving 7000 cycles are denoted by GT7K.

(f) Are there convincing differences among the blends?

(g) Which blends have a large chi-square contribution and differ from the others? Add the two chi-square contributions (LE7K and GT7K) to get the total chi-square contribution of a blend. How do those blends differ (better or worse)?

(h) If the blends that clearly differ are removed from the data, how do the remaining blends compare (subjective evaluation based on the contingency table analysis)?

The following computer output shows a one-way analysis of variance to compare the mean log life of the blends.

(i) Are there convincing differences among the blends? Are the sample sizes large enough so the F test is accurate enough? Explain.

(j) Examine the blend means. Which blends differ significantly from the others (better or worse)?

(k) Use the analysis of variance table to get an estimate of the standard deviation of the pooled data, ignoring blends. Compare this with the pooled estimate of the standard deviation within blends. Are the two estimates the same for practical purposes? If so, the alloy can be treated as a single homogeneous population, and blends can be ignored.

(l) Explain why Bartlett's test to compare (log) standard deviations of the blends is not suitable.

(m) In nonstatistical terms, write up all your conclusions in a form suitable for a department general manager.

	BLEND	----NUMBER---- OBSERVED	EXPECTED.	CHI-SQ CONTRIB
LE7K	13	3	3.084	0.002
LE7K	15	1	3.084	1.409
LE7K	59	2	2.056	0.002
LE7K	30	8	6.683	0.260
LE7K	10	10	11.309	0.152
LE7K	16	15	17.478	0.351
LE7K	07	6	7.197	0.199
LE7K	25	3	6.169	1.628
LE7K	11	9	11.309	0.472
LE7K	17	13	10.281	0.719
LE7K	20	26	23.647	0.234
LE7K	23	7	7.197	0.005
LE7K	12	5	4.113	0.192
LE7K	36	12	12.338	0.009
LE7K	34	9	12.338	0.903
LE7K	28	5	6.169	0.221
LE7K	24	9	8.225	0.073
LE7K	32	6	5.141	0.144
LE7K	33	9	5.141	2.897
LE7K	35	4	5.141	0.253
LE7K	38	6	5.141	0.144
LE7K	27	7	5.141	0.673
LE7K	37	2	4.627	1.491
LE7K	39	10	5.141	4.593
LE7K	40	3	4.113	0.301
LE7K	47	6	4.113	0.866
LE7K	50	5	4.627	0.030
GT7K	13	3	2.916	0.002
GT7K	15	5	2.916	1.490
GT7K	59	2	1.944	0.002
GT7K	30	5	6.317	0.275
GT7K	10	12	10.691	0.160
GT7K	16	19	16.522	0.372
GT7K	07	8	6.803	0.211
GT7K	25	9	5.831	1.722
GT7K	11	13	10.691	0.499
GT7K	17	7	9.719	0.761
GT7K	20	20	22.353	0.248
GT7K	23	7	6.803	0.006
GT7K	12	3	3.887	0.203
GT7K	36	12	11.662	0.010
GT7K	34	15	11.662	0.955
GT7K	28	7	5.831	0.234
GT7K	24	7	7.775	0.077
GT7K	32	4	4.859	0.152
GT7K	33	1	4.859	3.065
GT7K	35	6	4.859	0.268
GT7K	38	4	4.859	0.152
GT7K	27	3	4.859	0.711
GT7K	37	7	4.373	1.577
GT7K	39	0	4.859	4.859
GT7K	40	5	3.887	0.318
GT7K	47	2	3.887	0.916
GT7K	50	4	4.373	0.032
TOTAL		391	391.000	37.499

THE CHI-SQUARE STATISTIC TO TEST THAT EACH BELOW
HAS THE SAME PROPORTION IN EACH BLEND CATEGORY
HAS THE VALUE 37.499; THIS STATISTIC HAS 26
DEGREES OF FREEDOM.

THE PROBABILITY OF EXCEEDING THIS CHI-SQUARE
VALUE BY CHANCE ALONE IF THERE ARE NO
TRUE EFFECTS IS 6.7 PER CENT

95% CONFIDENCE LIMITS
FOR THE TRUE AVERAGE

BLEND	NO.	AVERAGE	LOWER	UPPER	STD. DEV.
13	5	3.820568	3.607575	4.033562	0.1570272
15	6	4.128080	3.933645	4.322515	0.2391549
59	4	3.989339	3.751205	4.227473	0.1953305
30	13	3.897007	3.764914	4.029100	0.2165405
10	22	3.934206	3.832665	4.035746	0.2069649
16	34	3.962591	3.880912	4.044270	0.3016540
07	14	4.005120	3.877832	4.132407	0.2745229
25	12	4.039750	3.902263	4.177236	0.2872952
11	22	4.026089	3.924548	4.127629	0.3029616
17	20	3.833650	3.727153	3.940147	0.1827026
20	46	3.909818	3.839596	3.980040	0.2625667
23	14	3.969889	3.842602	4.097177	0.2577815
12	8	3.868562	3.700176	4.036948	0.1605434
36	24	3.907170	3.809952	4.004387	0.2439618
34	24	3.977536	3.880319	4.074754	0.2407067
28	12	3.989117	3.851630	4.126603	0.2671224
24	16	3.847650	3.728583	3.966717	0.2357546
32	10	3.829747	3.679138	3.980356	0.1522724
33	10	3.787374	3.636765	3.937983	0.2203455
35	10	3.864816	3.714207	4.015425	0.1792929
38	10	3.840997	3.690388	3.991606	0.8554691E-01
27	10	3.804573	3.653964	3.955182	0.1035423
37	9	4.171735	4.012980	4.330491	0.3720517
39	10	3.708343	3.557734	3.858952	0.6933187E-01
40	8	3.893563	3.725177	4.061948	0.2324871
47	8	3.823750	3.655364	3.992135	0.2645411
50	9	3.880643	3.721887	4.039399	0.1921313
POOLED	390	3.920969	3.896852	3.945085	0.2421878

* ONE-WAY ANALYSIS OF VARIANCE

SOURCE OF VARIATION	DEG. OF FREEDOM	SUM OF SQUARES	MEAN SQUARE	F-TEST STATISTIC
BETWEEN BLENDS	26	2.9520013	0.11353851	1.9357036
ERROR	363	21.291731	0.58654907E-01	
TOTAL	389	24.243733		

11

Comparisons With Linear Estimators (Singly Censored and Complete Data)

This chapter presents linear methods for comparing samples by hypothesis tests. The topics include comparison of (1) a sample with a specified parameter of a distribution, (2) two samples, and (3) K samples. Another topic is pooled estimates for a parameter from a number of samples. The methods employ linear estimators of distribution parameters. So the methods apply **only** to one- or two-parameter distributions with location and scale parameters. Such distributions include the basic ones—the exponential, normal, lognormal, extreme value, and Weibull distributions. The necessary background for this chapter is in Chapter 7 on linear estimation and Chapter 10 for basic concepts on hypothesis testing.

An advantage of linear methods is that the calculations are easy. A disadvantage is that needed tables for computing linear estimates and confidence limits for most distributions are mostly limited to sample sizes of 20 or less. Also, most such tables are limited to complete and singly failure censored samples, but there are some tables for doubly and multiply failure censored samples. Moreover, many tables needed for exact linear confidence limits and hypothesis tests have not been developed. Therefore, approximate limits and tests are given below. In contrast, maximum likelihood methods of Chapter 12 require special computer programs, but more tables for singly failure censored samples are available.

Linear methods are exact for **failure** (Type II) censored data. Such data are less common than time (Type I) censored data. In practice, exact linear methods for failure censored data often provide adequate approximate analyses for time censored data. Maximum likelihood methods of Chapter 12 apply to failure and time censored data.

In this section, μ denotes a location parameter and σ a scale parameter. Suppose that the sample data consist of the first r order statistics $Y_{(1)} \leqslant \cdots \leqslant Y_{(r)}$ in a sample of size n, and suppose that the linear estimates are μ^* and σ^*, calculated as described in Chapter 7. These can be any linear estimators, but they will usually be taken to be the best linear unbiased estimators (BLUE). Then their variances and covariance have the form

$$\text{Var}(\mu^*) = A\sigma^2, \qquad \text{Var}(\sigma^*) = B\sigma^2, \qquad \text{Cov}(\mu^*, \sigma^*) = C\sigma^2,$$

where A, B, and C depend on r and n and are tabulated as described in Chapter 7.

This chapter explains linear methods for comparing exponential, normal, and extreme value distributions. Methods for normal data apply to the logs of lognormal data. Methods for extreme value data apply to the (base e) logs of Weibull data. The relationships between these pairs of distributions are explained in Chapter 2.

Singly censored samples can be subjectively compared with probability plots, described in Sections 4, 5, and 6 of Chapter 10 and in Chapter 3. Nonparametric comparisons of Chapter 10 can also be used for singly censored data, provided that the needed order statistics are observed. A combination of graphical and analytic methods is usually most effective.

Tests of fit based on singly censored samples are surveyed by Mann, Schafer, and Singpurwalla (1974), Lawless (1982, Chap. 9), and Bain (1978). Hawkins (1980) presents tests for outliers.

1. ONE-SAMPLE COMPARISONS

One often wants to assess whether a sample from an assumed distribution is consistent with (or else exceeds) a specified value of a parameter, percentile, reliability, or other quantity. For example, a reliability demonstration test for an exponential distribution assesses whether the mean life of a product exceeds a specified value. This section first describes appropriate hypothesis tests in general terms and then applies them to the basic distributions. Except for the exponential distribution, the operating characteristic curves (power functions) of most tests have not been derived for censored samples. This section summarizes material in Chapter 7.

Scale Parameter

Scale parameter test. Suppose that a linear estimate σ^* of a scale parameter σ is to be compared with a specified value σ_0. σ^* may be from a single sample or it may be a pooled estimate from a number of samples. For example, one might wish to test that a Weibull shape parameter equals 1. If a one- or two-sided $100\gamma\%$ confidence interval for the true σ encloses σ_0, then the sample is consistent with σ_0 at the $100\gamma\%$ confidence level. Also, for a demonstration test, if such a one-sided limit exceeds σ_0, then the value σ_0 (or better) has been demonstrated with $100\gamma\%$ confidence. Such exact and approximate limits based on σ^* appear in Chapter 7.

The variance of a BLUE σ^* is $\mathrm{Var}(\sigma^*) = B\sigma^2$. Linear two-sided approximate $100\gamma\%$ confidence limits are

$$\underline{\sigma} \simeq \sigma^*/\left[1-(B/9)+K_\gamma(B/9)^{1/2}\right]^3,$$

$$\tilde{\sigma} \simeq \sigma^*/\left[1-(B/9)-K_\gamma(B/9)^{1/2}\right]^3;$$

(1.1)

these employ the Wilson–Hilferty chi-square approximation, where K_γ is the $[100(1+\gamma)/2]$th standard normal percentile. To obtain a one-sided limit, replace K_γ by z_γ, the 100γth standard normal percentile, in the corresponding formula. These approximate limits apply to any scale parameter. Mann, Schafer, and Singpurwalla (1974, Ch. 5) justify these limits.

Exponential mean θ. The scale parameter of the exponential distribution is the mean θ. For a failure censored sample, the BLUE for θ is $\theta^* = [Y_{(1)} + \cdots + Y_{(r)} + (n-r)Y_{(r)}]/r$. $2\theta^*/\theta$ has a chi-square distribution with $2r$ degrees of freedom. Two-sided linear exact $100\gamma\%$ confidence limits for θ are

$$\underline{\theta} = 2r\theta^*/\chi^2[(1+\gamma)/2;2r], \qquad \tilde{\theta} = 2r\theta^*/\chi^2[(1-\gamma)/2;2r], \quad (1.2)$$

where $\chi^2(\delta;2r)$ is the 100δth chi-square percentile with $2r$ degrees of freedom. Section 2 of Chapter 7 gives examples of such limits for an exponential mean. Mann, Schafer, and Singpurwalla (1974, Sec. 6.3) extend such limits to the two-parameter exponential distribution. See Chapter 10.

Normal standard deviation σ and lognormal σ. The scale parameter of the (log) normal distribution is the (log) standard deviation σ. Chapter 7 presents exact and approximate confidence limits for σ. Also, (1.1) may be used for approximate confidence limits. Chapter 8 presents such confidence limits based on maximum likelihood estimates.

Extreme value scale parameter δ and Weibull shape parameter β. Chapter 7 presents exact confidence limits for the extreme value scale parameter δ. In addition, (1.1) can be used for approximate confidence limits. Limits for a Weibull shape parameter β come from (1) calculating confidence limits $\underline{\delta}$ and $\tilde{\delta}$ for the corresponding extreme value scale parameter δ from the base e logs of the Weibull data and then (2) calculating $\underline{\beta} = 1/\tilde{\delta}$ and $\tilde{\beta} = 1/\underline{\delta}$.

Insulating fluid example. Section 4 of Chapter 7 presents an example of time to breakdown data of an insulating fluid. Theory for such fluid says time to breakdown has an exponential distribution. The breakdown data at 34 kV yield a transformed BLUE and linear 90% confidence limits for the Weibull shape parameter of $\beta^* = 0.739$, $\underline{\beta} = 0.527$, and $\tilde{\beta} = 0.980$. The interval (barely) does not enclose $\beta = 1$, the value for an exponential distribution. This suggests that the exponential distribution is suspect. A β below 1 may indicate that the data come from a number of exponential distributions, owing to uncontrolled test conditions; Proschan (1963) describes such situations.

Other distribution scale parameters. For linear confidence limits for the scale parameters of other distributions, consult the recent indices by Ross and Tukey (1975) and by Joiner and others (1970, 1975).

Location Parameter

Location parameter test. Suppose that a linear estimate μ^* of a location parameter μ is to be compared with a specified value μ_0. μ^* may be from a single sample or it may be a pooled estimate from a number of samples, and $Var(\mu^*) = A\sigma^2$, where σ is the scale parameter. For example, one might wish to test that a normal mean has a specified value. If a one- or two-sided $100\gamma\%$ confidence interval for the true μ encloses μ_0, then the sample is consistent with μ_0 at the $100\gamma\%$ confidence level; otherwise, there is a statistically significant difference. Also, for a demonstration test, if such a one-sided limit is "better" than μ_0, then μ_0 has been demonstrated with $100\gamma\%$ confidence. Such exact and approximate intervals based on μ^* from singly censored samples appear in Chapter 7. Chapter 8 presents such limits based on maximum likelihood estimates.

If the number r of observed order statistics is large, the t-like statistic $t = (\mu^* - \mu)/(\sigma^* A^{1/2})$ approximately has a standard normal distribution. So linear two-sided approximate $100\gamma\%$ confidence limits for μ are

$$\underline{\mu} \cong \mu^* - K_\gamma A^{1/2}\sigma^*, \qquad \tilde{\mu} \cong \mu^* + K_\gamma A^{1/2}\sigma^*, \tag{1.3}$$

where K_γ is the $[100(1+\gamma)/2]$th standard normal percentile. To obtain a one-sided limit, replace K_γ by z_γ, the 100γth standard normal percentile, in the corresponding formula. These approximate intervals tend to be too narrow. Exact limits employ exact values in place of the $(K_\gamma A^{1/2})$.

Normal mean μ and lognormal μ. Chapter 7 presents exact and approximate confidence limits for μ from μ^* and σ^*. Also, (1.3) provides approximate limits for μ. Confidence limits for the corresponding normal mean are calculated from the base 10 logs of the lognormal data as described in Chapter 7. No OC curves for corresponding hypothesis tests have been developed.

Extreme value location parameter λ and Weibull scale parameter α. Chapter 7 presents exact and approximate confidence limits for λ and α from λ^* and δ^*. Also, (1.3) provides approximate limits for λ (and α). Confidence limits for a Weibull scale parameter α come from (1) calculating confidence limits $\underset{\sim}{\lambda}$ and $\tilde{\lambda}$ for the corresponding extreme value location parameter λ from the base e logs of the Weibull data and then (2) calculating $\underset{\sim}{\alpha} = \exp(\underset{\sim}{\lambda})$ and $\tilde{\alpha} = \exp(\tilde{\lambda})$, as described in Chapter 7.

Other distribution location parameters. For confidence limits for the location parameters of other distributions, consult the recent indices by Ross and Tukey (1975) and by Joiner and others (1970, 1975).

Percentiles

Linear confidence limits for a percentile are described in Chapter 7. Such limits have the same form as the limits for the location parameter, which is also a percentile (50% for the normal and 63.2% for the extreme value). However, such limits require a special table for each percentile. Limits based on maximum likelihood estimates are referenced in Chapter 8.

Reliabilities

Exact linear confidence limits for reliability have been tabulated for the (log) normal distribution by Nelson and Schmee (1979) for $n = 2(1)10$, $r = 2(1)n$. Fertig and Mann (1980) tabulate critical values for hypothesis tests for Weibull reliability for $n = 3(1)18$, $r = 3(1)n$, $1 - R = 1 \times 10^{-p}$ and 5×10^{-p}, $p = 1(1)5$. They also approximate the OC curve. There are confidence limits for these and other distributions in Chapter 7. Confidence limits and hypothesis tests based on maximum likelihood estimates are referenced in Chapters 8 and 12.

Sample Size

Sample size n and the observed number r of failures can be chosen as follows to achieve a desired width for a confidence interval for a population value. Consult a table of factors for the confidence limits, and, for selected n and r, calculate the (two-sided) interval length, which is a multiple of the scale parameter σ. For a one-sided interval, calculate the half-length, that is, the difference between the limit and the true value. Choose n and r that yield the desired length.

2. POOLED ESTIMATE OF A COMMON PARAMETER VALUE

Often one wishes to use a number of samples to estimate a parameter. In particular, the hypothesis tests in following sections employ such pooled estimates. This section describes how to calculate pooled estimates and confidence limits for such a parameter. This briefly repeats material from Chapter 7.

Pooled estimate. Suppose that $\theta_1^*,\ldots,\theta_K^*$ are statistically **independent** BLUEs of θ. Also, suppose that their variances can be expressed in terms of a **common** (unknown) scale parameter σ as $\operatorname{Var}(\theta_1^*)=D_1\sigma^2,\ldots,\operatorname{Var}(\theta_K^*)=D_K\sigma^2$, where the factors D_k are known. Then the pooled BLUE for θ is

$$\theta^* = D\big[(\theta_1^*/D_1)+ \cdots +(\theta_K^*/D_K)\big], \tag{2.1}$$

where

$$D=1/\big[(1/D_1)+ \cdots +(1/D_K)\big] \quad \text{and} \quad \operatorname{Var}(\theta^*)= D\sigma^2. \tag{2.2}$$

Before using such a pooled estimate or confidence limits, one should check that the samples are consistent with the assumption of a common θ value. The hypothesis tests in the following sections provide such checks.

Confidence limits. Confidence limits for such a parameter are calculated from (1.1) and (1.3). Such limits use the D value from (2.2).

Appliance cord example. Table 2.1 shows life data and BLUEs of normal parameters from three flex tests, each with 12 appliance cords. Each test ended before all cords failed. In the first test, one cord ran to failure at 176.8 hours; this cord will be treated as censored when the other two cords were stopped. So all samples are singly censored, with $n=12$ and $r=9$. The purpose of the analyses is to compare the three tests and two types of cord. Probability plots suggest that the normal distribution adequately fits the data. Tests 1 and 2 involve the same type of cord and should have the same true standard deviation (and mean). The calculation of their pooled estimate of a normal standard deviation σ_{12} of life is $B_{12}=1/[(1/0.0723)+ (1/0.0723)]=0.03615$ and $\sigma_{12}^* =0.03615[(35.9/0.0723)+(43.6/0.0723)]= 39.8$ hours. The calculation of their pooled estimate of a normal mean life

Table 2.1. Appliance Cord Life and Linear Estimates

Test k:	1	2	3	BLUE Coeff's	
Cord:	B6	B6	B7	μ^*	σ^*
Hours:	96.9	57.5	72.4	.0360	-.2545
	100.3	77.8	78.6	.0581	-.1487
	100.8	88.0	81.2	.0682	-.1007
	103.3	98.4	94.0	.0759	-.0633
	103.4	102.1	120.1	.0827	-.0308
	105.4	105.3	126.3	.0888	-.0007
	122.6	139.3	127.2	.0948	.0286
	151.3	143.9	128.7	.1006	.0582
	162.4	148.0	141.9	.3950	.5119
	162.7+	161.1+	164.1+	-	-
	163.1+	161.2+	164.1+	-	-
	176.8	161.2+	164.1+	-	-
μ_k^*:	132.9	124.0	122.0	$0.0926 \; \sigma^2 = \text{Var}(\mu^*)$	
σ_k^*:	35.9	43.6	35.7	$0.0723 \; \sigma^2 = \text{Var}(\sigma^*)$	
				$0.0152 \; \sigma^2 = \text{Cov}(\mu^*,\sigma^*)$	

_____

+ denotes a running time.

μ_{12} is $A_{12} = 1/[(1/0.0926)+(1/0.0926)]=0.0463$ and $\mu_{12}^* = 0.0463[(132.9/0.0926)+(124.0/0.0926)]=128.5$ hours. Two-sided approximate 90% confidence limits for σ_{12} from (1.1) are

$$\underline{\sigma}_{12} \cong 39.8/\left[1-(0.03615/9)+1.645(0.03615/9)^{1/2}\right]^3 = 29.2,$$

$$\tilde{\sigma}_{12} \cong 39.8/\left[1-(0.03615/9)-1.645(0.03615/9)^{1/2}\right]^3 = 56.1 \text{ hours.}$$

Two-sided approximate 90% confidence limits for μ_{12} from (1.3) are $\mu_{12} = 128.5 - 1.645(0.0463)^{1/2} = 114.4$ and $\tilde{\mu}_{12} = 128.5 + 14.1 = 142.6$ hours. Cord type B6 was the standard production cord; B7 was proposed as a cost improvement. Their comparison appears later.

3. TWO-SAMPLE COMPARISONS

This section explains how to use linear methods to compare **independent** samples from two populations. Samples may be compared for equality of location and scale parameters, percentiles, reliabilities, and other quantities. This section first presents general methods and then specific results for the basic distributions—exponential, normal, lognormal, extreme value, and Weibull distributions. Except for the exponential distribution, the OC curves (power functions) of most hypothesis tests with singly censored samples have not been derived.

Comparison of Scale Parameters

Confidence limits for their ratio. For $k = 1$, 2, suppose that σ_k^* is the BLUE of the scale parameter σ_k. Two-sided $100\gamma\%$ confidence limits for the ratio $\rho \equiv \sigma_1/\sigma_2$ are

$$\underset{\sim}{\rho} = (\sigma_1^*/\sigma_2^*)/v^*[(1+\gamma)/2], \qquad \tilde{\rho} = (\sigma_1^*/\sigma_2^*)/v^*[(1-\gamma)/2]; \quad (3.1)$$

$v^*(\delta)$ is the 100δth percentile of the standardized ratio

$$v^* = (\sigma_1^*/\sigma_1)/(\sigma_2^*/\sigma_2). \qquad (3.2)$$

If the confidence interval does not enclose 1, then σ_1 and σ_2 differ significantly at the $100(1-\gamma)\%$ level. (3.1) yields a one sided $100\gamma\%$ confidence limit when γ replaces $(1+\gamma)/2$ or $1-\gamma$ replaces $(1-\gamma)/2$. The distribution of v^* depends on the life distribution, the sample sizes, and censoring pattern—but not on the parameter values.

For samples with many failures, an approximation is

$$\underset{\sim}{\rho} \simeq (\sigma_1^*/\sigma_2^*)/F[(1+\gamma)/2; 2/B_2, 2/B_1],$$

$$\tilde{\rho} \simeq (\sigma_1^*/\sigma_2^*) \cdot F[(1+\gamma)/2; 2/B_1, 2/B_2], \qquad (3.3)$$

where $\mathrm{Var}(\sigma_k^*) = B_k \sigma_k^2$ and $F(\delta; a; b)$ is the 100δth F percentile with a degrees of freedom above and b below. $2/B_1$ and $2/B_2$ are usually not integers; note that they are reversed in the lower and upper limits.

Exponential means. Suppose that θ_1^* and θ_2^* are BLUEs for θ_1 and θ_2 and are based on the first r_1 and r_2 order statistics. Two-sided exact $100\gamma\%$ confidence limits for $\rho = \theta_1/\theta_2$ are

$$\underset{\sim}{\rho} = (\theta_1^*/\theta_2^*)/F[(1+\gamma)/2; 2r_2, 2r_1], \qquad \tilde{\rho} = (\theta_1^*/\theta_2^*)F[(1+\gamma)/2; 2r_1, 2r_2], \qquad (3.4)$$

where $F(\delta; a, b)$ is the 100δth F percentile with a degrees of freedom above and b below. Note that the degrees of freedom are reversed in the two limits. To get a one-sided $100\gamma\%$ limit, replace $(1+\gamma)/2$ by γ. See Chapter 10.

Normal standard deviations σ_k and lognormal σ_k.   The scale parameter of the normal distribution is the standard deviation σ. There are no exact tables of v^*. (3.3) may be used for approximate confidence limits. For lognormal σ_k, work with the log data, and use the methods for the normal distribution.

Appliance cord example.   Table 2.1 shows life data from three tests of appliance cords. The first and second tests involve the same type of cord. The test is used to check whether the two samples are consistent with the assumption of a common normal standard deviation. Linear approximate 90% confidence limits (3.4) for $\rho = \sigma_1/\sigma_2$ are

$$\rho \cong (35.9/43.6)/F[(1+0.90)/2;2/0.0723,2/0.0723] = 0.436,$$

$$\tilde{\rho} \cong (35.9/43.6) \cdot F[(1+0.90)/2;2/0.0723,2/0.0723] = 1.56.$$

This interval encloses 1; so the two standard deviations do not differ significantly. From Section 2, the pooled estimate is $\sigma_{12}^* = 39.8$ hours and $B_{12} = 0.03615$.

The same method is used to check whether the two types of cords have the same standard deviation. Approximate 90% confidence limits for $\rho = \sigma_{12}/\sigma_3$ are

$$\underline{\rho} \cong (39.8/43.6)/F[(1+0.90)/2;2/0.0723,2/0.03615] = 0.540,$$

$$\tilde{\rho} \cong (39.8/43.6) \cdot F[(1+0.90)/2;2/0.03615,2/0.0723] = 1.63.$$

This interval encloses 1; so the standard deviations of the two types of cords do not differ significantly.

Extreme value scale parameter and Weibull shape parameter.   There are no tables of the distribution of the ratio δ_1^*/δ_2^* of BLUEs of extreme value scale parameters. So (3.3) must be used for approximate confidence limits. For Weibull shape parameters β_k, use the base e logs of the data, and use the confidence limits (3.3) for the ratio of extreme value scale parameters. Then $\rho = \delta_1/\delta_2 = \beta_2/\beta_1$, where the subscripts are reversed. Chapter 12 describes exact tables for the ratio $\hat{\delta}_1/\hat{\delta}_2$ of maximum likelihood estimators.

Insulating fluid example.   Section 4 of Chapter 7 presents an example of time to breakdown data of an insulating fluid. Theory for such fluid says that the Weibull shape parameter has the same value at all test voltages. For 34 kV, the BLUE for the corresponding extreme value scale parameter is $\delta_{34}^* = 1.353$ calculated from the log data. For 36 kV, $\delta_{36}^* = 1.154$. Also, $\text{Var}(\delta_{34}^*) = 0.03502 \, \delta_{34}^2$ and $\text{Var}(\delta_{36}^*) = 0.04534 \, \delta_{36}^2$. Linear approximate two-sided 90% confidence limits for $\rho = \delta_{34}/\delta_{36}$ are

$$\underset{\sim}{\rho} \cong (1.353/1.154)/F[(1+0.90)/2; 2/0.04534, 2/0.03502] = 0.733,$$

$$\tilde{\rho} \cong (1.353/1.154) \cdot F[(1+0.90)/2; 2/0.03502, 2/0.04534] = 1.91.$$

These are the limits for the inverted ratio β_{36}/β_{34} of Weibull shape parameters. These limits enclose 1, so β_{34} and β_{36} do not differ significantly, consistent with theory.

Comparison of Location Parameters and Percentiles

Confidence limits for their difference. For $k = 1, 2$, suppose that the BLUE of a percentile y_k is y_k^*, and the BLUE of the scale parameter σ_k is σ_k^*. The location parameter is a particular percentile of a distribution. Suppose that σ^* is the pooled BLUE (Section 2). Assuming that $\sigma_1 = \sigma_2$, two-sided $100\gamma\%$ confidence limits for $\Delta = y_1 - y_2$ are

$$\underset{\sim}{\Delta} = (y_1^* - y_2^*) - d^*[(1+\gamma)/2]\sigma^*, \qquad \tilde{\Delta} = (y_1^* - y_2^*) - d^*[(1-\gamma)/2]\sigma^*;$$

$$(3.5)$$

$d^*(\varepsilon)$ is the 100εth percentile of the standardized difference

$$d^* = (y_1^* - y_2^*)/\sigma^*. \tag{3.6}$$

See (5.13) of Chapter 10. If the confidence interval does not enclose 0, then y_1 and y_2 differ significantly at the $100(1-\gamma)\%$ level. (3.5) yields a one-sided $100\gamma\%$ confidence limit when γ replaces $(1+\gamma)/2$ or $1-\gamma$ replaces $(1-\gamma)/2$. Generally, $d^*(\varepsilon) \neq -d^*(1-\varepsilon)$.

If there are no tables of d^*, then, for samples with many failures,

$$d^*(\varepsilon) \simeq z_\varepsilon (D_1 + D_2)^{1/2}, \tag{3.7}$$

where $\text{Var}(y_k^*) = D_k \sigma^2$ and z_ε is the 100εth standard normal percentile. D_k depends on the sample size, censoring, and the percentile. These approximate d^* percentiles tend to yield confidence intervals that are too short.

Normal means μ_k and lognormal μ_k. There are no exact tables of the statistic $d^* = (\mu_1^* - \mu_2^*)/\sigma^*$ of BLUEs of normal parameters. (3.5) and (3.7) may be used for approximate confidence limits for $\mu_1 - \mu_2$. For lognormal data, work with the base (10) logs of the data.

Appliance cord example. Table 2.1 shows life data from three tests of appliance cords. The preceding test is used to compare samples 1 and 2 (cord B6) with respect to means, which should be equal. Calculated above, the pooled estimate is $\sigma_{12}^* = 39.8$ hours. Approximate 90% confidence limits are $\Delta_{12} = (132.9 - 124.0) - 1.645(0.0463 + 0.0463)^{1/2}39.8 = -11.0$ and $\tilde{\Delta}_{12} = 8.9 - (-1.645)0.3043(39.8) = 28.8$ hours. These limits enclose zero; so the two means do not differ significantly.

The same method is used to check whether cords B6 and B7 have the same mean. The calculation of the pooled estimate of the standard deviation is $B = 1/[(1/0.03615) + (1/0.0723)] = 0.0482$ and $\sigma^* = 0.0482[(39.8/0.03615) + (35.7/0.0723)] = 38.4$ hours. Approximate 90% confidence limits are $\Delta = (128.5 - 122.0) - 1.645(0.0463 + 0.0926)^{1/2}38.4 = -15.0$ and $\tilde{\Delta} = 6.5 + 21.5 = 28.0$ hours. These limits enclose zero; so the means of the two cords do not differ significantly.

Extreme value location parameters λ_k and Weibull scale parameters α_k. There are no tables of the t-like statistic $d^* = (\lambda_1^* - \lambda_2^*)/\delta^*$ of BLUEs of extreme value parameters. So (3.5) and (3.7) must be used for approximate limits. Work with the (base e) logs of Weibull data; then $\lambda_k = \ln(\alpha_k)$. Chapter 12 describes exact tables for the t-like statistic $\hat{d} = (\hat{\lambda}_1 - \hat{\lambda}_2)/\hat{\delta}$ of maximum likelihood estimators.

Insulating fluid example. Section 4 of Chapter 7 presents an example of time to breakdown data of an insulating fluid. Theory for such fluid says that the Weibull scale parameter decreases with increasing test voltage. The preceding test is used to compare the samples at 34 and 36 kV with respect to Weibull scale parameters. The BLUEs are $\lambda_{34}^* = 2.531$ and $\lambda_{36}^* = 1.473$, where $\mathrm{Var}(\lambda_{34}^*) = 0.05890\delta^2$ and $\mathrm{Var}(\lambda_{36}^*) = 0.07481\delta^2$, and the pooled scale parameter estimate for all voltages is $\delta^* = 1.348$. Approximate 90% confidence limits are $\Delta = (2.531 - 1.473) - 1.645(0.05890 + 0.07481)^{1/2}1.348 = 0.247$ and $\tilde{\Delta} = 1.058 + 0.811 = 1.869$. These limits do not enclose 0. So the two Weibull scale parameters differ significantly at the 10% level, consistent with theory.

Comparison of Reliabilities

Under the assumption of unequal scale parameters, there are no tables for linear comparisons of two samples for equality of reliabilities. The usual procedure is to assume equality of the scale parameters and to test for equality of the location parameters. If they do not differ significantly, the reliabilities are regarded equal. This method is not the same as a direct comparison of reliabilities. Chapter 12 references some tables for such direct comparisons based on maximum likelihood estimates.

Sample Size

The sizes n_k and the numbers r_k of failures of the two samples can be chosen to achieve a desired width for a confidence interval for the difference (or ratio) of two population values. Consult a table of factors for the confidence limits, and, for selected n_k and r_k, calculate the (two-sided) interval length, which may be a multiple of the scale parameter. For a one-sided interval, calculate the half-length, that is, the difference between the limit and the true difference. Choose n_k and r_k that yield the desired length. Approximate formulas for confidence limits may be used.

4. *K*-SAMPLE COMPARISONS

This section shows how to compare K samples with respect to population location and scale parameters. The linear methods include (1) tests for equality of parameters by "analysis of variance" and (2) simultaneous confidence limits to compare all pairs of parameters. Suppose that $\theta_1^*, \ldots, \theta_K^*$ are statistically independent BLUEs for $\theta_1, \ldots, \theta_K$, which may be percentiles, location parameters, or scale parameters. $\mathrm{Var}(\theta_k^*) = D_k \sigma^2$ for $k = 1, \ldots, K$, where the D_k are known and σ is the unknown common scale parameter.

Test of Homogeneity

Hypothesis test. The following tests the homogeneity (equality) hypothesis that $\theta_1 = \cdots = \theta_K$ against the alternative that some $\theta_k \neq \theta_{k'}$. As in analysis of variance, the quadratic test statistic is

$$Q = \left(\sum_{k=1}^{K} (\theta_k^* - \theta^*)^2 / D_k \right) / \sigma^{*2}, \tag{4.1}$$

where θ^* and σ^* are the pooled estimates (2.1) for the common θ and σ. If the equality hypothesis is true, the distribution of Q is approximately chi square with $K - 1$ degrees of freedom. If the alternative is true, Q tends to have larger values. So the test is

1. If $Q \leqslant \chi^2(1 - \alpha; K - 1)$, accept the equality hypothesis.
2. If $Q > \chi^2(1 - \alpha; K - 1)$, reject the equality hypothesis

at the $100\alpha\%$ significance level. Here $\chi^2(1 - \alpha; K - 1)$ is the $100(1 - \alpha)$th chi-square percentile with $K-1$ degrees of freedom. The chi-square approximation is more precise the larger the observed number of failures in each sample.

If Q is statistically significant, examine the individual θ_k^* to see how they differ. Individual confidence limits for the θ_k aid the examination, particularly if plotted side by side.

The above test applies to location and scale parameters and to percentiles. But to compare scale parameters, it is better to use $\theta_k^* = \ln(\sigma_k^*)$ in (4.1). Then eliminate σ^{*2} from Q. The resulting Q is used the same way.

Exponential means.   There are no tables of the Q statistic (4.1) based on BLUEs of exponential means. So the above approximate test must be used, and one can work with $\ln(\theta_k^*)$, since the exponential mean is a scale parameter. Also, Bartlett's test (4.6) in Chapter 10 can be used; then one uses the number r_k of failures in place of the sample size n_k.

Normal and lognormal parameters.   There are no tables of the Q statistic (4.1) based on BLUEs of normal means or standard deviations. So the chi-square approximation must be used. For lognormal data, work with the (base 10) logs of data.

Appliance cord example.   Table 2.1 shows life data from three tests of appliance cords. The preceding test is used to compare the standard deviations of the three tests. The log estimates are $\theta_1^* = \ln(35.9) = 3.581$, $\theta_2^* = \ln(43.6) = 3.775$, and $\theta_3^* = \ln(35.7) = 3.575$, and their variance factors are $B_1 = B_2 = B_3 = 0.0723$. The calculation of the pooled estimate (2.1) is $B = 1/[(1/0.0723) + (1/0.0723) + (1/0.0723)] = 0.0241$ and $\theta^* = 0.0241[(3.581/0.0723)+(3.775/0.0723)+(3.575/0.0723)] = 3.6437$. Then $Q = [(3.581 - 3.6437)^2/0.0723] + [(3.775 - 3.6437)^2/0.0723] + [(3.575 - 3.6437)^2 /0.0723] = 0.358$. This is less than the corresponding 90th chi-square percentile $\chi^2(0.90; 3-1) = 4.605$. So the three standard deviations do not differ significantly, and it is reasonable to pool the three estimates.

The preceding test is also used to compare the means of the three tests. The BLUEs are $\mu_1^* = 132.9$, $\mu_2^* = 124.0$, and $\mu_3^* = 122.0$ and their variance factors are $A_1 = A_2 = A_3 = 0.0926$. The calculation of the pooled estimate is $A = 1/[(1/0.0926)+(1/0.0926)+(1/0.0926)] = 0.03087$ and $\mu^* = 0.03087$ $[(132.9/0.0926)+(124.0/0.0926)+(122.0/0.0926)] = 126.3$ hours. Also, $\sigma^* = 38.4$ hours. Then $Q = \{[(132.9 - 126.3)^2/0.0926] + [(124.0 - 126.3)^2/0.0926]+[(122.0-126.3)^2/0.0926]\}/(38.4)^2 = 0.493$. This is less than the corresponding 90th chi-square percentile $\chi^2(0.90; 3-1) = 4.605$. So the three means do not differ significantly, and the proposed cord appears comparable to the standard one.

Overall, the three tests are comparable. A more correct combined test for equality of the three distributions would simultaneously test for equality of the means and of the standard deviations. Such a Q statistic would take into account the correlation between μ_k^* and σ_k^*. Chapter 12 gives such a test based on maximum likelihood.

Weibull and extreme value parameters. There are no tables of the Q statistic (4.1) based on BLUEs of extreme value location or scale parameters. So the above approximate test must be used. For Weibull data, work with the (base e) logs of the data and the corresponding extreme value distributions.

Insulating fluid example. Section 4 of Chapter 7 presents time to breakdown data on an insulating fluid at different voltages. Theory for such fluid says that the Weibull shape parameter of the distribution of time to breakdown has the same value for the $K = 7$ voltages. For the log shape parameter estimates, (4.1) yields $Q = 8.02$ with $7 - 1 = 6$ degrees of freedom. Q is less than $\chi^2(0.95; 6) = 12.6$. So the shape parameters do not differ significantly, consistent with theory.

Simultaneous Pairwise Comparisons

All pairwise limits. For K parameters $\theta_1, \ldots, \theta_K$, there are $K(K-1)/2$ pairs of differences $\theta_k - \theta_{k'}$ or ratios $\theta_k / \theta_{k'}$. Simultaneous $100\gamma\%$ confidence limits enclose **all** true differences (or ratios) with probability γ. If such limits for a difference (ratio) do not enclose zero (one), there is a wholly significant difference between those two parameters. For a given γ, the simultaneous limits are wider than the limits for a single difference or ratio. The following gives approximate simultaneous limits based on BLUEs and singly censored samples.

Tables for exact simultaneous confidence limits from BLUEs exist only for exponential means. Section 4 of Chapter 10 describes these tables; they apply only to samples with the same number r of observed order statistics.

Simultaneous limits for scale parameters. Suppose that $\sigma_1^*, \ldots, \sigma_K^*$ are **independent** BLUEs for scale parameters $\sigma_1, \ldots, \sigma_K$, and $\mathrm{Var}(\sigma_1^*) = B_1 \sigma_1^2, \ldots, \mathrm{Var}(\sigma_K^*) = B_K \sigma_K^2$. Simultaneous linear two-sided approximate $100\gamma\%$ confidence limits for all ratios of pairs of scale parameters (k, $k' = 1, \ldots, K$) are

$$(\sigma_k^* / \sigma_{k'}^*) / F(\gamma'; 2/B_{k'}, 2/B_k) \lesssim \sigma_k / \sigma_{k'} \lesssim (\sigma_k^* / \sigma_{k'}^*) \cdot F(\gamma'; 2/B_k, 2/B_{k'}),$$

$$(4.2)$$

where $\gamma' = 1 - (1 - \gamma)K^{-1}(K - 1)^{-1}$ and $F(\gamma'; a, b)$ is the $100\gamma'$th F percentile with a degrees of freedom above and b below. Note that B_1 and B_2 are reversed in the lower and upper limits. These limits are the individual limits (3.1) evaluated for γ'. (4.2) gives one-sided simultaneous limits if $\gamma' = 1 - 2(1 - \gamma)K^{-1}(K - 1)^{-1}$. If the limits (3.1) are exact, rather than approximate, the simultaneous limits enclose all true ratios with a probability of at least γ. The probability is close to γ when K is small and γ is near 1. If such an interval does not enclose 1, those two scale parameters have a wholly significant difference.

Section 4 of Chapter 10 references tables of exact factors for such exponential means when all samples have the same number r of failures. Then r replaces the sample size n in (4.8) of Chapter 10. McCool (1975b) tabulates exact factors for such limits for Weibull shape parameters; they apply to ML estimates from samples with the same size and the same single censoring as described in Chapter 12.

Appliance cord example. Table 2.1 shows life data from three tests of appliance cords. The simultaneous intervals above are used to compare the three tests for equality of the normal standard deviations of life. For example, for linear two-sided approximate 90% simultaneous limits the calculation of the lower limit for σ_1/σ_2 is $\gamma' = 1 - (1 - 0.90)3^{-1}(3 - 1)^{-1} = 0.9833$ and $(35.9/43.6)/F(0.9833; 2/0.0723, 2/0.0723) = 1.93$. The limits are in Table 4.1. All intervals enclose 1. So the three standard deviations do not differ wholly significantly.

Simultaneous limits for location parameters and percentiles. Suppose that y_k^* is the BLUE for a percentile y_k for $k = 1, \ldots, K$. Also, suppose σ^* is the pooled BLUE for a **common** scale parameter and is based on samples k and k' or on all K **independent** samples. Simultaneous linear two-sided $100\gamma\%$ confidence limits for all differences $\Delta_{kk'} = y_k - y_{k'}$ are

$$\underset{\sim}{\Delta}_{kk'} = (y_k^* - y_{k'}^*) - q^*(\gamma)\sigma^*, \qquad \tilde{\Delta}_{kk'} = (y_k^* - y_{k'}^*) + q^*(\gamma)\sigma^*; \quad (4.3)$$

Table 4.1. Ratios of Standard Deviations and Simultaneous Limits for Appliance Cords

Ratio	Estimate	90% Limits	
		Lower	Upper
σ_1/σ_2	0.913	0.390	1.93
σ_2/σ_3	1.22	0.523	2.86
σ_3/σ_1	0.994	0.425	2.33

Table 4.2. Differences of Means and Simultaneous Limits for Appliance Cords

| | | 90% Limits | |
Difference	Estimate	Lower	Upper
$\mu_1-\mu_2$	8.9	-26.4	44.2
$\mu_2-\mu_3$	2.0	-33.3	37.3
$\mu_3-\mu_1$	-10.9	-46.2	24.4

$q^*(\gamma)$ is the 100γ th percentile of the maximum absolute standardized difference

$$q^* = \max_{k,k'} (|y_k^* - y_{k'}^*|/\sigma^*). \qquad (4.4)$$

If such an interval does not enclose 0, those two percentiles have a wholly significant difference.

There are no tables of q^*. Simultaneous two-sided approximate limits are given by (3.5), but $(1+\gamma)/2$ is replaced by $\gamma'=1-(1-\gamma)K^{-1}(K-1)^{-1}$. Such limits with exact $d^*(\gamma')$ enclose all true differences with a probability of at least γ. The probability is close to γ when K is small and γ is near 1. Chapter 12 references such exact tables for simultaneous limits based on maximum likelihood estimates.

Appliance cord example. Table 2.1 shows life data from three tests of appliance cords. The simultaneous intervals above are used to compare the three tests for equality of normal mean life. $\sigma^*=38.4$ hours is the overall pooled estimate. (4.3) provides simultaneous linear two-sided approximate 90% confidence limits. For example, for $\mu_1-\mu_2$, $\gamma'=1-(1-0.90)3^{-1}$ $(3-1)^{-1}=0.9833$, and $\Delta_{12}=(132.9-124.0)-z_{.9833}(0.0926+0.0926)^{1/2}38.4$ $=-26.4$ hours. Table 4.2 shows all limits. All intervals enclose 0. So the three means do not differ wholly significantly. Exact intervals would be wider than these approximate ones. Thus the data are consistent with the assumptions that the tests and the cords are comparable.

PROBLEMS

11.1. Insulating fluid. The table below shows estimates of the parameters of the extreme value distributions of ln time to breakdown of an insulating fluid at seven voltage stresses. The following analyses assess the equality of the seven extreme value scale parameters (Weibull shape), which are equal according to engineering theory.

(a) Calculate the pooled estimate of the extreme value scale parameter.

(b) Calculate the quadratic test statistic (4.1), using the scale parameter estimates. How many degrees of freedom does it have? Look up the 90 and 95% points of the corresponding chi-square distribution. Do the seven estimates differ statistically significantly? If so, how?

(c) How good is the chi-square approximation?

(d) Calculate the quadratic test statistic (4.1), using the ln estimates of the scale parameters and omitting the pooled scale parameter estimate from the denominator of (4.1). How many degrees of freedom does this statistic have? Look up the 90 and 95% points of the corresponding chi-square distribution. Do the seven estimates differ statistically significantly? If so, how?

(e) How good is this chi-square approximation?

(f) Would you expect the two test statistics to be roughly equal or not? Why?

k	1	2	3	4	5	6	7
n_k	3	5	11	15	19	15	8
λ_k^*	7.125	5.957	4.373	3.310	2.531	1.473	0.0542
δ_k^*	2.345	1.224	0.987	1.898	1.353	1.154	0.836
A_k	0.4029	0.2314	0.1025	0.0748	0.0589	0.0748	0.1420
B_k	0.3447	0.1667	0.0642	0.0453	0.0350	0.0453	0.0929
C_k	−0.0248	−0.0340	−0.0203	−0.0156	−0.0126	−0.0156	−0.0261

11.2. Appliance cord. Use the Weibull distribution for the following analyses of the appliance cord data in Table 2.1.

(a) Make separate Weibull plots of the data from the three tests and compare them visually for equality of the distributions. Are the distributions the same? Does the Weibull distribution fit adequately?

(b) Pool the data from the two tests on cord type B6 and make a Weibull plot. How does this plot compare with that for B7 cord?

(c) For each test, calculate the BLUEs for the corresponding extreme value parameters and give the variance and covariance factors.

(d) Calculate two-sided approximate 95% confidence limits for the ratio of the Weibull shape parameters for tests 1 and 2. Are the data consistent with the assumption that the two true shape parameters are equal?

(e) Calculate a pooled estimate of the common extreme value scale parameter for cord type B6, and calculate its variance factor.

(f) Calculate two-sided approximate 95% confidence limits for the ratio of the shape parameters for cords B6 and B7. Are the data consistent with the assumption that the true shape parameters of the two types of cords are equal?

(g) Calculate two-sided approximate 95% confidence limits for the difference between the extreme value location parameters for tests 1 and 2, using the pooled estimate of the extreme value scale parameter from (e). Give the interval for the corresponding Weibull scale parameters. Are the data consistent with the assumption that tests 1 and 2 have equal Weibull scale parameters?

(h) Repeat (g) for cords B6 and B7, using the pooled estimate of the common extreme value scale parameter for all three tests.

(i) Perform the test of homogeneity for the three extreme value scale parameters and state conclusions.

(j) Do (i) for the three extreme value location parameters.

(k) Calculate simultaneous (approximate) 90% confidence limits for all ratios of pairs of the three Weibull shape parameters, and state conclusions.

(l) Do (k) for all ratios of pairs of the three Weibull scale parameters.

(m) Explain why the analyses using the normal and Weibull distributions yield the same conclusions.

12

Maximum Likelihood Comparisons (Multiply Censored and Other Data)

This **advanced** chapter presents maximum likelihood (ML) methods for comparing samples. The topics include comparison of (1) a sample with a given distribution, (2) two samples, and (3) K samples. Other topics are pooled estimates for a parameter from a number of samples and advanced methods and theory for likelihood ratio tests. Needed background is in Chapter 10 on comparisons and in Chapter 8 on ML estimation. One can use the methods in this section without fully understanding their theoretical basis if one has a special computer program for the laborious calculations. Such programs include CENSOR of Meeker and Duke (1979), STATPAC of Nelson and Hendrickson (1972) and Nelson and others (1978), MLP of Ross and others (1976), CENS of Hahn and Miller (1968), and SURVREG of Preston and Clarkson (1980). Readers may wish to develop their own programs.

ML methods are **versatile**; they apply to most distributions and statistical models and to most types of data, including multiply censored, interval, and quantal-response data. In particular, available comparisons for multiply censored data employ ML methods. A disadvantage is that most ML methods require special computer programs. Also, many tables needed for exact ML confidence limits and hypothesis tests have not been developed. So approximate limits and tests are given below.

The ML methods for comparing distributions are applied to the exponential, normal, and extreme value distributions. The methods for normal data apply to the logs of lognormal data, and the methods for extreme value data apply to the (base e) logs of Weibull data. The relationships between these pairs of distributions are explained in Chapter 2.

Tables for exact ML confidence limits and hypothesis tests are strictly correct only for failure (Type II) censored samples. In practice, the tables often provide adequate approximations for time (Type I) censored data.

General methods and theory in this section are explained in terms of distributions with two parameters α and β. However, they extend to distributions with any number of parameters and to regression models where the distribution parameters are functions of independent variables and coefficients to be estimated from the data. For sample k, the ML estimates are $\hat{\alpha}_k$ and $\hat{\beta}_k$, and the maximum value of the sample log likelihood is $\hat{\mathcal{L}}_k$. Also, an estimate of the Fisher information matrix is $\hat{F}_k$, and an estimate of the covariance matrix of $\hat{\alpha}_k$ and $\hat{\beta}_k$ is $\hat{V}_k$. Chapter 8 describes the calculation of these estimates.

Multiply censored samples can be subjectively compared with hazard plots (Chapter 4). Singly censored and complete samples can be compared with probability plots (Chapter 3). A combination of such graphical methods and the ML methods presented here is usually most effective.

Nonparametric comparisons of multiply censored data are presented by Lawless (1982, Chap. 8), Kalbfleisch and Prentice (1980), Gross and Clark (1975), Elandt-Johnson and Johnson (1980), and Miller (1981), with biomedical applications.

1. ONE-SAMPLE COMPARISONS

One often wants to assess whether a sample surpasses (demonstration test) or is consistent with (hypothesis test) a specified value of a distribution parameter, percentile, reliability, or other quantity. For example, a reliability demonstration test for an exponential distribution may assess whether a product failure rate is below a specified value. This section describes hypothesis tests, confidence limits, and likelihood ratio tests for such comparisons. Essential background for this section is corresponding material in Chapter 8.

Confidence limits. Comparisons can employ ML confidence limits described in Chapter 8. In particular, if a $100\gamma\%$ confidence interval encloses the specified parameter value, the sample is **consistent** with that value. Also, if a one-sided confidence limit surpasses a specified value, the specified value is **demonstrated**. Chapter 8 gives examples of such comparisons. Also, Chapter 8 references exact tables for such intervals from singly censored and complete samples. Sample size can be chosen as described in Chapter 8 to achieve a confidence interval of desired length.

Hypothesis tests employing ML estimates.   Various authors have presented one-sample hypothesis tests based on ML estimates from censored data. For example, Spurrier and Wei (1980) give necessary tables for singly censored tests for an exponential mean and corresponding OC curves. Consult the references of Chapter 8 for such tests for particular distributions.

Likelihood ratio test.   Suppose that the specified value of the parameter θ is θ_0. Suppose that the equality hypothesis is $\theta = \theta_0$ and the alternative hypothesis is $\theta \neq \theta_0$. Let $\hat{\mathcal{L}}(\theta_0)$ denote the maximum value of the sample log likelihood when θ equals the specified value θ_0. Also, let $\hat{\mathcal{L}}(\hat{\theta})$ denote the maximum value of the sample log likelihood where $\hat{\theta}$ is the ML estimate of θ. Both likelihoods may depend on other parameters and are maximized with respect to them. The log likelihood ratio test statistic is

$$T = 2\left[\hat{\mathcal{L}}(\hat{\theta}) - \hat{\mathcal{L}}(\theta_0)\right]. \tag{1.1}$$

If the alternative is true, T tends to have larger values. So the test is

1.   If $T \leqslant t_{1-\alpha}$, accept the equality hypothesis.
2.   If $T > t_{1-\alpha}$, reject the equality hypothesis

at the $100\alpha\%$ significance level. Here $t_{1-\alpha}$ is the $100(1-\alpha)$th percentile of the distribution of T when $\theta = \theta_0$. Often $t_{1-\alpha}$ is not known, but a chi-square approximation applies to large samples with many observed failures. Then $t_{1-\alpha} \approx \chi^2(1-\alpha, 1)$, the $100(1-\alpha)$th chi-square percentile with one degree of freedom. Note that this is a two-sided test. Often the test can be made one sided by using 2α in place of α and rejecting only if $\hat{\theta}$ is above (below) θ_0. The two θ_0 values for which $T = \chi^2(1-\alpha, 1)$ are two sided (approximate) $100(1-\alpha)\%$ confidence limits for the true θ.

Insulating fluid example.   Section 5 of Chapter 7 presents an example of time to breakdown data on an insulating field. Theory for such fluid says that time to breakdown has an exponential distribution, that is, a Weibull distribution with a shape parameter of 1. Seven Weibull distributions with a common β were ML fitted to the data from the seven test voltages. The ML method of estimating a common parameter is explained in the next section.

The ML estimate of the common shape parameter is $\hat{\beta} = 0.799$, and the corresponding maximum log likelihood for that model is $\hat{\mathcal{L}}(\hat{\beta}) = -299.65$. For the model with seven exponential distributions ($\beta_0 = 1$), the maximum log likelihood is $\hat{\mathcal{L}}(1) = -303.10$. Since $T = 2[-299.65 - (-303.10)] = 6.90$ exceeds $\chi^2(0.99, 1) = 6.635$, $\hat{\beta} = 0.779$ is significantly below 1. So the distributions are not exponential, or the low shape parameter may come from experimental conditions that are not constant.

2. POOLED ESTIMATE OF A COMMON PARAMETER VALUE

Often one wishes to use a number of samples to estimate a parameter. In particular, the hypothesis tests in the following sections employ such pooled estimates. This section describes two pooled estimates and confidence limits for such a parameter: (1) the linearly pooled estimate and (2) the ML pooled estimate. A linearly pooled estimate is easier to calculate; a ML pooled estimate generally requires a special computer program.

Linearly pooled estimates. Suppose that $\hat{\theta}_1, \ldots, \hat{\theta}_K$ are statistically independent ML estimates of θ. Also, suppose that their (asymptotic) variances are $V_1, \ldots, V_K$. Then the linearly pooled estimate for θ is

$$\theta^* = V\left[(\hat{\theta}_1/V_1) + \cdots + (\hat{\theta}_K/V_K)\right], \tag{2.1}$$

where

$$V = 1/\left[(1/V_1) + \cdots + (1/V_K)\right]. \tag{2.2}$$

The V_k are often unknown and must be estimated from the data as described in Chapter 8. Most ML estimators are biased; so θ^* is usually a biased estimator. However, when each $\hat{\theta}_k$ comes from a large number of observed failure times, the cumulative distribution of θ^* is close to a normal one with a mean of θ and a variance V. Then θ^* is approximately unbiased. Sometimes $\hat{\theta}_k$ is a transformed ML estimate, for example, the (base e) log of a scale parameter or a ML estimate multiplied by an unbiasing factor.

Usually each $\hat{\theta}_k$ comes from a separate sample. Then a computer program for ML fitting can fit the distribution separately to each sample to get each $\hat{\theta}_k$. This is an advantage, since it is difficult to find computer programs that directly calculate a pooled ML estimate as described below.

Before using such a pooled estimate or confidence limits, one should check that the sampled populations do not have significantly different θ values. The hypothesis tests in following sections provide such checks.

Confidence limits.   Two-sided approximate $100\gamma\%$ confidence limits for θ are

$$\underset{\sim}{\theta} \cong \hat{\theta}^* - K_\gamma V^{1/2} \quad \text{and} \quad \tilde{\theta} \cong \hat{\theta}^* + K_\gamma V^{1/2}, \tag{2.3}$$

where K_γ is the $[100(1+\gamma)/2]$th standard normal percentile. The limits are more accurate the larger the number of observed failure times. Except for the exponential distribution (Chapter 8), there are no tables for exact confidence limits from pooled censored samples.

Insulating fluid example.   Section 5 of Chapter 7 presents time to breakdown data on an insulating fluid. Table 2.1 shows the logs of the ML estimates $\hat{\beta}_k$ of the shape parameters of the seven samples and the sample sizes n_k. For large samples, the log estimates $\hat{\theta}_k = \ln(\hat{\beta}_k)$ from complete samples have $V_k \cong 0.6079/n_k$. So, for $\hat{\theta}^* = \ln(\hat{\beta}^*), V \cong 0.6079/n$, where $n = n_1 + \cdots + n_K$, and $\ln(\hat{\beta}^*) = \frac{1}{76}[3(-0.606621) + \cdots + 8(0.309688)] = -0.199447$. The pooled estimate of the shape parameter is $\hat{\beta}^* = \exp(-0.199447) = 0.819$. Two-sided approximate 95% confidence limits for $\ln(\beta)$ are $\ln(\underset{\sim}{\beta}) = -0.199447 - 1.960(0.6079/76)^{1/2} = -0.374740$ and $\ln(\tilde{\beta}) = -0.024154$. Such limits for β are $\underset{\sim}{\beta} = \exp(-0.374740) = 0.687$ and $\tilde{\beta} = 0.976$, and the parameter differs significantly from 1. The large-sample theory may be crude, since the smallest sample size is 3; so the conclusion should be tentative, since $\tilde{\beta}$ is close to 1.

ML pooled estimate.   Suppose that $\mathcal{L}_1(\alpha_1, \beta_1), \ldots, \mathcal{L}_K(\alpha_K, \beta_K)$ are the log likelihoods of K statistically independent samples. Then the log likelihood of the combined samples is

$$\mathcal{L}(\alpha_1, \beta_1, \ldots, \alpha_K, \beta_K) = \mathcal{L}_1(\alpha_1, \beta_1) + \cdots + \mathcal{L}_K(\alpha_K, \beta_K). \tag{2.4}$$

Table 2.1. ML Estimates of Shape Parameters of Insulating Fluid

Voltage	n_k	$\ln(\hat{\beta}_k)$
26	3	-0.606621
28	5	-0.215485
30	11	0.057146
32	15	-0.577306
34	19	-0.260297
36	15	-0.117436
38	8	0.309688

Chapters 8 and 9 give formulas for such log likelihoods for various distributions and types of data. If some parameters have the same value, then they are set equal in the log likelihood. For example, if all of the β_k are equal to a common value β, the log likelihood is

$$\mathcal{L}_\beta(\alpha_1,\ldots,\alpha_K,\beta)=\mathcal{L}_1(\alpha_1,\beta)+\cdots+\mathcal{L}_K(\alpha_K,\beta). \qquad (2.5)$$

The ML estimates of the parameters in such a model are the values that maximize the corresponding log likelihood. For example, for the model with all β_k equal, the ML estimates of $\alpha_1,\ldots,\alpha_K,\beta$ are the parameter values that maximize (2.5). For a set of data, the ML estimates of α_k under a general model like (2.4) generally differ from those under a model with equal β_k parameters like (2.5). Also, the calculation of the ML estimates for (2.5) usually requires a special computer program such as those of Nelson and others (1978), Hahn and Miller (1968), and Meeker and Duke (1979). Such programs differ from the more available ones that ML fit a single distribution to a set of data.

Confidence limits. The general ML methods and theory from Chapter 8 apply to the model with equal parameter values. The theory gives the Fisher information matrix, the covariance matrix of the ML estimators, and approximate confidence limits.

Insulating fluid example. For the previous example, the pooled ML estimate of the common Weibull shape parameter is $\hat{\beta}=0.7994$; this differs slightly from the linearly pooled estimate $\hat{\beta}^*=0.819$. The corresponding scale parameter estimates are $\hat{\alpha}_{26}=1174$, $\hat{\alpha}_{28}=349.9$, $\hat{\alpha}_{30}=104.7$, $\hat{\alpha}_{32}=68.81$, $\hat{\alpha}_{34}=38.88$, $\hat{\alpha}_{36}=12.41$, $\hat{\alpha}_{38}=1.851$; these differ slightly from the scale parameter estimates from fitting a separate Weibull distribution (and shape parameter value) to the data from each test voltage. The maximum log likelihood is $\hat{\mathcal{L}}_\beta=-299.65$. The local estimate of the large-sample variance of $\hat{\beta}$ is $\text{var}(\hat{\beta})=0.005181$. This yields positive two-sided approximate 95% confidence limits $\beta=0.7994/\exp[1.960(0.005181)^{1/2}/0.7994]=0.670$ and $\tilde{\beta}=0.954$. These limits do not enclose 1. So as before, β is significantly different from 1, and the distribution is not exponential. Section 4 gives a test that the seven test voltages have a common β.

3. TWO-SAMPLE COMPARISONS

This section explains how to use ML methods to compare independent samples from two populations. Samples may be compared for equality of distribution parameters, percentiles, reliabilities, and other quantities. This section first presents general methods and then specific results for the basic distributions—exponential, normal, lognormal, extreme value, and Weibull distributions. Except for the exponential distribution, the OC curves (power functions) of most hypothesis tests for singly and multiply censored samples have not been derived. Sections 4 and 5 on K-sample comparisons give added methods that also apply to two-sample comparisons.

Comparison of Scale Parameters

Confidence limits. For $k = 1, 2$, suppose that $\hat{\sigma}_k$ is the ML estimator of the scale parameter σ_k. Two-sided $100\gamma\%$ confidence limits for the ratio $\rho = \sigma_1/\sigma_2$ are

$$\underset{\sim}{\rho} = (\hat{\sigma}_1/\hat{\sigma}_2)/\hat{v}[(1+\gamma)/2], \qquad \tilde{\rho} = (\hat{\sigma}_1/\hat{\sigma}_2)/\hat{v}[(1-\gamma)/2], \quad (3.1)$$

where $\hat{v}(\varepsilon)$ is the 100εth percentile of the standardized ratio

$$\hat{v} = (\hat{\sigma}_1/\sigma_1)/(\hat{\sigma}_2/\sigma_2). \tag{3.2}$$

$\hat{v}(\varepsilon)$ depends on sample sizes and censorings. If the confidence interval does not enclose 1, then σ_1 and σ_2 differ significantly at the $100(1-\gamma)\%$ level. (3.1) yields a one-sided $100\gamma\%$ confidence limit when γ replaces $(1+\gamma)/2$ or $(1-\gamma)$ replaces $(1-\gamma)/2$.

There are tables of $\hat{v}(\varepsilon)$ for certain distributions and failure censored samples; they are described below. For samples with many failures,

$$\hat{v}(\varepsilon) \simeq \exp\left[z_\varepsilon(B_1 + B_2)^{1/2}\right], \tag{3.3}$$

where $\text{Var}(\hat{\sigma}_k) = B_k\sigma_k^2$ and z_ε is the 100εth standard normal percentile. B_k depends on the sample size and censoring. If the B_k are not known, use $B_k \simeq \text{var}(\hat{\sigma}_k)/\hat{\sigma}_k^2$, where $\text{var}(\hat{\sigma}_k)$ is an estimate of $\text{Var}(\hat{\sigma}_k)$ as described in Chapter 8. $\text{Var}(\hat{\sigma}_k)$ here may be either the true or asymptotic variance.

Sample size. One may wish to prechoose sample sizes and observed numbers of failures so a (two-sided) interval has a desired length. Equation (3.3) or tables of exact factors for confidence limits can be used to calculate interval lengths for trial sample sizes and numbers of failures.

Exponential means. The ML estimate $\hat{\theta}_k$ of an exponential mean is identical to the BLUE θ_k^* for multiply failure censored samples. So the exact confidence limits in Section 3 of Chapter 11 apply.

Normal standard deviations σ_k and lognormal σ_k. Exact confidence limits for the ratio σ_1/σ_2 have not been derived for censored data. So (3.3) must be used for approximate confidence limits. For lognormal σ_k, work with the base 10 logs of the data, and use the methods for normal σ_k.

Snubber example. Table 3.1 shows multiply censored life test data on two snubber designs, a toaster component. The data are the number of toaster cycles (operations) to snubber failure (the toaster pops and throws the toast out). The basic question is, "How do the two normal life distributions compare?" The limits (3.1) compare the standard deviations. The ML estimates for the old and new designs are $\hat{\sigma}_O = 546.0$ and $\hat{\sigma}_N = 362.5$ cycles. The local estimates of their variances are $\mathrm{var}(\hat{\sigma}_O) = 9905$ and $\mathrm{var}(\hat{\sigma}_N) = 4023$; so $B_O \cong 9905/(546.0)^2 = 0.0332$ and $B_N \cong 4023/(362.5)^2 = 0.0306$. Two-sided approximate 95% confidence limits for $\rho = \sigma_O/\sigma_N$ are $\rho \cong (546.0/362.5)$ $/\exp[1.960(0.0332 + 0.0306)^{1/2}] = 0.92$ and $\tilde{\rho} \cong 2.48$. The limits enclose 1; so the standard deviations do not differ significantly. See Figure 3.4 of Chapter 5 for hazard plots of these data.

Extreme value scale parameters δ_k **and Weibull shape parameters** β_k. There are exact tables of percentiles $\hat{v}(\varepsilon)$ of the distribution of $\hat{v} = (\hat{\beta}_1/\beta_1)/(\hat{\beta}_2/\beta_2)$ for two Weibull shape parameters. The tables apply to pairs of complete or singly failure censored samples with the same size n and number r of failures; then $\hat{v}(1 - \varepsilon) = 1/\hat{v}(\varepsilon)$. McCool's (1970a) table covers $n = 10[r = 2(1)5, 7, 10]$, $n = 20[r = 3(1)7, 10, 15, 20]$, $n = 30[r = 3(1)10, 15, 30]$ and $\varepsilon = .90, .95$. Tables for $n = 5(r = 2, 3, 5)$, $n = 10(r = 2, 3, 5, 10)$, $n = 15(r = 2, 5(5)15)$, $n = 20[r = 2, 5(5)20]$, $n = 30[r = 2, 5(5)20, 30]$

Table 3.1. Cycles to Snubber Failure

Old Design				New Design			
90	410+	731	790+	45+	485+	608+	964
90	485	739	790+	47	485+	608+	1164+
90+	508	739+	790+	73	490	608+	1164+
190+	600+	739+	790+	136+	569+	608+	1164+
218+	600+	739+	790+	136+	571	608+	1164+
218+	600+	739+	855	136+	571+	608+	1164+
241+	600+	790	980	136+	575	608+	1164+
268	631	790+	980	136+	608	630	1164+
349+	631	790+	980+	145	608	670	1198
378+	631	790+	980+	190+	608+	670	1198+
378+	635	790+	980+	190+	608+	731+	1300+
410	658	790+	980+	281+	608+	838	1300+
410	658+	790+	980+	311	608+	964	1300+
				417+	608+		

+ denotes running time without snubber failure.

are given by McCool (1975b) for $\varepsilon = .90$, by McCool (1978a) for six values of ε, and by McCool (1974) for 21 values of ε. Thoman and Bain's (1969) table for complete samples covers $n = 5(1)20(2)80, 90, 100, 120$ and $\varepsilon = .60, .70(.05).95, .98$; they also plot the OC functions of level 0.05 and 0.10 tests for selected n's. These tables apply to extreme value scale parameters; then one uses the relationship between the Weibull and extreme value distributions. Use (3.3) outside the range of these tables.

Comparison of Location Parameters and Percentiles

Confidence limits. For $k = 1, 2$, suppose that the ML estimator of a percentile y_k is $\hat{y}_k$, and the ML estimator of the scale parameter σ_k is $\hat{\sigma}_k$. The location parameter is a particular percentile of a distribution. Suppose that $\hat{\sigma}$ is a (ML or linearly) pooled estimate (Section 2), assuming that $\sigma_1 = \sigma_2$. Two-sided $100\gamma\%$ confidence limits for $\Delta = y_1 - y_2$ are

$$\underline{\Delta} = (\hat{y}_1 - \hat{y}_2) - \hat{d}[(1+\gamma)/2]\hat{\sigma}, \qquad \tilde{\Delta} = (\hat{y}_1 - \hat{y}_2) - \hat{d}[(1-\gamma)/2]\hat{\sigma}, \quad (3.4)$$

where $\hat{d}(\varepsilon)$ is the 100εth percentile of the standardized difference

$$\hat{d} = (\hat{y}_1 - \hat{y}_2 - y_1 + y_2)/\hat{\sigma}, \tag{3.5}$$

a "t-like" statistic. If the confidence interval does not enclose 0, then y_1 and y_2 differ significantly at the $[100(1-\gamma)]\%$ level. (3.4) yields a one-sided $100\gamma\%$ confidence limit when γ replaces $(1+\gamma)/2$ or $1-\gamma$ replaces $(1-\gamma)/2$.

There are tables of $\hat{d}(\varepsilon)$ for certain distributions and failure censored samples; they are described below. The $\hat{d}$ distributions depend on the sizes and censorings of both samples. Also, the distributions are not symmetric; that is, $\hat{d}(1-\varepsilon) \neq -\hat{d}(\varepsilon)$. If both samples have many failures,

$$\hat{d}(\varepsilon) \cong z_\varepsilon (D_1 + D_2)^{1/2}, \tag{3.6}$$

where $\text{Var}(\hat{y}_k) = D_k \sigma_k^2$ and z_ε is the 100εth standard normal percentile. If the D_k are not known, they can be approximated from $D_k \cong \text{var}(\hat{y}_k)/\hat{\sigma}_k^2$, where $\text{var}(\hat{y}_k)$ is an estimate of $\text{Var}(\hat{y}_k)$ and is obtained as described in Chapter 8.

Sample size. One may wish to prechoose sample sizes and observed numbers of failures so a (two-sided) interval has a desired length. Equation (3.6) or tables of exact factors for confidence limits can be used to calculate interval lengths for trial sample sizes and numbers of failures.

Normal means μ_k and percentiles y_k and lognormal μ_k. There are no tables of percentiles of the statistic $\hat{d} = (\hat{y}_1 - \hat{y}_2 - y_1 + y_2)/\hat{\sigma}$ of ML estimators of normal parameters from censored samples. So (3.6) must be used for approximate confidence limits for $\mu_1 - \mu_2$ and $y_1 - y_2$. For lognormal data, work with the (base 10) logs of the data.

Snubber example. Table 3.1 shows life test data on old and new snubber designs. (3.4) is used to compare the means of their normal life distributions. The ML estimates of the means are $\hat{\mu}_O = 1128$ and $\hat{\mu}_N = 908$ cycles, and the linearly pooled estimate is $\hat{\sigma}^* = 450.4$ cycles. The calculations for two-sided approximate 95% confidence limits for $\mu_O - \mu_N$ employ $\text{var}(\hat{\mu}_O) = 15,191, \text{var}(\hat{\mu}_N) = 5814, \hat{\sigma}_O = 546.0$, and $\hat{\sigma}_N = 362.5$. Then $A_O \cong 15,191/(546.0)^2 = 0.05096$, $A_N \cong 5814/(362.5)^2 = 0.04424$, and $\hat{d}(0.975) \cong -\hat{d}(0.025) \cong 1.960(0.05096 + 0.04424)^{1/2} = 0.6047$. Finally, $\Delta \cong (1128 - 908) - 0.6047(450.4) = -52$ and $\tilde{\Delta} \cong 492$ cycles. These limits enclose zero; so the means do not differ significantly. See Figure 3.4 of Chapter 5 for hazard plots of the two samples.

Extreme value location parameters λ_k and percentiles y_k; Weibull scale parameters α_k and percentiles t_k. There are exact tables of 100εth percentiles $\hat{u}(\varepsilon)$ of $\hat{u} = \exp(\hat{d}) = (\hat{t}_1/\hat{t}_2)(\hat{\beta}_1 + \hat{\beta}_2)/2$ for Weibull percentiles t_1 and t_2 when $\beta_1 = \beta_2$. The tables apply to pairs of complete or singly failure censored samples with the same size n and number r of failures; then $\hat{u}(\varepsilon) = 1/\hat{u}(1 - \varepsilon)$. McCool's (1970a) table covers t_{10} and n and r values in the above paragraph on Weibull shape parameters. McCool's (1974) table covers t_{01}, t_{10}, and t_{50} and n and r values in that same paragraph. Excerpts from this table appear in McCool (1975b, 1978b). Thoman and Bain's (1969) table for the scale parameter t_{632} covers complete samples with the n values in that same paragraph. These tables apply to extreme value location parameters; then one uses the relationship between the Weibull and extreme value distributions. Schafer and Sheffield (1976) give exact tables of 100γth percentiles of $\hat{\beta}[\ln(\hat{\alpha}_1) - \ln(\hat{\alpha}_2)]$, where the ML estimate of the common β is $\hat{\beta}$ and that of the scale parameter α_k is $\hat{\alpha}_k$. Their tables cover complete samples of size $n = 5(1)20(4)40(10)100$ and ε and $1 - \varepsilon = .50, .60, .70(.05).95, .975$. Use (3.6) outside the range of the tables.

4. K-SAMPLE COMPARISONS

This section explains ML methods for comparing K samples with respect to population parameters. The methods include (1) quadratic test statistics for equality of K parameters, (2) simultaneous confidence intervals for all pairs of K parameters, and (3) likelihood ratio (LR) tests for equality of K

parameters. LR tests appear in Section 5; they are versatile and apply to most statistical distributions and models and to most types of data; however, they are advanced.

Suppose that $\hat{\theta}_1, \ldots, \hat{\theta}_K$ are statistically independent ML estimators for $\theta_1, \ldots, \theta_K$ and $\mathrm{var}(\hat{\theta}_k)$ is an estimate of the (asymptotic) $\mathrm{Var}(\hat{\theta}_k)$ for $k = 1, \ldots, K$.

Quadratic Test of Homogeneity

Test. The following tests the homogeneity (equality) hypothesis that $\theta_1 = \cdots = \theta_K$ against the alternative that some $\theta_k \neq \theta_{k'}$. The quadratic test statistic is

$$Q = \sum_{k=1}^{K} \left(\hat{\theta}_k - \hat{\theta} \right)^2 / \mathrm{var}(\hat{\theta}_k), \qquad (4.1)$$

where $\hat{\theta}$ is a pooled estimate for a common θ (Section 8.2). Also, where appropriate, Q can be calculated by (4.1) of Chapter 11, but with ML estimates. If the hypothesis is true, the distribution of Q is approximately chi square with $K - 1$ degrees of freedom. If the alternative is true, Q tends to have larger values. So the approximate test is

1. If $Q \leqslant \chi^2(1 - \alpha, K - 1)$, accept the equality hypothesis.
2. If $Q > \chi^2(1 - \alpha, K - 1)$, reject the equality hypothesis

at the $100\alpha\%$ significance level; here $\chi^2(1 - \alpha, K - 1)$ is the $100(1 - \alpha)$th chi-square percentile with $K - 1$ degrees of freedom. The chi-square approximation is more precise the larger the number of failures in each sample. There are no tables of the exact distribution of Q for censored samples. McCool (1977, 1978b) investigates the performance of this test for comparing Weibull shape and scale parameters.

If Q is statistically significant, examine the individual $\hat{\theta}_k$ to see how they differ. Individual confidence limits for the $\hat{\theta}_k$ help one to see the significant differences, particularly, if limits are plotted side by side.

Snubber example. Table 3.1 shows data on cycles to failure of two snubber designs. The test above is used to compare the normal means of the old and new designs. The estimates are $\hat{\mu}_O = 1128$, $\hat{\mu}_N = 908$, $\mathrm{var}(\hat{\mu}_O) = 15{,}191$, and $\mathrm{var}(\hat{\mu}_N) = 5814$. The linearly pooled estimate (2.1) is $\hat{\mu}^* = 969$. The test statistic is $Q = [(1128 - 969)^2 / 15{,}191] + [(908 - 969)^2 / 5814] = 2.28$. Since $Q = 2.28 < 3.841 = \chi^2(0.95, 1)$, the two means do not differ significantly.

All Pairwise Comparisons

For K parameters $\theta_1, \ldots, \theta_K$, there are $K(K-1)/2$ pairs of differences $\theta_k - \theta_{k'}$ or ratios $\theta_k / \theta_{k'}$. Simultaneous $100\gamma\%$ confidence limits enclose all true differences (or ratios) with probability γ. Such limits are wider than a $100\gamma\%$ interval for a single difference (or ratio). If such limits for a difference (ratio) do not enclose zero (one), there is a wholly significant difference between those two parameters. The following gives approximate simultaneous limits based on independent ML estimators and references tables for exact limits.

Simultaneous limits for scale parameters. Suppose that $\hat{\sigma}_k$ is the ML estimator for the scale parameter σ_k for $k = 1, \ldots, K$. Simultaneous two-sided $100\gamma\%$ confidence limits for all ratios $\rho_{kk'} = \sigma_k / \sigma_{k'}$ are

$$\underline{\rho}_{kk'} = (\hat{\sigma}_k / \hat{\sigma}_{k'}) / \hat{w}(\gamma) \quad \text{and} \quad \tilde{\rho}_{kk'} = (\hat{\sigma}_k / \hat{\sigma}_{k'}) \cdot \hat{w}(\gamma), \qquad (4.2)$$

where $\hat{w}(\gamma)$ is the 100γth percentile of the maximum standardized ratio

$$\hat{w} = \max_{k,k'} \left[(\hat{\sigma}_k / \sigma_k) / (\hat{\sigma}_{k'} / \sigma_{k'}) \right]. \qquad (4.3)$$

If such an interval does not enclose 1, those two scale parameters have a wholly significant difference.

Section 4 of Chapter 10 references tables of $\hat{w}(\gamma)$ for exponential means when all samples have the same number r of failures. Then r replaces the sample size n in (4.9) of Chapter 10. McCool (1974) tabulates $\hat{w}(\gamma)$ for Weibull shape parameters for 21 values of γ, $K = 2(1)10$, and n and r have selected values, which are given in Section 3. Excerpts from this table appear in McCool (1975b, 1978b). His tables also apply to extreme value scale parameters. There are no such tables for (log) normal standard deviations and censored samples.

Simultaneous two-sided approximate $100\gamma\%$ confidence limits are given by (3.1) for a single ratio, but replace $(1+\gamma)/2$ by $\gamma' = 1 - (1-\gamma)K^{-1}(K-1)^{-1}$. Such limits with exact $\hat{v}(\gamma')$ enclose all true ratios with a probability of at least γ. The probability is close to γ when K is small and γ is near 1.

Insulating oil example. Sixty times to breakdown of an insulating oil were measured at each of six test conditions. Engineering theory says that time to oil breakdown has a Weibull distribution with the same shape parameter value at each test condition. The above simultaneous intervals are used to test for equality of the shape parameters. The ML estimates of the six shape parameters are $10.84, 12.22, 12.53, 13.32, 14.68,$ and 16.45. The maximum ratio of shape parameter estimates is $\hat{w} = 16.45/10.84 = 1.518$. For simultaneous approximate 95% confidence limits, $\gamma' = 1 - (1 - 0.95)6^{-1}5^{-1} = 0.9983$, and $\hat{w}(0.95) \cong \hat{v}(0.9966) \cong \exp\{z_{0.9983}[(0.6079/60) + (0.6079/60)]^{1/2}\} = 1.518$. Since $\hat{w} \cong \hat{w}(0.95)$, the shape parameters just differ wholly significantly at the 5% level. One can examine a plot of the estimates to see how they differ.

Simultaneous limits for location parameters and percentiles. Suppose that $\hat{y}_k$ is the ML estimator for y_k, the location parameter or a percentile for $k = 1, \ldots, K$. Also, suppose that $\hat{\sigma}$ is a pooled estimate of a common scale parameter (Section 2) and is based on samples k and k' or on all K samples. Simultaneous two-sided $100\gamma\%$ confidence limits for all differences $\Delta_{kk'} = y_k - y_{k'}$ are

$$\underset{\sim}{\Delta}_{kk'} = (\hat{y}_k - \hat{y}_{k'}) - \hat{q}(\gamma)\hat{\sigma}, \qquad \tilde{\Delta}_{kk'} = (\hat{y}_k - \hat{y}_{k'}) + \hat{q}(\gamma)\hat{\sigma}, \qquad (4.4)$$

where $\hat{q}(\gamma)$ is the 100γth percentile of the maximum absolute standardized difference

$$\hat{q} = \max_{k,k'} \left(|\hat{y}_k - \hat{y}_{k'} - y_k + y_{k'}|/\hat{\sigma} \right). \qquad (4.5)$$

If such an interval does not enclose 0, those two percentiles have a wholly significant difference.

McCool (1974) tabulates $\exp[\hat{q}(\gamma)]$ for Weibull percentiles $t_{.01}, t_{.10}$, and $t_{.50}$ and singly censored samples for 21 values of γ, $K = 2(1)10$ and n and r with the selected values given in Section 3. McCool (1975) gives the same table for just $\gamma = 0.90$. These tables also apply to corresponding extreme value percentiles. There are no such tables for (log) normal percentiles and censored samples.

Simultaneous two-sided approximate $100\gamma\%$ confidence limits are given by (3.4) for a single difference, but $(1 + \gamma)/2$ is replaced by $\gamma' = 1 - (1 - \gamma)K^{-1}(K - 1)^{-1}$. These limits employ the Bonferroni inequality (Miller, 1966); such limits with exact $\hat{d}(\gamma')$ enclose all true differences with a probability of at least γ. The probability is close to γ when K is small and γ is near 1.

5. LIKELIHOOD RATIO TESTS

The following likelihood ratio tests apply to most statistical distributions and models and to most types of data. The presentation covers (1) K-sample tests that a particular parameter has the same value in all K distributions and (2) K-sample tests that the K distributions are identical (that is, corresponding parameters all have the same value). This advanced material depends on Chapter 8.

Equality of K Parameters

The problem. For $k = 1, \ldots, K$, suppose that sample k comes from a distribution with parameters α_k and β_k. Two parameters are used here, but the test below applies to distributions and models with any number of parameters. We test the equality (homogeneity) hypothesis that $\beta_1 = \cdots = \beta_K$ against the alternative that some $\beta_k \neq \beta_{k'}$. Suppose that the log likelihood of sample k is $\mathcal{L}_k(\alpha_k, \beta_k)$ and has a maximum $\hat{\mathcal{L}}_k$ at the ML estimates $\hat{\alpha}_k$ and $\hat{\beta}_k$. The combined log likelihood when all $\beta_k = \beta$ (a common value) is $\mathcal{L}_\beta(\alpha_1, \ldots, \alpha_K, \beta) = \mathcal{L}_1(\alpha_1, \beta) + \cdots + \mathcal{L}_K(\alpha_K, \beta)$, since the samples are assumed to be independent. Calculation of its maximum $\hat{\mathcal{L}}_\beta$ and the ML estimates $\hat{\alpha}'_1, \ldots, \hat{\alpha}'_K, \hat{\beta}$ generally requires a special computer program such as those of Nelson and others (1978), Hahn and Miller (1968), Schafer and Sheffield (1976), and Meeker and Duke (1979).

The test. The (log) likelihood ratio test statistic is

$$T = 2\left(\hat{\mathcal{L}} - \hat{\mathcal{L}}_\beta\right) = 2\left(\hat{\mathcal{L}}_1 + \cdots + \hat{\mathcal{L}}_K - \hat{\mathcal{L}}_\beta\right). \tag{5.1}$$

Under the equality hypothesis, T has a distribution that is approximately chi square with $K - 1$ degrees of freedom. Under the alternative, T tends to have larger values. So the approximate test is

1. If $T \leqslant \chi^2(1 - \alpha, K - 1)$, accept the equality hypothesis.
2. If $T > \chi^2(1 - \alpha, K - 1)$, reject the equality hypothesis

at the $100\alpha\%$ significance level; here $\chi^2(1 - \alpha, K - 1)$ is the $100(1 - \alpha)$th chi-square percentile with $K - 1$ degrees of freedom. The approximation is more precise the larger the number of failures in each sample. For large samples (asymptotically), the T statistic equals the Q statistic (4.1).

If T is significant, examine the estimates and confidence limits for the β_k to see how they differ.

Equality of Poisson occurrence rates. This example applies the likelihood ratio test to Poisson data. Suppose that $Y_1, \ldots, Y_K$ are independent Poisson counts, where Y_k has an occurrence rate λ_k and length of observation t_k. We derive the likelihood ratio test of the equality hypothesis $\lambda_1 = \cdots = \lambda_K$ against the alternative some $\lambda_k \neq \lambda_{k'}$. The log likelihood for the Y_k under the alternative is

$$\mathcal{L}(\lambda_1, \ldots, \lambda_K) = \sum_{k=1}^{K} \mathcal{L}_k(\lambda_k) = \sum_{k=1}^{K} \left[-\lambda_k t_k + Y_k \ln(\lambda_k t_k) - \ln(Y_k!) \right].$$

$\mathcal{L}$ and the $\mathcal{L}_k$ are maximized by the separate ML estimates $\hat{\lambda}_1 = Y_1/t_1, \ldots, \hat{\lambda}_K = Y_K/t_K$, and

$$\hat{\mathcal{L}} = \sum_{k=1}^{K} \hat{\mathcal{L}}_k = \sum_{k=1}^{K} \left[-Y_k + Y_k \ln(Y_k) - \ln(Y_k!) \right].$$

The log likelihood for the Y_k under the equality hypothesis is

$$\mathcal{L}_\lambda(\lambda) = \sum_{k=1}^{K} \mathcal{L}_k(\lambda) = \sum_{k=1}^{K} \left[-\lambda t_k + Y_k \ln(\lambda t_k) - \ln(Y_k!) \right].$$

Let $Y = (Y_1 + \cdots + Y_K)$ and $t = (t_1 + \cdots + t_K)$. Then $\hat{\lambda} = Y/t$ maximizes $\mathcal{L}_\lambda$, and

$$\hat{\mathcal{L}}_\lambda = -Y + \sum_{k=1}^{K} \left[Y_k \ln(\hat{\lambda} t_k) - \ln(Y_k!) \right].$$

The log likelihood ratio test statistic is

$$T = 2(\hat{\mathcal{L}} - \hat{\mathcal{L}}_\lambda) = 2 \sum_{k=1}^{K} Y_k \ln(\hat{\lambda}_k / \hat{\lambda}). \tag{5.2}$$

Under the equality hypothesis, the distribution of T is approximately chi square with $K - 1$ degrees of freedom.

Tree and bare wire example. This example uses the preceding Poisson test of homogeneity. Power line failure data are $Y_1 = 12$ tree-wire failures in $t_1 = 467.9$ 1000 ft·years of exposure and $Y_2 = 69$ bare wire failures in $t_2 = 1079.6$ 1000 ft·years. Here $\hat{\lambda}_1 = 12/467.9 = 0.0256$, $\hat{\lambda}_2 = 69/1079.6 = 0.0639$, $\hat{\lambda} = (12 + 69)/(467.9 + 1079.6) = 0.0523$ failures per 1000 ft·years. Then $T = 2[12 \cdot \ln(0.0256/0.0523) + 69 \cdot \ln(0.0639/0.0523)] = 10.42$; this has $K - 1 = 2 - 1 = 1$ degree of freedom. $\chi^2(0.99, 1) = 6.64$ and $\chi^2(0.999, 1) = 10.83$. So the observed failure rates differ very significantly at almost the 0.1% level.

Insulating fluid example. Chapter 11 presents an example of time to breakdown data on an insulating fluid. The data provide a numerical example of the likelihood ratio test. Theory says that time to breakdown has a Weibull distribution with a common shape parameter β at all test voltages. To test this, we use the maximum log likelihoods of the seven samples, each fitted with a separate Weibull distribution, namely, $\hat{\mathcal{L}}_{26} = -23.72$, $\hat{\mathcal{L}}_{28} = -34.38$, $\hat{\mathcal{L}}_{30} = -58.59$, $\hat{\mathcal{L}}_{32} = -65.74$, $\hat{\mathcal{L}}_{34} = -68.39$, $\hat{\mathcal{L}}_{36} = -37.69$, $\hat{\mathcal{L}}_{38} = -6.76$. We also use the maximum log likelihood when Weibull distributions with a common shape parameter are fitted to the data, namely, $\hat{\mathcal{L}}_{\beta} = -299.65$. The test statistic is $T = 2[-23.72 - 34.38 - 58.59 - 65.74 - 68.39 - 37.69 - 6.76 - (-299.65)] = 8.76$. Since $T < 12.59 = \chi^2(1 - 0.05, 7 - 1)$, the seven Weibull shape parameters do not differ significantly. So the pooled β estimate appears satisfactory. The maximum log likelihoods were calculated with the STATPAC program of Nelson and others (1978). Bilikam, Moore, and Petrick (1979) give tables of exact percentage points $\gamma = .8(.05).95$ of T for $K = 2$ complete samples with sizes $n_1, n_2 = 10(10)40$.

Equality of K Parameters Assuming That Other Parameters Are Equal

The problem. The likelihood ratio test for equality of parameters can be used when some other parameters are assumed to be equal. For example, one can test the equality hypothesis $\alpha_1 = \cdots = \alpha_K$ under the assumption that all $\beta_k = \beta$. For example, the likelihood ratio test for equality of K normal means assuming that the standard deviations are equal is the usual one-way analysis of variance test for complete data in Chapter 6.

For K distributions with parameters α_k, β_k, suppose that $\mathcal{L}_k(\alpha_k, \beta_k)$ are the log likelihoods of K statistically independent samples. Under the assumption that all $\beta_k = \beta$, the combined sample log likelihood under the equality hypothesis all $\alpha_k = \alpha$ is $\mathcal{L}_{\alpha\beta}(\alpha, \beta) = \mathcal{L}_1(\alpha, \beta) + \cdots + \mathcal{L}_K(\alpha, \beta)$. Suppose that the maximum is $\hat{\mathcal{L}}_{\alpha\beta}$ at the ML estimates $\hat{\alpha}$ and $\hat{\beta}$; these are easy to obtain by fitting a single distribution to the pooled data treated as a single sample. Under the assumption that all $\beta_k = \beta$, the combined sample log likelihood under the alternative is $\mathcal{L}_{\beta}(\alpha_1, \ldots, \alpha_K, \beta) = \mathcal{L}_1(\alpha_1, \beta) + \cdots + \mathcal{L}_K(\alpha_K, \beta)$. Calculation of its maximum $\hat{\mathcal{L}}_{\beta}$ and the ML estimates $\hat{\alpha}_1, \ldots, \hat{\alpha}_K, \hat{\beta}'$ generally requires a special computer program such as those of Nelson and others (1978), Hahn and Miller (1968), Schafer and Sheffield (1976), and Meeker and Duke (1979).

The test. The test statistic is

$$T = 2\left(\hat{\mathcal{L}}_{\beta} - \hat{\mathcal{L}}_{\alpha\beta}\right). \tag{5.3}$$

Under the equality hypothesis, T has a distribution that is approximately chi square with $K - 1$ degrees of freedom. Under the alternative, T tends to have larger values. So the approximate test is

1. If $T \leqslant \chi^2(1-\alpha, K-1)$, accept the equality hypothesis.
2. If $T > \chi^2(1-\alpha, K-1)$, reject the equality hypothesis

at the $100(1-\alpha)\%$ significance level; here $\chi^2(1-\alpha, K-1)$ is the $100(1-\alpha)$th chi-square percentile with $K-1$ degrees of freedom. The approximation is more exact the larger the number of failures in each sample. McCool (1979) tabulates the distribution of T for singly censored samples with all $n_k = n$ and all $r_k = r$. He also gives an improvement for the chi-square approximation.

If T is significant, examine the estimates and confidence limits for the α_k to see how they differ.

The equality hypothesis $\alpha_1 = \cdots = \alpha_K$ can be tested either with or without the assumption that some other parameters are equal. The assumption should be used only if physical theory and a hypothesis test indicate it is appropriate. Use of the assumption, if valid, usually yields a more sensitive test of the equality hypothesis.

Snubber example. Table 3.1 shows life test data on old and new snubber designs. The means of normal life distributions are compared with the above test under the assumption that the standard deviations are equal. The maximum log likelihood under the equality hypothesis is $\hat{\mathcal{L}}_{\mu\sigma} = -286.95$; this value and $\hat{\mu} = 1020$ and $\hat{\sigma} = 462$ come from pooling the two samples and fitting a single normal distribution to that pooled sample. Under the alternative, $\hat{\mathcal{L}}_\sigma = -286.66$, $\hat{\mu}_O = 974$, $\hat{\mu}_N = 1061$, and $\hat{\sigma}' = 458$. The test statistic is $T = 2[-286.66 - (-286.95)] = 0.58$. This is less than $\chi^2(1-0.05, 2-1) = 3.84$. So the two means do not differ significantly. Calculation of $\hat{\mathcal{L}}_\sigma, \hat{\mu}_O, \hat{\mu}_N$, and $\hat{\sigma}'$ was performed with STATPAC of Nelson and others (1978).

K Distributions Are Identical

The problem. For $k = 1, \ldots, K$, suppose that sample k comes from a distribution with parameters α_k and β_k. Two parameters are used here, but the test below applies to distributions and models with any number of parameters. We test the hypothesis of identical distributions, that is, $\alpha_1 = \cdots = \alpha_K$ and $\beta_1 = \cdots = \beta_K$. The alternative is some $\alpha_k \neq \alpha_{k'}$ and/or some $\beta_k \neq \beta_{k'}$. Suppose that the log likelihood of sample k is $\mathcal{L}_k(\alpha_k, \beta_k)$ and has a maximum $\hat{\mathcal{L}}_k$ at the ML estimates $\hat{\alpha}_k$ and $\hat{\beta}_k$. The combined log likelihood when all $\alpha_k = \alpha$ and all $\beta_k = \beta$ is $\mathcal{L}_{\alpha\beta}(\alpha, \beta) = \mathcal{L}_1(\alpha, \beta) + \cdots + \mathcal{L}_K(\alpha, \beta)$. Suppose that its maximum is $\hat{\mathcal{L}}_{\alpha\beta}$ at the ML estimates $\hat{\alpha}$ and $\hat{\beta}$; these are easy to obtain by fitting a single distribution to the pooled data treated as a single big sample.

The test.   The test statistic is

$$T = 2\left(\hat{\mathcal{L}}_1 + \cdots + \hat{\mathcal{L}}_K - \hat{\mathcal{L}}_{\alpha\beta}\right). \tag{5.4}$$

Under the equality hypothesis, T has a distribution that is approximately chi square with $2K - 2$ degrees of freedom. Under the alternative, T tends to have larger values. So the approximate test is

1.   If $T \leqslant \chi^2(1 - \alpha, 2K - 2)$, accept the equality hypothesis.
2.   If $T > \chi^2(1 - \alpha, 2K - 2)$, reject the equality hypothesis

at the $100\alpha\%$ significance level; here $\chi^2(1 - \alpha, 2K - 2)$ is the $100(1 - \alpha)$th chi-square percentile with $2K - 2$ degrees of freedom. The approximation is more precise the larger the number of failures in each sample.

An alternative analysis could consist of a likelihood ratio test of the equality hypothesis $\alpha_1 = \cdots = \alpha_K$ and a separate test of the equality hypothesis $\beta_1 = \cdots = \beta_K$. The combination of these two tests is not quite the same as the above test for identical distributions. In practice, one may wish to use all three tests, pairwise comparisons, and separate confidence intervals.

Snubber example.   Table 3.1 shows life test data on old and new snubber designs. The test above is used to assess whether the two normal life distributions are identical. Under the hypothesis of identical distributions, $\hat{\mathcal{L}}_{\mu\sigma} = -286.95$, and the common parameter estimates are $\hat{\mu} = 1020$ and $\hat{\sigma} = 462$; these come from fitting a single normal distribution to the pooled samples. Separate normal fits to the two samples yield $\hat{\mathcal{L}}_O = -138.60$, $\hat{\mu}_O = 1128$, $\hat{\sigma}_O = 546$, $\hat{\mathcal{L}}_N = -146.75$, $\hat{\mu}_N = 908$, and $\hat{\sigma}_N = 362$. The test statistic is $T = 2[-138.60 - 146.75 - (-286.95)] = 3.20$. This is less than $\chi^2(1 - 0.05, 2 \cdot 2 - 2) = 5.99$. So the two distributions (means and standard deviations together) do not differ significantly. The estimates were calculated with the STATPAC program of Nelson and others (1978).

6.   GENERAL THEORY FOR LIKELIHOOD RATIO AND RELATED TESTS

This technical section informally presents advanced theory for likelihood ratio tests. Rao (1973) and Wilks (1962) rigorously present the theory. The tests have certain asymptotic optimum properties (e.g., locally most powerful), and they generally have good properties for small sample sizes. The tests are described below in terms of a complete sample, but they apply to censored and other types of samples with the appropriate sample likelihood.

This section presents (1) the general model, (2) the null hypothesis model, (3) sample likelihoods, (4) likelihood ratio, (5) likelihood ratio test, (6) approximate test, (7) the OC function and consistency, (8) test that K parameters are equal, (9) test that K distributions are identical, (10) Rao's equivalent test, and (11) Wald's equivalent test. Chapter 8 is needed background.

Problem Statement

The general model. Suppose that a sample has n observations $Y_1, \ldots, Y_n$ with a joint parametric density $f_\Theta(y_1, \ldots, y_n)$, continuous or discrete; the parameter Θ has r components $(\theta_1, \ldots, \theta_r)$ in some r-dimensional parameter space Ω, an open subset of Euclidean r-space. The distribution f_Θ and the parameter space Ω are assumed given; they are the general model for the problem.

Example A. $Y_1, \ldots, Y_n$ are independent observations from a normal distribution with mean μ and variance σ^2. Ω consists of all points (μ, σ^2) where $-\infty < \mu < \infty$ and $\sigma^2 > 0$; Ω is the upper half-plane in Figure 6.1a.

Example B. $Y_1, \ldots, Y_n$ are independent Bernoulli trials with probability p that $Y_i = 1$ and probability $1 - p$ that $Y_i = 0$. Ω consists of the p values satisfying $0 < p < 1$. Ω is the open unit interval in Figure 6.1b.

Example C. Y_1 ands Y_2 are independent binomial observations, where Y_k is the number of successes in n_k trials, each with success probability p_k, $k = 1, 2$. Ω consists of all points (p_1, p_2) where $0 < p_1 < 1$ and $0 < p_2 < 1$; Ω is the open unit square in Figure 6.1c.

The null hypothesis model. We test a hypothesis that the true Θ is in some subspace ω of Ω. The alternative is that the true Θ is in the subset $\Omega - \omega$. ω is the subspace of the null hypothesis, and its dimension r' must be less than r. The null hypothesis specifies a constrained model for the problem.

Example A. For the null hypothesis $\mu = \mu_0$ where μ_0 is given, ω consists of all points (μ_0, σ^2) where $\sigma^2 > 0$; ω is the half-line in Figure 6.2a. Here ω has $r' = 1$ dimension.

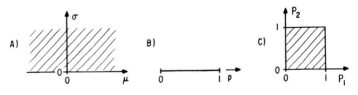

Figure 6.1. Parameter spaces Ω.

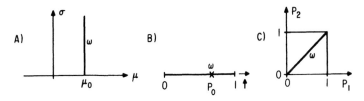

Figure 6.2. Hypothesis spaces ω.

Example B. For the null hypothesis $p = p_0$, where p_0 is given, ω consists of the single point p_0 on the unit interval as shown in Figure 6.2b. Here ω has $r' = 0$ dimensions.

Example C. For the null hypothesis $p_1 = p_2$, ω consists of all points (p_1, p_2) where $p_1 = p_2$; ω is the line segment from $(0,0)$ to $(1,1)$ as shown in Figure 6.2c. Here ω has $r' = 1$ dimension.

The subspace ω of the null hypothesis is usually a (hyper) plane, line, or point in Ω. Common forms of ω are the following.

1. $\theta_1 = \theta_{10}, \ldots, \theta_c = \theta_{c0}$ are given constants, and $\theta_{c+1}, \ldots, \theta_r$ may have any allowed values. Examples A and B are of this type. Here ω is an $(r - c)$-dimensional subspace (a hyperplane) in Ω. The dimension of ω is the number of free parameters $\theta_{c+1}, \ldots, \theta_r$ under the null hypothesis.

2. $\theta_1 = \theta_2 = \cdots = \theta_d$, the equality hypothesis, and $\theta_{d+1}, \ldots, \theta_r$ may have any values. Example C is of this type. Here ω is an $(r - d + 1)$-dimensional subspace (a hyperplane) of Ω. The dimensions of ω is the number of free parameters under the null hypothesis, namely, $\theta_{d+1}, \ldots, \theta_r$, and the common value for $\theta_1 = \cdots = \theta_d$.

Many null hypotheses can be expressed in either common form by reparametrizing the model. For example, suppose that $Y_{k1}, \ldots, Y_{kn_k}$ are independent observations from a normal distribution with mean μ_k and standard deviation $\sigma_k, k = 1, 2$. Ω is the four-dimensional space of points $(\mu_1, \sigma_1, \mu_2, \sigma_2)$ where $-\infty < \mu_k < \infty$ and $\sigma_k > 0, k = 1, 2$. The null hypothesis that the $100P$th percentiles are equal is $\mu_1 + z_P\sigma_1 = \mu_2 + z_P\sigma_2$, a three-dimensional hyperplane. Suppose that the distributions are parametrized and written in terms of the percentiles $y_{Pk} = \mu_k + z_P\sigma_k$ and standard deviations $\sigma_k, k = 1, 2$. Then Ω is the equivalent four-dimensional space of points $(y_{P1}, \sigma_1, y_{P2}, \sigma_2)$ where $-\infty < y_{Pk} < \infty$ and $\sigma_k > 0, k = 1, 2$. Also, then the null hypothesis is $y_{P1} = y_{P2}$, which is common form (2). Further reparametrization by replacing y_{P2} by $\Delta = y_{P1} - y_{P2}$ yields a null hypothesis $\Delta = 0$, which is common form (1). In general, forms (1) and (2) are equivalent through such a reparametrization.

Later results generally hold for a subspace ω of the null hypothesis that is expressible in common form (1) or (2).

A null hypothesis subspace ω can also be specified in terms of c equality constraints:

$$h_1(\theta_1,\ldots,\theta_r)=0,\ldots, h_c(\theta_1,\ldots,\theta_r)=0; \tag{6.1}$$

these must have continuous first partial derivatives. For Example C, another possible null hypothesis is $p_1=p_2^2$ or $p_1-p_2^2=0$. For models with such constraints, the method of Lagrange multipliers may be useful for finding ML estimates.

Note that the null hypothesis model must be a special case of the general model. The likelihood ratio test does not apply to a test between, say, a normal distribution and an exponential distribution. It does apply to a test between a Weibull distribution and an exponential distribution, since the exponential distribution is a Weibull distribution with a shape parameter of 1.

Sample likelihoods. We use the sample likelihood (probability model) for $Y_1,\ldots, Y_n$ (assumed only for simplicity here to be observed values):

$$L(y_1,\ldots, y_n; \mathbf{\Theta})= f_{\mathbf{\Theta}}(y_1,\ldots, y_n), \tag{6.2}$$

where $\mathbf{\Theta}$ is in the parameter space Ω or in the subspace ω of the null hypothesis.

Example A.   Under Ω, $L(y_1,\ldots, y_n; \mu, \sigma^2)=(2\pi\sigma^2)^{-1/2}\exp[-(y_1-\mu)^2/(2\sigma^2)]\cdots(2\pi\sigma^2)^{-1/2}\exp[-(y_n-\mu)^2/(2\sigma^2)]$. Under ω, $L(y_1,\ldots, y_n; \mu_0, \sigma^2)=(2\pi\sigma^2)^{-1/2}\exp[-(y_1-\mu_0)^2/(2\sigma^2)]\cdots(2\pi\sigma^2)^{-1/2}\exp[-(y_n-\mu_0)^2/(2\sigma^2)]$.

Example B.   Under Ω, $L(y_1,\ldots, y_n; p)= p^{y_1}(1-p)^{1-y_1}\cdots p^{y_n}(1-p)^{1-y_n}$. Under ω, $L(y_1,\ldots, y_n; p_0)= p_0^{y_1}(1-p_0)^{1-y_1}\cdots p_0^{y_n}(1-p_0)^{1-y_n}$.

Example C.   Under Ω, $L(y_1, y_2; p_1, p_2)=\binom{n_1}{y_1}p_1^{y_1}(1-p_1)^{n_1-y_1}\binom{n_2}{y_2}p_2^{y_2}(1-p_2)^{n_2-y_2}$. Under ω, $p_1=p_2=p$ and $L(y_1, y_2; p)=\binom{n_1}{y_1}\binom{n_2}{y_2}p^{y_1+y_2}(1-p)^{n_1+n_2-y_1-y_2}$.

Likelihood Ratio Test

Likelihood ratio.   For observations $y_1,\ldots, y_n$, the **likelihood ratio** (LR) for testing the null hypothesis ω is

$$\lambda=\lambda(y_1,\ldots,y_n)\equiv\left[\max_{\mathbf{\Theta}\text{ in }\omega} L(y_1,\ldots,y_n; \mathbf{\Theta})\right]\bigg/\left[\max_{\mathbf{\Theta}\text{ in }\Omega} L(y_1,\ldots,y_n; \mathbf{\Theta})\right]. \tag{6.3}$$

λ is a function of $y_1,\ldots,y_n$. It is a statistic, since it does not depend on the unknown true Θ value. The Θ value that maximizes the numerator (denominator) likelihood is the ML estimate $\hat{\Theta}(y_1,\ldots,y_n)$ with respect to ω (Ω). Denote the numerator maximum value by $\hat{L}_\omega$ and the denominator maximum value by $\hat{L}_\Omega$; they are usually found by the calculus method of equating to zero the derivative of the likelihood with respect to each parameter and solving the resulting equations as described in Chapter 8 to get the parameter estimates.

Example A.   Under Ω, the ML estimates are $\hat{\mu} = \bar{y}$ and $\hat{\sigma}^2 = \Sigma_{i=1}^n (y_i - \bar{y})^2/n$. So

$$\hat{L}_\Omega = \max_\Omega L\left(y_1,\ldots, y_n; \mu, \sigma^2\right)$$

$$= \left(2\pi \sum_{i=1}^n (y_i - \bar{y})^2/n\right)^{-n/2} \exp\left[\sum_{i=1}^n (y_i - \bar{y})^2 \Big/ \left(2 \sum_{i=1}^n (y_i - \bar{y})^2/n\right)\right]$$

$$= (2\pi)^{-n/2} \left(\sum_{i=1}^n (y_i - \bar{y})^2/n\right)^{-n/2} e^{-n/2}.$$

Under ω, $\mu = \mu_0$ and $\hat{\sigma}^2 = \Sigma_{i=1}^n (y_i - \mu_0)^2/n$. This maximizes

$$L_\omega\left(y_1,\ldots, y_n; \mu_0, \sigma^2\right) = (2\pi\sigma^2)^{-n/2} \exp\left(- \sum_{i=1}^n (y_i - \mu_0)^2/(2\sigma^2)\right).$$

So

$$\hat{L}_\omega = \left(2\pi \sum_{i=1}^n (y_i - \mu_0)^2/n\right)^{-n/2} e^{-n/2},$$

and the likelihood ratio is

$$\lambda = \left(\sum_{i=1}^n (y_i - \bar{y})^2/n\right)^{n/2} \Big/ \left(\sum_{i=1}^n (y_i - \mu_0)^2/n\right)^{n/2}. \qquad (6.4)$$

Example B.   Under Ω, $\hat{p} = y/n$, where $y = y_1 + \cdots + y_n$ is the number of successes. Under ω, $p = p_0$, since p_0 is the only allowed value. Then $\hat{L}_\Omega = (y/n)^y [1 - (y/n)]^{n-y}$ and $\hat{L}_\omega = p_0^y (1 - p_0)^{n-y}$. The likelihood ratio is

$$\lambda = p_0^y (1 - p_0)^{n-y} \Big/ \left\{(y/n)^y [1 - (y/n)]^{n-y}\right\}. \qquad (6.5)$$

Example C.   Under Ω, the ML estimates are $\hat{p}_1 = y_1/n_1$ and $\hat{p}_2 = y_2/n_2$. Then

$$\hat{L}_\Omega = \binom{n_1}{y_1}(y_1/n_1)^{y_1}[1-(y_1/n_1)]^{n_1-y_1}\binom{n_2}{y_2}(y_2/n_2)^{y_2}[1-(y_2/n_2)]^{n_2-y_2}.$$

Under ω, $\hat{p} = (y_1 + y_2)/(n_1 + n_2)$ and

$$\hat{L}_\omega = \binom{n_1}{y_1}\binom{n_2}{y_2}\left(\frac{y_1+y_2}{n_1+n_2}\right)^{y_1+y_2}\left(1-\frac{y_1+y_2}{n_1+n_2}\right)^{(n_1+n_2)-(y_1+y_2)}.$$

The likelihood ratio is

$$\lambda = [(y_1+y_2)/(n_1+n_2)]^{y_1+y_2}\{1-[(y_1+y_2)/(n_1+n_2)]\}^{n_1+n_2-(y_1+y_2)}$$
$$\bigg/ \{(y_1/n_1)^{y_1}[1-(y_1/n_1)]^{n_1-y_1}(y_2/n_2)^{y_2}[1-(y_2/n_2)]^{n_2-y_2}\}. \quad (6.6)$$

Likelihood ratio test.   For different samples, the likelihood ratio λ would take on different values between 0 and 1. That is, it is a function of random variables $Y_1, \ldots, Y_n$, and it is a random variable. An observed λ value near 1 indicates that the corresponding values of $Y_1, \ldots, Y_n$ are likely under ω, and a λ value near 0 indicates that the $Y_1, \ldots, Y_n$ are not likely under ω compared to under $\Omega - \omega$. This suggests the test

1.   If $\lambda \leqslant C_\alpha$, reject ω (null hypothesis).
2.   If $\lambda > C_\alpha$, accept ω.

This is called the **likelihood ratio test**. The C_α is chosen so

$$\max_{\boldsymbol{\Theta}\text{ in }\omega} P_{\boldsymbol{\Theta}}(\text{reject }\omega) = \max_{\boldsymbol{\Theta}\text{ in }\omega} P_{\boldsymbol{\Theta}}(\lambda \leqslant C_\alpha) = \alpha;$$

that is, the test has level α.

To find such, a C_α, one must know the sampling distribution of λ. Two methods can be used to obtain C_α:

1.   Show that λ is a function of a statistic U with a known distribution and that the likelihood ratio test defines the same critical (rejection) region as a test based on U.

2.   Use an asymptotic (large-sample) approximation to the distribution of λ under the null hypothesis to get an approximate critical region.

Example A. Method (1) can be used to rewrite the likelihood ratio as

$$\lambda = \left(\sum_{i=1}^{n} (y_i - \bar{y})^2 \Big/ \sum_{i=1}^{n} [(y_i - \bar{y}) + (\bar{y} - \mu_0)]^2 \right)^{n/2}$$

$$= \left\{ \sum_i (y_i - \bar{y})^2 \Big/ \left[\sum_i (y_i - \bar{y})^2 + n(\bar{y} - \mu_0)^2 \right] \right\}^{n/2}$$

$$= \left\{ 1 + [t^2/(n-1)] \right\}^{-n/2}, \tag{6.7}$$

where

$$t = (\bar{y} - \mu_0) n^{1/2} \Big/ \left[\sum_i (y_i - \bar{y})^2 / (n-1) \right]^{1/2}$$

is the t statistic. Since λ is a monotone function of $|t|$, the critical region for the likelihood ratio test

$$\lambda = \left(1 + \frac{1}{n-1} t^2 \right)^{-n/2} < C_\alpha$$

is equivalent to the critical region $|t| > t(1 - 0.5\alpha, n - 1) = (n-1)(C_\alpha^{-2/n} - 1)$, where $t(1 - 0.5\alpha, n - 1)$ is the $100(1 - 0.5\alpha)$th t percentile with $n - 1$ degrees of freedom. Thus the likelihood ratio test is equivalent to the usual two-sided t-test for testing $\mu = \mu_0$ versus $\mu \neq \mu_0$ when σ^2 is unknown. Rather than work with λ and C_α, it is more convenient to work with the equivalent t statistic. They both provide the same test, that is, the same critical region and hence the same OC function.

Often it is more convenient to work with the maximum log likelihoods $\hat{\ell}_\Omega = \ln(\hat{L}_\Omega)$ and $\hat{\ell}_\omega = \ln(\hat{L}_\omega)$ and the equivalent statistic $T = -2\ln(\lambda) = 2(\hat{\ell}_\Omega - \hat{\ell}_\omega)$.

Figure 6.3 adds insight to the likelihood ratio test. There a sample log likelihood $\ell(\alpha, \beta)$ is a function of all points (α, β) in Ω, which is the horizontal plane; $r = 2$ for the general model. The subspace ω for the null hypothesis $\beta = \beta_0$ is the labeled line in Figure 6.3; $r' = 1$ for the null hypothesis subspace. Figure 6.3 shows ML estimates $\hat{\alpha}, \hat{\beta}$ under Ω and the maximum log likelihood $\hat{\ell}_\Omega$. Figure 6.3 also shows the ML estimate $\hat{\alpha}', \beta_0$ under ω and the maximum log likelihood $\hat{\ell}_\omega$. If $\hat{\ell}_\Omega$ is much above $\hat{\ell}_\omega$, the general model fits the data much better than the null hypothesis model; this corresponds to a large value of $T = -2 \cdot \ln(\lambda) = 2(\hat{\ell}_\Omega - \hat{\ell}_\omega)$. For different

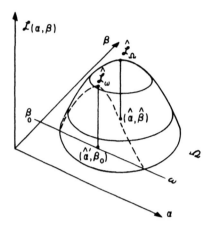

Figure 6.3. Likelihood function and optima.

samples, the log likelihood $\mathcal{L}(\alpha, \beta)$ differs, and thus so do $\hat{\alpha}, \hat{\beta}, \hat{\alpha}', \hat{\mathcal{L}}_\Omega, \hat{\mathcal{L}}_\omega$, and T.

Approximate test.   Method (1) does not work on Examples B and C. The asymptotic method does; it is presented next. First let r denote the number of dimensions of the parameter space Ω and let r' denote the number of dimensions of the null hypothesis space ω. It is useful to note that r is the number of parameters that must be estimated in the general model, and r' is the number that must be estimated in the null hypothesis model.

Example A.   As Figure 6.1a shows, Ω is a half-plane and has $r=2$ dimensions. As Figure 6.2a shows, ω is a semi-infinite line and has $r'=1$ dimension.

Example B.   As Figure 6.1b shows, Ω is a line segment and has $r=1$ dimension. As Figure 6.2b shows, ω is a point and has $r'=0$ dimensions.

Example C.   As Figure 6.1c shows, Ω is a square and has $r=2$ dimensions. As Figure 6.2c shows, ω is a line segment and has $r'=1$ dimension.

The large-sample distribution of $\ln(\lambda)$ is given by the following "theorem".

Theorem.   Under certain mild regularity conditions on f_θ, ω, and Ω, when $(r-r')>0$, the asymptotic distribution (large sample size) of twice the log likelihood ratio statistic

$$T = 2 \cdot \ln(\lambda) = 2(\hat{\mathcal{L}}_\Omega - \hat{\mathcal{L}}_\omega) \tag{6.8}$$

is chi square with $(r-r')$ degrees of freedom under the null hypothesis ω.

Here $\hat{\ell}_\Omega = \ln(\hat{L}_\Omega)$ and $\hat{\ell}_\omega = \ln(\hat{L}_\omega)$ are the maximum log likelihoods. Rao (1973) and Wilks (1962) state such regularity conditions, which are satisfied in most applications.

Under the conditions of the theorem, an approximate level α test of the null hypothesis ω is

1. If $T = 2(\hat{\ell}_\Omega - \hat{\ell}_\omega) \leqslant \chi^2(1-\alpha, r-r')$, accept ω.
2. If $T = 2(\hat{\ell}_\Omega - \hat{\ell}_\omega) > \chi^2(1-\alpha, r-r')$, reject ω.

Here $\chi^2(1-\alpha, r-r')$ is the $100(1-\alpha)$th chi-square percentile with $r-r'$ degrees of freedom.

There is no simple rule of thumb that indicates when the chi-square approximation is satisfactory. This means that the true level of the test differs from α. As a practical matter, one often has only the approximation and must use it, since it is better than no test. Then marginally significant results are not convincing.

The approximation can sometimes be improved by transforming parameters to reparametrize the model. For example, $\theta'_k = g_k(\theta_1, \ldots, \theta_K)$ for $k = 1, \ldots, K$. The functions $g_k(\ )$ are chosen to make the distribution of $\hat{\theta}'_k$ closer to normal. Typical transformations are $\sigma' = \ln(\sigma)$ for $\sigma > 0$ and $p' = \ln[p/(1-p)]$ for $0 < p < 1$. Such transformations do not affect the value of the likelihood ratio test statistic.

Example B. Let $Y = Y_1 + \cdots + Y_n$. Then

$$T = -2\{ Y \cdot \ln(p_0) + (n-Y)\ln(1-p_0)$$
$$- Y \cdot \ln(Y/n) - (n-Y)\ln[1-(Y/n)]\}$$
$$= 2((n-Y)\ln\{[1-(Y/n)]/(1-p_0)\} + Y \cdot \ln[(Y/n)/p_0]). \quad (6.9)$$

For large n, T is approximately chi-square distributed with $r - r' = 1 - 0 = 1$ degree of freedom under ω. For large samples, this is equivalent to the usual chi-square statistic for testing $p = p_0$ against $p \neq p_0$, namely, $Q = [(Y - np_0)^2/(np_0)] + \{[(n-Y) - n(1-p_0)]^2/[n(1-p_0)]\}$, which also has one degree of freedom.

Example C. Let $Y = Y_1 + Y_2$ and $n = n_1 + n_2$. Then

$$T = -2\{ Y \cdot \ln(Y/n) + (n-Y)\ln[1-(Y/n)]$$
$$- Y_1 \ln(Y_1/n_1) - (n_1-Y_1)\ln[1-(Y_1/n_1)] \quad (6.10)$$
$$- Y_2 \ln(Y_2/n_2) - (n_2-Y_2)\ln[1-(Y_2/n_2)]\}.$$

For n_1 and n_2 large, T is approximately chi-square distributed with $(2-1)=1$ degree of freedom under the null hypothesis ω ($p_1 = p_2$).

Example A.   Here

$$T = -2\left[(n/2)\ln\left(\sum_{i=1}^{n}(Y_i-\bar{Y})^2/n\right) - (n/2)\ln\left(\sum_{i=1}^{n}(Y_i-\mu_0)^2/n\right)\right].$$

(6.11)

For large n, T is approximately chi-square distributed with $r-r'=2-1=1$ degree of freedom under the null hypothesis. Since the exact distribution of the equivalent t statistic is known, this approximation is unnecessary.

OC function and consistency.   The OC function $P_{\Theta}\{\lambda \leqslant C_\alpha\}$ gives the performance of a likelihood ratio test as a function of Θ in Ω. Sometimes the distribution of λ is known or the test is equivalent to a known one, as in Example A; then one can get the OC function. If, as in Examples B and C, the sampling distribution of T is not known, one cannot find the exact OC function. The following theorem assures that the test is good for a large sample.

First we need a definition. A test of level α of the null hypothesis ω against the alternative $\Omega - \omega$ is called a **consistent test** if $P_{\Theta}\{\text{reject } \omega\} \to 1$ as the sample size $n \to \infty$ for any Θ in $\Omega - \omega$. This just says that the test is almost sure to reject the null hypothesis when it is false (i.e., Θ is in $\Omega - \omega$) if the sample size is large enough.

For one example, consider $Y_1,\ldots,Y_n$ from a normal distribution with unknown mean μ and known standard deviation σ_0. Consider testing ω: $\mu = \mu_0$ against $\mu \neq \mu_0$. The level α test is: if $|\bar{Y}-\mu_0| \leqslant K_{1-\alpha}\sigma_0/n^{1/2}$, accept ω, and if $|\bar{Y}-\mu_0| > K_{1-\alpha}\sigma_0/n^{1/2}$, reject ω. Then $P_\mu\{\text{reject } \omega\} = P_\mu\{|\bar{Y}-\mu_0| > K_{1-\alpha}\sigma_0/n^{1/2}\} = \Phi\{[\mu_0 - K_{1-\alpha}(\sigma_0/n^{1/2}) - \mu](n^{1/2}/\sigma_0)\} + 1 - \Phi\{[\mu_0 + K_{1-\alpha}(\sigma_0/n^{1/2}) - \mu](n^{1/2}/\sigma_0)\}$. For $\mu \neq \mu_0$, $P_\mu\{\text{reject } \omega\} \to 1$ as $n \to \infty$. So the test is consistent.

The following theorem states a general result on the consistency of LR tests.

Theorem.   The likelihood ratio test is consistent under some mild conditions on f_{Θ}, Ω, and ω. That is, if T_n is the log likelihood ratio based on sample size n, then $P_{\Theta}\{T_n > \chi^2(1-\alpha, r-r')\} \to 1$ as $n \to \infty$ for any Θ in $\Omega - \omega$. Moreover, no other test achieves this limit faster.

Consequently, the likelihood ratio test is said to be "asymptotically" uniformly most powerful. (This last sentence is not intended to be precise.) Rao (1973) and Wilks (1962) state such conditions and prove the theorem.

Test that K parameters are equal. The following applies the above general theory to get the methods in Section 5 for testing for equality of K parameter values. Suppose that there are K independent samples, and sample k comes from a distribution with parameters α_k, β_k. The parameter space Ω consists of all allowed points $(\alpha_1, \beta_1, \ldots, \alpha_K, \beta_K)$; Ω has $r = 2K$ dimensions. The subspace ω for the null (equality) hypothesis that $\beta_1 = \cdots = \beta_K = \beta$ consists of all allowed points $(\alpha_1, \beta, \ldots, \alpha_K, \beta)$; ω has $r' = K + 1$ dimensions.

Suppose that sample k has log likelihood $\mathcal{L}_k(\alpha_k, \beta_k)$. Then the log likelihood for the general model is $\mathcal{L}_\Omega(\alpha_1, \beta_1, \ldots, \alpha_K, \beta_K) = \mathcal{L}_1(\alpha_1, \beta_1) + \cdots + \mathcal{L}_K(\alpha_K, \beta_K)$. Also, the log likelihood for the null hypothesis model is $\mathcal{L}_\omega(\alpha_1, \ldots, \alpha_K, \beta) = \mathcal{L}_1(\alpha_1, \beta) + \cdots + \mathcal{L}_K(\alpha_K, \beta)$. Under Ω, the ML estimates α_k, β_k maximize $\mathcal{L}_k(\alpha_k, \beta_k)$; that is, the distribution is separately fitted to sample k to obtain $\hat{\alpha}_k, \hat{\beta}_k$. Then the maximum log likelihood is $\hat{\mathcal{L}}_\Omega = \mathcal{L}_1(\hat{\alpha}_1, \hat{\beta}_1) + \cdots + \mathcal{L}_K(\hat{\alpha}_K, \hat{\beta}_K)$. Under ω, the ML estimates $\hat{\alpha}'_1, \ldots, \hat{\alpha}'_K, \hat{\beta}$ maximize $\mathcal{L}_\omega(\alpha_1, \ldots, \alpha_K, \beta)$, and its maximum value is $\hat{\mathcal{L}}_\omega$. The test statistic is $T = 2(\hat{\mathcal{L}}_\Omega - \hat{\mathcal{L}}_\omega)$ and has $r - r' = 2K - (K + 1) = K - 1$ degrees of freedom, as stated in Section 5.

Under Ω, the Fisher and covariance matrices for the $\hat{\alpha}_k, \hat{\beta}_k$ are block diagonal, where $\mathrm{Cov}(\hat{\alpha}_k, \hat{\alpha}_{k'}) = \mathrm{Cov}(\hat{\beta}_k, \hat{\beta}_{k'}) = \mathrm{Cov}(\hat{\alpha}_k, \hat{\beta}_{k'}) = 0$ for $k \neq k'$.

McCool (1977, 1978a, b, 1979) investigates this test for comparing Weibull scale parameters. He gives percentiles of T for $K = 2(1)5, 10, n = 5$ ($r = 3, 5$), $n = 10 (r = 5, 10)$. He also gives the OC curves for $K = 2, 10$ and the same n and r.

Test that K distributions are identical. The following applies the above general theory to get the methods in Section 5 for testing that K distributions are identical. Suppose that there are K independent samples and sample k comes from a distribution with parameters α_k, β_k. The parameter space Ω consists of all allowed points $(\alpha_1, \beta_1, \ldots, \alpha_K, \beta_K)$; Ω has $r = 2K$ dimensions. The null (equality) hypothesis is $\alpha_1 = \cdots = \alpha_K$ and $\beta_1 = \cdots = \beta_K$. Its subspace ω consists of all allowed points $(\alpha, \beta, \ldots, \alpha, \beta)$, where α and β are the common values; ω has $r' = 2$ dimensions.

Suppose that sample k has log likelihood $\mathcal{L}_k(\alpha_k, \beta_k)$. Then the log likelihood for the general model is $\mathcal{L}_\Omega(\alpha_1, \beta_1, \ldots, \alpha_K, \beta_K) = \mathcal{L}_1(\alpha_1, \beta_1) + \cdots + \mathcal{L}_K(\alpha_K, \beta_K)$. The log likelihood for the null hypothesis model is $\mathcal{L}_\omega(\alpha, \beta) = \mathcal{L}_1(\alpha, \beta) + \cdots + \mathcal{L}_K(\alpha, \beta)$. Under Ω, the ML estimates $\hat{\alpha}_k, \hat{\beta}_k$ maximize $\mathcal{L}_k(\alpha_k, \beta_k)$; that is, the distribution is separately fitted to sample k to obtain $\hat{\alpha}_k, \hat{\beta}_k$. Then the maximum log likelihood is $\hat{\mathcal{L}}_\Omega = \mathcal{L}_1(\hat{\alpha}_1, \hat{\beta}_1) + \cdots + \mathcal{L}_K(\hat{\alpha}_K, \hat{\beta}_K)$. Under ω, the ML estimates $\hat{\alpha}, \hat{\beta}$ maximize $\mathcal{L}_\omega(\alpha, \beta)$, and its maximum value is $\hat{\mathcal{L}}_\omega$, where a single distribution is fitted to the pooled data from all K samples to get $\hat{\alpha}, \hat{\beta}$, since $\mathcal{L}(\alpha, \beta)$ is the likelihood for a single distribution and the pooled data. The test statistic is $T = 2(\hat{\mathcal{L}}_\Omega - \hat{\mathcal{L}}_\omega)$ and has $r - r' = 2K - 2$ degrees of freedom.

Under Ω, the Fisher and covariance matrices for the $\hat{\alpha}_k, \hat{\beta}_k$ are block diagonal. Under ω, the Fisher and covariance matrices for $\hat{\alpha}$ and $\hat{\beta}$ are 2×2.

Related Tests

Rao's equivalent test.   Rao (1973, p. 418) gives the following test statistic; it is asymptotically equivalent to the likelihood ratio statistic. Suppose that the $r \times 1$ column vector of scores is

$$\mathbf{S}(\mathbf{\Theta}) = (\partial \mathcal{L} / \partial \theta_1, \ldots, \partial \mathcal{L} / \partial \theta_r)', \tag{6.12}$$

where $\mathcal{L}$ is the sample log likelihood and $\mathbf{\Theta} = (\theta_1, \ldots, \theta_r)$ in Ω is the vector of the r parameters under the general model. Denote the $r \times r$ Fisher matrix under Ω by

$$\mathbf{F}(\mathbf{\Theta}) = \left\{ - E\left[\partial^2 \mathcal{L} / \partial \theta_i \partial \theta_j \right] \right\}$$

$$= \left\{ - E\left[(\partial \mathcal{L} / \partial \theta_i)(\partial \mathcal{L} / \partial \theta_j) \right] \right\}, \qquad i, j = 1, \ldots, r. \tag{6.13}$$

Also, suppose that $\hat{\mathbf{\Theta}} = (\hat{\theta}_1, \ldots, \hat{\theta}_r)'$ is the $r \times 1$ column vector of ML estimates under the null hypothesis model ω; for example, some $\hat{\theta}_i$ will be constants or equal to other $\hat{\theta}_i$ or functions of other $\hat{\theta}_j$. Rao's statistic for testing the null hypothesis that Ω is in ω is the quadratic form

$$R = \mathbf{S}'(\hat{\mathbf{\Theta}})\left[\mathbf{F}(\hat{\mathbf{\Theta}}) \right]^{-1} \mathbf{S}(\hat{\mathbf{\Theta}}). \tag{6.14}$$

Under the null hypothesis, R is asymptotically equal to the log likelihood ratio statistic (6.8) and to Wald's test statistic (6.19). That is, R then has a chi-square distribution with $r - r'$ degrees of freedom. R is not convenient for multiply censored data, since the expectations (6.13) are difficult to calculate for such data. In practice one can use the local estimate of the Fisher information matrix as it does not use the expectations. This is done in the SURVREG program of Preston and Clarkson (1980). R employs only the ML estimates under the null hypothesis; thus, use of R avoids the labor of calculating the ML estimates for the general model.

Example on equality of Poisson λ_k.   The following is an example of Rao's test. Suppose that $Y_1, \ldots, Y_K$ are independent Poisson counts, where λ_k is the occurrence rate and t_k the length of observation, $k = 1, \ldots, K$. The null (equality) hypothesis is $\lambda_1 = \cdots = \lambda_K$. The sample log likelihood under

Ω is

$$\mathcal{L}(\lambda_1,\ldots,\lambda_K)= \sum_{k=1}^{K} \left[-\lambda_k t_k + Y_k \ln(\lambda_k t_k) - \ln(Y_k!)\right].$$

The kth score is $\partial\mathcal{L}/\partial\lambda_k = -t_k + (Y_k/\lambda_k), k=1,\ldots, K$. Under the null hypothesis, $\overset{*}{\lambda}_1 = \cdots = \overset{*}{\lambda}_K = \hat{\lambda} = (Y_1 + \cdots + Y_K)/(t_1 + \cdots + t_K)$, and the vector of scores is

$$\mathbf{s}(\overset{*}{\lambda}_1,\ldots,\overset{*}{\lambda}_K) = \left[-t_1 + (Y_1/\hat{\lambda}),\ldots, -t_K + (Y_K/\hat{\lambda})\right]'.$$

The terms in the Fisher matrix are

$$E\{-\partial^2\mathcal{L}/\partial\lambda_k^2\} = E\{Y_k/\lambda_k^2\} = \lambda_k t_k/\lambda_k^2 = t_k/\lambda_k,$$

$$E\{-\partial^2\mathcal{L}/\partial\lambda_k\partial\lambda_{k'}\} = E\{0\}=0 \qquad \text{for } k \neq k'.$$

This Fisher matrix is diagonal. Its estimate under the null hypothesis has $\overset{*}{\lambda}_k = \hat{\lambda}$ in place of λ_k. Then Rao's test statistic is

$$R = \left[(Y_1/\hat{\lambda}) - t_1,\ldots,(Y_K/\hat{\lambda}) - t_K\right] \begin{bmatrix} t_1/\hat{\lambda} & & 0 \\ & \ddots & \\ 0 & & t_K/\hat{\lambda} \end{bmatrix}^{-1} \begin{bmatrix} (Y_1/\hat{\lambda}) - t_1 \\ \vdots \\ (Y_K/\hat{\lambda}) - t_K \end{bmatrix}$$

$$= \sum_{k=1}^{K} (Y_k - \hat{\lambda}t_k)^2/(\hat{\lambda}t_k). \tag{6.15}$$

This is the chi-square test statistic (quadratic form) for equality of Poisson occurrence rates (Chapter 10). Under the null hypothesis, R has an asymptotic distribution that is chi square with $K-1$ degrees of freedom.

Wald's equivalent test. Rao (1973, p. 419) gives Wald's test statistic, which is asymptotically equivalent to the likelihood ratio statistic. Suppose that the subspace ω of the null hypothesis is specified by c constraints

$$h_1(\theta_1,\ldots,\theta_r)=0,\ldots, h_c(\theta_1,\ldots,\theta_r)=0. \tag{6.16}$$

The test uses the $r \times c$ matrix of partial derivatives

$$\mathbf{H}(\boldsymbol{\Theta})= \{\partial h_i/\partial\theta_j\}, \qquad i=1,\ldots, c, \quad j=1,\ldots, r, \tag{6.17}$$

which depends on Θ. Suppose that $\hat{\Theta} = (\hat{\theta}_1, \ldots, \hat{\theta}_r)'$ are the ML estimates under Ω, the general model, and their asymptotic covariance matrix is $\Sigma_{\hat{\Theta}}(\Theta)$, which depends on Θ. $\hat{h} = h(\hat{\Theta})$ denotes the $c \times 1$ vector of constraints evaluated at $\hat{\Theta}$. The asymptotic covariance matrix of $h(\hat{\Theta})$ is the $c \times c$ matrix

$$\Sigma_{\hat{h}}(\Theta) = H'(\Theta)\Sigma_{\hat{\Theta}}(\Theta)H(\Theta). \tag{6.18}$$

Wald's statistic for testing the null hypothesis (6.16) is

$$W = h'(\hat{\Theta})\left(\Sigma_{\hat{h}}(\hat{\Theta})\right)^{-1}h(\hat{\Theta}). \tag{6.19}$$

This is a quadratic form in the observed values of the constraints and is a measure of how close they are to zero. Under the null hypothesis (6.16), W is asymptotically equal to the log likelihood ratio statistic (6.8) and to Rao's test statistic (6.14); that is, W has a chi square distribution with degrees of freedom. W is convenient to use with multiply censored data. Then one can use the local estimate of $\Sigma_{\hat{\Theta}}(\Theta)$ in place of the ML estimate $\Sigma_{\hat{\Theta}}(\hat{\Theta})$ in (6.18). W employs only the ML estimates under the general model. This is convenient, as it is often difficult to calculate estimates for the null hypothesis model without special computer programs.

Snubber example.   Table 3.1 shows life test data on old and new types of snubbers, which are assumed to have normal life distributions. One comparison of the two types is a test of the hypothesis ω: $\mu_O = \mu_N$, assuming $\sigma_O = \sigma_N = \sigma$ against the alternative Ω: $\mu_O \neq \mu_N$, assuming $\sigma_O = \sigma_N = \sigma$. Expressed as a constraint, the null hypothesis is

$$h_1(\mu_O, \mu_N, \sigma) = \mu_O - \mu_N = 0.$$

The partial derivatives are

$$\partial h_1/\partial \mu_O = 1, \qquad \partial h_1/\partial \mu_N = -1, \qquad \partial h_1/\partial \sigma = 0.$$

The matrix of partial derivatives is (the column vector) $H = (1 \quad -1 \quad 0)'$. The ML estimates under Ω are $\hat{\mu}_O = 974.3$, $\hat{\mu}_N = 1061.3$, and $\hat{\sigma} = 458.4$. The local estimate of the covariance matrix of the ML estimates under Ω is

$$\Sigma_{\hat{\Theta}} = \begin{array}{ccc} \hat{\mu}_O & \hat{\mu}_N & \hat{\sigma} \\ \begin{bmatrix} 7930.61 & 1705.42 & 2325.45 \\ 1705.42 & 8515.93 & 2435.20 \\ 2325.45 & 2435.20 & 3320.58 \end{bmatrix} \end{array}.$$

The estimate of the 1×1 covariance matrix (6.18) of the constraint estimate is $\Sigma_{\hat{h}}^* = (1 \quad -1 \quad 0)\Sigma_{\hat{\theta}}^*(1 \quad -1 \quad 0)' = 13035.7$. The Wald statistic is $W = (974.3 - 1061.3)'(13035.7)^{-1}(974.3 - 1061.3) = 0.58$. Under $\mu_O = \mu_N$, the distribution of W is approximately chi square with one degree of freedom. Since $W < \chi^2(0.90, 1) = 2.706$, the two means do not differ significantly. The ML estimates and local estimate of their covariance matrix were obtained with STATPAC of Nelson and others (1978).

PROBLEMS

12.1. Power line outages. Use the data on the four longest power lines in Table 1.1 of Chapter 10.

(a) Use the chi-square test to compare their outage rates.

(b) Use the log likelihood ratio test to compare their outage rates.

(c) Calculate the pooled ML estimate of the common outage rate.

(d) Calculate two-sided (approximate) 95% confidence limits for the common outage rate.

(e) Calculate (exact) two-sided (approximate) 90% prediction limits for the total number of outages on the four lines in a coming year.

12.2 LR test for K binomial samples. Suppose that Y_k has a binomial distribution with sample size n_k and proportion p_k, $k = 1, \ldots, K$. All Y_k are statistically independent of each other.

(a) Derive the (log) likelihood ratio test statistic for equality $p_1 = \cdots = p_K$.

(b) How many degrees of freedom does the approximate chi-square distribution for this statistic have?

(c) Apply the test to the capacitor data of Chapter 10.

(d) Do you think that the asymptotic theory is adequate for the capacitor data? Explain why.

(e) Apply the test to the appliance component data of Problem 10.5, doing parts (b) through (e) of the problem.

(f) Do you think that the asymptotic theory is adequate for the appliance component data? Explain why.

12.3 Exponential means—singly censored. Suppose that one observes the first r_k order statistics of a sample of size n_k from an exponential distribution with mean θ_k, $k = 1, \ldots, K$, where the samples are statistically independent.

(a) Derive the test statistic of the likelihood ratio test for equality $\theta_1 = \cdots = \theta_K$.

(b) Use this test on the data in Table 2.1 of Chapter 7. Determine a simple approximate and satisfactory way of handling left and multiply censored samples there.

(c) Do you think that the asymptotic theory is adequate for this application? Explain why.

(d) Plot the samples on Weibull paper. Is a formal hypothesis test needed?

12.4* Exponential means—multiply censored. Do problem 10.5 for samples that are multiply time censored on the right.

12.5 Singly censored normal samples. Suppose that sample k is singly time censored at y_0 and comes from a normal distribution with mean μ_k and standard deviation σ_k, $k = 1, \ldots, K$. Suppose that sample k has r_k failures among its n_k units. Derive the (log) likelihood test for equality of the distributions as follows.

(a) For the appliance data in Table 2.1 of Chapter 11, make normal probability plots of the three samples. Are the plots convincing enough that a formal test for equality is not needed? State what the plots show.

(b) Write the combined log likelihood for the model with K different distributions.

(c) Write the log likelihood for a single common normal distribution with mean μ and standard deviation σ.

(d) Calculate the separate ML estimates for the three distributions. Use tables of Cohen (1961) or (easier) of Schmee and Nelson (1977). Evaluate the corresponding maximum log likelihood.

(e) Calculate the ML estimates for a common distribution. Use the same table used in (d). Evaluate the corresponding maximum log likelihood.

(f) Calculate the (log) likelihood ratio test statistic. How many degrees of freedom does its approximate chi-square distribution have?

(g) Test for equality of the three distributions.

(h) Do you think that the asymptotic theory is adequate for the cord data? Explain why.

12.6 Multiply censored normal samples. The appliance cord data in Table 2.1 of Chapter 11 are analyzed as follows to compare the samples

*Asterisk denotes laborious or difficult.

from the three tests. The following table summarizes the results of ML fitting a normal distribution to the separate and combined samples.

Data	$\hat{\mathcal{L}}$	$\hat{\mu}$	$\hat{\sigma}$	var($\hat{\mu}$)	var($\hat{\sigma}$)	cov($\hat{\mu}, \hat{\sigma}$)
Test 1	−51.73	132.26	34.7	105.182	65.077	8.302
Test 2	−49.87	126.64	44.3	177.912	123.098	24.568
Test 3	−49.40	127.60	41.5	155.712	110.140	29.491
All, common σ and μ	−151.37	128.75	40.3	48.324	32.452	5.619
All, common σ	−151.27	below	40.0	—	32.158	—

$\hat{\mu}_1 = 132.94$, $\hat{\mu}_2 = 125.81$, $\hat{\mu}_3 = 127.32$

(a) Calculate the log likelihood ratio test statistic for equality of the three standard deviations, assuming the means differ. Compare it with an appropriate chi-square percentile and state your conclusions.

(b) Calculate the log likelihood ratio test statistic for equality of the three means, assuming the standard deviations are equal. Compare it with an appropriate chi-square percentile and state your conclusions.

(c) Calculate the log likelihood ratio test statistic for equality of the three means and of the three standard deviations. Compare it with an appropriate chi-square percentile and state your conclusions.

(d) Tests 1 and 2 employed one type of cord and Test 3 another type. Explain in detail the analyses (fittings and test statistics) you would use to compare the two cord types.

(e*) Use Wald's test to compare the 1% points of the three distributions, assuming the means and standard deviations may differ.

12.7 Insulating oil. An experiment on breakdown voltage of an insulating oil used two electrode diameters and three rates of rise of voltage (in V/second)—six test conditions in all. Theory assumes that each distribution of breakdown voltage is Weibull, the scale parameter α_k depends on the test condition, but the shape is a constant β. For each test condition, there were 60 breakdown voltages. Separate Weibull fits to the six data sets yielded the following.

Condition	Diameter (in.)	Rate	$\hat{\alpha}_k$	$\hat{\beta}_k$	$\hat{\mathcal{L}}_k$
1	1	10	44.567	10.839	−174.70
2	1	100	50.677	12.526	−176.59
3	1	1000	60.055	13.316	−181.65
4	3	10	39.695	12.220	−165.58
5	3	100	46.248	16.453	−155.44
6	3	1000	50.860	14.681	−169.96

For the model with a common shape parameter, the results are

Condition	1	2	3	4	5	6
$\hat{\alpha}_k$	44.928	50.741	60.010	39.772	45.992	50.711

$\hat{\beta} = 12.997, \hat{\mathcal{L}}_\beta = -1029.04.$

(a) Calculate the log likelihood ratio test statistic for equality of the six shape parameters.

(b) How many degrees of freedom does this statistic have?

(c) Do the shape parameters differ statistically significantly? In Section 4, they have a wholly significant difference at the 5% level.

(d) Make a normal probability plot of the $\hat{\beta}_k$ and determine how the shape parameters differ.

12.8* Nonparametric comparison with grouped data and progressive censoring. For motivation, first see Problem 9.6, where interval data on K groups of transformers were pooled to fit a nonparametric multinomial model to the data. The following (log) likelihood ratio test compares the K groups for equality of corresponding proportions failing in each year. Suppose that the groups differ. Then, for group k, π_{km} denotes the expected proportion failing in year m ($m = 1, \ldots, k$), and $\pi_{k,k+1}$ denotes the expected proportion surviving the latest year (k) in service. Similarly, y_{km} denotes the observed number failing in year m, and $y_{k,k+1}$ denotes the observed number surviving the latest year (k) in service.

(a) For each transformer group, plot the sample cumulative distribution function on the same plotting paper. (Connect plotted points so that the plot is easier to see.) Do the distributions differ convincingly? How? Confidence limits for the cumulative distributions may help.

(b) Write the general expression for the log likelihood for the multinomial model for group k, $k = 1, \ldots, K$.

(c) Write the general total log likelihood for all groups (assuming that they differ).

(d) Derive the formulas for the ML estimates for the proportions π_{km}, $k = 1, \ldots, K$, $m = 1, \ldots, k + 1$.

(e) Evaluate the estimates (c) for the transformer data of Problem 9.6.

(f) Give the general expression for the maximum log likelihood for differing groups, and evaluate it for the transformer data.

(g) Use the results of Problem 9.6 to calculate the maximum log likelihood for the transformer data under the model there, with a common proportion failing in each year, all $\pi_{km} = \pi$.

(h) Calculate the log likelihood ratio statistic for testing for equality of group proportions within each year, $\pi_{1m} = \pi_{2m} = \cdots = \pi_{km}$.

(i) Give a general formula for the number of degrees of freedom of the test statistic, and evaluate it for the transformer data.

(j) Test whether the groups differ significantly. How do they differ?

(k) Do you think that the asymptotic theory is adequate for the transformer data? Explain why.

12.9 Circuit breaker. Use the data from Problem 3.4 and the results from Problem 9.7. Compare the two samples of size 18 from the new design using the likelihood ratio test, as follows. Use a computer program, if available.

(a) Assume that both samples come from the same exponential distribution with (unknown) true mean θ_0. Write the sample log likelihood $\mathcal{L}(\theta)$ for the combined samples.

(b) Calculate the ML estimate $\hat{\theta}$ for the assumed common exponential distribution and the corresponding maximum log likelihood $\mathcal{L}(\hat{\theta})$.

(c) Assume that the sample with failures comes from an exponential distribution with unknown true mean η_0, and the other sample has a true other proportion π_0 failed by 15,000 cycles. (There is no need to assume that the distribution is exponential.) Write the sample log likelihood $\mathcal{L}(\eta, \pi)$ for the two samples.

(d) Calculate the ML estimates $\hat{\eta}$ and $\hat{\pi}$ and the corresponding maximum log likelihood.

(e) Calculate the log likelihood ratio test statistic. How many degrees of freedom does it have? Do the samples differ statistically significantly?

(f) In your opinion, is the asymptotic theory crude or adequate for these samples? Explain.

13

Survey of Other Topics

This chapter briefly surveys some topics not covered in this book. Each survey briefly states the aims of the topic and gives key references. The survey is limited to statistical and probabilistic methods for reliability and life data analysis. So the survey omits reliability management, reliability physics, handbook data on component failure rates and life, and similar topics.

The survey includes the following.

1. System reliability models and their analyses.
2. Maintainability models and data analyses.
3. Availability models and data analyses.
4. Estimation of system reliability from component data.
5. Bayesian methods in reliability.
6. Reliability demonstration and acceptance tests.
7. Reliability growth (Duane plots).
8. Renewal theory and data analysis.
9. Life as a function of other variables.
10. Accelerated life testing.
11. Depreciation, replacement, and maintenance policies.
12. Books with reliability and life data analysis.
13. Abstracts, bibliographies, and indices on reliability and life data analysis.
14. Journals with papers on reliability and life data analysis.
15. Computerized literature searches.

558

1. SYSTEM RELIABILITY MODELS AND THEIR ANALYSES

Reliability books are concerned with methods for determining the reliability of a system of components from the component reliabilities. Such methods, which are described below, help designers to assess whether a design meets reliability requirements. Also, they show how design reliability can be improved through better components and other system designs, for example, including redundant components. Such reliability analyses can be repeated for different definitions of system failure and operating environments. Also, failure can be catastrophic or result from performance dropping below a specified value. Major approaches to system reliability analyses described below are FMEA, fault tree and other analyses, coherent structures, common mode failures, and simulation. These methods generally do not involve statistical data analysis, but some involve probability modeling.

FMEA. Failure Mode and Effects Analysis is a simplified approach to system reliability analysis. It involves identifying (1) each failure mode of a component and its cause, (2) the effect of such a failure on the component, and (3) the effect on the system performance or safety. This method can include combinations of component failure modes that cause system failure. Also, it assumes that the failure modes are statistically independent. For these reasons, FMEA overestimates system reliability. Its main value is that it helps one identify components that contribute most to unreliability. Shooman (1968, Sec. 3.7) and the General Electric reliability manual (1975, Sec. 5) describe FMEA in more detail.

Fault tree and other analyses. Fault tree analysis has two goals: (1) to identify the minimal cuts (the distinct combinations of component failures that cause system failure) and (2) to calculate system reliability. The analysis involves developing a model for system reliability in terms of a fault tree that expresses which combinations of component failures produce system failure. Fault tree analysis provides systematic methods for determining fault trees, minimal cuts, and system reliabilities. These methods are described in the following references.

Barlow, R. E., Fussell, J. B., and Singpurwalla, N. D., Eds. (1975), *Reliability and Fault Tree Analysis—Theoretical and Applied Aspects of System Reliability and Safety Assessment*, Society for Industrial and Applied Mathematics, 33 S. 17th Street, Philadelphia, PA 19103.

Fussell, J. B. (1973), "Fault Tree Analysis—Concepts and Techniques," presented at the NATO Advanced Study Institute of Generic Techniques of System Reliability Assessment, Liverpool, England, July 1973, Aerojet Nuclear Co., National Reactor Testing Station, Idaho Falls, ID 83401.

Fussel, J. B., Henry, E. B., and Marshall, N. H. (1974), "MOCUS—A Computer Program to Obtain Minimal Cut Sets from Fault Trees," Aerojet Nuclear Company Report ANCR-1156, Idaho Falls, ID 83401.

Fussell, J. B., and Burdick, G. R., Eds. (1977), *Nuclear Systems Reliability Engineering and Risk Assessment*, Proceedings of the 1977 International Conference on Nuclear Systems Reliability Engineering and Risk Assessment, Society for Industrial and Applied Math, 33 S. 17th St., Philadelphia, PA 19103.

General Electric Company Corporate Research and Development (1975), *Reliability Manual for Liquid Metal Fast Breeder Reactors*, Vols. 1 and 2, General Electric Company Corporate Research and Development Report SRD-75-064; see References for source. Section 6 has fault tree analysis.

Pradip, P., and Spector, M. (1975),"Computerized Fault Tree Analysis: TREEL and MICSUP," Operations Research Center Report ORC75-3, University of California, Berkeley, CA 94720.

Vesley, W. E., and Narum, R. E. (1970), "PREP and KITT: Computer Codes for the Automatic Evaluation of a Fault Tree," Idaho Nuclear Corp. Report IN-1349, Idaho Falls, ID 83401.

Wheeler, D. B., Hsuan, J. S., Duersch, R. R., and Roe, G. M. (1977), "Fault Tree Analysis Using Bit Manipulation," *IEEE Trans. Reliab.* **R-26**, 95–99.

The KITT and PREP programs for fault tree analysis are best known and widely available; more recent programs can handle larger trees and are faster. Both analytic and Monte Carlo methods are used to analyze fault trees. Fault tree analysis is surveyed by General Electric (1975, Sec. 6).

System reliability may also be analyzed by means of reliability block diagrams and Boolean algebra statements. Tillman and others (1977) review optimization techniques for system reliability.

Coherent structure analysis. Coherent structure functions are another means of representing a model for system reliability. A coherent structure function is a binary function that takes on the value 1 (0) for system success (failure). It is a function of the component indicator variables. Such a variable is 1 (0) for component success (failure). The aim of such models and analyses is to determine minimal cut sets and to calculate system reliability from component reliabilities. Theory for coherent structures appears in Barlow and Proschan (1975).

Common mode failure analysis. Most reliability analyses of large systems involve dividing the system into subsystems that are physically separate and analyzing the subsystems separately. Often such subsystems are not completely statistically independent, and common mode failure analysis attempts to take into account the common failure modes. For example, two separate safety shut-down systems for a nuclear reactor may fail together when an earthquake occurs. The General Electric reliability manual (1975, Sec. 7) describes common mode failure analysis.

Simulation. Often the reliability model for a system is too complex to handle by analytic methods. Then one may use Monte Carlo simulation on a computer to "build and run" a system many times to obtain estimates of system reliability and other measures of performance. For example, reliability simulation of Apollo moon flights were run; 1000 simulated flights took 10 hours on an IBM 7094. General methods for simulation appear in books, for example,

Fishman, G. S. (1978), *Principles of Discrete Event Simulation*, Wiley, New York.

Hammersley, J. M., and Hanscomb, D. C. (1964), *Monte Carlo Methods*, Methuen Monograph, Wiley, New York.

Kleijnen, J. P. C. (1975), *Statistical Techniques in Simulation*, Vols. 1 and 2, Dekker, New York.

Examples of applications to reliability analysis appear in General Electric (1975, Sec. 8).

Component failure rates. There are many handbooks on component failure rates. Some of the best known are

MILITARY HANDBOOK 217C, "Reliability Prediction of Electronic Equipment," 9 April 1979. Available from Naval Publications and Forms Center, 5801 Tabor Ave., Philadelphia, PA 19120.

Government–Industry Data Exchange Program (GIDEP) Reliability–Maintainability Data Summaries. Available from GIDEP Operations Center, Corona, CA 91720. This includes electronic and mechanical components.

INSPEC (1981) *Electronic Reliability Data: A Guide to Selected Components*, ISBN 852962 40 1, 225 pp, $300, from INSPEC, IEEE Service Center, 445 Hoes Lane, Piscataway, NJ 08854.

2. MAINTAINABILITY MODELS AND DATA ANALYSES

Repairable systems need to be designed for easy maintenance, that is, fast and cheap repair. Analytic methods for maintainability deal with models and distributions for the time to repair. Maintainability demonstration deals with sampling plans and data analyses to estimate the distribution of time to repair. For example, demonstration plans compare the mean and percentiles of the distribution of time to repair with specified values. The methods of this book apply to such data. The references below specifically treat maintainability.

Bird, G. T. (1969), "MIL-STD-471, Maintainability Demonstration," *J. Qual. Technol.* **1**, 134–148.

Blanchard, Jr., B. S., and Lowery, E. E. (1969), *Maintainability Principles and Practices*, McGraw-Hill, New York.

Department of Defense (1973), "MIL-STD-471A, Maintainability Demonstration," (March 1973). Available from the Commanding Officer, Naval Publications and Forms Center, 5801 Tabor Ave., Attn. NPFC 105, Philadelphia, PA 19120.

Gertsbakh, I. B. (1977), *Models of Preventive Maintenance*, North Holland, New York.

Goldman, A., and Slattery, T. (1964), *Maintainability: A Major Element of System Effectiveness*, Wiley, New York.

3.   AVAILABILITY MODELS AND DATA ANALYSES

Availability is a commonly used measure of performance of repairable systems. Such systems are working (up) of failed and awaiting or undergoing repair (down). The long run fraction or the time that a system is up is called the **steady state availability.** Models and analyses for this availability are like those for reliability. Availability is determined by the system's distributions of time to failure and time to repair. If μ (ν) is the mean time to failure (repair), the steady state availability is $A = \mu/(\mu + \nu)$; Sandler (below) calls this system reliability. This availability is used for engineering design. The probability of a system being up at a given instant in time is called the **instantaneous availability.** This availability is determined from queuing theory for finite source models and Markov chains. There are some methods for analyzing failure and repair time data to get availability information; Nelson (below) references such work. A few references on availability follow. Also, many of the reliability books listed in Section 11 contain some availability methods.

Barlow, R. E., and Proschan, F. (1965), *Mathematical Theory of Reliability*, Wiley, New York.

Jaiswal, N. K. (1966), "Finite Source Queuing Models," Case Institute of Technology Technical Memo No. 45, Cleveland, OH.

Nelson, W. (1970), "A Statistical Prediction Interval for Availability," *IEEE Trans. Reliab.* **R-19**, 179–182.

Sandler, G. H. (1963), *System Reliability Engineering*, Prentice-Hall, Englewood Cliffs, NJ.

4.   ESTIMATION OF SYSTEM RELIABILITY FROM COMPONENT DATA

A common problem is to calculate an estimate and confidence limits for system reliability from component reliability data. Such an analysis employs a model for the system reliability and distributions for component life to analyze such data. Most theoretical work involves either binary (success–failure) data on reliability or exponential component life distributions. Such work is surveyed by Mann, Schafer, and Singpurwalla (1974, Chap. 10). This field continues to develop, and the bibliographies and journals listed later give the latest methods.

5.   BAYESIAN METHODS IN RELIABILITY

Engineers who use Bayesian methods do so because these methods provide a formal means of including subjective information on product or component reliability in a reliability analysis of a system (or component). Such methods are used for probabilistic analysis of system reliability and for data analysis.

The analyst assigns (1) noninformative or (2) subjective distributions called "priors" to the parameters of the model for system reliability or performance. The priors express the uncertainties in the parameter values. The theory combines actual data with such priors to get new parameter distributions called "posteriors." The posteriors yield estimates and Bayesian confidence limits for the parameters. The prior distribution combines with the data and produces more precise estimates than would the data alone. Of course, the validity of the results of a Bayesian analysis strongly depends on the validity of the model and prior distributions. Bayesian methods are surveyed by Mann, Schafer, and Singpurwalla (1974, Chap. 8) and by General Electric (1975, Sec. 9). Locks (1973, Chap. 7) gives an introduction to Bayesian methods. This field continues to develop rapidly. Consult books, bibliographies, and journals listed in the following sections for the latest methods. Also see Tsokos and Shimi (1977).

6. RELIABILITY DEMONSTRATION AND ACCEPTANCE TESTS

Reliability demonstration and acceptance tests are used to assess whether product performance meets a specification. The specification usually requires that some parameter of the distribution of product performance surpass a specified value. For example, the mean life of an exponential distribution is to be above a specified value. Test plans specify the sample size, test time, test statistic, and criteria for the product to pass or fail the test. Many plans have been given in standards, books, and the literature dealing with reliability and quality control. Pabst (1975) reviews a number of standards for such tests. Such reviews, a regular feature in the *Journal of Quality Technology*, are generally more readable than the original standards. Mann, Schafer, and Singpurwalla (1974, Ch. 6) present the statistical theory of many such tests. Chapter 10 here briefly describes such demonstration tests and references standards containing such tests.

Selected standards follow.

MIL-STD-690B, "Failure Rate Sampling Plans and Procedures."

MIL-STD-781C, "Reliability Tests: Exponential Distribution."

MIL-HDBK-108, "Sampling Procedures and Tables for Life and Reliability Testing (Based on Exponential Distribution)."

TR-3, "Sampling Procedures and Tables for Life and Reliability Testing Based on the Weibull Distribution (Mean Life Criterion)."

TR-4, "Sampling Procedures and Tables for Life and Reliability Testing Based on the Weibull Distribution (Hazard Rate Criterion)."

TR-6, "Sampling Procedures and Tables for Life and Reliability Testing Based on the Weibull Distribution (Reliable Life Criterion)."

MIL-S-19500D, "General Specification for Semiconductor Devices."

These are available from the U.S. Government Printing Office, Washington, DC 20402.

7. RELIABILITY GROWTH (DUANE PLOTS)

In many development programs, hardware reliability grows during design, development, testing, and actual use. Such growth results from continuing engineering effort to improve the design, manufacture, and operation of repairable hardware. Reliability managers often need predictions of hardware reliability that will result from continued engineering effort. Reliability growth models and data analysis methods provide such predictions, confidence limits for the true reliability (or Poisson failure rate), and prediction limits for future numbers of failures. Such an analysis usually involves fitting to failure data an equation for the failure rate as a function of time. The best known analysis is the Duane (1964) plot, which is described by Codier (below). Gross and Clark (1975, Chap. 5) and Crow (below) present and reference some recent work. MIL-HDBK-189 (below) explains how to use the latest reliability growth methods.

Codier, E. O. (1968), "Reliability Growth in Real Life," *Proceedings of the 1968 Annual Symposium on Reliability*, pp. 458–469.

Crow, L. H. (1975), "Reliability Analysis for Complex, Repairable Systems," U.S. Army Material Systems Analysis Activity, Technical Report No. 138, Aberdeen Proving Ground, MD 21005. Also available as document AD-020296 from the Defense Documentation Center, Defense Logistics Agency, Cameron Station, Alexandria, VA 22314.

Crow, L. H. (1977), "Confidence Interval Procedures for Reliability Growth Analysis," U.S. Army Material Systems Analysis Activity, Technical Report No. 197, Aberdeen Proving Ground, MD 21005. Also available as document AD-A044788 from Defense Documentation Center, Defense Logistics Agency, Cameron Station, Alexandria, VA 22314.

Duane, J. T. (1964), "Learning Curve Approach to Reliability Monitoring," *IEEE Trans. Aerosp.* **2**, No. 2, 563–566.

MIL-HDBK-189, "Reliability Growth Management," Department of Defense, Washington, DC (1981). Available from the Commanding Officer, Naval Publications and Forms Center, 5801 Tabor Ave., Philadelphia, PA 19120.

8. RENEWAL THEORY AND DATA ANALYSIS

Many systems and products have components that are replaced (immediately) when they fail, and the replacement continues in service. For a fleet of such systems, one is often interested in the expected number of replacements as a function of calendar time; this function is called the "renewal function." It is useful for forecasting and planning needed numbers of replacements. Cox (1962), Parzen (1962, Chap. 5), and journals present probabilistic models and theory for renewal problems. The theory gives, for example, the relationship between the component life distribution and the renewal function.

One has **renewal data** if one knows only how long the system has been running when a component fails and one does not know if it is an original or a replacement. That is, for a failed or a running component, one knows only the time since the system first went into service. Few methods for analysis of renewal data have been developed; for examples, see the parametric maximum likelihood approach of Bassin (1969) and the non-parametric approaches of Klega (1976) and of Trindade and Haugh (1978).

9. LIFE AS A FUNCTION OF OTHER VARIABLES

Product life may depend on variables arising in design, manufacture, operation, etc. Such so-called independent variables may be qualitative (categories, for example, vendors A, B, and C), and they may be quantitative (for example, operating temperature). This section briefly reviews models and analyses for such data.

Models. A general model for such situations consists of a life distribution whose parameters are functions of the variables. The unknown values of coefficients in the functions are estimated from data. Nelson and Hendrickson (1972) describe such models in general terms. Examples of such models are referenced in Section 10 on accelerated testing. References that follow generally present models entirely with qualitative or else quantitative variables, but both types of variables can be in a model.

Qualitative variables. Analysis of variance relationships are commonly used to express a life distribution parameter in terms of qualitative variables. Such relationships are described in general by Box, Hunter, and Hunter (1978), Mendenhall (1968), and Scheffé (1959). The data analysis methods of these books do not apply to censored data. Examples of the use of such relationships with censored life data are given by Sampford and Taylor (1959), Zahn (1975), and Zelen (1969).

Quantitative variables. Linear regression relationships are commonly used to express a life distribution parameter in terms of quantitative variables. Such relationships are described in general by Draper and Smith (1981) and Neter and Wasserman (1974). Applications to censored life data are given by Cox (1972), Lieblein and Zelen (1956), and Nelson and Hahn (1972). Further applications are referenced in Section 10. Kalbfleisch and Prentice (1980) and Elandt-Johnson and Johnson (1980) present the Cox (1972) proportional hazards model and data analyses in detail for biomedical applications. Lawless (1982) treats parametric models (his Chapter 6) and the nonparametric Cox model (his Chapter 7).

10. ACCELERATED LIFE TESTING

Accelerated life testing of products and materials is used to get information quickly on their life distributions. Such testing involves subjecting the test units to conditions that are more severe than normal. This results in shorter lives than would be observed under normal conditions. Accelerated test conditions are typically produced by testing units at high levels of temperature, voltage, pressure, vibration, cycling rate, load, etc., or some combination of these. The use of certain accelerating or stress variables is a well-established engineering practice for many products and materials. In other fields, similar problems involve estimating a relationship between life and variables that affect life.

The data obtained at the more severe or accelerated conditions are extrapolated by means of an appropriate model to the normal conditions to obtain an estimate of the life distribution under normal conditions. Such testing provides a savings in time and cost compared with testing at normal conditions. Indeed, for many products and materials, life at normal conditions is so long that testing at those conditions is completely out of the question.

Nelson (1974a), General Electric (1975, Sec. 11), Little and Jebe (1975), and Yurkowsky and others (1967) survey statistical methods for planning and analyzing accelerated tests where units are subjected to high stresses. In particular, they give references on how to analyze such data before all test units fail. This important advance makes it possible (1) to terminate a test before all units fail, resulting in a savings of time and cost, and (2) to test at lower stresses so that extrapolation is reduced. Also, the references show how to properly analyze such data with different failure modes. Previously, it was not known how to use test data with a mix of failure modes to estimate the life distribution at design conditions. Meeker (1979) gives a bibliography on accelerated testing, including statistical methods and applications.

Metal fatigue under high stress, temperature, and other variables is an important application of accelerated testing. References on statistical methods include Little and Jebe (1975), Weibull (1961), and ASTM (1963).

A variety of statistical methods and special problems are presented by Nelson (1971, 1973, 1974a,b, 1975), Nelson and Hahn (1972), and Hahn and Nelson (1971).

11. DEPRECIATION, REPLACEMENT, AND MAINTENANCE POLICIES

Some references on these topics follow.

Barlow, R. E., and Proschan, F. (1965), *Mathematical Theory of Reliability*, Wiley, New York, Chapters 3 and 4.

Glasser, G. J. (1969), "Planned Replacement: Some Theory and Its Application," *J. Qual. Technol.* **1**, 110–119.

Jorgenson, D. W., McCall, J. J., Radner, R. (1967), *Optimal Replacement Policy*, Rand McNally, Chicago, IL.

12. BOOKS WITH RELIABILITY AND LIFE DATA ANALYSIS

Below is a list of books on mathematical methods for reliability and for life data analysis. The list omits books on reliability management, physics of failure, and other nonmathematical methods. There are few books on life data analysis, and most reliability books present only a few simple data analyses. Reliability is a rapidly changing field, and many early books (say, before 1974) lack important new developments. Many of the following books have been reviewed in the journals listed in Section 14.

Amstadter, B. L. (1971), *Reliability Mathematics — Fundamentals; Practices; Procedures*, McGraw-Hill, New York.

Bain, L. J. (1978), *Statistical Analysis of Reliability and Life-Testing Models*, Dekker, New York.

Barlow, R. E., and Proschan, F. (1965), *Mathematical Theory of Reliability*, Wiley, New York.

Barlow, R. E., and Proschan, F. (1975), *Statistical Theory of Reliability and Life Testing*, Holt, Rinehart, and Winston, New York.

Bazovsky, I. (1961), *Reliability Theory & Practice*, Prentice-Hall, Englewood Cliffs, NJ.

Becker, P. W., and Jensen, F. (1977), *Design of Systems and Circuits — for Maximum Reliability or Maximum Production Yield*, McGraw-Hill, New York.

Billington, R. (1970), *Power System Reliability Evaluation*, Gordon and Breach, New York.

Bompas-Smith, J. H. (1973), *Mechanical Survival: The Use of Reliability Data*, McGraw-Hill, Maidenhead, Berkshire, England.

Bury, K. V. (1975), *Statistical Models in Applied Science*, Wiley, New York.

Calabro, S. R. (1962), *Reliability Principles and Practices*, McGraw-Hill, New York.

Chiang, C. L. (1968), *Introduction to Stochastic Processes in Biostatistics*, Wiley, New York.

David, H. A. (1981), *Order Statistics*, 2nd Ed., Wiley, New York.

Dummer, G. W., and Winton, R. C. (1968), *An Elementary Guide to Reliability*, Pergamon, Elmsford, NY.

Elandt-Johnson, R. C., and Johnson, N. L. (1980), *Survival Models and Data Analysis*, Wiley, New York.

Enrick, N. L. (1967), *Quality Control and Reliability*, Industrial Press, New York.

Gertsbakh, I. B., and Kordonskiy, Kh. B. (1969), *Models of Failure*, Springer-Verlag, New York.

Green, A. E., and Bourne, A. J. (1972), *Reliability Technology*, Wiley-Interscience, New York.

Gross, A. J., and Clark, V. A. (1975), *Survival Distributions: Reliability Applications in the Medical Sciences*, Wiley, New York.

Gumbel, E. J. (1958), *Statistics of Extremes*, Columbia University Press, New York.

Hahn, G. J., and Shapiro, S. S. (1967), *Statistical Models in Engineering*, Wiley, New York.

Halpern, S. (1978), *The Assurance Sciences: An Introduction to Quality Control and Reliability*, Prentice-Hall, Englewood Cliffs, NJ.

Haviland, R. P. (1964), *Engineering Reliability and Long Life Design*, Van Nostrand, Princeton, NJ.

Henley, E. J., and Kumamoto, H. (1980), *Reliability Engineering and Risk Assessment*, Prentice-Hall, New York.

Ireson, W. G., Ed. (1966), *Reliability Handbook*, McGraw-Hill, New York.

Johnson, L. G. (1964), *The Theory and Technique of Variation Research*, Elsevier, New York.

Kalbfleisch, J. D., and Prentice, R. L. (1980), *The Statistical Analysis of Failure Time Data*, Wiley, New York.

Kapur, K. C., and Lamberson, L. R. (1977), *Reliability in Engineering Design*, Wiley, New York.

Kaufman, A., Grouchko, D., and Cruon, R. (1977), *Mathematical Models for the Study of Reliability of Systems*, Academic, New York.

King, J. R. (1971), *Probability Plots for Decision Making*, Industrial Press, New York.

Kivenson, G. (1971), *Durability and Reliability in Engineering Design*, Hayden, New York.

Kulldorff, G. (1961), *Estimation from Grouped and Partially Grouped Samples*, Wiley, New York.

Landers, R. R. (1963), *Reliability and Product Assurance*, Prentice-Hall, Englewood Cliffs, NJ.

Lawless, J. F. (1982), *Statistical Models and Methods for Lifetime Data*, Wiley, New York.

Lee, E. (1980), *Statistical Methods for Survival Data Analysis*, Lifetime Learning, Belmont, CA.

Lipson, C., and Sheth, N. C. (1973), *Statistical Design and Analysis of Engineering Experiments*, McGraw-Hill, New York.

Little, R. E., and Jebe, E. H. (1975), *Statistical Design of Fatigue Experiments*, Halstead, New York.

Lloyd, D. K., and Lipow, M. (1977), *Reliability: Management, Methods and Mathematics*, 2nd ed., McGraw-Hill, New York.

Locks, M. O. (1973), *Reliability, Maintainability, and Availability Assessment*, Hayden, Rochelle Park, NJ.

Lowrance, W. W. (1976), *Of Acceptable Risk: Science and the Determination of Safety*, William Kaufman, Los Altos, CA.

Mann, N. R., Schafer, R. E., and Singpurwalla, N. D. (1974), *Methods for Statistical Analysis of Reliability and Life Data*, Wiley, New York.

Miller, R. (1981), *Survival Analysis*, Wiley, New York.

Pieruschka, E. (1963), *Principles of Reliability*, Prentice-Hall, Englewood Cliffs, NJ.

Polovko, A. M. (1968), *Fundamentals of Reliability Theory*, Academic, New York.

Proschan, F., and Serfling, R. J., Eds. (1973), *Reliability and Biometry — Statistical Analysis of Lifelength*, Society for Industrial and Applied Mathematics, 33 South 17th St., Philadelphia, PA 19103.

Rau, J. J. (1970), *Optimization and Probability in Systems Engineering*, Van Nostrand Reinhold, New York.

Roberts, N. H. (1964), *Mathematical Methods in Reliability Engineering*, McGraw-Hill, New York.

Rowe, W. D. (1977), *An Anatomy of Risk*, Wiley, New York.

Sandler, G. H. (1963), *System Reliability Engineering*, Prentice-Hall, Englewood Cliffs, NJ.

Shooman, M. L. (1968), *Probabilistic Reliability: An Engineering Approach*, McGraw-Hill, New York.

Sinha, S. K., and Kale, B. K. (1980), *Life Testing and Reliability Estimation*, Halstead, New York.

Smith, D. J. (1972), *Reliability Engineering*, Barnes and Noble, New York.

Tsokos, C. P., and Shimi, I. N., Eds. (1977), *The Theory and Application of Reliability with Emphasis on Bayesian and Nonparametric Methods*, Vol. I—*Theory*, Vol. II—*Application*, Academic, New York.

Von Alven, W. H., Ed. (1964), *Reliability Engineering*, ARINC Research Corporation, Prentice-Hall, Englewood Cliffs, NJ.

Weibull, W. (1961), *Fatigue Testing and the Analysis of Results*, Pergamon, New York.

Zelen, M., Ed. (1963), *Statistical Theory of Reliability*, University of Wisconsin Press, Madison, WI.

13. ABSTRACTS, BIBLIOGRAPHIES, AND INDICES ON RELIABILITY AND LIFE DATA ANALYSIS

The following abstracts, bibliographies, and indices reference methods for reliability and life data analysis. Also, the yearly indices of journals list papers. Applications of such methods are not listed here. Reliability and life data analysis are rapidly developing fields; so many references over 10 years old may be out of date. There are bibliographies in many of the books listed in Section 12.

Buckland, W. R. (1964), *Statistical Assessment of Life Characteristic: a Bibliographic Guide*, Hafner, London. Papers grouped by topic with annotations.

Dolby, J. L., and Tukey, J. W. (1973), *The Statistics Cum Index*, R&D Press, Los Altos, CA.

The Engineering Index.

Govindarajulu, Z. (1964), "A Supplement to Mendenhall's Bibliography on Life Testing and Related Topics," *J. Am. Stat. Assoc.* **59**, 1231–1291. Papers listed by author and with subject labels.

Gross, A. J., and Clark, V. A. (1975), *Survival Distributions: Reliability Applications in the Biomedical Sciences*, Wiley, New York. The recent bibliography is annotated.

International Journal of Abstracts on Statistical Methods in Industry.

International Journal of Abstracts — Statistical Theory & Methods.

Joiner, B. L. Ed. (1975), *Current Index to Statistics: Applications, Methods and Theory*, American Statistical Association and the Institute of Mathematical Statistics. By author and keywords in title; issued yearly since 1975.

Joiner, B. L., Laubscher, N. F., Brown, E. S., and Levy, B. (1970), *An Author and Permuted Title Index to Selected Statistical Journals*, National Bureau of Standards Special Publication 321. Available from the U. S. Government Printing Office, Washington, DC 20402. By author and by key words in title; covers selected journals starting different years through 1969.

Kendall, M. G., and Doig, A. G. (1962), *Bibliography of Statistical Literature 1950–1958*, Hafner, London.

Kendall, M. G., and Doig, A. G. (1965), *Bibliography of Statistical Literature 1940–1949*, Hafner, London.

Kendall, M. G., and Doig, A. G. (1968), *Bibliography of Statistical Literature Pre-1940*, Hafner, London.

Lancaster, H. O. (1971), "A Bibliography of Statistical Bibliographies: A Fourth List," *Rev. Int. Stat. Inst.* **39**, No. 1.

Mathematical Reviews.

Mendenhall, W. (1958), "A Bibliography of Life Testing and Related Topics," *Biometrika* **45**, 521–543. Papers listed by author and with subject labels.

Quality Control Abstracts.

Reliability Abstracts and Technical Reviews, National Aeronautics and Space Administration, Washington, DC.

Ross, I. C., and Tukey, J. W., (1975), *Index to Statistics and Probability*, Vol. 2—*Citation Index*, Vol. 3—*Permuted Titles: A–K*, Vol. 4—*Permuted Titles L–Z*, Vol. 5—*Locations and Authors*, Vol. 6—*Permuted Index to Minimum Abbreviation*.

Schaefer, B. K. (1979), *Using the Mathematical Literature: A Practical Guide*, Dekker, New York.

Science Citation Index. Useful for locating more recent papers that reference known papers on a topic.

Statistical Theory & Methods Abstracts, Oliver & Boyd, Tweeddale Court, 14 High St., Edinburgh 1, Scotland.

14. JOURNALS WITH PAPERS ON RELIABILITY AND LIFE DATA ANALYSIS

The following journals contain papers on methods for reliability and life data analysis. Many engineering, biomedical, and other journals contain papers on applications of such methods.

American Society for Quality Control Technical Conference Transactions.

American Statistician.

Annals of Mathematical Statistics—Author, subject, and citation indices vols. 1–31 (1960).

Annals of Probability. Successor to *Annals of Mathematical Statistics*.

Annals of Reliability and Maintainability

Annals of Statistics. Successor to *Annals of Mathematical Statistics*.

Applied Statistics.

Berkeley Symposium.

Biometrics—Author and subject indices for vols. 1–20 (1965).

Biometrika—Subject index for vols. 1–37 (1950), author indices for vols. 1–48 (1961) and 49–56 (1969).

Evaluation Engineering.

Industrial Quality Control—Subject index for vols I–X (1954).

IEEE Transactions on Power Apparatus and Systems.

IEEE Transactions on Reliability.

Journal of Quality Technology. Successor to *Industrial Quality Control*.

Journal of the American Statistical Association—Author index for vols. 51–60 (1966).

Journal of the Operations Research Society of America—Author, subject, and title indices for vols. 1–15 (1967).

Journal of Statistical Planning and Inference.

Journal of the Royal Statistical Society, Series B—Methodological.

Naval Research Logistics Quarterly.

Operations Research Quarterly.

Proceedings of the Annual Reliability and Maintainability Symposium.

The Q R Journal—Theory and Practice, Methods and Management.

Quality Assurance.

Quality Progress.

THE RELIA-COM.

Sankhya, Series A and Series B.

Skandinavisk Aktuarietidskrift (many papers in English)—Index for vols. 1–40.

Technometrics—Author and subject indices for vols. 1–7 (1966), 8–20 (1978).

Trabajos de Estadistica e Investigaciones Operativas (some papers in English).

15. COMPUTERIZED LITERATURE SEARCHES

Computerized literature searches of data bases are now available through many libraries. Searches may be made by key words in paper titles, author names, year of publication, journal, etc. Some data bases relevant to life data analysis follow.

Biosis Previews of BioScience Information Service.   This provides comprehensive worldwide coverage of research in the life sciences, including anatomy, bacteriology, biochemistry, genetics, immunology, microbiology, physiology, toxicology, virology, and zoology. It includes all citations from *Biological Abstracts* and *Bioresearch Index*. 1969–present.

CA Condensation/Casia of Chemical Abstracts Service (CAS).   This is the merger of two CAS files covering the world's chemical literature, including biochemistry, analytical and physical chemistry, applied chemistry, and chemical engineering. 1970–present.

Compendex of Engineering Index.   This corresponds to *Engineering Index Monthly*. It covers civil–environmental–geological engineering, mining–metals–petroleum–fuel engineering, mechanical–automotive–nuclear–aerospace engineering, electrical–electronics–control engineering, chemical–agricultural–food engineering, and industrial engineering, management, mathematics, physics, and instruments, including approximately 1500 serials and over 900 monographic publications.

DOE Energy Data Base, U.S. Department of Energy.   This covers all unclassified energy information processed at the DOE Technical Information Center. It includes all literature announced in *DOE Energy Research Abstracts, Energy Abstracts for Policy Analysis, Power Reactor Docket Information, Atomindex,* and the *Solar, Geothermal, and Fossil Energy Updates.* January 1974–present.

Inspec of Institution of Electrical Engineers. This provides worldwide coverage of the literature in physics and electrical and electronics engineering. Source documents are primarily journal articles; however, government reports, patents, and monographs are included. It corresponds in coverage to *Physics Abstracts*, *Electrical and Electronics Abstracts*, and *Computer and Control Abstracts*. January 1969–present.

Scisearch of Institute for Scientific Information, Philadelphia, PA. This is a multidisciplinary index to the literature of science and technology. It contains all records published in *Science Citation Index* (SCI) and additional records from the *Current Contents* series of publications that are not included in the printed version of SCI. Indexed journals are carefully selected on the basis of several criteria, including citation analysis, resulting in the inclusion of 90% of the world's significant scientific and technical literature. Citation indexing allows retrieval of newly published articles through the subject relationships established by an author's reference to prior articles. Scisearch covers every area of the pure and applied sciences. Scisearch indexes all significant items (articles, reports of meetings, letters, editorials, correction notices, etc.) from about 2600 major scientific and technical journals. In addition, the Scisearch file for 1974–1975 includes approximately 38,000 items from *Current Contents — Clinical Practice*. Beginning January 1, 1976, all items from *Current Contents — Engineering, Technology, and Applied Science* and *Current Contents — Agriculture, Biology, and Environmental Sciences* that are not presently covered in the printed SCI will be included each month. This expanded coverage will add approximately 58,000 items per year to the Scisearch file.

PROBLEMS

13.1. Compile a bibliography on one of the following topics.

(a) Renewal data analysis.

(b) Stress-strength interference.

(c) Applications to a particular engineering problem, fatigue of metals, endurance of dielectrics, or whatever.

(d) Interval data analysis.

(e) Failure modes and effects analysis (FMEA).

(f) Estimation of system reliability from component data.

(g) Size effect on specimen life.

(h) Accelerated testing.

(i) Goodness of fit tests, particularly for censored data.

(j)   The method of minimum chi-square fitting.

(k)   Nonparametric methods for multiply censored data.

(l)   Availability methods.

(m)   Maintainability methods.

(n)   Competing risks, including dependent failure modes.

(o)   A topic of your choice.

APPENDIX
TABLES

Appendix A1. Standard Normal Cumulative Distribution Function $\Phi(u)$

u	·00	·01	·02	·03	·04	·05	·06	·07	·08	·09
− ·0	·5000	·4960	·4920	·4880	·4840	·4801	·4761	·4721	·4681	·4641
− ·1	·4602	·4562	·4522	·4483	·4443	·4404	·4364	·4325	·4286	·4247
− ·2	·4207	·4168	·4129	·4090	·4052	·4013	·3974	·3936	·3897	·3859
− ·3	·3821	·3783	·3745	·3707	·3669	·3632	·3594	·3557	·3520	·3483
− ·4	·3446	·3409	·3372	·3336	·3300	·3264	·3228	·3192	·3156	·3121
− ·5	·3085	·3050	·3015	·2981	·2946	·2912	·2877	·2843	·2810	·2776
− ·6	·2743	·2709	·2676	·2643	·2611	·2578	·2546	·2514	·2483	·2451
− ·7	·2420	·2389	·2358	·2327	·2297	·2266	·2236	·2206	·2177	·2148
− ·8	·2119	·2090	·2061	·2033	·2005	·1977	·1949	·1922	·1894	·1867
− ·9	·1841	·1814	·1788	·1762	·1736	·1711	·1685	·1660	·1635	·1611
−1·0	·1587	·1562	·1539	·1515	·1492	·1469	·1446	·1423	·1401	·1379
−1·1	·1357	·1335	·1314	·1292	·1271	·1251	·1230	·1210	·1190	·1170
−1·2	·1151	·1131	·1112	·1093	·1075	·1056	·1038	·1020	·1003	·09853
−1·3	·09680	·09510	·09342	·09176	·09012	·08851	·08691	·08534	·08379	·08226
−1·4	·08076	·07927	·07780	·07636	·07493	·07353	·07215	·07078	·06944	·06811
−1·5	·06681	·06552	·06426	·06301	·06178	·06057	·05938	·05821	·05705	·05592
−1·6	·05480	·05370	·05262	·05155	·05050	·04947	·04846	·04746	·04648	·04551
−1·7	·04457	·04363	·04272	·04182	·04093	·04006	·03920	·03836	·03754	·03673
−1·8	·03593	·03515	·03438	·03362	·03288	·03216	·03144	·03074	·03005	·02938
−1·9	·02872	·02807	·02743	·02680	·02619	·02559	·02500	·02442	·02385	·02330
−2·0	·02275	·02222	·02169	·02118	·02068	·02018	·01970	·01923	·01876	·01831
−2·1	·01786	·01743	·01700	·01659	·01618	·01578	·01539	·01500	·01463	·01426
−2·2	·01390	·01355	·01321	·01287	·01255	·01222	·01191	·01160	·01130	·01101
−2·3	·01072	·01044	·01017	$·0^2$9903	$·0^2$9642	$·0^2$9387	$·0^2$9137	$·0^2$8894	$·0^2$8656	$·0^2$8424
−2·4	$·0^2$8198	$·0^2$7976	$·0^2$7760	$·0^2$7549	$·0^2$7344	$·0^2$7143	$·0^2$6947	$·0^2$6756	$·0^2$6569	$·0^2$6387
−2·5	$·0^2$6210	$·0^2$6037	$·0^2$5868	$·0^2$5703	$·0^2$5543	$·0^2$5386	$·0^2$5234	$·0^2$5085	$·0^2$4940	$·0^2$4799
−2·6	$·0^2$4661	$·0^2$4527	$·0^2$4396	$·0^2$4269	$·0^2$4145	$·0^2$4025	$·0^2$3907	$·0^2$3793	$·0^2$3681	$·0^2$3573
−2·7	$·0^2$3467	$·0^2$3364	$·0^2$3264	$·0^2$3167	$·0^2$3072	$·0^2$2980	$·0^2$2890	$·0^2$2803	$·0^2$2718	$·0^2$2635
−2·8	$·0^2$2555	$·0^2$2477	$·0^2$2401	$·0^2$2327	$·0^2$2256	$·0^2$2186	$·0^2$2118	$·0^2$2052	$·0^2$1988	$·0^2$1926
−2·9	$·0^2$1866	$·0^2$1807	$·0^2$1750	$·0^2$1695	$·0^2$1641	$·0^2$1589	$·0^2$1538	$·0^2$1489	$·0^2$1441	$·0^2$1395
−3·0	$·0^2$1350	$·0^2$1306	$·0^2$1264	$·0^2$1223	$·0^2$1183	$·0^2$1144	$·0^2$1107	$·0^2$1070	$·0^2$1035	$·0^2$1001
−3·1	$·0^3$9676	$·0^3$9354	$·0^3$9043	$·0^3$8740	$·0^3$8447	$·0^3$8164	$·0^3$7888	$·0^3$7622	$·0^3$7364	$·0^3$7114
−3·2	$·0^3$6871	$·0^3$6637	$·0^3$6410	$·0^3$6190	$·0^3$5976	$·0^3$5770	$·0^3$5571	$·0^3$5377	$·0^3$5190	$·0^3$5009
−3·3	$·0^3$4834	$·0^3$4665	$·0^3$4501	$·0^3$4342	$·0^3$4189	$·0^3$4041	$·0^3$3897	$·0^3$3758	$·0^3$3624	$·0^3$3495
−3·4	$·0^3$3369	$·0^3$3248	$·0^3$3131	$·0^3$3018	$·0^3$2909	$·0^3$2803	$·0^3$2701	$·0^3$2602	$·0^3$2507	$·0^3$2415
−3·5	$·0^3$2326	$·0^3$2241	$·0^3$2158	$·0^3$2078	$·0^3$2001	$·0^3$1926	$·0^3$1854	$·0^3$1785	$·0^3$1718	$·0^3$1653
−3·6	$·0^3$1591	$·0^3$1531	$·0^3$1473	$·0^3$1417	$·0^3$1363	$·0^3$1311	$·0^3$1261	$·0^3$1213	$·0^3$1166	$·0^3$1121
−3·7	$·0^3$1078	$·0^3$1036	$·0^4$9961	$·0^4$9574	$·0^4$9201	$·0^4$8842	$·0^4$8496	$·0^4$8162	$·0^4$7841	$·0^4$7532
−3·8	$·0^4$7235	$·0^4$6948	$·0^4$6673	$·0^4$6407	$·0^4$6152	$·0^4$5906	$·0^4$5669	$·0^4$5442	$·0^4$5223	$·0^4$5012
−3·9	$·0^4$4810	$·0^4$4615	$·0^4$4427	$·0^4$4247	$·0^4$4074	$·0^4$3908	$·0^4$3747	$·0^4$3594	$·0^4$3446	$·0^4$3304
−4·0	$·0^4$3167	$·0^4$3036	$·0^4$2910	$·0^4$2789	$·0^4$2673	$·0^4$2561	$·0^4$2454	$·0^4$2351	$·0^4$2252	$·0^4$2157
−4·1	$·0^4$2066	$·0^4$1978	$·0^4$1894	$·0^4$1814	$·0^4$1737	$·0^4$1662	$·0^4$1591	$·0^4$1523	$·0^4$1458	$·0^4$1395
−4·2	$·0^4$1335	$·0^4$1277	$·0^4$1222	$·0^4$1168	$·0^4$1118	$·0^4$1069	$·0^4$1022	$·0^5$9774	$·0^5$9345	$·0^5$8934
−4·3	$·0^5$8540	$·0^5$8163	$·0^5$7801	$·0^5$7455	$·0^5$7124	$·0^5$6807	$·0^5$6503	$·0^5$6212	$·0^5$5934	$·0^5$5668
−4·4	$·0^5$5413	$·0^5$5169	$·0^5$4935	$·0^5$4712	$·0^5$4498	$·0^5$4294	$·0^5$4098	$·0^5$3911	$·0^5$3732	$·0^5$3561
−4·5	$·0^5$3398	$·0^5$3241	$·0^5$3092	$·0^5$2949	$·0^5$2813	$·0^5$2682	$·0^5$2558	$·0^5$2439	$·0^5$2325	$·0^5$2216
−4·6	$·0^5$2112	$·0^5$2013	$·0^5$1919	$·0^5$1828	$·0^5$1742	$·0^5$1660	$·0^5$1581	$·0^5$1506	$·0^5$1434	$·0^5$1366
−4·7	$·0^5$1301	$·0^5$1239	$·0^5$1179	$·0^5$1123	$·0^5$1069	$·0^5$1017	$·0^6$9680	$·0^6$9211	$·0^6$8765	$·0^6$8339
−4·8	$·0^6$7933	$·0^6$7547	$·0^6$7178	$·0^6$6827	$·0^6$6492	$·0^6$6173	$·0^6$5869	$·0^6$5580	$·0^6$5304	$·0^6$5042
−4·9	$·0^6$4792	$·0^6$4554	$·0^6$4327	$·0^6$4111	$·0^6$3906	$·0^6$3711	$·0^6$3525	$·0^6$3348	$·0^6$3179	$·0^6$3019

From A. Hald, *Statistical Tables and Formulas*, Wiley, New York, 1952, Table II. Reproduced by permission.

Appendix A1. Standard Normal Cumulative Distribution Function $\Phi(u)$
(*Continued*)

u	·00	·01	·02	·03	·04	·05	·06	·07	·08	·09
·0	·5000	·5040	·5080	·5120	·5160	·5199	·5239	·5279	·5319	·5359
·1	·5398	·5438	·5478	·5517	·5557	·5596	·5636	·5675	·5714	·5753
·2	·5793	·5832	·5871	·5910	·5948	·5987	·6026	·6064	·6103	·6141
·3	·6179	·6217	·6255	·6293	·6331	·6368	·6406	·6443	·6480	·6517
·4	·6554	·6591	·6628	·6664	·6700	·6736	·6772	·6808	·6844	·6879
·5	·6915	·6950	·6985	·7019	·7054	·7088	·7123	·7157	·7190	·7224
·6	·7257	·7291	·7324	·7357	·7389	·7422	·7454	·7486	·7517	·7549
·7	·7580	·7611	·7642	·7673	·7703	·7734	·7764	·7794	·7823	·7852
·8	·7881	·7910	·7939	·7967	·7995	·8023	·8051	·8078	·8106	·8133
·9	·8159	·8186	·8212	·8238	·8264	·8289	·8315	·8340	·8365	·8389
1·0	·8413	·8438	·8461	·8485	·8508	·8531	·8554	·8577	·8599	·8621
1·1	·8643	·8665	·8686	·8708	·8729	·8749	·8770	·8790	·8810	·8830
1·2	·8849	·8869	·8888	·8907	·8925	·8944	·8962	·8980	·8997	·90147
1·3	·90320	·90490	·90658	·90824	·90988	·91149	·91309	·91466	·91621	·91774
1·4	·91924	·92073	·92220	·92364	·92507	·92647	·92785	·92922	·93056	·93189
1·5	·93319	·93448	·93574	·93699	·93822	·93943	·94062	·94179	·94295	·94408
1·6	·94520	·94630	·94738	·94845	·94950	·95053	·95154	·95254	·95352	·95449
1·7	·95543	·95637	·95728	·95818	·95907	·95994	·96080	·96164	·96246	·96327
1·8	·96407	·96485	·96562	·96638	·96712	·96784	·96856	·96926	·96995	·97062
1·9	·97128	·97193	·97257	·97320	·97381	·97441	·97500	·97558	·97615	·97670
2·0	·97725	·97778	·97831	·97882	·97932	·97982	·98030	·98077	·98124	·98169
2·1	·98214	·98257	·98300	·98341	·98382	·98422	·98461	·98500	·98537	·98574
2·2	·98610	·98645	·98679	·98713	·98745	·98778	·98809	·98840	·98870	·98899
2·3	·98928	·98956	·98983	·$9^2$0097	·$9^2$0358	·$9^2$0613	·$9^2$0863	·$9^2$1106	·$9^2$1344	·$9^2$1576
2·4	·$9^2$1802	·$9^2$2024	·$9^2$2240	·$9^2$2451	·$9^2$2656	·$9^2$2857	·$9^2$3053	·$9^2$3244	·$9^2$3431	·$9^2$3613
2·5	·$9^2$3790	·$9^2$3963	·$9^2$4132	·$9^2$4297	·$9^2$4457	·$9^2$4614	·$9^2$4766	·$9^2$4915	·$9^2$5060	·$9^2$5201
2·6	·$9^2$5339	·$9^2$5473	·$9^2$5604	·$9^2$5731	·$9^2$5855	·$9^2$5975	·$9^2$6093	·$9^2$6207	·$9^2$6319	·$9^2$6427
2·7	·$9^2$6533	·$9^2$6636	·$9^2$6736	·$9^2$6833	·$9^2$6928	·$9^2$7020	·$9^2$7110	·$9^2$7197	·$9^2$7282	·$9^2$7365
2·8	·$9^2$7445	·$9^2$7523	·$9^2$7599	·$9^2$7673	·$9^2$7744	·$9^2$7814	·$9^2$7882	·$9^2$7948	·$9^2$8012	·$9^2$8074
2·9	·$9^2$8134	·$9^2$8193	·$9^2$8250	·$9^2$8305	·$9^2$8359	·$9^2$8411	·$9^2$8462	·$9^2$8511	·$9^2$8559	·$9^2$8605
3·0	·$9^2$8650	·$9^2$8694	·$9^2$8736	·$9^2$8777	·$9^2$8817	·$9^2$8856	·$9^2$8893	·$9^2$8930	·$9^2$8965	·$9^2$8999
3·1	·$9^3$0324	·$9^3$0646	·$9^3$0957	·$9^3$1260	·$9^3$1553	·$9^3$1836	·$9^3$2112	·$9^3$2378	·$9^3$2636	·$9^3$2886
3·2	·$9^3$3129	·$9^3$3363	·$9^3$3590	·$9^3$3810	·$9^3$4024	·$9^3$4230	·$9^3$4429	·$9^3$4623	·$9^3$4810	·$9^3$4991
3·3	·$9^3$5166	·$9^3$5335	·$9^3$5499	·$9^3$5658	·$9^3$5811	·$9^3$5959	·$9^3$6103	·$9^3$6242	·$9^3$6376	·$9^3$6505
3·4	·$9^3$6631	·$9^3$6752	·$9^3$6869	·$9^3$6982	·$9^3$7091	·$9^3$7197	·$9^3$7299	·$9^3$7398	·$9^3$7493	·$9^3$7585
3·5	·$9^3$7674	·$9^3$7759	·$9^3$7842	·$9^3$7922	·$9^3$7999	·$9^3$8074	·$9^3$8146	·$9^3$8215	·$9^3$8282	·$9^3$8347
3·6	·$9^3$8409	·$9^3$8469	·$9^3$8527	·$9^3$8583	·$9^3$8637	·$9^3$8689	·$9^3$8739	·$9^3$8787	·$9^3$8834	·$9^3$8879
3·7	·$9^3$8922	·$9^3$8964	·$9^4$0039	·$9^4$0426	·$9^4$0799	·$9^4$1158	·$9^4$1504	·$9^4$1838	·$9^4$2159	·$9^4$2468
3·8	·$9^4$2765	·$9^4$3052	·$9^4$3327	·$9^4$3593	·$9^4$3848	·$9^4$4094	·$9^4$4331	·$9^4$4558	·$9^4$4777	·$9^4$4988
3·9	·$9^4$5190	·$9^4$5385	·$9^4$5573	·$9^4$5753	·$9^4$5926	·$9^4$6092	·$9^4$6253	·$9^4$6406	·$9^4$6554	·$9^4$6696
4·0	·$9^4$6833	·$9^4$6964	·$9^4$7090	·$9^4$7211	·$9^4$7327	·$9^4$7439	·$9^4$7546	·$9^4$7649	·$9^4$7748	·$9^4$7843
4·1	·$9^4$7934	·$9^4$8022	·$9^4$8106	·$9^4$8186	·$9^4$8263	·$9^4$8338	·$9^4$8409	·$9^4$8477	·$9^4$8542	·$9^4$8605
4·2	·$9^4$8665	·$9^4$8723	·$9^4$8778	·$9^4$8832	·$9^4$8882	·$9^4$8931	·$9^4$8978	·$9^5$0226	·$9^5$0655	·$9^5$1066
4·3	·$9^5$1460	·$9^5$1837	·$9^5$2199	·$9^5$2545	·$9^5$2876	·$9^5$3193	·$9^5$3497	·$9^5$3788	·$9^5$4066	·$9^5$4332
4·4	·$9^5$4587	·$9^5$4831	·$9^5$5065	·$9^5$5288	·$9^5$5502	·$9^5$5706	·$9^5$5902	·$9^5$6089	·$9^5$6268	·$9^5$6439
4·5	·$9^5$6602	·$9^5$6759	·$9^5$6908	·$9^5$7051	·$9^5$7187	·$9^5$7318	·$9^5$7442	·$9^5$7561	·$9^5$7675	·$9^5$7784
4·6	·$9^5$7888	·$9^5$7987	·$9^5$8081	·$9^5$8172	·$9^5$8258	·$9^5$8340	·$9^5$8419	·$9^5$8494	·$9^5$8566	·$9^5$8634
4·7	·$9^5$8699	·$9^5$8761	·$9^5$8821	·$9^5$8877	·$9^5$8931	·$9^5$8983	·$9^6$0320	·$9^6$0789	·$9^6$1235	·$9^6$1661
4·8	·$9^6$2067	·$9^6$2453	·$9^6$2822	·$9^6$3173	·$9^6$3508	·$9^6$3827	·$9^6$4131	·$9^6$4420	·$9^6$4696	·$9^6$4958
4·9	·$9^6$5208	·$9^6$5446	·$9^6$5673	·$9^6$5889	·$9^6$6094	·$9^6$6289	·$9^6$6475	·$9^6$6652	·$9^6$6821	·$9^6$6981

Appendix A2   Standard Normal Percentiles z_P

$100P\%$	z_P	$100P\%$	z_P
10^{-4}	-4.753	50	0.
10^{-3}	-4.265	60	0.253
0.01	-3.719	70	0.524
0.1	-3.090	75	0.675
0.5	-2.576	80	0.842
1.0	-2.326	90	1.282
2.0	-2.054	95	1.645
2.5	-1.960	97.5	1.960
5	-1.645	98	2.054
10	-1.282	99	2.326
20	-0.842	99.5	2.576
25	-0.675	99.9	3.090
30	-0.524		
40	-0.253		

Appendix A3. Chi-Square Percentiles $\chi^2(P; \nu)$

ν \ P	0.005	0.010	0.025	0.050	0.100	0.250	0.500
1	0.00004	0.00016	0.00098	0.00393	0.01579	0.1015	0.4549
2	0.0100	0.0201	0.0506	0.1026	0.2107	0.5754	1.386
3	0.0717	0.1148	0.2158	0.3518	0.5844	1.213	2.366
4	0.2070	0.2971	0.4844	0.7107	1.064	1.923	3.357
5	0.4117	0.5543	0.8312	1.145	1.610	2.675	4.351
6	0.6757	0.8721	1.2373	1.635	2.204	3.455	5.348
7	0.9893	1.239	1.690	2.167	2.833	4.255	6.346
8	1.344	1.646	2.180	2.733	3.490	5.071	7.344
9	1.735	2.088	2.700	3.325	4.168	5.899	8.343
10	2.156	2.558	3.247	3.940	4.865	6.737	9.342
11	2.603	3.053	3.816	4.575	5.578	7.584	10.34
12	3.074	3.571	4.404	5.226	6.304	8.438	11.34
13	3.565	4.107	5.009	5.892	7.041	9.299	12.34
14	4.075	4.660	5.629	6.571	7.790	10.17	13.34
15	4.601	5.229	6.262	7.261	8.547	11.04	14.34
16	5.142	5.812	6.908	7.962	9.312	11.91	15.34
17	5.697	6.408	7.564	8.672	10.09	12.79	16.34
18	6.265	7.015	8.231	9.390	10.86	13.68	17.34
19	6.844	7.633	8.907	10.12	11.65	14.56	18.34
20	7.434	8.260	9.591	10.85	12.44	15.45	19.34
21	8.034	8.897	10.28	11.59	13.24	16.34	20.34
22	8.643	9.542	10.98	12.34	14.04	17.24	21.34
23	9.260	10.20	11.69	13.09	14.85	18.14	22.34
24	9.886	10.86	12.40	13.85	15.66	19.04	23.34
25	10.52	11.52	13.12	14.61	16.47	19.94	24.34
26	11.16	12.20	13.84	15.38	17.29	20.84	25.34
27	11.81	12.88	14.57	16.15	18.11	21.75	26.34
28	12.46	13.56	15.31	16.93	18.94	22.66	27.34
29	13.12	14.26	16.05	17.71	19.77	23.57	28.34
30	13.79	14.95	16.79	18.49	20.60	24.48	29.34
40	20.71	22.16	24.43	26.51	29.05	33.66	39.34
50	27.99	29.71	32.36	34.76	37.69	42.94	49.33
60	35.53	37.48	40.48	43.19	46.46	52.29	59.33
70	43.28	45.44	48.76	51.74	55.33	61.70	69.33
80	51.17	53.54	57.15	60.39	64.28	71.14	79.33
90	59.20	61.75	65.65	69.13	73.29	80.62	89.33
100	67.33	70.06	74.22	77.93	82.36	90.13	99.33

Appendix A3. Chi-Square Percentiles $\chi^2(P; \nu)$ (*Continued*)

ν \ P	0.750	0.900	0.950	0.975	0.990	0.995	0.999
1	1.323	2.706	3.841	5.024	6.635	7.879	10.83
2	2.773	4.605	5.991	7.378	9.210	10.60	13.82
3	4.108	6.251	7.815	9.348	11.34	12.84	16.27
4	5.385	7.779	9.488	11.14	13.28	14.86	18.47
5	6.626	9.236	11.07	12.83	15.09	16.75	20.52
6	7.841	10.64	12.59	14.45	16.81	18.55	22.46
7	9.037	12.02	14.07	16.01	18.48	20.28	24.32
8	10.22	13.36	15.51	17.53	20.09	21.96	26.12
9	11.39	14.68	16.92	19.02	21.67	23.59	27.88
10	12.55	15.99	18.31	20.48	23.21	25.19	29.59
11	13.70	17.28	19.68	21.92	24.72	26.76	31.26
12	14.85	18.55	21.03	23.34	26.22	28.30	32.91
13	15.98	19.81	22.36	24.74	27.69	29.82	34.53
14	17.12	21.06	23.68	26.12	29.14	31.32	36.12
15	18.25	22.31	25.00	27.49	30.58	32.80	37.70
16	19.37	23.54	26.30	28.85	32.00	34.27	39.25
17	20.49	24.77	27.59	30.19	33.41	35.72	40.79
18	21.60	25.99	28.87	31.53	34.81	37.16	42.31
19	22.72	27.20	30.14	32.85	36.19	38.58	43.82
20	23.83	28.41	31.41	34.17	37.57	40.00	45.32
21	24.93	29.62	32.67	35.48	38.93	41.40	46.80
22	26.04	30.81	33.92	36.78	40.29	42.80	48.27
23	27.14	32.01	35.17	38.08	41.64	44.18	49.73
24	28.24	33.20	36.42	39.36	42.98	45.56	51.18
25	29.34	34.38	37.65	40.65	44.31	46.93	52.62
26	30.43	35.56	38.89	41.92	45.64	48.29	54.05
27	31.53	36.74	40.11	43.19	46.96	49.64	55.48
28	32.62	37.92	41.34	44.46	48.28	50.99	56.89
29	33.71	39.09	42.56	45.72	49.59	52.34	58.30
30	34.80	40.26	43.77	46.98	50.89	53.67	59.70
40	45.62	51.80	55.76	59.34	63.69	66.77	73.40
50	56.33	63.17	67.50	71.42	76.15	79.49	86.66
60	66.98	74.40	79.08	83.30	88.38	91.95	99.61
70	77.58	85.53	90.53	95.02	100.4	104.2	112.3
80	88.13	96.58	101.9	106.6	112.3	116.3	124.8
90	98.65	107.6	113.1	118.1	124.1	128.3	137.2
100	109.1	118.5	124.3	129.6	135.8	140.2	149.4

Appendix A4. t-Distribution Percentiles $t(P; \nu)$

ν \ P	0.750	0.900	0.950	0.975	0.990	0.995	0.999	0.9995
1	1.000	3.078	6.314	12.706	31.821	63.657	318.31	636.62
2	0.816	1.886	2.920	4.303	6.965	9.925	22.326	31.598
3	0.765	1.638	2.353	3.182	4.541	5.841	10.213	12.924
4	0.741	1.533	2.132	2.776	3.747	4.604	7.173	8.610
5	0.727	1.476	2.015	2.571	3.365	4.032	5.893	6.869
6	0.718	1.440	1.943	2.447	3.143	3.707	5.208	5.959
7	0.711	1.415	1.895	2.365	2.998	3.499	4.785	5.408
8	0.706	1.397	1.860	2.306	2.896	3.355	4.501	5.041
9	0.703	1.383	1.833	2.262	2.821	3.250	4.297	4.781
10	0.700	1.372	1.812	2.228	2.764	3.169	4.144	4.587
11	0.697	1.363	1.796	2.201	2.718	3.106	4.025	4.437
12	0.695	1.356	1.782	2.179	2.681	3.055	3.930	4.318
13	0.694	1.350	1.771	2.160	2.650	3.012	3.852	4.221
14	0.692	1.345	1.761	2.145	2.624	2.977	3.787	4.140
15	0.691	1.341	1.753	2.131	2.602	2.947	3.733	4.073
16	0.690	1.337	1.746	2.120	2.583	2.921	3.686	4.015
17	0.689	1.333	1.740	2.110	2.567	2.898	3.646	3.965
18	0.688	1.330	1.734	2.101	2.552	2.878	3.610	3.922
19	0.688	1.328	1.729	2.093	2.539	2.861	3.579	3.883
20	0.687	1.325	1.725	2.086	2.528	2.845	3.552	3.850
21	0.686	1.323	1.721	2.080	2.518	2.831	3.527	3.819
22	0.686	1.321	1.717	2.074	2.508	2.819	3.505	3.792
23	0.685	1.319	1.714	2.069	2.500	2.807	3.485	3.767
24	0.685	1.318	1.711	2.064	2.492	2.797	3.467	3.745
25	0.684	1.316	1.708	2.060	2.485	2.787	3.450	3.725
26	0.684	1.315	1.706	2.056	2.479	2.779	3.435	3.707
27	0.684	1.314	1.703	2.052	2.473	2.771	3.421	3.690
28	0.683	1.313	1.701	2.048	2.467	2.763	3.408	3.674
29	0.683	1.311	1.699	2.045	2.462	2.756	3.396	3.659
30	0.683	1.310	1.697	2.042	2.457	2.750	3.385	3.646
40	0.681	1.303	1.684	2.021	2.423	2.704	3.307	3.551
60	0.679	1.296	1.671	2.000	2.390	2.660	3.232	3.460
120	0.677	1.289	1.658	1.980	2.358	2.617	3.160	3.373
∞	0.674	1.282	1.645	1.960	2.326	2.576	3.090	3.291

From N. L. Johnson and F. C. Leone, *Statistics and Experimental Design in Engineering and the Physical Sciences*, 2nd ed., Wiley, New York, 1977, Vol. 1, p. 466. Reproduced by permission of the publisher and the Biometrika Trustees. Use $t(1 - P; \nu) = - t(P; \nu)$ for small percentiles.

Appendix A5a. *F*-Distribution 90 Percent Points $F(0.90; \nu_1, \nu_2)$

ν_2 \ ν_1	1	2	3	4	5	6	7	8	9
1	39.864	49.500	53.593	55.833	57.241	58.204	58.906	59.439	59.858
2	8.5263	9.0000	9.1618	9.2434	9.2926	9.3255	9.3491	9.3668	9.3805
3	5.5383	5.4624	5.3908	5.3427	5.3092	5.2847	5.2662	5.2517	5.2400
4	4.5448	4.3246	4.1908	4.1073	4.0506	4.0098	3.9790	3.9549	3.9357
5	4.0604	3.7797	3.6195	3.5202	3.4530	3.4045	3.3679	3.3393	3.3163
6	3.7760	3.4633	3.2888	3.1808	3.1075	3.0546	3.0145	2.9830	2.9577
7	3.5894	3.2574	3.0741	2.9605	2.8833	2.8274	2.7849	2.7516	2.7247
8	3.4579	3.1131	2.9238	2.8064	2.7265	2.6683	2.6241	2.5893	2.5612
9	3.3603	3.0065	2.8129	2.6927	2.6106	2.5509	2.5053	2.4694	2.4403
10	3.2850	2.9245	2.7277	2.6053	2.5216	2.4606	2.4140	2.3772	2.3473
11	3.2252	2.8595	2.6602	2.5362	2.4512	2.3891	2.3416	2.3040	2.2735
12	3.1765	2.8068	2.6055	2.4801	2.3940	2.3310	2.2828	2.2446	2.2135
13	3.1362	2.7632	2.5603	2.4337	2.3467	2.2830	2.2341	2.1953	2.1638
14	3.1022	2.7265	2.5222	2.3947	2.3069	2.2426	2.1931	2.1539	2.1220
15	3.0732	2.6952	2.4898	2.3614	2.2730	2.2081	2.1582	2.1185	2.0862
16	3.0481	2.6682	2.4618	2.3327	2.2438	2.1783	2.1280	2.0880	2.0553
17	3.0262	2.6446	2.4374	2.3077	2.2183	2.1524	2.1017	2.0613	2.0284
18	3.0070	2.6239	2.4160	2.2858	2.1958	2.1296	2.0785	2.0379	2.0047
19	2.9899	2.6056	2.3970	2.2663	2.1760	2.1094	2.0580	2.0171	1.9836
20	2.9747	2.5893	2.3801	2.2489	2.1582	2.0913	2.0397	1.9985	1.9649
21	2.9609	2.5746	2.3649	2.2333	2.1423	2.0751	2.0232	1.9819	1.9480
22	2.9486	2.5613	2.3512	2.2193	2.1279	2.0605	2.0084	1.9668	1.9327
23	2.9374	2.5493	2.3387	2.2065	2.1149	2.0472	1.9949	1.9531	1.9189
24	2.9271	2.5383	2.3274	2.1949	2.1030	2.0351	1.9826	1.9407	1.9063
25	2.9177	2.5283	2.3170	2.1843	2.0922	2.0241	1.9714	1.9292	1.8947
26	2.9091	2.5191	2.3075	2.1745	2.0822	2.0139	1.9610	1.9188	1.8841
27	2.9012	2.5106	2.2987	2.1655	2.0730	2.0045	1.9515	1.9091	1.8743
28	2.8939	2.5028	2.2906	2.1571	2.0645	1.9959	1.9427	1.9001	1.8652
29	2.8871	2.4955	2.2831	2.1494	2.0566	1.9878	1.9345	1.8918	1.8568
30	2.8807	2.4887	2.2761	2.1422	2.0492	1.9803	1.9269	1.8841	1.8490
40	2.8354	2.4404	2.2261	2.0909	1.9968	1.9269	1.8725	1.8289	1.7929
60	2.7914	2.3932	2.1774	2.0410	1.9457	1.8747	1.8194	1.7748	1.7380
120	2.7478	2.3473	2.1300	1.9923	1.8959	1.8238	1.7675	1.7220	1.6843
∞	2.7055	2.3026	2.0838	1.9449	1.8473	1.7741	1.7167	1.6702	1.6315

From C. A. Bennett and N. L. Franklin, *Statistical Analysis in Chemistry and the Chemical Industry*, New York, 1954, pp. 702–705. Reproduced by permission of the publisher and the Biometrika Trustees.

Appendix A5a. F-Distribution 90 Percent Points $F(0.90; \nu_1, \nu_2)$
(Continued)

ν_2 \ ν_1	10	12	15	20	24	30	40	60	120	∞
1	60.195	60.705	61.220	61.740	62.002	62.265	62.529	62.794	63.061	63.328
2	9.3916	9.4081	9.4247	9.4413	9.4496	9.4579	9.4663	9.4746	9.4829	9.4913
3	5.2304	5.2156	5.2003	5.1845	5.1764	5.1681	5.1597	5.1512	5.1425	5.1337
4	3.9199	3.8955	3.8689	3.8443	3.8310	3.8174	3.8036	3.7896	3.7753	3.7607
5	3.2974	3.2682	3.2380	3.2067	3.1905	3.1741	3.1573	3.1402	3.1228	3.1050
6	2.9369	2.9047	2.8712	2.8363	2.8183	2.8000	2.7812	2.7620	2.7423	2.7222
7	2.7025	2.6681	2.6322	2.5947	2.5753	2.5555	2.5351	2.5142	2.4928	2.4708
8	2.5380	2.5020	2.4642	2.4246	2.4041	2.3830	2.3614	2.3391	2.3162	2.2926
9	2.4163	2.3789	2.3396	2.2983	2.2768	2.2547	2.2320	2.2085	2.1843	2.1592
10	2.3226	2.2841	2.2435	2.2007	2.1784	2.1554	2.1317	2.1072	2.0818	2.0554
11	2.2482	2.2087	2.1671	2.1230	2.1000	2.0762	2.0516	2.0261	1.9997	1.9721
12	2.1878	2.1474	2.1049	2.0597	2.0360	2.0115	1.9861	1.9597	1.9323	1.9036
13	2.1376	2.0966	2.0532	2.0070	1.9827	1.9576	1.9315	1.9043	1.8759	1.8462
14	2.0954	2.0537	2.0095	1.9625	1.9377	1.9119	1.8852	1.8572	1.8280	1.7973
15	2.0593	2.0171	1.9722	1.9243	1.8990	1.8728	1.8454	1.8168	1.7867	1.7551
16	2.0281	1.9854	1.9399	1.8913	1.8656	1.8388	1.8108	1.7816	1.7507	1.7182
17	2.0009	1.9577	1.9117	1.8624	1.8362	1.8090	1.7805	1.7506	1.7191	1.6856
18	1.9770	1.9333	1.8868	1.8368	1.8103	1.7827	1.7537	1.7232	1.6910	1.6567
19	1.9557	1.9117	1.8647	1.8142	1.7873	1.7592	1.7298	1.6988	1.6659	1.6308
20	1.9367	1.8924	1.8449	1.7938	1.7667	1.7382	1.7083	1.6768	1.6433	1.6074
21	1.9197	1.8750	1.8272	1.7756	1.7481	1.7193	1.6890	1.6569	1.6228	1.5862
22	1.9043	1.8593	1.8111	1.7590	1.7312	1.7021	1.6714	1.6389	1.6042	1.5668
23	1.8903	1.8450	1.7964	1.7439	1.7159	1.6864	1.6554	1.6224	1.5871	1.5490
24	1.8775	1.8319	1.7831	1.7302	1.7019	1.6721	1.6407	1.6073	1.5715	1.5327
25	1.8658	1.8200	1.7708	1.7175	1.6890	1.6589	1.6272	1.5934	1.5570	1.5176
26	1.8550	1.8090	1.7596	1.7059	1.6771	1.6468	1.6147	1.5805	1.5437	1.5036
27	1.8451	1.7989	1.7492	1.6951	1.6662	1.6356	1.6032	1.5686	1.5313	1.4906
28	1.8359	1.7895	1.7395	1.6852	1.6560	1.6252	1.5925	1.5575	1.5198	1.4784
29	1.8274	1.7808	1.7306	1.6759	1.6465	1.6155	1.5825	1.5472	1.5090	1.4670
30	1.8195	1.7727	1.7223	1.6673	1.6377	1.6065	1.5732	1.5376	1.4989	1.4564
40	1.7627	1.7146	1.6624	1.6052	1.5741	1.5411	1.5056	1.4672	1.4248	1.3769
60	1.7070	1.6574	1.6034	1.5435	1.5107	1.4755	1.4373	1.3952	1.3476	1.2915
120	1.6524	1.6012	1.5450	1.4821	1.4472	1.4094	1.3676	1.3203	1.2646	1.1926
∞	1.5987	1.5458	1.4871	1.4206	1.3832	1.3419	1.2951	1.2400	1.1686	1.0000

Interpolation should be carried out using the reciprocals of the degrees of freedom.

Appendix A5b. F-Distribution 95 Percent Points $F(0.95; \nu_1, \nu_2)$

ν_2 \ ν_1	1	2	3	4	5	6	7	8	9
1	161.45	199.50	215.71	224.58	230.16	233.99	236.77	238.88	240.54
2	18.513	19.000	19.164	19.247	19.296	19.330	19.353	19.371	19.385
3	10.128	9.5521	9.2766	9.1172	9.0135	8.9406	8.8868	8.8452	8.8123
4	7.7086	6.9443	6.5914	6.3883	6.2560	6.1631	6.0942	6.0410	5.9988
5	6.6079	5.7861	5.4095	5.1922	5.0503	4.9503	4.8759	4.8183	4.7725
6	5.9874	5.1433	4.7571	4.5337	4.3874	4.2839	4.2066	4.1468	4.0990
7	5.5914	4.7374	4.3468	4.1203	3.9715	3.8660	3.7870	3.7257	3.6767
8	5.3177	4.4590	4.0662	3.8378	3.6875	3.5806	3.5005	3.4381	3.3881
9	5.1174	4.2565	3.8626	3.6331	3.4817	3.3738	3.2927	3.2296	3.1789
10	4.9646	4.1028	3.7083	3.4780	3.3258	3.2172	3.1355	3.0717	3.0204
11	4.8443	3.9823	3.5874	3.3567	3.2039	3.0946	3.0123	2.9480	2.8962
12	4.7472	3.8853	3.4903	3.2592	3.1059	2.9961	2.9134	2.8486	2.7964
13	4.6672	3.8056	3.4105	3.1791	3.0254	2.9153	2.8321	2.7669	2.7144
14	4.6001	3.7389	3.3439	3.1122	2.9582	2.8477	2.7642	2.6987	2.6458
15	4.5431	3.6823	3.2874	3.0556	2.9013	2.7905	2.7066	2.6408	2.5876
16	4.4940	3.6337	3.2389	3.0069	2.8524	2.7413	2.6572	2.5911	2.5377
17	4.4513	3.5915	3.1968	2.9647	2.8100	2.6987	2.6143	2.5480	2.4943
18	4.4139	3.5546	3.1599	2.9277	2.7729	2.6613	2.5767	2.5102	2.4563
19	4.3808	3.5219	3.1274	2.8951	2.7401	2.6283	2.5435	2.4768	2.4227
20	4.3513	3.4928	3.0984	2.8661	2.7109	2.5990	2.5140	2.4471	2.3928
21	4.3248	3.4668	3.0725	2.8401	2.6848	2.5727	2.4876	2.4205	2.3661
22	4.3009	3.4434	3.0491	2.8167	2.6613	2.5491	2.4638	2.3965	2.3419
23	4.2793	3.4221	3.0280	2.7955	2.6400	2.5277	2.4422	2.3748	2.3201
24	4.2597	3.4028	3.0088	2.7763	2.6207	2.5082	2.4226	2.3551	2.3002
25	4.2417	3.3852	2.9912	2.7587	2.6030	2.4904	2.4047	2.3371	2.2821
26	4.2252	3.3690	2.9751	2.7426	2.5868	2.4741	2.3883	2.3205	2.2655
27	4.2100	3.3541	2.9604	2.7278	2.5719	2.4591	2.3732	2.3053	2.2501
28	4.1960	3.3404	2.9467	2.7141	2.5581	2.4453	2.3593	2.2913	2.2360
29	4.1830	3.3277	2.9340	2.7014	2.5454	2.4324	2.3463	2.2782	2.2229
30	4.1709	3.3158	2.9223	2.6896	2.5336	2.4205	2.3343	2.2662	2.2107
40	4.0848	3.2317	2.8387	2.6060	2.4495	2.3359	2.2490	2.1802	2.1240
60	4.0012	3.1504	2.7581	2.5252	2.3683	2.2540	2.1665	2.0970	2.0401
120	3.9201	3.0718	2.6802	2.4472	2.2900	2.1750	2.0867	2.0164	1.9588
∞	3.8415	2.9957	2.6049	2.3719	2.2141	2.0986	2.0096	1.9384	1.8799

584

Appendix A5b. **_F_-Distribution 95 Percent Points $F(0.95; \nu_1, \nu_2)$**
(_Continued_)

ν_1 \ ν_2	10	12	15	20	24	30	40	60	120	∞
1	241.88	243.91	245.95	248.01	249.05	250.09	251.14	252.20	253.25	254.32
2	19.396	19.413	19.429	19.446	19.454	19.462	19.471	19.479	19.487	19.496
3	8.7855	8.7446	8.7029	8.6602	8.6385	8.6166	8.5944	8.5720	8.5494	8.5265
4	5.9644	5.9117	5.8578	5.8025	5.7744	5.7459	5.7170	5.6878	5.6581	5.6281
5	4.7351	4.6777	4.6188	4.5581	4.5272	4.4957	4.4638	4.4314	4.3984	4.3650
6	4.0600	3.9999	3.9381	3.8742	3.8415	3.8082	3.7743	3.7398	3.7047	3.6688
7	3.6365	3.5747	3.5108	3.4445	3.4105	3.3758	3.3404	3.3043	3.2674	3.2298
8	3.3472	3.2840	3.2184	3.1503	3.1152	3.0794	3.0428	3.0053	2.9669	2.9276
9	3.1373	3.0729	3.0061	2.9365	2.9005	2.8637	2.8259	2.7872	2.7475	2.7067
10	2.9782	2.9130	2.8450	2.7740	2.7372	2.6996	2.6609	2.6211	2.5801	2.5379
11	2.8536	2.7876	2.7186	2.6464	2.6090	2.5705	2.5309	2.4901	2.4480	2.4045
12	2.7534	2.6866	2.6169	2.5436	2.5055	2.4663	2.4259	2.3842	2.3410	2.2962
13	2.6710	2.6037	2.5331	2.4589	2.4202	2.3803	2.3392	2.2966	2.2524	2.2064
14	2.6021	2.5342	2.4630	2.3879	2.3487	2.3082	2.2664	2.2230	2.1778	2.1307
15	2.5437	2.4753	2.4035	2.3275	2.2878	2.2468	2.2043	2.1601	2.1141	2.0658
16	2.4935	2.4247	2.3522	2.2756	2.2354	2.1938	2.1507	2.1058	2.0589	2.0096
17	2.4499	2.3807	2.3077	2.2304	2.1898	2.1477	2.1040	2.0584	2.0107	1.9604
18	2.4117	2.3421	2.2686	2.1906	2.1497	2.1071	2.0629	2.0166	1.9681	1.9168
19	2.3779	2.3080	2.2341	2.1555	2.1141	2.0712	2.0264	1.9796	1.9302	1.8780
20	2.3479	2.2776	2.2033	2.1242	2.0825	2.0391	1.9938	1.9464	1.8963	1.8432
21	2.3210	2.2504	2.1757	2.0960	2.0540	2.0102	1.9645	1.9165	1.8657	1.8117
22	2.2967	2.2258	2.1508	2.0707	2.0283	1.9842	1.9380	1.8895	1.8380	1.7831
23	2.2747	2.2036	2.1282	2.0476	2.0050	1.9605	1.9139	1.8649	1.8128	1.7570
24	2.2547	2.1834	2.1077	2.0267	1.9838	1.9390	1.8920	1.8424	1.7897	1.7331
25	2.2365	2.1649	2.0889	2.0075	1.9643	1.9192	1.8718	1.8217	1.7684	1.7110
26	2.2197	2.1479	2.0716	1.9898	1.9464	1.9010	1.8533	1.8027	1.7488	1.6906
27	2.2043	2.1323	2.0558	1.9736	1.9299	1.8842	1.8361	1.7851	1.7307	1.6717
28	2.1900	2.1179	2.0411	1.9586	1.9147	1.8687	1.8203	1.7689	1.7138	1.6541
29	2.1768	2.1045	2.0275	1.9446	1.9005	1.8543	1.8055	1.7537	1.6981	1.6377
30	2.1646	2.0921	2.0148	1.9317	1.8874	1.8409	1.7918	1.7396	1.6835	1.6223
40	2.0772	2.0035	1.9245	1.8389	1.7929	1.7444	1.6928	1.6373	1.5766	1.5089
60	1.9926	1.9174	1.8364	1.7480	1.7001	1.6491	1.5943	1.5343	1.4673	1.3893
120	1.9105	1.8337	1.7505	1.6587	1.6084	1.5543	1.4952	1.4290	1.3519	1.2539
∞	1.8307	1.7522	1.6664	1.5705	1.5173	1.4591	1.3940	1.3180	1.2214	1.0000

Appendix A6. Poisson Cumulative Distribution Function $F(y)$ for Mean μ

y	0.001	0.005	0.010	0.015	0.020	0.025
0	0.9990 0050	0.9950 1248	0.9900 4983	0.9851 1194	0.9801 9867	0.9753 099
1	0.9999 9950	0.9999 8754	0.9999 5033	0.9998 8862	0.9998 0264	0.9996 927
2	1.0000 0000	0.9999 9998	0.9999 9983	0.9999 9945	0.9999 9868	0.9999 974
3	-	1.0000 0000	1.0000 0000	1.0000 0000	0.9999 9999	1.0000 000
4	-	-	-	-	1.0000 0000	1.0000 000

y	0.030	0.035	0.040	0.045	0.050	0.055
0	0.970 446	0.965 605	0.960 789	0.955 997	0.951 229	0.946 485
1	0.999 559	0.999 402	0.999 221	0.999 017	0.998 791	0.998 542
2	0.999 996	0.999 993	0.999 990	0.999 985	0.999 980	0.999 973
3	1.000 000	1.000 000	1.000 000	1.000 000	1.000 000	1.000 000

y	0.060	0.065	0.070	0.075	0.080	0.085
0	0.941 765	0.937 067	0.932 394	0.927 743	0.923 116	0.918 512
1	0.998 270	0.997 977	0.997 661	0.997 324	0.996 966	0.996 586
2	0.999 966	0.999 956	0.999 946	0.999 934	0.999 920	0.999 904
3	0.999 999	0.999 999	0.999 999	0.999 999	0.999 998	0.999 998
4	1.000 000	1.000 000	1.000 000	1.000 000	1.000 000	1.000 000

y	0.090	0.095	0.100	0.200	0.300	0.400
0	0.913 931	0.909 373	0.904 837	0.818 731	0.740 818	0.670 320
1	0.996 185	0.995 763	0.995 321	0.982 477	0.963 064	0.938 448
2	0.999 886	0.999 867	0.999 845	0.998 852	0.996 401	0.992 074
3	0.999 997	0.999 997	0.999 996	0.999 943	0.999 734	0.999 224
4	1.000 000	1.000 000	1.000 000	0.999 998	0.999 984	0.999 939
5				1.000 000	0.999 999	0.999 996
6					1.000 000	1.000 000

y	0.500	0.600	0.700	0.800	0.900	1.000
0	0.606 531	0.548 812	0.496 585	0.449 329	0.406 570	0.367 879
1	0.909 796	0.878 099	0.844 195	0.808 792	0.772 482	0.735 759
2	0.985 612	0.976 885	0.965 858	0.952 577	0.937 143	0.919 699
3	0.998 248	0.996 642	0.994 247	0.990 920	0.986 541	0.981 012
4	0.999 828	0.999 606	0.999 214	0.998 589	0.997 656	0.996 340
5	0.999 986	0.999 961	0.999 910	0.999 816	0.999 657	0.999 406
6	0.999 999	0.999 997	0.999 991	0.999 979	0.999 957	0.999 917
7	1.000 000	1.000 000	0.999 999	0.999 998	0.999 995	0.999 990
8			1.000 000	1.000 000	1.000 000	0.999 999
9						1.000 000

From Donald B. Owen, *Handbook of Statistical Tables*, © 1962. U.S. Department of Energy. Published by Addison-Wesley Publishing Company, Inc., Reading Massachusetts. Table 9.3, pp. 260–261, "Poisson Table." Reprinted with permission of the publisher.

Appendix A6. Poisson Cumulative Distribution Function $F(y)$ for Mean μ
(Continued)

y	1.20	1.40	1.60	1.80	2.00	2.50	3.00	3.50
0	0.3012	0.2466	0.2019	0.1653	0.1353	0.0821	0.0498	0.0302
1	0.6626	0.5918	0.5249	0.4628	0.4060	0.2873	0.1991	0.1359
2	0.8795	0.8335	0.7834	0.7306	0.6767	0.5438	0.4232	0.3208
3	0.9662	0.9463	0.9212	0.8913	0.8571	0.7576	0.6472	0.5366
4	0.9923	0.9857	0.9763	0.9636	0.9473	0.8912	0.8153	0.7254
5	0.9985	0.9968	0.9940	0.9896	0.9834	0.9580	0.9161	0.8576
6	0.9997	0.9994	0.9987	0.9974	0.9955	0.9858	0.9665	0.9347
7	1.0000	0.9999	0.9997	0.9994	0.9989	0.9958	0.9881	0.9733
8		1.0000	1.0000	0.9999	0.9998	0.9989	0.9962	0.9901
9				1.0000	1.0000	0.9997	0.9989	0.9967
10						0.9999	0.9997	0.9990
11						1.0000	0.9999	0.9997
12							1.0000	0.9999
13								1.0000

y	4.00	4.50	5.00	6.00	7.00	8.00	9.00	10.00
0	0.0183	0.0111	0.0067	0.0025	0.0009	0.0003	0.0001	0.0000
1	0.0916	0.0611	0.0404	0.0174	0.0073	0.0030	0.0012	0.0005
2	0.2381	0.1736	0.1247	0.0620	0.0296	0.0138	0.0062	0.0028
3	0.4335	0.3423	0.2650	0.1512	0.0818	0.0424	0.0212	0.0103
4	0.6288	0.5321	0.4405	0.2851	0.1730	0.0996	0.0550	0.0293
5	0.7851	0.7029	0.6160	0.4457	0.3007	0.1912	0.1157	0.0671
6	0.8893	0.8311	0.7622	0.6063	0.4497	0.3134	0.2068	0.1301
7	0.9489	0.9134	0.8666	0.7440	0.5987	0.4530	0.3239	0.2202
8	0.9786	0.9597	0.9319	0.8472	0.7291	0.5925	0.4577	0.3328
9	0.9919	0.9829	0.9682	0.9161	0.8305	0.7166	0.5874	0.4579
10	0.9972	0.9933	0.9863	0.9574	0.9015	0.8159	0.7060	0.5830
11	0.9991	0.9976	0.9945	0.9799	0.9467	0.8881	0.8030	0.6968
12	0.9997	0.9992	0.9980	0.9912	0.9730	0.9362	0.8758	0.7916
13	0.9999	0.9997	0.9993	0.9964	0.9872	0.9658	0.9261	0.8645
14	1.0000	0.9999	0.9998	0.9986	0.9943	0.9827	0.9585	0.9165
15		1.0000	0.9999	0.9995	0.9976	0.9918	0.9780	0.9513
16			1.0000	0.9998	0.9990	0.9963	0.9889	0.9730
17				0.9999	0.9996	0.9984	0.9947	0.9857
18				1.0000	0.9999	0.9993	0.9976	0.9928
19						0.9997	0.9989	0.9965
20					1.0000	0.9999	0.9996	0.9984
21						1.0000	0.9998	0.9993
22							0.9999	0.9997
23							1.0000	0.9999
24								1.0000

Appendix A7.   Binomial Cumulative Distribution Function $F(y)$

n	y	p 0.05	0.10	0.15	0.20	0.25	0.30	0.35	0.40	0.45	0.50
2	2	0.9025	0.8100	0.7225	0.6400	0.5625	0.4900	0.4225	0.3600	0.3025	0.2500
	1	0.9975	0.9900	0.9775	0.9600	0.9375	0.9100	0.8755	0.8400	0.7975	0.7500
3	0	0.8574	0.7290	0.6141	0.5120	0.4219	0.3430	0.2746	0.2160	0.1664	0.1250
	1	0.9928	0.9720	0.9392	0.8960	0.8438	0.7840	0.7182	0.6480	0.5748	0.5000
	2	0.9999	0.9990	0.9966	0.9920	0.9844	0.9730	0.9571	0.9360	0.9089	0.8750
4	0	0.8145	0.6561	0.5220	0.4096	0.3164	0.2401	0.1785	0.1296	0.0915	0.0625
	1	0.9860	0.9477	0.8905	0.8192	0.7383	0.6517	0.5630	0.4752	0.3910	0.3125
	2	0.9995	0.9963	0.9880	0.9728	0.9492	0.9163	0.8735	0.8208	0.7585	0.6875
	3	1.0000	0.9999	0.9995	0.9984	0.9961	0.9919	0.9850	0.9744	0.9590	0.9375
5	0	0.7738	0.5905	0.4437	0.3277	0.2373	0.1681	0.1160	0.0778	0.0503	0.0312
	1	0.9774	0.9185	0.8352	0.7373	0.6328	0.5282	0.4284	0.3370	0.2562	0.1875
	2	0.9988	0.9914	0.9734	0.9421	0.8965	0.8369	0.7648	0.6826	0.5931	0.5000
	3	1.0000	0.9995	0.0078	0.9933	0.9844	0.9692	0.9460	0.9130	0.8688	0.8125
	4	1.0000	1.0000	0.9999	0.9997	0.9990	0.9976	0.9947	0.9898	0.9815	0.9688
6	0	0.7351	0.5314	0.3771	0.2621	0.1780	0.1176	0.0754	0.0467	0.0277	0.0156
	1	0.9672	0.8857	0.7765	0.6554	0.5339	0.4202	0.3191	0.2333	0.1636	0.1094
	2	0.9978	0.9842	0.9527	0.9011	0.8306	0.7443	0.6471	0.5443	0.4415	0.3438
	3	0.9999	0.9987	0.9941	0.9830	0.9624	0.9295	0.8826	0.8208	0.7447	0.6562
	4	1.0000	0.9999	0.9996	0.9984	0.9954	0.9891	0.9777	0.9590	0.9308	0.8906
	5	1.0000	1.0000	1.0000	0.9999	0.9998	0.9993	0.9982	0.9959	0.9917	0.9844
7	0	0.6983	0.4783	0.3206	0.2097	0.1335	0.0824	0.0490	0.0280	0.0152	0.0078
	1	0.9556	0.8503	0.7166	0.5767	0.4449	0.3294	0.2338	0.1586	0.1024	0.0625
	2	0.9962	0.9743	0.9262	0.8520	0.7564	0.6471	0.5323	0.4199	0.3164	0.2266
	3	0.9998	0.9973	0.9879	0.9667	0.9294	0.8740	0.8002	0.7102	0.6083	0.5000
	4	1.0000	0.9998	0.9988	0.9953	0.9871	0.9712	0.9444	0.9037	0.8471	0.7734
	5	1.0000	1.0000	0.9999	0.9996	0.9987	0.9962	0.9910	0.9812	0.9643	0.9375
	6	1.0000	1.0000	1.0000	1.0000	0.9999	0.9998	0.9994	0.9984	0.9963	0.9922
8	0	0.6634	0.4305	0.2725	0.1678	0.1001	0.0576	0.0319	0.0168	0.0084	0.0039
	1	0.9428	0.8131	0.6572	0.5033	0.3671	0.2553	0.1691	0.1064	0.0632	0.0352
	2	0.9942	0.9619	0.8948	0.7969	0.6785	0.5518	0.4278	0.3154	0.2201	0.1445
	3	0.9996	0.9950	0.9786	0.9437	0.8862	0.8059	0.7064	0.5941	0.4770	0.3633
	4	1.0000	0.9996	0.9971	0.9896	0.9727	0.9420	0.8939	0.8263	0.7396	0.6367
	5	1.0000	1.0000	0.9998	0.9988	0.9958	0.9887	0.9747	0.9502	0.9115	0.8555
	6	1.0000	1.0000	1.0000	0.9999	0.9996	0.9987	0.9964	0.9915	0.9819	0.9648
	7	1.0000	1.0000	1.0000	1.0000	1.0000	0.9999	0.9998	0.9993	0.9983	0.9961
9	0	0.6302	0.3874	0.2316	0.1342	0.0751	0.0404	0.0207	0.0101	0.0046	0.0020
	1	0.9288	0.7748	0.5995	0.4362	0.3003	0.1960	0.1211	0.0705	0.0385	0.0195
	2	0.9916	0.9470	0.8591	0.7382	0.6007	0.4628	0.3373	0.2318	0.1495	0.0898
	3	0.9994	0.9917	0.9661	0.9144	0.8343	0.7297	0.6089	0.4826	0.3614	0.2539
	4	1.0000	0.9991	0.9944	0.9804	0.9511	0.9012	0.8283	0.7334	0.6214	0.5000
	5	1.0000	0.9999	0.9994	0.9969	0.9900	0.9747	0.9464	0.9006	0.8342	0.7461
	6	1.0000	1.0000	1.0000	0.9997	0.9987	0.9957	0.9888	0.9750	0.9502	0.9102
	7	1.0000	1.0000	1.0000	1.0000	0.9999	0.9996	0.9986	0.9962	0.9909	0.9805
	8	1.0000	1.0000	1.0000	1.0000	1.0000	1.0000	0.9999	0.9997	0.9992	0.9980

From Irwin Miller and John E. Freund, *Probability and Statistics for Engineers*, 2nd ed., © 1977, pp. 477–479, 481. Reprinted by permission of Prentice-Hall, Inc., Englewood Cliffs, N.J.

Appendix A7. Binomial Cumulative Distribution Function $F(y)$
(Continued)

n	y	0.05	0.10	0.15	0.20	0.25	0.30	0.35	0.40	0.45	0.50
10	0	0.5987	0.3487	0.1969	0.1074	0.0563	0.0282	0.0135	0.0060	0.0025	0.0010
	1	0.9139	0.7361	0.5443	0.3758	0.2440	0.1493	0.0860	0.0464	0.0232	0.0107
	2	0.9885	0.9298	0.8202	0.6778	0.5256	0.3828	0.2616	0.1673	0.0996	0.0547
	3	0.9990	0.9872	0.9500	0.8791	0.7759	0.6496	0.5138	0.3823	0.2660	0.1719
	4	0.9999	0.9984	0.9901	0.9672	0.9219	0.8497	0.7515	0.6331	0.5044	0.3770
	5	1.0000	0.9999	0.9986	0.9936	0.9803	0.9527	0.9051	0.8338	0.7384	0.6230
	6	1.0000	1.0000	0.9999	0.9991	0.9965	0.9894	0.9740	0.9452	0.8980	0.8281
	7	1.0000	1.0000	1.0000	0.9999	0.9996	0.9984	0.9952	0.9877	0.9726	0.9453
	8	1.0000	1.0000	1.0000	1.0000	1.0000	0.9999	0.9995	0.9983	0.9955	0.9893
	9	1.0000	1.0000	1.0000	1.0000	1.0000	1.0000	1.0000	0.9999	0.9997	0.9990
15	0	0.4633	0.2059	0.0874	0.0352	0.0134	0.0047	0.0016	0.0005	0.0001	0.0000
	1	0.8290	0.5490	0.3186	0.1671	0.0802	0.0353	0.0142	0.0052	0.0017	0.0005
	2	0.9638	0.8159	0.6042	0.3980	0.2361	0.1268	0.0617	0.0271	0.0107	0.0037
	3	0.9945	0.9444	0.8227	0.6482	0.4613	0.2969	0.1727	0.0905	0.0424	0.0176
	4	0.9994	0.9873	0.9383	0.8358	0.6865	0.5155	0.3519	0.2173	0.1204	0.0592
	5	0.9999	0.9978	0.9832	0.9389	0.8516	0.7216	0.5643	0.4032	0.2608	0.1509
	6	1.0000	0.9997	0.9964	0.9819	0.9434	0.8689	0.7548	0.6098	0.4522	0.3036
	7	1.0000	1.0000	0.9996	0.9958	0.9827	0.9500	0.8868	0.7869	0.6535	0.5000
	8	1.0000	1.0000	0.9999	0.9992	0.9958	0.9848	0.9578	0.9050	0.8182	0.6964
	9	1.0000	1.0000	1.0000	0.9999	0.9992	0.9963	0.9876	0.9662	0.9231	0.8491
	10	1.0000	1.0000	1.0000	1.0000	0.9999	0.9993	0.9972	0.9907	0.9745	0.9408
	11	1.0000	1.0000	1.0000	1.0000	1.0000	0.9999	0.9995	0.9981	0.9937	0.9824
	12	1.0000	1.0000	1.0000	1.0000	1.0000	1.0000	0.9999	0.9997	0.9989	0.9963
	13	1.0000	1.0000	1.0000	1.0000	1.0000	1.0000	1.0000	1.0000	0.9999	0.9995
	14	1.0000	1.0000	1.0000	1.0000	1.0000	1.0000	1.0000	1.0000	1.0000	1.0000
20	0	0.3585	0.1216	0.0388	0.0115	0.0032	0.0008	0.0002	0.0000	0.0000	0.0000
	1	0.7358	0.3917	0.1756	0.0692	0.0243	0.0076	0.0021	0.0005	0.0001	0.0000
	2	0.9245	0.6769	0.4049	0.2061	0.0913	0.0355	0.0121	0.0036	0.0009	0.0002
	3	0.9841	0.8670	0.6477	0.4114	0.2252	0.1071	0.0444	0.0160	0.0049	0.0013
	4	0.9974	0.9568	0.8298	0.6296	0.4148	0.2375	0.1182	0.0510	0.0189	0.0059
	5	0.9997	0.9887	0.9327	0.8042	0.6172	0.4164	0.2454	0.1256	0.0553	0.0207
	6	1.0000	0.9976	0.9781	0.9133	0.7858	0.6080	0.4166	0.2500	0.1299	0.0577
	7	1.0000	0.9996	0.9941	0.9679	0.8982	0.7723	0.6010	0.4159	0.2520	0.1316
	8	1.0000	0.9999	0.9987	0.9900	0.9591	0.8867	0.7624	0.5956	0.4143	0.2517
	9	1.0000	1.0000	0.9998	0.9974	0.9861	0.9520	0.8782	0.7553	0.5914	0.4119
	10	1.0000	1.0000	1.0000	0.9994	0.9961	0.9829	0.9468	0.8725	0.7507	0.5881
	11	1.0000	1.0000	1.0000	0.9999	0.9991	0.9949	0.9804	0.9435	0.8692	0.7483
	12	1.0000	1.0000	1.0000	1.0000	0.9998	0.9987	0.9940	0.9790	0.9420	0.8684
	13	1.0000	1.0000	1.0000	1.0000	1.0000	0.9997	0.9985	0.9935	0.9786	0.9423
	14	1.0000	1.0000	1.0000	1.0000	1.0000	1.0000	0.9997	0.9984	0.9936	0.9793
	15	1.0000	1.0000	1.0000	1.0000	1.0000	1.0000	1.0000	0.9997	0.9985	0.9941
	16	1.0000	1.0000	1.0000	1.0000	1.0000	1.0000	1.0000	1.0000	0.9997	0.9987
	17	1.0000	1.0000	1.0000	1.0000	1.0000	1.0000	1.0000	1.0000	1.0000	0.9998
	18	1.0000	1.0000	1.0000	1.0000	1.0000	1.0000	1.0000	1.0000	1.0000	1.0000

Appendix A8. Confidence Limits for Binomial Proportion p with 95% confidence (two-sided); 97.5% (one-sided).

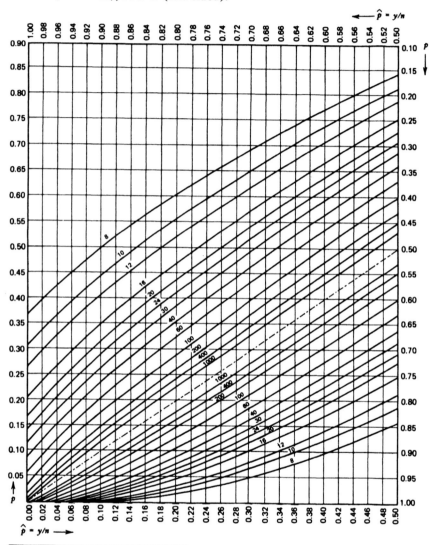

The numbers printed along the curves indicate the sample size n. For a given y/n, p and $\bar{p}$ are read from (or interpolated between) the appropriate lower and upper curves.

From G. E. P. Box, W. G. Hunter, and J. S. Hunter, *Statistics for Experimenters*, Wiley, New York, 1978, pp. 642–643. Reproduced by permission of the publisher and the Biometrika Trustees.

Appendix A8. Confidence Limits for Binomial Proportion p with 99% confidence (two-sided); 99.5% (one-sided).

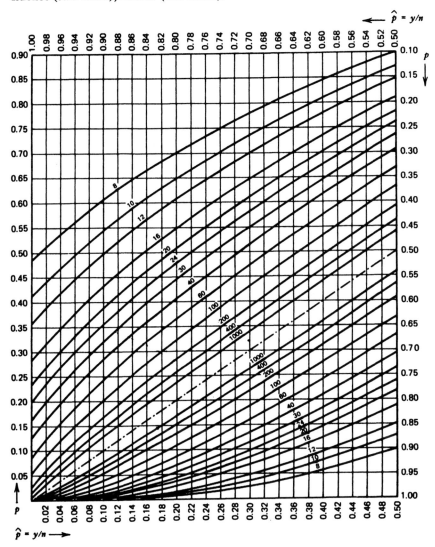

Appendix A9. Plotting Positions $100(i - 0.5)/n$

n

i	6	7	8	9	10	11	12	13	14	15	16	17	18	19	20
1	8.3	7.1	6.2	5.6	5.0	4.5	4.2	3.8	3.6	3.3	3.1	2.9	2.8	2.6	2.5
2	25.0	21.4	18.7	16.7	15.0	13.6	12.5	11.5	10.7	10.0	9.4	8.8	8.3	7.9	7.5
3	41.7	35.7	31.2	27.8	25.0	22.7	20.8	19.2	17.9	16.7	15.6	14.7	13.9	13.2	12.5
4	58.3	50.0	43.7	38.9	35.0	31.8	29.2	26.9	25.0	23.3	21.9	20.6	19.4	18.4	17.5
5	75.0	64.3	56.2	50.0	45.0	40.9	37.5	34.6	32.1	30.0	28.1	26.5	25.0	23.7	22.5
6	91.7	78.6	68.7	61.1	55.0	50.0	45.8	42.3	39.3	36.7	34.4	32.4	30.6	28.9	27.5
7		92.9	81.2	72.2	65.0	59.1	54.2	50.0	46.4	43.3	40.6	38.2	36.1	34.2	32.5
8			93.7	83.3	75.0	68.2	62.5	57.7	53.6	50.0	46.9	44.1	41.7	39.5	37.5
9				94.4	85.0	77.3	70.8	65.4	60.7	56.7	53.1	50.0	47.2	44.7	42.5
10					95.0	86.4	79.2	73.1	67.9	63.3	59.4	55.9	52.8	50.0	47.5
11						95.5	87.5	80.8	75.0	70.0	65.6	61.8	58.3	55.3	52.5
12							95.8	88.5	82.1	76.7	71.9	67.6	63.9	60.5	57.5
13								96.2	89.3	83.3	78.1	73.5	69.4	65.8	62.5
14									96.4	90.0	84.4	79.4	75.0	71.1	67.5
15										96.7	90.6	85.3	80.6	76.3	72.5
16											96.9	91.2	86.1	81.6	77.5
17												97.1	91.7	86.8	82.5
18													97.2	92.1	87.5
19														97.4	92.5
20															97.5

n

i	21	22	23	24	25	26	27	28	29	30	31	32	33	34	35
1	2.4	2.3	2.2	2.1	2.0	1.9	1.9	1.8	1.7	1.7	1.6	1.6	1.5	1.5	1.4
2	7.1	6.8	6.5	6.2	6.0	5.8	5.6	5.4	5.2	5.0	4.8	4.7	4.5	4.4	4.3
3	11.9	11.4	10.9	10.4	10.0	9.6	9.3	8.9	8.6	8.3	8.1	7.8	7.6	7.4	7.1
4	16.7	15.9	15.2	14.6	14.0	13.5	13.0	12.5	12.1	11.7	11.3	10.9	10.6	10.3	10.0
5	21.4	20.5	19.6	18.7	18.0	17.3	16.7	16.1	15.5	15.0	14.5	14.1	13.6	13.2	12.9

															i
15.7	16.2	16.7	17.2	17.7	18.3	19.0	19.6	20.4	21.2	22.0	22.9	23.9	25.0	26.2	6
18.6	19.1	19.7	20.3	21.0	21.7	22.4	23.2	24.1	25.0	26.0	27.1	28.3	29.5	31.0	7
21.4	22.1	22.7	23.4	24.2	25.0	25.9	26.8	27.8	28.8	30.0	31.2	32.6	34.1	35.7	8
24.3	25.0	25.8	26.6	27.4	28.3	29.3	30.4	31.5	32.7	34.0	35.4	37.0	38.6	40.5	9
27.1	27.9	28.8	29.7	30.6	31.7	32.8	33.9	35.2	36.5	38.0	39.6	41.3	43.2	45.2	10
30.0	30.9	31.8	32.8	33.9	35.0	36.2	37.5	38.9	40.4	42.0	43.7	45.7	47.7	50.0	11
32.9	33.8	34.8	35.9	37.1	38.3	39.7	41.1	42.6	44.2	46.0	47.9	50.0	52.3	54.8	12
35.7	36.8	37.9	39.1	40.3	41.7	43.1	44.6	46.3	48.1	50.0	52.1	54.3	56.8	59.5	13
38.6	39.7	40.9	42.2	43.5	45.0	46.6	48.2	50.0	51.9	54.0	56.2	58.7	61.4	64.3	14
41.4	42.6	43.9	45.3	46.8	48.3	50.0	51.8	53.7	55.8	58.0	60.4	63.0	65.9	69.0	15
44.3	45.6	47.0	48.4	50.0	51.7	53.4	55.4	57.4	59.6	62.0	64.6	67.4	70.5	73.8	16
47.1	48.5	50.0	51.6	53.2	55.0	56.9	58.9	61.1	63.5	66.0	68.7	71.7	75.0	78.6	17
50.0	51.5	53.0	54.7	56.5	58.3	60.3	62.5	64.8	67.3	70.0	72.9	76.1	79.5	83.3	18
52.9	54.4	56.1	57.8	59.7	61.7	63.8	66.1	68.5	71.2	74.0	77.1	80.4	84.1	88.1	19
55.7	57.4	59.1	60.9	62.9	65.0	67.2	69.6	72.2	75.0	78.0	81.2	84.8	88.6	92.9	20
58.6	60.3	62.1	64.1	66.1	68.3	70.7	73.2	75.9	78.8	82.0	85.4	89.1	93.2	97.6	21
61.4	63.2	65.2	67.2	69.4	71.7	74.1	76.8	79.6	82.7	86.0	89.6	93.5	97.7		22
64.3	66.2	68.2	70.3	72.6	75.0	77.6	80.4	83.3	86.5	90.0	93.7	97.8			23
67.1	69.1	71.2	73.4	75.8	78.3	81.0	83.9	87.0	90.4	94.0	97.9				24
70.0	72.1	74.2	76.6	79.0	81.7	84.5	87.5	90.7	94.2	98.0					25
72.9	75.0	77.3	79.7	82.3	85.0	87.9	91.1	94.4	98.1						26
75.7	77.9	80.3	82.8	85.5	88.3	91.4	94.6	98.1							27
78.6	80.9	83.3	85.9	88.7	91.7	94.8	98.2								28
81.4	83.8	86.4	89.1	91.9	95.0	98.3									29
84.3	86.8	89.4	92.2	95.2	98.3										30
87.1	89.7	92.4	95.3	98.4											31
90.0	92.6	95.5	98.4												32
92.9	95.6	98.5													33
95.7	98.5														34
98.6															35

Appendix A10. Hazard Values $100/K$

k	100/k	k	100/k	k	100/k	k	100/k
1	100.00	51	1.96	101	0.99	151	0.66
2	50.00	52	1.92	102	0.98	152	0.66
3	33.33	53	1.89	103	0.97	153	0.65
4	25.00	54	1.85	104	0.96	154	0.65
5	20.00	55	1.82	105	0.95	155	0.65
6	16.67	56	1.79	106	0.94	156	0.64
7	14.29	57	1.75	107	0.93	157	0.64
8	12.50	58	1.72	108	0.93	158	0.63
9	11.11	59	1.69	109	0.92	159	0.63
10	10.00	60	1.67	110	0.91	160	0.62
11	9.09	61	1.64	111	0.90	161	0.62
12	8.33	62	1.61	112	0.89	162	0.62
13	7.69	63	1.59	113	0.88	163	0.61
14	7.14	64	1.56	114	0.88	164	0.61
15	6.67	65	1.54	115	0.87	165	0.61
16	6.25	66	1.52	116	0.86	166	0.60
17	5.88	67	1.49	117	0.85	167	0.60
18	5.56	68	1.47	118	0.85	168	0.60
19	5.26	69	1.45	119	0.84	169	0.59
20	5.00	70	1.43	120	0.83	170	0.59
21	4.76	71	1.41	121	0.83	171	0.58
22	4.55	72	1.39	122	0.82	172	0.58
23	4.35	73	1.37	123	0.81	173	0.58
24	4.17	74	1.35	124	0.81	174	0.57
25	4.00	75	1.33	125	0.80	175	0.57
26	3.85	76	1.32	126	0.79	176	0.57
27	3.70	77	1.30	127	0.79	177	0.56
28	3.57	78	1.28	128	0.78	178	0.56
29	3.45	79	1.27	129	0.78	179	0.56
30	3.33	80	1.25	130	0.77	180	0.56
31	3.23	81	1.23	131	0.76	181	0.55
32	3.12	82	1.22	132	0.76	182	0.55
33	3.03	83	1.20	133	0.75	183	0.55
34	2.94	84	1.19	134	0.75	184	0.54
35	2.86	85	1.18	135	0.74	185	0.54
36	2.78	86	1.16	136	0.74	186	0.54
37	2.70	87	1.15	137	0.73	187	0.53
38	2.63	88	1.14	138	0.72	188	0.53
39	2.56	89	1.12	139	0.72	189	0.53
40	2.50	90	1.11	140	0.71	190	0.53
41	2.44	91	1.10	141	0.71	191	0.52
42	2.38	92	1.09	142	0.70	192	0.52
43	2.33	93	1.08	143	0.70	193	0.52
44	2.27	94	1.06	144	0.69	194	0.52
45	2.22	95	1.05	145	0.69	195	0.51
46	2.17	96	1.04	146	0.68	196	0.51
47	2.13	97	1.03	147	0.68	197	0.51
48	2.08	98	1.02	148	0.68	198	0.51
49	2.04	99	1.01	149	0.67	199	0.50
50	2.00	100	1.00	150	0.67	200	0.50

Appendix A11a. Coefficients of BLUEs for Normal μ and σ

$n = 2$

$n - r$		$y_{(1)}$	$y_{(2)}$
0	μ	.5000	.5000
	σ	−.8862	.8862

$n = 3$

$n - r$		$y_{(1)}$	$y_{(2)}$	$y_{(3)}$
0	μ	.3333	.3333	.3333
	σ	−.5908	.0000	.5908
1	μ	.0000	1.0000	
	σ	−1.1816	1.1816	

$n = 4$

$n - r$		$y_{(1)}$	$y_{(2)}$	$y_{(3)}$	$y_{(4)}$
0	μ	.2500	.2500	.2500	.2500
	σ	−.4539	−.1102	.1102	.4539
1	μ	.1161	.2408	.6431	
	σ	−.6971	−.1268	.8239	
2	μ	−.4056	1.4056		
	σ	−1.3654	1.3654		

$n = 5$

$n - r$		$y_{(1)}$	$y_{(2)}$	$y_{(3)}$	$y_{(4)}$	$y_{(5)}$
0	μ	.2000	.2000	.2000	.2000	.2000
	σ	−.3724	−.1352	.0000	.1352	.3724
1	μ	.1252	.1830	.2147	.4771	
	σ	−.5117	−.1668	.0274	.6511	
2	μ	−.0638	.1498	.9139		
	σ	−.7696	−.2121	.9817		
3	μ	−.7411	1.7411			
	σ	−1.4971	1.4971			

$n = 6$

$n - r$		$y_{(1)}$	$y_{(2)}$	$y_{(3)}$	$y_{(4)}$	$y_{(5)}$	$y_{(6)}$
0	μ	.1667	.1667	.1667	.1667	.1667	.1667
	σ	−.3175	−.1386	−.0432	.0432	.1386	.3175
1	μ	.1183	.1510	.1680	.1328	.3799	
	σ	−.4097	−.1685	−.0406	.0740	.5448	
2	μ	.0185	.1226	.1761	.6828		
	σ	−.5528	−.2091	−.0290	.7909		
3	μ	−.2159	.0649	1.1511			
	σ	−.8244	−.2760	1.1004			
4	μ	−1.0261	2.0261				
	σ	−1.5988	1.5988				

$n = 7$

$n - r$		$y_{(1)}$	$y_{(2)}$	$y_{(3)}$	$y_{(4)}$	$y_{(5)}$	$y_{(6)}$	$y_{(7)}$
0	μ	.1429	.1429	.1429	.1429	.1429	.1429	.1429
	σ	−.2778	−.1351	−.0625	.0000	.0625	.1351	.2778
1	μ	.1088	.1295	.1400	.1487	.1571	.3159	
	σ	−.3440	−.1610	−.0681	.0114	.0901	.4716	
2	μ	.0465	.1072	.1375	.1626	.5462		
	σ	−.4370	−.1943	−.0718	.0321	.6709		
3	μ	−.0738	.0677	.1375	.8686			
	σ	−.5848	−.2428	−.0717	.8994			
4	μ	−.3474	−.0135	1.3609				
	σ	−.8682	−.3269	1.1951				
5	μ	−1.2733	2.2733					
	σ	−1.6812	1.6812					

From A. E. Sarhan and B. G. Greenberg, "Estimation of Location and Scale Parameters by Order Statistics from Singly and Doubly Censored Samples, Parts I and II," *Ann. Math. Statist.*, Vol. 27 (1956), pp. 427–451 and Vol. 29 (1958), pp. 79–105, with permission of Heebok Park, Treasurer of the Inst. of Math. Statist. Photocopied from E. Lee, *Statistical Methods for Survival Data Analysis*, 1980, with permission of Lifetime Learning Publications, Belmont, CA.

Appendix A11a. Coefficients of BLUEs for Normal μ and σ (*Continued*)

$n = 8$

$n - r$		$y_{(1)}$	$y_{(2)}$	$y_{(3)}$	$y_{(4)}$	$y_{(5)}$	$y_{(6)}$	$y_{(7)}$	$y_{(8)}$
0	μ	.1250	.1250	.1250	.1250	.1250	.1250	.1250	.1250
	σ	−.2476	−.1294	−.0713	−.0230	.0230	.0713	.1294	.2476
1	μ	.0997	.1139	.1208	.1265	.1318	.1370	.2704	
	σ	−.2978	−.1515	−.0796	−.0200	.0364	.0951	.4175	
2	μ	.0569	.0962	.1153	.1309	.1451	.4555		
	σ	−.3638	−.1788	−.0881	−.0132	.0570	.5868		
3	μ	−.0167	.0677	.1084	.1413	.6993			
	σ	−.4586	−.2156	−.0970	.0002	.7709			
4	μ	−.1549	.0176	.1001	1.0372				
	σ	−.6110	−.2707	−.1061	.9878				
5	μ	−.4632	−.0855	1.5487					
	σ	−.9045	−.3690	1.2735					
6	μ	−1.4915	2.4915						
	σ	−1.7502	1.7502						

$n = 9$

$n - r$		$y_{(1)}$	$y_{(2)}$	$y_{(3)}$	$y_{(4)}$	$y_{(5)}$	$y_{(6)}$	$y_{(7)}$	$y_{(8)}$	$y_{(9)}$
0	μ	.1111	.1111	.1111	.1111	.1111	.1111	.1111	.1111	.1111
	σ	−.2237	−.1233	−.0751	−.0360	.0000	.0360	.0751	.1233	.2237
1	μ	.0915	.1018	.1067	.1106	.1142	.1177	.1212	.2365	
	σ	−.2633	−.1421	−.0841	−.0370	.0062	.0492	.0954	.3757	
2	μ	.0602	.0876	.1006	.1110	.1204	.1294	.3909		
	σ	−.3129	−.1647	−.0938	−.0364	.0160	.0678	.5239		
3	μ	.0104	.0660	.0923	.1133	.1320	.5860			
	σ	−.3797	−.1936	−.1048	−.0333	.0317	.6797			
4	μ	−.0731	.0316	.0809	.1199	.8408				
	σ	−.4766	−.2335	−.1181	−.0256	.8537				
5	μ	−.2272	−.0284	.0644	1.1912					
	σ	−.6330	−.2944	−.1348	1.0622					
6	μ	−.5664	−.1521	1.7185						
	σ	−.9355	−.4047	1.3402						
7	μ	−1.6868	2.6868							
	σ	−1.8092	1.8092							

$n = 10$

$n - r$		$y_{(1)}$	$y_{(2)}$	$y_{(3)}$	$y_{(4)}$	$y_{(5)}$	$y_{(6)}$	$y_{(7)}$	$y_{(8)}$	$y_{(9)}$	$y_{(10)}$
0	μ	.1000	.1000	.1000	.1000	.1000	.1000	.1000	.1000	.1000	.1000
	σ	−.2044	−.1172	−.0763	−.0436	−.0142	.0142	.0436	.0763	.1172	.2044
1	μ	.0843	.0921	.0957	.0986	.1011	.1036	.1060	.1085	.2101	
	σ	−.2364	−.1334	−.0851	−.0465	−.0119	.0215	.0559	.0937	.3423	
2	μ	.0605	.0804	.0898	.0972	.1037	.1099	.1161	.3424		
	σ	−.2753	−.1523	−.0947	−.0488	−.0077	.0319	.0722	.4746		
3	μ	.0244	.0636	.0818	.0962	.1089	.1207	.5045			
	σ	−.3252	−.1758	−.1058	−.0502	−.0006	.0469	.6107			
4	μ	−.0316	.0383	.0707	.0962	.1185	.7078				
	σ	−.3930	−.2063	−.1192	−.0501	.0111	.7576				
5	μ	−.1240	−.0016	.0549	.0990	.9718					
	σ	−.4919	−.2491	−.1362	−.0472	.9243					
6	μ	−.2923	−.0709	.0305	1.3327						
	σ	−.6520	−.3150	−.1593	1.1263						
7	μ	−.6596	−.2138	1.8734							
	σ	−.9625	−.4357	1.3981							
8	μ	−1.8634	2.8634								
	σ	−1.8608	1.8608								

$n = 11$

$n-r$		$y_{(1)}$	$y_{(2)}$	$y_{(3)}$	$y_{(4)}$	$y_{(5)}$	$y_{(6)}$	$y_{(7)}$	$y_{(8)}$	$y_{(9)}$	$y_{(10)}$	$y_{(11)}$
0	μ	.0909	.0909	.0909	.0909	.0909	.0909	.0909	.0909	.0909	.0909	.0909
	σ	-.1883	-.1115	-.0760	-.0481	-.0234	.0000	.0234	.0481	.0760	.1115	.1883
1	μ	.0781	.0841	.0869	.0891	.0910	.0928	.0945	.0963	.0982	.1891	
	σ	-.2149	-.1256	-.0843	-.0519	-.0233	.0038	.0309	.0593	.0911	.3149	
2	μ	.0592	.0744	.0814	.0869	.0917	.0962	.1005	.1049	.3047		
	σ	-.2463	-.1417	-.0934	-.0555	-.0220	.0095	.0409	.0736	.4349		
3	μ	.0320	.0609	.0741	.0845	.0935	.1020	.1101	.4430			
	σ	-.2852	-.1610	-.1038	-.0589	-.0194	.0178	.0545	.5562			
4	μ	-.0082	.0415	.0642	.0820	.0974	.1116	.6116				
	σ	-.3357	-.1854	-.1163	-.0621	-.0146	.0299	.6842				
5	μ	-.0698	.0128	.0504	.0797	.1049	.8220					
	σ	-.4045	-.2175	-.1317	-.0647	-.0061	.8246					
6	μ	-.1702	-.0323	.0303	.0786	1.0937						
	σ	-.5053	-.2627	-.1519	-.0657	.9857						
7	μ	-.3516	-.1104	-.0016	1.4636							
	σ	-.6687	-.3331	-.1807	1.1825							
8	μ	-.7445	-.2712	2.0157								
	σ	-.9862	-.4630	1.4402								
9	μ	-2.0245	3.0245									
	σ	-1.9065	1.9065									

$n = 12$

$n-r$		$y_{(1)}$	$y_{(2)}$	$y_{(3)}$	$y_{(4)}$	$y_{(5)}$	$y_{(6)}$	$y_{(7)}$	$y_{(8)}$	$y_{(9)}$	$y_{(10)}$	$y_{(11)}$	$y_{(12)}$
0	μ	.0833	.0833	.0833	.0833	.0833	.0833	.0833	.0833	.0833	.0833	.0833	.0833
	σ	-.1748	-.1061	-.0749	-.0506	-.0294	-.0097	.0097	.0294	.0506	.0749	.1061	.1748
1	μ	.0726	.0775	.0796	.0813	.0828	.0842	.0855	.0868	.0882	.0896	.1719	
	σ	-.1972	-.1185	-.0827	-.0548	-.0305	-.0079	.0142	.0367	.0608	.0881	.2919	
2	μ	.0574	.0693	.0747	.0789	.0825	.0859	.0891	.0923	.0956	.2745		
	σ	-.2232	-.1324	-.0911	-.0590	-.0310	-.0050	.0203	.0461	.0733	.4020		
3	μ	.0360	.0581	.0682	.0759	.0827	.0888	.0948	.1006	.3950			
	σ	-.2545	-.1487	-.1007	-.0633	-.0308	-.0007	.0286	.0582	.5119			
4	μ	.0057	.0428	.0595	.0724	.0836	.0938	.1036	.5386				
	σ	-.2937	-.1686	-.1119	-.0678	-.0296	.0058	.0400	.6259				
5	μ	-.0382	.0210	.0477	.0684	.0861	.1022	.7128					
	σ	-.3448	-.1939	-.1255	-.0726	-.0267	.0155	.7479					
6	μ	-.1048	-.0109	.0313	.0637	.0915	.9292						
	σ	-.4146	-.2274	-.1428	-.0774	-.0210	.8833						
7	μ	-.2125	-.0609	.0070	.0589	1.2075							
	σ	-.5171	-.2749	-.1659	-.0820	1.0399							
8	μ	-.4059	-.1472	-.0321	1.5852								
	σ	-.6836	-.3493	-.1996	1.2324								
9	μ	-.8225	-.3249	2.1474									
	σ	-1.0075	-.4874	1.4948									
10	μ	-2.1728	3.1728										
	σ	-1.9474	1.9474										

Appendix A11b. Variance and Covariance Factors for BLUEs for Normal μ^* and σ^*

n		0	1	2	3	4	5	6	7	8	9	10
									$n-r$			
2	Var (μ^*)	.5000										
	Var (σ^*)	.5708										
	Cov (μ^*,σ^*)	0										
3	Var (μ^*)	.3333	.4487									
	Var (σ^*)	.2755	.6378									
	Cov (μ^*,σ^*)	0	.2044									
4	Var (μ^*)	.2500	.2870	.5130								
	Var (σ^*)	.1801	.3021	.6730								
	Cov (μ^*,σ^*)	0	.0672	.3567								
5	Var (μ^*)	.2000	.2177	.2839	.6112							
	Var (σ^*)	.1333	.1948	.3181	.6957							
	Cov (μ^*,σ^*)	0	.0330	.1234	.4749							
6	Var (μ^*)	.1667	.1769	.2068	.2999	.7186						
	Var (σ^*)	.1057	.1428	.2044	.3292	.7119						
	Cov (μ^*,σ^*)	0	.0195	.0624	.1702	.5705						
7	Var (μ^*)	.1429	.1494	.1660	.2071	.3248	.8264					
	Var (σ^*)	.0875	.1123	.1493	.2114	.3375	.7243					
	Cov (μ^*,σ^*)	0	.0128	.0375	.0881	.2099	.6503					
8	Var (μ^*)	.1250	.1295	.1399	.1623	.2138	.3541	.9310				
	Var (σ^*)	.0746	.0924	.1171	.1542	.2168	.3441	.7342				
	Cov (μ^*,σ^*)	0	.0090	.0250	.0538	.1106	.2442	.7186				
9	Var (μ^*)	.1111	.1144	.1214	.1352	.1629	.2241	.3854	1.0313			
	Var (σ^*)	.0650	.0784	.0960	.1207	.1581	.2212	.3494	.7423			
	Cov (μ^*,σ^*)	0	.0067	.0178	.0362	.0684	.1305	.2743	.7781			
10	Var (μ^*)	.1000	.1025	.1075	.1167	.1336	.1664	.2366	.4174	1.1269		
	Var (σ^*)	.0576	.0681	.0813	.0989	.1237	.1613	.2248	.3539	.7491		
	Cov (μ^*,σ^*)	0	.0051	.0132	.0260	.0465	.0816	.1483	.3011	.8306		
11	Var (μ^*)	.0909	.0929	.0966	.1031	.1143	.1342	.1718	.2504	.4493	1.2179	
	Var (σ^*)	.0517	.0601	.0704	.0836	.1013	.1262	.1640	.2279	.3577	.7550	
	Cov (μ^*,σ^*)	0	.0041	.0102	.0195	.0336	.0559	.0936	.1644	.3251	.8777	
12	Var (μ^*)	.0833	.0849	.0878	.0926	.1004	.1136	.1363	.1784	.2650	.4809	1.3044
	Var (σ^*)	.0469	.0538	.0620	.0723	.0855	.1033	.1283	.1663	.2305	.3610	.7601
	Cov (μ^*,σ^*)	0	.0033	.0081	.0152	.0254	.0406	.0645	.1045	.1791	.3469	.9202

Appendix A12a. Coefficients of the BLUEs for the Extreme Value Distribution λ and δ

i	n	r	$a(i;n,r)$	$b(i;n,r)$
1	2	2	0.0836269	-0.7213475
2	2	2	0.9163731	0.7213475
1	3	2	-0.3777001	-0.8221011
2	3	2	1.3777000	0.8221011
1	3	3	0.0879664	-0.3747251
2	3	3	0.2557135	-0.2558160
3	3	3	0.6563201	0.6305411
1	4	2	-0.7063194	-0.8690149
2	4	2	1.7063194	0.8690149
1	4	3	-0.0801058	-0.4143997
2	4	3	0.0604318	-0.3258576
3	4	3	1.0196739	0.7402573
1	4	4	0.0713800	-0.2487965
2	4	4	0.1536799	-0.2239193
3	4	4	0.2639426	-0.0859033
4	4	4	0.5109975	0.5586191
1	5	2	-0.9598627	-0.8962840
2	5	2	1.9598626	0.8962840
1	5	3	-0.2101147	-0.4343423
2	5	3	-0.0860216	-0.3642452
3	5	3	1.2961362	0.7985875
1	5	4	-0.0153827	-0.2730335
2	5	4	0.0519625	-0.2499443
3	5	4	0.1520766	-0.1491092
4	5	4	0.8113436	0.6720870
1	5	5	0.0583501	-0.1844827
2	5	5	0.1088237	-0.1816557
3	5	5	0.1676093	-0.1304549
4	5	5	0.2462824	-0.0065338
5	5	5	0.4189344	0.5031271
1	6	2	-1.1655650	-0.9141358
2	6	2	2.1655650	0.9141358
1	6	3	-0.3153967	-0.4466018
2	6	3	-0.2034317	-0.3886493
3	6	3	1.5188283	0.8352511
1	6	4	-0.0865380	-0.2858649
2	6	4	-0.0280567	-0.2654763
3	6	4	0.0649474	-0.1858689
4	6	4	1.0496472	0.7372101
1	6	5	0.0057312	-0.2015427
2	6	5	0.0465760	-0.1972715
3	6	5	0.1002434	-0.1536128
4	6	5	0.1722854	-0.0645867
5	6	5	0.6751639	0.6170138
1	6	6	0.0488670	-0.1458073
2	6	6	0.0835217	-0.1495343
3	6	6	0.1210537	-0.1267241
4	6	6	0.1656198	-0.0731993
5	6	6	0.2254881	0.0359918
6	6	6	0.3554496	0.4592732
1	7	2	-1.3382740	-0.9267370
2	7	2	2.3382740	0.9267370
1	7	3	-0.4036103	-0.4549547
2	7	3	-0.3012125	-0.4055742
3	7	3	1.7048228	0.8605289
1	7	4	-0.1463256	-0.2940405
2	7	4	-0.0940681	-0.2760193
3	7	4	-0.0070967	-0.2101609
4	7	4	1.2474903	0.7802208
1	7	5	-0.0392570	-0.2110152
2	7	5	-0.0043641	-0.2064593
3	7	5	0.0458324	-0.1691176
4	7	5	0.1134204	-0.0991828
5	7	5	0.8843683	0.6857749
1	7	6	0.0137303	-0.1586850
2	7	6	0.0417973	-0.1608703
3	7	6	0.0758807	-0.1396394
4	7	6	0.1175849	-0.0950700
5	7	6	0.1721153	-0.0176462
6	7	6	0.5799915	0.5719108
1	7	7	0.0418411	-0.1201405
2	7	7	0.0673314	-0.1258588
3	7	7	0.0937470	-0.1148675
4	7	7	0.1232231	-0.0873391
5	7	7	0.1585900	-0.0361916
6	7	7	0.2062600	0.0606980
7	7	7	0.3090075	0.4236996
1	8	2	-1.4869219	-0.9361095
2	8	2	2.4869219	0.9361095
1	8	3	-0.4793981	-0.4610279
2	8	3	-0.3848212	-0.4180145
3	8	3	1.8642193	0.8790424
1	8	4	-0.1977185	-0.2997588
2	8	4	-0.1502017	-0.2836886
3	8	4	-0.0684910	-0.2274858
4	8	4	1.4164112	0.8109332
1	8	5	-0.0781440	-0.2172247
2	8	5	-0.0474155	-0.2127596
3	8	5	-0.0000857	-0.1802817
4	8	5	0.0637119	-0.1225159
5	8	5	1.0619333	0.7328019
1	8	6	-0.0172425	-0.1661332
2	8	6	0.0065283	-0.1674873
3	8	6	0.0380245	-0.1482977
4	8	6	0.0779914	-0.1105318
5	8	6	0.1292040	-0.0499901
6	8	6	0.7654942	0.6424402
1	8	7	0.0168081	-0.1302929
2	8	7	0.0375935	-0.1347891
3	8	7	0.0612335	-0.1238688
4	8	7	0.0888741	-0.0990771
5	8	7	0.1224279	-0.0571224
6	8	7	0.1654593	0.0108656
7	8	7	0.5076035	0.5342847
1	8	8	0.0364852	-0.1019365
2	8	8	0.0561317	-0.1080740
3	8	8	0.0759037	-0.1027278
4	8	8	0.0971419	-0.0871624
5	8	8	0.1211747	-0.0589284
6	8	8	0.1501999	-0.0111245
7	8	8	0.1894276	0.0757666
8	8	8	0.2735352	0.3941872
1	9	2	-1.6172796	-0.9433541
2	9	2	2.6172795	0.9433541
1	9	3	-0.5457604	-0.4656486
2	9	3	-0.4577309	-0.4275495
3	9	3	2.0034912	0.8931981

From John S. White, *Industrial Mathematics*, Vol. 14, Part 1, 1964, pp. 21–60. Reproduced by permission of J. S. White, who developed the table, and R. Schmidt, editor of *Industrial Mathematics*, where the table first appeared.

Appendix A12a. Coefficients of the BLUEs for the Extreme Value Distribution λ and δ

(Continued)

i	r	n	a(i;n,r)	b(i;n,r)
1	4	9	-0.2427102	-0.3140002
2	4	9	-0.9898152	-0.2895314
3	4	9	-0.1218960	-0.2404933
4	4	9	1.5635877	0.8340249
1	5	9	-0.1122653	-0.2217028
2	5	9	-0.0846628	-0.2174080
3	5	9	-0.0398298	-0.1887179
4	5	9	-0.0206119	-0.1394352
5	5	9	1.2161461	0.7672640
1	6	9	-0.0446124	-0.1711693
2	6	9	-0.0239090	-0.1720278
3	6	9	-0.0057259	-0.1546900
4	6	9	0.0439637	-0.1219925
5	6	9	0.0924579	-0.0720775
6	6	9	0.9263739	0.6919572
1	7	9	-0.0057741	-0.1363684
2	7	9	0.0117905	-0.1400394
3	7	9	0.0335898	-0.1297226
4	7	9	0.0599713	-0.1076490
5	7	9	0.0921887	-0.0723188
6	7	9	0.1325194	-0.0193719
7	7	9	0.6757144	0.6054702
1	8	9	0.0177962	-0.1101884
2	8	9	0.0336652	-0.1154096
3	8	9	0.0515769	-0.1097441
4	8	9	0.0713590	-0.0950005
5	8	9	0.0942696	-0.0700075
6	8	9	0.1218432	-0.0312302
7	8	9	0.1569098	0.0292244
8	8	9	0.4522801	0.5023558
1	9	9	0.0322910	-0.0883912
2	9	9	0.0479566	-0.0943693
3	9	9	0.0633995	-0.0919651
4	9	9	0.0795686	-0.0826548
5	9	9	0.0972179	-0.0655738
6	9	9	0.1173565	-0.0379772
7	9	9	0.1417891	0.0064858
8	9	9	0.1748819	0.0852032
1	2	10	-1.7332827	-0.9491221
2	2	10	2.7332827	0.9491221
1	3	10	-0.6047400	-0.4692847
2	3	10	-0.5222903	-0.4350929
3	3	10	2.1270303	0.9043776
1	4	10	-0.2826778	-0.3072783
2	4	10	-0.2420734	-0.2941358
3	4	10	-0.1690791	-0.2506307
4	4	10	1.6938303	0.8520448
1	5	10	-0.1426067	-0.2250678
2	5	10	-0.1174612	-0.2209984
3	5	10	-0.0748446	-0.1953226
4	5	10	-0.0174033	-0.1523120
5	5	10	1.3523158	0.7937008
1	6	10	-0.0690324	-0.1748478
2	6	10	-0.0506448	-0.1753912
3	6	10	-0.0225702	-0.1596414
4	6	10	0.0140717	-0.1308280
5	6	10	0.0601887	-0.0882729
6	6	10	1.0679869	0.7289813
1	7	10	-0.0260821	-0.1405755
2	7	10	-0.0108783	-0.1436594
3	7	10	0.0095210	-0.1340341
4	7	10	0.0348369	-0.1142583
5	7	10	0.0658392	-0.0837641
6	7	10	0.1040540	-0.0401921
7	7	10	0.8227093	0.6564835
1	8	10	0.0006317	-0.1152690
2	8	10	0.0143203	-0.1197883
3	8	10	0.0304554	-0.1142025
4	8	10	0.0492565	-0.1005983
5	8	10	0.0713719	-0.0785228
6	8	10	0.0978952	-0.0460265
7	8	10	0.1306530	0.0008844
8	8	10	0.6054161	0.5735230
1	9	10	0.0178349	-0.0952545
2	9	10	0.0308561	-0.1005502
3	9	10	0.0445900	-0.0977581
5	9	10	0.0763356	-0.0727481
6	9	10	0.0956763	-0.0486079
7	9	10	0.1188411	-0.0128578
8	9	10	0.1481415	0.0415222
9	9	10	0.4081592	0.4748591
1	10	10	0.0289290	-0.0779399
2	10	10	0.0417478	-0.0835515
3	10	10	0.0541930	-0.0827706
4	10	10	0.0669876	-0.0770208
5	10	10	0.0806178	-0.0660648
6	10	10	0.0956359	-0.0486710
7	10	10	0.1128684	-0.0221794
8	10	10	0.1338452	0.0192100
9	10	10	0.1623082	0.0911583
10	10	10	0.2228670	0.3478297
1	2	11	-1.8377307	-0.9538235
2	2	11	2.8377307	0.9538235
1	3	11	-0.6577871	-0.4722220
2	3	11	-0.5801607	-0.4412107
3	3	11	2.2379478	0.9134327
1	4	11	-0.3186053	-0.3098909
2	4	11	-0.2806371	-0.2978600
3	4	11	-0.2112831	-0.2587592
4	4	11	1.8105255	0.8665101
1	5	11	-0.1698921	-0.2277048
2	5	11	-0.1467420	-0.2238631
3	5	11	-0.1061114	-0.2006363
4	5	11	-0.0513661	-0.1624608
5	5	11	1.4741115	0.8146650
1	6	11	-0.0910322	-0.1776687
2	6	11	-0.0744627	-0.1780023
3	6	11	-0.0477433	-0.1636020
4	6	11	-0.0125799	-0.1378512
5	6	11	0.0314153	-0.1007180
6	6	11	1.1944028	0.7578422
1	7	11	-0.0444511	-0.1437042
2	7	11	-0.0310560	-0.1463524
3	7	11	-0.0117979	-0.1373925

Block 1

i	r	n	a(i;n,r)	b(i;n,r)
5	7	11	0.0424182	-0.0926953
6	7	11	0.0788225	-0.0555804
7	7	11	0.9535072	0.6952474
1	8	11	-0.0150243	-0.1188596
2	8	11	-0.0032013	-0.1228351
3	8	11	0.0118120	-0.1174590
4	8	11	0.0298233	-0.1049450
5	8	11	0.0512298	-0.0852557
6	8	11	0.0768116	-0.0852781
7	8	11	0.1078325	-0.0187425
8	8	11	0.7407162	0.6253750
1	9	11	-0.0043670	-0.0995823
2	9	11	0.0154366	-0.1043068
3	9	11	0.0279638	-0.1014022
4	9	11	0.0421240	-0.0927166
5	9	11	0.0582942	-0.0770311
6	9	11	0.0770601	-0.0570311
7	9	11	0.0993329	-0.0271922
8	9	11	0.1266006	0.0148710
9	9	11	0.5488207	0.5455931
1	10	11	0.0174317	-0.0837528
2	10	11	0.0281914	-0.0888527
3	10	11	0.0392646	-0.0877099
4	10	11	0.0510690	-0.0818786
5	10	11	0.0639625	-0.0713651
6	10	11	0.0783871	-0.0554233
7	10	11	0.0949853	-0.0324598
8	10	11	0.1148044	-0.0005783
9	10	11	0.1397781	0.0499838
10	10	11	0.3721260	0.4508800
1	11	11	0.0261799	-0.0696435
2	11	11	0.0368858	-0.0748304
3	11	11	0.0471592	-0.0749774
4	11	11	0.0575780	-0.0713809
5	11	11	0.0684852	-0.0640708
6	11	11	0.0802218	-0.0524642
7	11	11	0.0932342	-0.0352839
8	11	11	0.1082257	-0.0100318
9	11	11	0.1265219	0.0286042

Block 2

i	r	n	a(i;n,r)	b(i;n,r)
10	11	11	0.1513848	0.0948691
11	11	11	0.2041233	0.3292097
1	2	12	-1.9326838	-0.9577292
2	2	12	2.9326838	-0.9577292
1	3	12	-0.7059671	-0.4746448
2	3	12	-0.6325588	-0.4462730
3	3	12	2.3385259	0.9209178
1	4	12	-0.3512178	-0.3120236
2	4	12	-0.3155127	-0.3009354
3	4	12	-0.2494171	-0.2654253
4	4	12	1.9161476	0.8783843
1	5	12	-0.1946622	-0.2298303
2	5	12	-0.1731727	-0.2262054
3	5	12	-0.1343335	-0.2050051
4	5	12	-0.0820210	-0.1706756
5	5	12	1.5841895	0.8317164
1	6	12	-0.1110243	-0.1799074
2	6	12	-0.0959246	-0.1800965
3	6	12	-0.0704054	-0.1668469
4	6	12	-0.0366109	-0.1435706
5	6	12	0.0054727	-0.1106083
6	6	12	1.3084924	0.7810298
1	7	12	-0.0611825	-0.1461377
2	7	12	-0.0492184	-0.1484513
3	7	12	-0.0309288	-0.1401000
4	7	12	-0.0074584	-0.1238187
5	7	12	0.0213293	-0.0998649
6	7	12	0.0561330	-0.0674908
7	7	12	1.0713258	0.7258634
1	8	12	-0.0293475	-0.1215721
2	8	12	-0.0189783	-0.1251165
3	8	12	-0.0048692	-0.1199910
4	8	12	0.0124567	-0.1084511
5	8	12	0.0331956	-0.0907082
6	8	12	0.0578972	-0.0661294
7	8	12	0.0874668	-0.0333351
8	8	12	0.8621786	0.6653035
1	9	12	-0.0080605	-0.1026948
2	9	12	0.0015064	-0.1069507

Block 3

i	r	n	a(i;n,r)	b(i;n,r)
3	9	12	0.0131077	-0.1040492
4	9	12	0.0266268	-0.0958851
5	9	12	0.0422903	-0.0826430
6	9	12	0.0605311	-0.0637936
7	9	12	0.0820263	-0.0381598
8	9	12	0.1078161	-0.0036644
9	9	12	0.6741558	0.5978406
1	10	12	0.0065957	-0.0874933
2	10	12	0.0157917	-0.0921338
3	10	12	0.0258664	-0.0908158
4	10	12	0.0369782	-0.0851487
5	10	12	0.0493695	-0.0753004
6	10	12	0.0633846	-0.0608340
7	10	12	0.0795237	-0.0407555
8	10	12	0.0985502	-0.0132751
9	10	12	0.1217001	0.0248324
10	10	12	0.5022398	0.5209243
1	11	12	0.0168287	-0.0746398
2	11	12	0.0258986	-0.0794387
3	11	12	0.0350553	-0.0792736
4	11	12	0.0446466	-0.0755164
5	11	12	0.0549243	-0.0683231
6	11	12	0.0661658	-0.0573406
7	11	12	0.0787381	-0.0417422
8	11	12	0.0931854	-0.0200137
9	11	12	0.1104001	0.0559065
10	11	12	0.1320313	0.1320313
11	11	12	0.3421256	0.4297430
1	12	12	0.0238938	-0.0629056
2	12	12	0.0329842	-0.0676705
3	12	12	0.0416281	-0.0683572
4	12	12	0.0503027	-0.0661223
5	12	12	0.0592658	-0.0611125
6	12	12	0.0687474	-0.0530529
7	12	12	0.0790177	-0.0412780
8	12	12	0.0904553	-0.0245480
9	12	12	0.1036731	-0.0005340
10	12	12	0.1198380	0.0356552
11	12	12	0.1418326	0.0970860
12	12	12	0.1883611	0.3128398

Appendix A12b. Variance and Covariance Factors for BLUEs for Extreme Value λ and δ

r	n	A(n, r)	C(n, r)	B(n, r)	r	n	A(n, r)	C(n, r)	B(n, r)
2	2	0.6595468	0.0643216	0.7118574	5	10	0.3210449	0.1462586	0.2154736
2	3	0.9160385	0.4682464	0.8183653	6	10	0.2143651	0.0734417	0.1657706
3	3	0.4028637	-0.0247719	0.3447117	7	10	0.1615247	0.0312775	0.1321255
2	4	1.3340190	0.7720299	0.8670220	8	10	0.1340605	0.0052601	0.1074787
3	4	0.4331573	0.1180273	0.3922328	9	10	0.1197760	-0.0113587	0.0881441
4	4	0.2934587	-0.0346903	0.2252829	10	10	0.1129729	-0.0219764	0.0715730
2	5	1.7891718	1.0115593	0.8950462	2	11	4.2754239	1.8605901	0.9535829
3	5	0.5293953	0.2353742	0.4168157	3	11	1.3394275	0.6622444	0.4644704
4	5	0.2918142	0.0385707	0.2537910	4	11	0.6258965	0.3207513	0.3010332
5	5	0.2313953	-0.0339905	0.1666472	5	11	0.3571655	0.1722376	0.2189574
2	6	2.2440054	1.2082247	0.9132926	6	11	0.2339774	0.0940754	0.1693639
3	6	0.6529407	0.3332487	0.4321160	7	11	0.1711050	0.0482322	0.1359374
4	6	0.3237185	0.1020225	0.2697164	8	11	0.1370347	0.0194672	0.1116516
5	6	0.2236065	0.0105329	0.1861065	9	11	0.1180946	0.0006384	0.0929336
6	6	0.1911738	-0.0313731	0.1319602	10	11	0.1076972	-0.0119594	0.0776696
2	7	2.6856914	1.3746095	0.9261257	11	11	0.1025087	-0.0203275	0.0641736
3	7	0.7879540	0.4167048	0.4426123	2	12	4.6309696	1.9536097	0.9575278
4	7	0.3726466	0.1569581	0.2801583	3	12	1.4714833	0.7093955	0.4675528
5	7	0.2352926	0.0504482	0.1975663	4	12	0.6927236	0.3524030	0.3039034
6	7	0.1827014	-0.0014908	0.1462713	5	12	0.3948335	0.1960075	0.2217941
7	7	0.1629283	-0.0286029	0.1090962	6	12	0.2557970	0.1130176	0.1722580
2	8	3.1099904	1.5185785	0.9356461	7	12	0.1832612	0.0638719	0.1389599
3	8	0.9270270	0.4892374	0.4502769	8	12	0.1428230	0.0326676	0.1148810
4	8	0.4307225	0.2050898	0.2875948	9	12	0.1194292	0.0119220	0.0964838
5	8	0.2580472	0.0859328	0.2053689	10	12	0.1057367	-0.0022798	0.0817536
6	8	0.1865817	0.0259555	0.1550330	11	12	0.0978899	-0.0121362	0.0693731
7	8	0.1550975	-0.0071836	0.1201520	12	12	0.0938212	-0.0188937	0.0581499
8	8	0.1419827	-0.0260831	0.0929162					
2	9	3.5159953	1.6453318	0.9429907					
3	9	1.0664362	0.5532675	0.4561257					
4	9	0.4937336	0.2477853	0.2931801					
5	9	0.2874813	0.1176612	0.2110851					
6	9	0.1980218	0.0508391	0.1611722					
7	9	0.1556183	0.0128437	0.1271266					
8	9	0.1350590	-0.0099918	0.1017627					
9	9	0.1258228	-0.0238813	0.0808757					
2	10	3.9041486	1.7584720	0.9488296					
3	10	1.2042414	0.6105167	0.4607383					
4	10	0.5592714	0.2860785	0.2975366					

From John S. White, *Industrial Mathematics*, Vol. 14, Part 1, 1964, pp. 21–60. Reproduced by permission of J. S. White, who developed the table, and R. Schmidt, editor of *Industrial Mathematics*, where the table first appeared.

References

General Electric reports and reprints marked with an asterisk are available from the Technical Information Exchange, General Electric Company Corporate Research & Development, Schenectady, NY 12345.

Abramowitz, M., and Stegun, I. A. (1964), *Handbook of Mathematical Functions with Formulas, Graphs, and Mathematical Tables*, National Bureau of Standards, *Applied Mathematics Series 55*. For sale by the Superintendent of Documents, U.S. Government Printing Office, Washington, DC 20402. Price $6.50.

Aitchison, J., and Brown, J. A. C. (1957), *The Lognormal Distribution*, Cambridge University Press, New York and London.

Albrecht, P. F., and Campbell, H. E. (1972), "Reliability Analysis of Distribution Equipment Failure Data," Electric Utilities Systems Engineering Department, General Electric Company, Schenectady, NY.

Albrecht, P. F., Nelson, W., and Ringlee, R. J. (1968), "Discussion of 'Determination and Analysis of Data for Reliability Studies,' by Alton D. Patton," *IEEE Trans. Power Appar. Syst.* **PAS-87**, 84–100.

American Society for Testing and Materials (1963), "A Guide for Fatigue Testing and the Statistical Analysis of Fatigue Data," Special Technical Publication No. 91-A, 2nd ed., 1916 Race St., Philadelphia, PA.

Aroian, L. A. (1976), "Applications of the Direct Method in Sequential Analysis," *Technometrics* **18**, 301–306.

Association for the Advancement of Medical Instrumentation (1975), "AAMI Implantable Pacemaker Standard (Proposed)," Document AAMI IP-P 10/75 from AAMI, P.O. Box 460, Springfield, VA 22150. Price $20

Bailey, R. C., Homer, L. D., and Summe, J. P. (1977), "A Proposal for the Analysis of Kidney Graft Survival Data," *Transplantation* **24**, 309–315.

Bain, L. J. (1978), *Statistical Analysis of Reliability and Life-Testing Models: Theory and Methods*, Dekker, New York.

Barlow, R. E., and Proschan, F. (1965), *Mathematical Theory of Reliability*, Wiley, New York.

Barlow, R. E., and Proschan, F. (1975), *Statistical Theory of Reliability and Life Testing — Probability Models*, Holt, Rhinehart, and Winston, New York.

Barnett, V., and Lewis, T. (1978), *Outliers in Statistical Data*, Wiley, New York.

Bartholomew, D. J. (1957), "Testing for Departure from the Exponential Distribution," *Biometrika* **44**, 253–257.

Bartholomew, D. J. (1963), "The Sampling Distribution of an Estimate Arising in Life Testing," *Technometrics* **5**, 361–374.

Bartree, E. M. (1968), *Engineering Experimental Design Fundamentals*, Prentice-Hall, Englewood Cliffs, NJ.

Bassin, W. M. (1969), "Increasing Hazard Functions and Overhaul Policy," in *Proceedings of the 1969 Annual Symposium on Reliability*, IEEE, 345 East 47th St., New York, NY 10017, pp. 173–180.

Basu, A. P., and Ghosh, J. K. (1980), "Asymptotic Properties of a Solution to the Likelihood Equation with Applications to Life Testing," *J. Am. Stat. Assoc.* **75**, 410–414.

Benard, A., and Bos-Levenbach, E. D. (1953), "The Plotting of Observations on Probability Paper," *Statistica Neerlandica* **7**, 163–173. In Dutch.

Berkson, J. (1953), "A Statistically Precise and Relatively Simple Method of Estimating the Bioassay with Quantal Response, Based on the Logistic Function," *J. Am. Stat. Assoc.* **48**, 565–599.

Bilikam, J. E., Moore, A. H., and Petrick, G. (1979), "Small *k*-Sample Maximum Likelihood Ratio Tests for Change of a Weibull Shape Parameter," *IEEE Trans. Reliab.* **R-28**, 47–50.

Billman, B. R., Antle, C. E., and Bain, L. J. (1972), "Statistical Inference from Censored Weibull Samples," *Technometrics* **14**, 831–840.

Birnbaum, Z. W. (1979), "On the Mathematics of Competing Risks," DHEW Publication No. (PHS)79-1351. For sale by the Superintendent of Documents, U.S. Government Printing Office, Washington, DC 20402.

Bishop, Y. M. M., Fienberg, S. E., and Holland, P. W. (1975), *Discrete Multivariate Analysis*, Massachusetts Institute of Technology Press, Cambridge, MA.

Block, H. W. (1975), "Continuous Multivariate Exponential Extensions," in *Reliability and Fault Tree Analysis*, R. E. Barlow, J. B. Fussell, and N. D. Singpurwalla, Eds., SIAM, 117 S. 17th St., Philadelphia, PA.

Block, H. W., and Savits, T. H. (1981), "Multivariate Distributions in Reliability Theory and Life Testing." Technical Report No. 81-13, Institute for Statistics and Applications, Department of Mathematics and Statistics, University of Pittsburgh, Pittsburgh, PA 15260.

Boardman, T. J., and Kendell, P. J. (1970), "Estimation in Compound Exponential Failure Models," *Technometrics* **12**, 891–900.

Box, G. E. P., Hunter, W. G., and Hunter, J. S. (1978), *Statistics for Experimenters*, Wiley, New York.

Brownlee, K. A. (1965), *Statistical Theory and Methodology in Science and Engineering*, 2nd ed., Wiley, New York.

Buckland, W. R. (1964), *Statistical Assessment of Life Characteristic—A Bibliographic Guide*, Hafner, New York.

Chernoff, H., and Lieberman, G. J. (1954), "Use of Normal Probability Paper," *J. Am. Stat. Assoc.* **49**, 778–785.

Chiang, C. L. (1968), *Introduction to Stochastic Processes in Biostatistics*, Wiley, New York.

Cochran, W. G. (1977), *Sampling Techniques*, 3rd ed., Wiley, New York.

Cohen, Jr., A. C. (1961), "Tables for Maximum Likelihood Estimates: Singly Truncated and Singly Censored Samples," *Technometrics* **3**, 535–541.

Cohen, Jr., A. C. (1963), "Progressively Censored Samples in Life Testing," *Technometrics* **5**, 327–339.

Cohen, Jr., A. C. (1965), "Maximum Likelihood Estimation in the Weibull Distribution Based on Complete and on Censored Samples," *Technometrics* **7**, 579–588.

Cohen, A. C. (1966), Query 18 "Life Testing and Early Failure," *Technometrics* **8**, 539–545.

Cohen, A. C. (1975), "Multi-Censored Sampling in the Three-Parameter Weibull Distribution," *Technometrics* **17**, 347–351.

Cohen, A. C. (1976), "Progressively Censored Sampling in the Three Parameter Log-Normal Distribution," *Technometrics* **18**, 99–103.

Cohen, J. (1977), *Statistical Power Analysis for the Behavioral Sciences*, rev. ed., Academic, New York.

Cox, D. R. (1959), "The Analysis of Exponentially Distributed Life-Times with Two Types of Failures," *J. R. Stat. Soc. B* **21**, 411–421.

Cox, D. R. (1962), *Renewal Theory*, Wiley, New York.

Cox, D. R. (1970), *Analysis of Binary Data*, Halsted, New York.

Cox, D. R. (1972), "Regression Models and Life Tables," *J. R. Stat. Soc. B* **34**, 187–220.

Cox, D. R., and Lewis, P. A. (1966), *The Statistical Analysis of Series of Events*, Methuen, London.

Cox, D. R., and Miller, H. D. (1965), *The Theory of Stochastic Processes*, Wiley, New York.

Cramer, H. (1945), *Mathematical Methods of Statistics*, Princeton University Press, Princeton, NJ.

Crawford, D. E. (1970), "Analysis of Incomplete Life Test Data on Motorettes," *Insul./Circuits* **16**, 43–48.

David, H. A. (1970), *Order Statistics*, Wiley, New York. Updated 2nd edition (1981).

David, H. A., and Moeschberger, M. L. (1979), *The Theory of Competing Risks*, Griffin's Statistical Monograph No. 39, Methuen, London.

Dixon, W. J., Ed. (1974), *BMD Biomedical Computer Programs*, University of California Press, Los Angeles. Available from the UCLA Student Store, 380 Westwood Blvd., Los Angeles, CA. Price $8.25.

Dixon, W. J., and Brown, M. B. (1977), *BMDP-77: Biomedical Computer Programs P-Series*, University of California Press, Berkeley.

Dixon, W. J., and Massey, Jr., F. J. (1969), *Introduction to Statistical Analysis*, 3rd ed., McGraw-Hill, New York.

Dodge, H. F., and Romig, H. G. (1959), *Sampling Inspection Tables*, 2nd ed., Wiley, New York.

Draper, N., and Smith, H. (1966), *Applied Regression Analysis*, Wiley, New York. Updated 2nd edition (1981).

Dubey, S. D. (1965), "Asymptotic Properties of Several Estimators of Weibull Parameters," *Technometrics* **7**, 423–434.

Dubey, S. D. (1967), "Some Percentile Estimators for Weibull Parameters," *Technometrics* **9**, 119–129.

Easterling, R. G. (1975), "Reliability Estimation and Sensitivity Testing," *Microelectron. Reliab.* **14**, 141–152.

Ehrenfeld, S. (1962), "Some Experimental Design Problems in Attribute Life Testing," *J. Am. Stat. Assoc.* **57**, 668–679.

Elandt-Johnson, R. C., and Johnson, N. L. (1980), *Survival Models and Data Analysis*, Wiley, New York.

Ellison, B. E. (1964), "Statistical Techniques for the Univariate Normal Distribution," Lockheed Missiles and Space Division Report 6-74-64-9, Sunnyvale, CA.

Engelhardt, M. (1975), "Simple Linear Estimation of the Parameters of the Logistic Distribution from a Complete or Censored Sample," *J. Am. Stat. Assoc.* **70**, 899–902.

Epstein, B., and Sobel, M. (1953), "Life Testing," *J. Am. Stat. Assoc.* **48**, 486–502.

Everitt, B. S., and Hand, D. J. (1981), *Finite Mixture Distributions*, Methuen, New York.

Farewell, V. T., and Prentice, R. L. (1979), "A Study of Distributional Shape in Life Testing," to be published.

Feder, P. I. (1972), "Studentized Range Graph Paper—A New Tool for Extracting Information from Data," General Electric Company Corporate Research and Development TIS Report 72CRD193.*

Feller, W. G. (1961), *An Introduction to Probability Theory and Its Applications*, 2nd ed., Wiley, New York.

Fertig, K. W., and Mann, N. R. (1980), "Life-Test Sampling Plans for Two-Parameter Weibull Populations," *Technometrics* **22**, 165–177.

Fertig, K. W., Meyer, M. E., and Mann, N. R. (1980), "On Constructing Prediction Intervals for Samples from a Weibull or Extreme Value Distribution," *Technometrics* **22**, 567–573.

Finney, D. J. (1968), *Probit Analysis*, 3rd ed., Cambridge University Press, New York and London.

Fleiss, J. L. (1973), *Statistical Methods for Rates and Proportions*, Wiley, New York.

Friedman, L., and Gertsbakh, I. B. (1980), "Maximum Likelihood Estimation in a Minimum-Type Model with Exponential and Weibull Failure Modes," *J. Am. Stat. Assoc.* **75**, 460–465.

Galambos, J. (1978), *The Asymptotic Theory of Extreme Order Statistics*, Wiley-Interscience, New York.

General Electric Company (1962), *Tables of the Individual and Cumulative Terms of Poisson Distribution*, Van Nostrand, Princeton, NJ.

General Electric Company (1975), "Reliability Manual for Liquid Metal Fast Breeder Reactors," Vol. 1 and 2, General Electric Company Corporate Research and Development Report SRD-75-064, Schenectady, NY 12345. Available from Mr. Richard Gilchrist, Fast Breeder Reactor Department, GE Company, DeGuigne Dr., Sunnyvale, CA 94086.

General Electric Information Services Company (1970), "Reliability Analysis by Weibull Distribution," Publication FT910373, Documentation Unit, 401 N. Washington St., Rockville, MD 20850.

General Electric Information Services Company (1979), "STATSYSTEM Users Guide," GE Information Services Company Publication 5707.12. Available from your service representative and 401 N. Washington St., Rockville, MD 20850.

Gibbons, J. D. (1976), *Nonparametric Methods for Quantitative Analysis*, Holt, Rinehart, and Winston, New York.

Glasser, M. (1965), "Regression Analysis with Censored Data," *Biometrics* **21**, 300–307.

Goldberger, A. A. (1962), "Best Linear Unbiased Prediction in the Generalized Linear Regression Model," *J. Am. Stat. Assoc.* **57**, 369–375.

Goldsmith, P. L. (1968), "The Construction of Life Distributions from Garment Wear-Trials," *Statistician* **18**, 355–375.

Golub, A., and Grubbs, F. E. (1956), "Analysis of Sensitivity Experiments When the Levels of Stimulus Cannot Be Controlled," *J. Am. Stat. Assoc.* **51**, 257–265; correction **51**, 650.

Govindarajulu, Z. (1964), "A Supplement to Mendenhall's Bibliography on Life Testing and Related Topics," *J. Am. Stat. Assoc.* **59**, 1231–1291.

Grant, E. L., and Leavenworth, R. S. (1980), *Statistical Quality Control*, 5th ed., McGraw-Hill, New York.

Greenwood, J. A., and Hartley, H. O. (1962), *Guide to Tables in Mathematical Statistics*, Princeton University Press, Princeton, NJ.

Gross, A. J., and Clark, V. A. (1975), *Survival Distributions: Reliability Applications in the Biomedical Sciences*, Wiley, New York.

Grubbs, F. E. (1969), "Procedures for Detecting Outlying Observations in Samples," *Technometrics* **11**, 1–21.

Gumbel, E. J. (1958), *Statistics of Extremes*, Columbia University Press, New York.

Gupta, S. S., Qureishi, A. S., and Shah, B. K. (1967), "Best Linear Unbiased Estimators of the Parameters of the Logistic Distribution Using Order Statistics," *Technometrics* **9**, 43–56.

Hahn, G. J. (1970), "Statistical Intervals for a Normal Population," *J. Qual. Technol.* **2**, "Part I—Tables, Examples, and Applications," 115–125, "Part II—Computations and Discussion," 195–206.

Hahn, G. J., and Miller, J. M. (1968), "Time-Sharing Computer Programs for Estimating Parameters of Several Normal Populations and for Regression Estimation from Censored Data," General Electric Research & Development Center TIS Report 68-C-366.*

Hahn, G. J., and Nelson, W. (1973), "A Survey of Prediction Intervals and Their Application," *J. Qual. Technol.* **5**, 178–188.

Hahn, G. J., and Nelson, Wayne (1971), "Graphical Analysis of Incomplete Accelerated Life Test Data," *Insul./Circuits* **17**, 79–84.

Hahn, G. J., and Shapiro, S. S. (1967), *Statistical Models in Engineering*, Wiley, New York.

Hahn, G. J., Nelson, W. B., and Cillay, C. (1975), "STATSYSTEM—A User-Oriented Interactive System for Statistical Data Analysis," in *1975 Statistical Computing Section Proceedings of the American Statistical Association*, American Statistical Association, 806 15th St., N.W., Washington, DC 20005, pp. 118–123.

Haight, F. A. (1967), *Handbook of the Poisson Distribution*, Wiley, New York.

Hanson, D. L., and Koopmans, L. H. (1964), "Tolerance Limits for the Class of Distributions with Increasing Hazard Rates," *Ann. Math. Stat.* **35**, 1561–1575.

Harter, H. L. (1964), *New Tables of the Incomplete Gamma-Function Ratio and of Percentage Points of the Chi-Square and Beta Distributions*, Aerospace Research Laboratory, Wright-Patterson Air Force Base, OH. Available from the Superintendent of Documents, U.S. Government Printing Office, Washington, DC 20402.

Harter, H. L. (1977a), "A Survey of the Literature on the Size Effect on Material Strength," Report No. AFFDL-TR-77-11, Air Force Flight Dynamics Laboratory AFSC, Wright-Patterson Air Force Base, OH 45433.

Harter, H. L. (1977b), *A Chronological Annotated Bibliography on Order Statistics*, Volume 1: *Pre-1950*, Air Force Flight Dynamics Laboratory, Wright-Patterson Air Force Base, OH 45433.

Harter, H. L. (1978), "A Bibliography of Extreme-Value Theory," *Int. Stat. Rev.* **46**, 279–306.

Harter, H. L., and Moore, A. H. (1966), "Iterative Maximum-Likelihood Estimation of the Parameters of Normal Populations from Singly and Doubly Censored Samples," *Biometrika* **53**, 205–213.

Harter, H. L., and Moore, A. H. (1968), "Maximum Likelihood Estimation, from Double Censored Samples, of the Parameters of the First Asymptotic Distribution of Extreme Values," *J. Am. Stat. Assoc.* **63**, 889–901.

Harvard Computation Laboratory (1955), *Tables of the Cumulative Binomial Probability Distribution*, Harvard University Press, Cambridge, MA.

Hawkins, D. W. (1980), *Identification of Outliers*, Methuen, New York.

Herd, G. R. (1960), "Estimation of Reliability from Incomplete Data," in *Proceedings of the 6th National Symposium on Reliability and Quality Control*, IEEE, 345 East 47th St., New York, NY 10017, pp. 202–217.

Herman, R. J., and Patell, R. K. N. (1971), "Maximum Likelihood Estimation for Multi-Risk Model," *Technometrics* **13**, 385–396.

Hjorth, U. (1980), "A Reliability Distribution with Increasing, Decreasing, Constant, and Bathtub-Shaped Failure Rates," *Technometrics* **22**, 99–107.

Hoadley, B. (1970), "Strategies for Removing Telephones When Service Is Disconnected," presented at the Joint Statistical Meeting, Detroit, MI.

Hoadley, B. (1971), "Asymptotic Properties of Maximum Likelihood Estimators for the Independent Not Identically Distributed Case," *Ann. Math. Stat.* **42**, 1977–1991.

Hoel, P. G. (1960), *Elementary Statistics*, Wiley, New York.

Hollander, M. and Wolfe, D. (1973), *Nonparametric Statistical Methods*, Wiley, New York.

IMSL (1975), "Library 2 Reference Manual," 5th ed. (November 1975), International Mathematical and Statistical Libraries, Inc., 6th Floor, GNB Building, 7500 Bellaire Boulevard, Houston, TX 77036.

Jacoby, S. L. S., Kowalik, J. S., and Pizzo, J. T. (1972), *Iterative Methods for Nonlinear Optimization Problems*, Prentice-Hall, Englewood Cliffs, NJ.

Jaeger, C. M., and Pennock, J. L. (1957), "Estimating the Service Life of Household Goods by Actuarial Methods," *J. Am. Stat. Assoc.* **52**, 175–185.

Jensen, F., and Petersen, N. E. (1982), *Burn-in: An Engineering Approach to the Design and Analysis of Burn-in Procedures*, Wiley, New York.

Johnson, L. G. (1964), *The Statistical Treatment of Fatigue Experiments*, Elsevier, New York.

Johnson, N. L., and Kotz, S. (1969), *Distributions in Statistics — Discrete Distributions*, Wiley, New York.

Johnson, N. L., and Kotz, S. (1970), *Distributions in Statistics: Continuous Univariate Distributions*, Vols. 1 and 2, Houghton-Mifflin, Boston.

Joiner, B. L., Ed. (1975), *Current Index to Statistics: Applications, Methods and Theory*, issued yearly by the American Statistical Association, 806 15th St., N.W., Washington, DC 20005.

Joiner, B. L., Laubscher, N. F., Brown, E. S., and Levy, B. (1970), *An Author and Permuted Title Index to Selected Statistical Journals*, U.S. Department of Commerce, National Bureau of Standards Special Publication 321.

Kalbfleisch, J. D., and Prentice, R. L. (1980), *The Statistical Analysis of Failure Time Data*, Wiley, New York.

Kaminsky, K. S., and Nelson, P. I. (1975), "Best Linear Unbiased Prediction of Order Statistics in Location and Scale Families," *J. Am. Stat. Assoc.* **70**, 145–150.

Kao, J. H. K. (1959), "A Graphical Estimation of Mixed Weibull Parameters in Life-Testing of Electron Tubes," *Technometrics* **1**, 389–407.

Kaplan, E. L., and Meier, P. (1958), "Nonparametric Estimation from Incomplete Observations," *J. Am. Stat. Assoc.* **53**, 457–481.

Karn, M. N. (1931), "An Inquiry into Various Death-Rates and the Comparative Influence of Certain Diseases on the Duration of Life," *Ann. Eugen.* **4**, 279–326.

Kennedy, Jr., W. J., and Gentle, J. E. (1980), *Statistical Computing*, Dekker, New York.

King, J. R. (1971), *Probability Charts for Decision Making*, Industrial Press, New York.

Kirkpatrick, R. L. (1970), "Confidence Limits for a Percent Defective Characterized by Two Specification Limits," *J. Qual. Technol.* **2**, 150–155.

Kelga, V. (1976), "Empirical Failure Distribution of Identical Components of Series System Determined from System Failure Data," *Acta Tech. CSAV* **12**, 112–117.

Kodlin, D. (1967), "A New Response Time Distribution," *Biometrics* **23**, 227–239.

Kosambi, D. D. (1966), "Scientific Numismatics," *Sci. Am.* **214**, 102–111.

Kpedekpo, G. M. K. (1969), "Working Life Tables for Males in Ghana, 1960," *J. Am. Stat. Assoc.* **64**, 102–110.

Krane, S. A. (1963), "Analysis of Survival Data by Regression Techniques," *Technometrics* **5**, 161–174.

Kulldorff, G. (1961), *Estimation from Grouped and Partially Grouped Samples*, Wiley, New York.

Lagakos, S. W. (1979), "General Right Censoring and Its Impact on the Analysis of Survival Data," *Biometrics* **35**, 139–156.

Lancaster, H. O. (1969), *The Chi-Squared Distribution*, Wiley, New York.

Lawless, J. F. (1971), "A Prediction Problem Concerning Samples from the Exponential Distribution, with Applications to Life Testing," *Technometrics* **13**, 725–730.

Lawless, J. F. (1972a), "On Prediction Intervals for Samples from the Exponential Distribution and Prediction Limits for System Survival," *Sankhya B* **34**, 1–14.

Lawless, J. F. (1972b), "Conditional Confidence Interval Estimation for the Parameters of the Weibull Distribution," *Util. Math.* **2**, 71–87.

Lawless, J. F. (1973), "On the Estimation of Safe Life When the Underlying Life Distribution Is Weibull," *Technometrics* **15**, 857–865.

Lawless, J. F. (1974), "On Prediction of Survival Time for Individual Systems," *IEEE Trans. Reliab.* **R-23**, 235–241.

Lawless, J. F. (1975), "Construction of Tolerance Bounds for the Extreme-Value and Weibull Distributions," *Technometrics* **17**, 255–261.

Lawless, J. F. (1978), "Confidence Interval Estimation for the Weibull and Extreme Value Distributions," *Technometrics* **20**, 355–368.

Lawless, J. F. (1982), *Statistical Models and Methods for Lifetime Data*, Wiley, New York.

Lee, L., and Thompson, Jr., W. A. (1974), "Results on Failure Time and Pattern for the Series System," in *Reliability and Biometry*, F. Proschan and R. Serfling, Eds., SIAM, 117 S. 17th St., Philadelphia, PA, pp. 291–302.

Lehmann, E. L. (1959), *Testing Statistical Hypotheses*, Wiley, New York.

Lehmann, E. L. (1975), *Nonparametrics: Statistical Methods Based on Ranks*, Holden-Day, San Francisco, CA.

Lemon, G. H., and Wattier, J. B. (1976), "Confidence and 'A' and 'B' Allowable Factors for the Weibull Distribution," *IEEE Trans. Reliab.* **R-25**, 16–19.

Lieberman, G. J. (1958), "Tables for One-Sided Statistical Tolerance Limits," *Indus. Qual. Control* **14**, 7–9.

Lieberman, G., and Owen, D. B. (1961), *Tables of the Hypergeometric Probability Distribution*, Stanford University Press, Stanford, CA.

Lieblein, J. (1954), "A New Method of Analyzing Extreme Value Data," National Advisory Commission for Aeronautics, Technical Note 3053.

Lieblein, J., and Zelen, M. (1956), "Statistical Investigation of the Fatigue Life of Deep-Groove Ball Bearings," *J. Res. Natl Bur. Standards* **57**, 273–316.

Little, R. E., and Jebe, E. H. (1975), *Statistical Design of Fatigue Experiments*, Halstead, New York.

Locks, M. O. (1973), *Reliability, Maintainability, and Availability Assessment*, Hayden, Rochelle Park, NJ.

Locks, M. O., Alexander, M. J., and Byars, B. J. (1963), "New Tables of the Non-Central *t*-Distribution," Aerospace Research Laboratory Technical Report No. 63-19, Wright-Patterson Air Force Base, OH 45433.

Mace, A. E. (1964), *Sample Size Determination*, Reinhold, New York.

Mann, N. R. (1968), "Point and Interval Estimation Procedures for the Two-Parameter Weibull and Extreme-Value Distributions," *Technometrics* **10**, 231–256.

Mann, N. R. (1970a), "Estimation of Location and Scale Parameters under Various Models of Censoring and Truncation," Aerospace Research Laboratories Report ARL 70-0026, Wright-Patterson Air Force Base, OH 45433.

Mann, N. R. (1970b), "Warranty Periods Based on Three Ordered Sample Observations from a Weibull Population," *IEEE Trans. Reliab.* **R-19**, 167–171.

Mann, N. R. (1971), "Best Linear Invariant Estimation for Weibull Parameters under Progressive Censoring," *Technometrics* **13**, 521–533.

Mann, N. R., and Fertig, K. W. (1973), "Tables for Obtaining Confidence Bounds and Tolerance Bounds Based on Best Linear Invariant Estimates of Parameters of the Extreme-Value Distribution," *Technometrics* **15**, 87–101.

Mann, N. R., and Saunders, S. C. (1969), "On Evaluation of Warranty Assurance When Life Has a Weibull Distribution," *Biometrika* **56**, 615–625.

Mann, N. R., Fertig, K. W., and Scheuer, E. M. (1971), "Confidence and Tolerance Bounds and a New Goodness-of-Fit Test for the Two Parameter Weibull or Extreme Value Distribution (with Tables for Censored Samples of Size 3(1)25)," Aerospace Research Laboratories Report ARL 71-0077, Office of Aerospace Research, Wright-Patterson Air Force Base, OH 45433.

Mann, N. R., Schafer, R. E., and Singpurwalla, N. D. (1974), *Methods for Statistical Analysis of Reliability and Life Data*, Wiley, New York.

Mayer, Jr., J. E. (1970) "Prediction of Ball Bearing Assembly Life from the Two-Parameter Weibull Life of the Components," in *Annals of Reliability and Maintainability — 1970*, Vol. 9, *Assurance Technology Spinoffs*, Society of Automative Engineers, 400 Commonwealth Dr., Warrendale, PA 15096, pp. 549–561.

Maxwell, A. (1961), *Analyzing Qualitative Data*, Wiley, New York.

McCool, J. I. (1965), "The Construction of Good Linear Unbiased Estimates from the Best Linear Estimates for a Smaller Sample Size," *Technometrics* **7**, 543–552.

McCool, J. I. (1970a), "Inference on Weibull Percentiles and Shape Parameter from Maximum-Likelihood Estimates," *IEEE Trans. Reliab.* **R-19**, 2–9.

McCool, J. I. (1970b), "Evaluating Weibull Endurance Data by the Method of Maximum Likelihood," *Am. Soc. Lubr. Eng.* **13**, 189–202.

McCool, J. I. (1974), "Inferential Techniques for Weibull Populations," Aerospace Research Laboratories Report ARL TR 74-0180, available from National Technical Information Services Clearinghouse, Springfield, VA 22151, publication AD A 009 645.

McCool, J. I. (1975a), "Inferential Techniques for Weibull Populations II," Aerospace Research Laboratories Report ARL TR 75-0233, available from National Technical Information Services Clearinghouse, Springfield, VA 22151.

McCool, J. I. (1975b), "Multiple Comparison for Weibull Parameters," *IEEE Trans. Reliab.* **R-24**, 186–192.

McCool, J. I. (1976), "Estimation of Weibull Parameters with Competing-Mode Censoring," *IEEE Trans. Reliab.* **R-25**, 25–31.

McCool, J. I. (1977), "Analysis of Variance for Weibull Populations," in *Theory and Applications of Reliability*, C. Tsokos and I. Shimi, Eds., Academic, New York.

McCool, J. I. (1978a), "Competing Risk and Multiple Comparison Analysis for Bearing Fatigue Tests," *Am. Soc. Lubr. Eng. Trans.* **21**, 271–284.

McCool, J. I. (1978b), "The Comparative Power of the Likelihood Ratio and a Shape Parameter Ratio Test for the Equality of Weibull Scale Parameters," SKF Industries, Inc., Report AL78P025, King of Prussia, PA 19406.

McCool, J. I. (1979), "Analysis of Single Classification Experiments Based on Censored Samples from the Two-Parameter Weibull Distribution," *J. Stat. Plan. Infer.* **3**, 39–68.

Meeker, W. Q. (1979), *Bibliography on Accelerated Testing*, Statistics Department, Iowa State University, Ames, IA.

Meeker, W. Q. (1980), "Maximum Likelihood Percentile Estimates and Confidence Limits from Grouped Weibull Data," private communication, Statistics Department, Iowa State University, Ames, IA.

Meeker, Jr., W. Q., and Duke, S. (1979), "CENSOR—A User-Oriented Computer Program for Life Data Analysis," Department of Statistics, Iowa State University, Ames, IA.

Meeker, Jr., W. Q., and Hahn, G. J. (1977), "Asymptotically Optimum Over-Stress Tests to Estimate the Survival Probability at a Condition with a Low Expected Failure Probability," *Technometrics* **19**, 381–399.

Meeker, W. Q., and Nelson, W. B. (1974), "Tables for the Weibull and Smallest Extreme Value Distributions," General Electric Company Corporate Research & Development TIS Report 74CRD230. Also published in *Relia-Com Review* **1**, Fall 1976.*

Meeker, Jr., W. Q., and Nelson, W. (1977), "Weibull Variances and Confidence Limits by Maximum Likelihood for Singly Censored Data," *Technometrics* **19**, 473–476.

Mendenhall, W. (1958), "A Bibliography of Life Testing and Related Topics," *Biometrika* **45**, 521–543.

Mendenhall, W. (1968), *Introduction to Linear Models and the Design and Analysis of Experiments*, Wadsworth, Belmont, CA.

Mendenhall, W., Ott, L., and Scheaffer, R. L. (1971), *Elementary Survey Sampling*, Wadsworth, Belmont, CA.

Menon, M. V. (1963), "Estimation of the Shape and Scale Parameters of the Weibull Distribution," *Technometrics* **5**, 175–182.

Michael, J. R., and Schucany, W. R. (1979), "A New Approach to Testing Goodness of Fit for Censored Samples," *Technometrics* **21**, 435–441.

Miller, R. (1966), *Simultaneous Statistical Inference*, McGraw-Hill, New York. 2nd Ed. (1981), Springer-Verlag, New York.

Miller, R. (1981), *Survival Analysis*, Wiley, New York.

Moeschberger, M. L. (1974), "Life Tests under Dependent Competing Causes of Failure," *Technometrics* **16**, 39–47.

Moeschberger, M. L., and David, H. A. (1971), "Life Tests under Competing Causes of Failure and the Theory of Competing Risks," *Biometrics* **27**, 909–933.

Molina, E. C. (1949), *Poisson's Exponential Binomial Limit*, Van Nostrand, Princeton, NJ.

Morrison, D. G. (1976), *Multivariate Statistical Methods*, 2nd ed., McGraw-Hill, New York.

Myers, R. H. (1971), *Response Surface Methodology*, Allyn and Bacon, Boston.

Nadas, A. (1969), "A Graphical Procedure for Estimating All Parameters of a Life Distribution in the Presence of Two Dependent Death Mechanisms, Each Having a Lognormally Distributed Killing Time ," private communication from the author at the IBM Corporation, East Fishkill Facility, Hopewell Junction, NY.

National Bureau of Standards (1950), *Tables of the Binomial Probability Distribution*, National Bureau of Standards, Applied Mathematics Series 6, U.S. Government Printing Office, Washington, DC 20402.

National Bureau of Standards (1953), *Probability Tables for the Analysis of Extreme-Value Data*, Applied Mathematics Series 22, U.S. Government Printing Office, Washington, DC 20402.

Natrella, M. G. (1963), *Experimental Statistics*, National Bureau of Standards Handbook 91, U.S. Government Printing Office, Washington, DC 20402.

Nelson, W. (1970), "Confidence Intervals for the Ratio of Two Poisson Means and Poisson Predictor Intervals," *IEEE Trans. Reliab.* **R-19**, 42–49.

Nelson, W. (1971), "Analysis of Accelerated Life Test Data," *IEEE Trans. Electr. Insulat.*: "Part I. The Arrhenius Model and Graphical Methods," **EI-6** (December 1971) 165–181; "Part II. Numerical Methods and Test Planning," **EI-7** (March 1972) 36–55; "Part III. Product Comparisons and Checks on the Validity of the Model and Data," **EI-7** (June 1972) 99–119; "Table of Contents," **EI-7** (September 1972) 158–159.

Nelson, W. (1972a), "Charts for Confidence Limits and Tests for Failure Rates," *J. Qual. Technol.* **4**, 190–195.

Nelson, W. (1972b), "Theory and Application of Hazard Plotting for Censored Failure Data," *Technometrics* **14**, 945–966.

Nelson, W. (1973), "Analysis of Residuals from Censored Data," *Technometrics* **15**, 697–715.

Nelson, W. (1974a), "A Survey of Methods for Planning and Analyzing Accelerated Life Tests," *IEEE Trans. Electr. Insul.* **EI-9**, 12–18.

Nelson, W. (1974b), "Analysis of Accelerated Life Test Data with a Mix of Failure Modes by Maximum Likelihood," General Electric Company Corporate Research & Development TIS Report 74CRD160.*

Nelson, W. (1975), "Graphical Analysis of.Accelerated Life Test Data with a Mix of Failure Modes," *IEEE Trans. Reliab.* **R-24**, 230–237.

Nelson, W. (1977), "Optimum Life Tests with Grouped Data from an Exponential Distribution," *IEEE Trans. Reliab.* **R-26**, 226–231.

Nelson, W. (1979), "How to Do Data Analysis with Simple Plots," The ASQC Basic References in Quality Control: Statistical Techniques, E. J. Dudewicz, Ed., American Society for Quality Control, Milwaukee, WI. 53203.

Nelson, W., and Hahn, G. J. (1972), "Linear Estimation of a Regression Relationship from Censored Data—Part I. Simple Methods and Their Application," *Technometrics* **14**, 247–269.

Nelson, W. B., and Hendrickson, R. (1972), "1972 User Manual for STATPAC—A General Purpose Package for Data Analysis and for Fitting Statistical Models to Data," General Electric Company Corporate Research & Development TIS Report 72GEN009.*

Nelson, W. B., and Schmee, J. (1976), "Confidence Limits for *Percentiles* of (Log) Normal Life Distributions from Small Singly Censored Samples and Best Linear Unbiased Estimates,"

General Electric Company Corporate Research & Development TIS Report 76CRD222.*

Nelson, W. B., and Schmee, J. (1977a), "Confidence Limits for *Reliabilities* of (Log) Normal Life Distributions from Small Singly Censored Samples and Best Linear Unbiased Estimates," General Electric Company Corporate Research & Development TIS Report 76CRD259.*

Nelson, W. B., and Schmee, J. (1977b), "Prediction Limits for the Last Failure Time of a (Log) Normal Sample from Early Failure Times," General Electric Company Corporate Research and Development TIS Report 77CRD189.* Also, in *IEEE Trans. Reliab.* **R-30**, Dec. 1981.

Nelson, W., and Schmee, J. (1979), "Inference for (Log) Normal Life Distributions from Small Singly Censored Samples and BLUEs," *Technometrics* **21**, 43–54.

Nelson, W., and Thompson, V. C. (1971), "Weibull Probability Papers," *J. Qual. Technol.* **3**, 45–50.

Nelson, W. B., Morgan, C. B., and Caporal, P. (1978), "1979 STATPAC Simplified—A Short Introduction to How to Run STATPAC, a General Statistical Package for Data Analysis," General Electric Company Corporate Research & Development TIS Report 78CRD276.* STATPAC may be obtained on license through Mr. Keith Burk, Technology Marketing Operation, GE Corporate Research & Development, 120 Erie Blvd., Schenectady, NY 12345.

Neter, J., and Wasserman, W. (1974), *Applied Linear Statistical Models*, Richard D. Irwin, Homewood, IL.

Odeh, R. E., and Owen, D. B. (1980), *Tables for Normal Tolerance Limits, Sampling Plans, and Screening*, Dekker, New York.

Olivier, C., and Neff, R. K. (1976), *LOGLIN 1.0 Users Guide*, Program Librarian, Health Sciences Computing Facility, Harvard School of Public Health, Boston, MA 02115.

Owen, D. B. (1962), *Handbook of Statistical Tables*, Addison-Wesley, Reading, MA.

Owen, D. B. (1968), "A Survey of Properties and Applications of the Noncentral *t*-Distribution," *Technometrics* **10**, 445–478.

Pabst, Jr., W. R., Ed. (1975), *Standards and Specifications*, Publication 103, American Society for Quality Control, 161 W. Wisconsin Ave., Milwaukee, WI 53203. Price $15.00.

Parzen, E. (1960), *Modern Probability Theory and Its Applications*, Wiley, New York.

Parzen, E. (1962), *Stochastic Processes*, Holden-Day, San Francisco.

Patil, G. P., and Joshi, S. W. (1968), *A Dictionary and Bibliography of Discrete Distributions*, Hafner, New York.

Pearson, E. S., and Hartley, H. O. (1954), *Biometrika Tables for Statisticians*, Vol I, Cambridge University Press, New York.

Peto, R. (1973), "Experimental Survival Curves for Interval-Censored Data," *Appl. Stat.* **22**, 86–91.

Plackett, R. L. (1974), *Analysis of Categorical Data*, Hafner, New York.

Powell, M. J. D. (1964), "An Efficient Method for Finding the Minimum of a Function of Several Variables without Calculating Derivatives," *Comput. J.* **7**, 155–162.

Prentice, R. L. (1974), "A Log Gamma Model and Its Maximum Likelihood Estimation," *Biometrika* **61**, 539–544.

Preston, D. L., and Clarkson, D. B. (1980), "SURVREG: An Interactive Program for Regression Analysis with Censored Survival Data," *Amer. Statist. Assoc. 1980 Proceedings of the Section on Statistical Computing.*

Prince, M. (1967), "Life Table Analysis of Prime Time Programs on a Television Network," *Am. Stat.* **21**, 21–23.

Proschan, F. (1963), "Theoretical Explanation of Observed Decreasing Failure Rate," *Technometrics* **5**, 375–383.

Proschan, F., and Sullo, P. (1976), "Estimating the Parameters of a Multivariate Exponential Distribution," *J. Am. Stat. Assoc.* **71**, 465–472.

Pugh, E. L. (1963), "The Best Estimate of Reliability in the Exponential Case," *Oper. Res.* **11** 57–61.

Rao, C. R. (1973), *Linear Statistical Inference and Its Applications*, 2nd ed., Wiley, New York.

Regal, R. R., and Larntz, K. (1978), "Likelihood Methods for Testing Group Problem Solving Models with Censored Data," *Psychometrika* **43**, 353–366.

Resnikoff, G. J., and Lieberman, G. J. (1957), *Tables of the Noncentral t-Distribution*, Stanford University Press, Stanford, CA.

Romig, H. G. (1953), *50–100 Binomial Tables*, Wiley, New York.

Ross, G. J. S. (1970), "The Efficient Use of Function Minimization in Non-Linear Maximum-Likelihood Estimation," *Appl. Stat.* **19**, 205–221.

Ross, G. J. S., Kempton, R. A., Laukner, F. B., Payne, R. W., Hawkins, D., and White, R. (1976), "Maximum Likelihood Program (MPL)—Prospectus 1976," available from The Programs Secretary, Statistics Department, Rothamsted Experimental Station, Harpenden, Herts AL52JQ, England.

Ross, I. C., and Tukey, J. W. (1975), *Index to Statistics and Probability*, R&D Press, Los Altos, CA.

Sampford, M. R., and Taylor, J. (1959), "Censored Observations in Randomized Block Experiments," *J. R. Stat. Soc.*, *B*, **21**, 214–237.

Sarhan, A. E., and Greenberg, B. G., Eds. (1962), *Contributions to Order Statistics*, Wiley, New York.

Schafer, R. E., and Sheffield, T. S. (1976), "On Procedures for Comparing Two Weibull Populations," *Technometrics* **18**, 231–235.

Scheffé, H. (1959), *The Analysis of Variance*, Wiley, New York.

Schilling, E. G. (1982), *Acceptance Sampling in Quality Control*, Dekker, New York.

Schmee, J., and Nelson, W. B. (1976), "Confidence Limits for *Parameters* of (Log) Normal Life Distributions from Small Singly Censored Samples by Maximum Likelihood," General Electric Company Corporate Research & Development TIS Report 76CRD218.*

Schmee, J., and Nelson, W. B. (1977), "Estimates and Approximate Confidence Limits for (Log) Normal Life Distributions from Singly Censored Samples by Maximum Likelihood," General Electric Company Corporate Research and Development TIS Report 76CRD250.*

Schmee, J., and Nelson, W. B. (1979), "Predicting from Early Failures the Last Failure Time of a (Log) Normal Sample," *IEEE Trans. Reliab.* **R-28**, 23–26.

Shapiro, S. S. (1980), "How to Test Normality and Other Distributional Assumptions," Vol. 3 of the ASQC Basic References in Quality Control: Statistical Techniques, from ASQC, 161 W. Wisconsin Ave., Milwaukee, WI 53203. Price $10.25.

Sheesley, J. H. (1975), "A Computer Program to Perform Probit Analysis on Data from Sensitivity Experiments," General Electric Lighting Research & Technical Services Operation Report No. 1300-1202, Nela Park, Cleveland, OH 44112.

Shenton, S. H., and Bowman, K. O. (1977), *Maximum Likelihood Estimation in Small Samples*, Griffin Monograph No. 38, Macmillian, New York.

Shooman, M. L. (1968), *Probabilistic Reliability: An Engineering Approach*, McGraw-Hill, New York.

Spurrier, J. D., and Wei, L. J. (1980), "A Test of the Parameter of the Exponential Distribution in the Type I Censoring Case," *J. Am. Stat. Assoc.* **75**, 405–409.

TEAM (Technical and Engineering Aids for Management), "Catalogue and Price List," Box 25, Tamworth, NH 03886.

Thoman, D. R., and Bain, L. J. (1969), "Two Sample Tests in the Weibull Distribution," *Technometrics* **11**, 805–815.

Thoman, D. R., Bain, L. J., and Antle, C. E. (1969), "Inferences on the Parameters of the Weibull Distribution," *Technometrics* **11**, 445–460.

Thoman, D. R., Bain, L. J., and Antle, C. E. (1970), "Maximum Likelihood Estimation, Exact Confidence Intervals for Reliability and Tolerance Limits in the Weibull Distribution," *Technometrics* **12**, 363–371.

Thomas, D. R., and Wilson, W. M. (1972), "Linear Order Statistic Estimation for the Two-Parameter Weibull and Extreme-Value Distributions from Type II Progressively Censored Samples," *Technometrics* **14**, 679–691.

Tillman, F. A., Hwang, C-L, and Kuo, W. (1977), "Optimization Techniques for System Reliability with Redundancy—A Review," *IEEE Trans. Reliab.* **R-26**, 148–155.

Todhunter, I. (1949), *A History of the Mathematical Theory of Probability*, Chelsea, New York.

Trindade, D. C, and Haugh, L. D. (1978), "Nonparametric Estimation of a Lifetime Distribution via the Renewal Function," Available from Dr. Trindade, IBM Technology Division, Essex Junction, VT 05452. Submitted for publication.

Tsokos, C. P., and Shimi, I. N., Eds. (1977), *The Theory and Application of Reliability with Emphasis on Bayesian and Nonparametric Methods*, Vol. I — *Theory*, Vol. II — *Application*, Academic, New York.

Turnbull, B. W. (1976), "The Empirical Distribution Function with Arbitrarily Grouped, Censored, and Truncated Data," *J. R. Stat. Soc.* **38**, 290–295.

Wagner, S. S., and Altman, S. A. (1973), "What Time Do the Baboons Come Down from the Trees? (An Estimation Problem)," *Biometrics* **28**, 623–635.

Weibull, W. (1951), "A Statistical Distribution Function of Wide Applicability," *J. Appl. Mech.* **18**, 293–297.

Weibull, W. (1961), *Fatigue Testing and the Analysis of Results*, Pergamon, New York.

Weintraub, S. (1963), *Tables of the Cumulative Binomial Probability Distribution for Small Values of p*, Free Press of Glencoe, New York.

White, J. S. (1964), "Least-Squares Unbiased Censored Linear Estimation for the Log Weibull (Extreme-Value) Distribution," *J. Indus. Math. Soc.* **14**, 21–60.

White, J. S. (1967), "The Moments of Log-Weibull Order Statistics," Research Publication GMP-717, Research Laboratories, General Motors Corporation, Warren, MI.

Wilde, D. J., and Beightler, C. (1967), *Foundations of Optimization*, Prentice-Hall, Englewood Cliffs, NJ.

Wilks, S. S. (1962), *Mathematical Statistics*, Wiley, New York.

Williamson, E., and Bretherton, M. H. (1963), *Tables of the Negative Binomial Probability Distribution*, Wiley, New York.

Wingo, D. R. (1973), "Solution of the Three-Parameter Weibull Equations by Constrained Modified Quasilinearization (Progressively Censored Samples)," *IEEE Trans. Reliab.* **R-22**, 96–102.

Yurkowsky, W., Schafer, R. E., and Finkelstein, J. M. (1967), "Accelerated Testing Technology," Rome Air Development Center Technical Report No. RADC-TR-67-420, Griffiss Air Force Base, NY.

Zahn, D. A. (1975), "On the Analysis of Factorial Experiments with Type I Censored Data," Florida State University Statistics Report M315, Department of Statistics, Tallahassee, FL.

Zelen, M. (1969), "Factorial Experiments in Life Testing," *Technometrics* **1**, 269–288.

Index

Many phrases are listed according to the main noun, rather than the first word, for example: Distribution, normal; Data, multiply censored; Comparisons, binomial. For distribution properties, consult: Distribution, name. Many data analysis methods are under the name of the distribution, for example: Normal analysis, confidence interval for mean.

WILEY SERIES IN PROBABILITY AND STATISTICS
ESTABLISHED BY WALTER A. SHEWHART AND SAMUEL S. WILKS

Editors: *David J. Balding, Noel A. C. Cressie, Nicholas I. Fisher,*
Iain M. Johnstone, J. B. Kadane, Geert Molenberghs. Louise M. Ryan,
David W. Scott, Adrian F. M. Smith, Jozef L. Teugels
Editors Emeriti: *Vic Barnett, J. Stuart Hunter, David G. Kendall*

The *Wiley Series in Probability and Statistics* is well established and authoritative. It covers many topics of current research interest in both pure and applied statistics and probability theory. Written by leading statisticians and institutions, the titles span both state-of-the-art developments in the field and classical methods.

Reflecting the wide range of current research in statistics, the series encompasses applied, methodological and theoretical statistics, ranging from applications and new techniques made possible by advances in computerized practice to rigorous treatment of theoretical approaches.

This series provides essential and invaluable reading for all statisticians, whether in academia, industry, government, or research.

*Now available in a lower priced paperback edition in the Wiley Classics Library.

BELSLEY, KUH, and WELSCH · Regression Diagnostics: Identifying Influential Data and Sources of Collinearity

BENDAT and PIERSOL · Random Data: Analysis and Measurement Procedures, *Third Edition*

BERRY, CHALONER, and GEWEKE · Bayesian Analysis in Statistics and Econometrics: Essays in Honor of Arnold Zellner

BERNARDO and SMITH · Bayesian Theory

BHAT and MILLER · Elements of Applied Stochastic Processes, *Third Edition*

BHATTACHARYA and JOHNSON · Statistical Concepts and Methods

BHATTACHARYA and WAYMIRE · Stochastic Processes with Applications

BILLINGSLEY · Convergence of Probability Measures, *Second Edition*

BILLINGSLEY · Probability and Measure, *Third Edition*

BIRKES and DODGE · Alternative Methods of Regression

BLISCHKE AND MURTHY (editors) · Case Studies in Reliability and Maintenance

BLISCHKE AND MURTHY · Reliability: Modeling, Prediction, and Optimization

BLOOMFIELD · Fourier Analysis of Time Series: An Introduction, *Second Edition*

BOLLEN · Structural Equations with Latent Variables

BOROVKOV · Ergodicity and Stability of Stochastic Processes

BOULEAU · Numerical Methods for Stochastic Processes

BOX · Bayesian Inference in Statistical Analysis

BOX · R. A. Fisher, the Life of a Scientist

BOX and DRAPER · Empirical Model-Building and Response Surfaces

*BOX and DRAPER · Evolutionary Operation: A Statistical Method for Process Improvement

BOX, HUNTER, and HUNTER · Statistics for Experimenters: An Introduction to Design, Data Analysis, and Model Building

BOX and LUCEÑO · Statistical Control by Monitoring and Feedback Adjustment

BRANDIMARTE · Numerical Methods in Finance: A MATLAB-Based Introduction

BROWN and HOLLANDER · Statistics: A Biomedical Introduction

BRUNNER, DOMHOF, and LANGER · Nonparametric Analysis of Longitudinal Data in Factorial Experiments

BUCKLEW · Large Deviation Techniques in Decision, Simulation, and Estimation

CAIROLI and DALANG · Sequential Stochastic Optimization

CHAN · Time Series: Applications to Finance

CHATTERJEE and HADI · Sensitivity Analysis in Linear Regression

CHATTERJEE and PRICE · Regression Analysis by Example, *Third Edition*

CHERNICK · Bootstrap Methods: A Practitioner's Guide

CHERNICK and FRIIS · Introductory Biostatistics for the Health Sciences

CHILÈS and DELFINER · Geostatistics: Modeling Spatial Uncertainty

CHOW and LIU · Design and Analysis of Clinical Trials: Concepts and Methodologies, *Second Edition*

CLARKE and DISNEY · Probability and Random Processes: A First Course with Applications, *Second Edition*

*COCHRAN and COX · Experimental Designs, *Second Edition*

CONGDON · Applied Bayesian Modelling

CONGDON · Bayesian Statistical Modelling

CONOVER · Practical Nonparametric Statistics, *Second Edition*

COOK · Regression Graphics

COOK and WEISBERG · Applied Regression Including Computing and Graphics

COOK and WEISBERG · An Introduction to Regression Graphics

CORNELL · Experiments with Mixtures, Designs, Models, and the Analysis of Mixture Data, *Third Edition*

COVER and THOMAS · Elements of Information Theory

*Now available in a lower priced paperback edition in the Wiley Classics Library.

*Now available in a lower priced paperback edition in the Wiley Classics Library.

GROSS and HARRIS · Fundamentals of Queueing Theory, *Third Edition*
*HAHN and SHAPIRO · Statistical Models in Engineering
HAHN and MEEKER · Statistical Intervals: A Guide for Practitioners
HALD · A History of Probability and Statistics and their Applications Before 1750
HALD · A History of Mathematical Statistics from 1750 to 1930
HAMPEL · Robust Statistics: The Approach Based on Influence Functions
HANNAN and DEISTLER · The Statistical Theory of Linear Systems
HEIBERGER · Computation for the Analysis of Designed Experiments
HEDAYAT and SINHA · Design and Inference in Finite Population Sampling
HELLER · MACSYMA for Statisticians
HINKELMAN and KEMPTHORNE: · Design and Analysis of Experiments, Volume 1:
    Introduction to Experimental Design
HOAGLIN, MOSTELLER, and TUKEY · Exploratory Approach to Analysis
    of Variance
HOAGLIN, MOSTELLER, and TUKEY · Exploring Data Tables, Trends and Shapes
*HOAGLIN, MOSTELLER, and TUKEY · Understanding Robust and Exploratory
    Data Analysis
HOCHBERG and TAMHANE · Multiple Comparison Procedures
HOCKING · Methods and Applications of Linear Models: Regression and the Analysis
    of Variance, *Second Edition*
HOEL · Introduction to Mathematical Statistics, *Fifth Edition*
HOGG and KLUGMAN · Loss Distributions
HOLLANDER and WOLFE · Nonparametric Statistical Methods, *Second Edition*
HOSMER and LEMESHOW · Applied Logistic Regression, *Second Edition*
HOSMER and LEMESHOW · Applied Survival Analysis: Regression Modeling of
    Time to Event Data
HUBER · Robust Statistics
HUBERTY · Applied Discriminant Analysis
HUNT and KENNEDY · Financial Derivatives in Theory and Practice
HUSKOVA, BERAN, and DUPAC · Collected Works of Jaroslav Hajek—
    with Commentary
IMAN and CONOVER · A Modern Approach to Statistics
JACKSON · A User's Guide to Principle Components
JOHN · Statistical Methods in Engineering and Quality Assurance
JOHNSON · Multivariate Statistical Simulation
JOHNSON and BALAKRISHNAN · Advances in the Theory and Practice of Statistics: A
    Volume in Honor of Samuel Kotz
JUDGE, GRIFFITHS, HILL, LÜTKEPOHL, and LEE · The Theory and Practice of
    Econometrics, *Second Edition*
JOHNSON and KOTZ · Distributions in Statistics
JOHNSON and KOTZ (editors) · Leading Personalities in Statistical Sciences: From the
    Seventeenth Century to the Present
JOHNSON, KOTZ, and BALAKRISHNAN · Continuous Univariate Distributions,
    Volume 1, *Second Edition*
JOHNSON, KOTZ, and BALAKRISHNAN · Continuous Univariate Distributions,
    Volume 2, *Second Edition*
JOHNSON, KOTZ, and BALAKRISHNAN · Discrete Multivariate Distributions
JOHNSON, KOTZ, and KEMP · Univariate Discrete Distributions, *Second Edition*
JUREČKOVÁ and SEN · Robust Statistical Procedures: Aymptotics and Interrelations
JUREK and MASON · Operator-Limit Distributions in Probability Theory
KADANE · Bayesian Methods and Ethics in a Clinical Trial Design
KADANE AND SCHUM · A Probabilistic Analysis of the Sacco and Vanzetti Evidence
KALBFLEISCH and PRENTICE · The Statistical Analysis of Failure Time Data, *Second
    Edition*

*Now available in a lower priced paperback edition in the Wiley Classics Library.

*Now available in a lower priced paperback edition in the Wiley Classics Library.

McLACHLAN · Discriminant Analysis and Statistical Pattern Recognition

McLACHLAN and KRISHNAN · The EM Algorithm and Extensions

McLACHLAN and PEEL · Finite Mixture Models

McNEIL · Epidemiological Research Methods

MEEKER and ESCOBAR · Statistical Methods for Reliability Data

MEERSCHAERT and SCHEFFLER · Limit Distributions for Sums of Independent Random Vectors: Heavy Tails in Theory and Practice

*MILLER · Survival Analysis, *Second Edition*

MONTGOMERY, PECK, and VINING · Introduction to Linear Regression Analysis, *Third Edition*

MORGENTHALER and TUKEY · Configural Polysampling: A Route to Practical Robustness

MUIRHEAD · Aspects of Multivariate Statistical Theory

MULLER and STOYAN · Comparison Methods for Stochastic Models and Risks

MURRAY · X-STAT 2.0 Statistical Experimentation, Design Data Analysis, and Nonlinear Optimization

MURTHY, XIE, and JIANG · Weibull Models

MYERS and MONTGOMERY · Response Surface Methodology: Process and Product Optimization Using Designed Experiments, *Second Edition*

MYERS, MONTGOMERY, and VINING · Generalized Linear Models. With Applications in Engineering and the Sciences

NELSON · Accelerated Testing, Statistical Models, Test Plans, and Data Analyses

NELSON · Applied Life Data Analysis

NEWMAN · Biostatistical Methods in Epidemiology

OCHI · Applied Probability and Stochastic Processes in Engineering and Physical Sciences

OKABE, BOOTS, SUGIHARA, and CHIU · Spatial Tesselations: Concepts and Applications of Voronoi Diagrams, *Second Edition*

OLIVER and SMITH · Influence Diagrams, Belief Nets and Decision Analysis

PALTA · Quantitative Methods in Population Health: Extensions of Ordinary Regressions

PANKRATZ · Forecasting with Dynamic Regression Models

PANKRATZ · Forecasting with Univariate Box-Jenkins Models: Concepts and Cases

*PARZEN · Modern Probability Theory and Its Applications

PEÑA, TIAO, and TSAY · A Course in Time Series Analysis

PIANTADOSI · Clinical Trials: A Methodologic Perspective

PORT · Theoretical Probability for Applications

POURAHMADI · Foundations of Time Series Analysis and Prediction Theory

PRESS · Bayesian Statistics: Principles, Models, and Applications

PRESS · Subjective and Objective Bayesian Statistics, *Second Edition*

PRESS and TANUR · The Subjectivity of Scientists and the Bayesian Approach

PUKELSHEIM · Optimal Experimental Design

PURI, VILAPLANA, and WERTZ · New Perspectives in Theoretical and Applied Statistics

PUTERMAN · Markov Decision Processes: Discrete Stochastic Dynamic Programming

*RAO · Linear Statistical Inference and Its Applications, *Second Edition*

RAUSAND and HØYLAND · System Reliability Theory: Models, Statistical Methods, and Applications, *Second Edition*

RENCHER · Linear Models in Statistics

RENCHER · Methods of Multivariate Analysis, *Second Edition*

RENCHER · Multivariate Statistical Inference with Applications

RIPLEY · Spatial Statistics

RIPLEY · Stochastic Simulation

ROBINSON · Practical Strategies for Experimenting

ROHATGI and SALEH · An Introduction to Probability and Statistics, *Second Edition*

*Now available in a lower priced paperback edition in the Wiley Classics Library.

ROLSKI, SCHMIDLI, SCHMIDT, and TEUGELS · Stochastic Processes for Insurance and Finance

ROSENBERGER and LACHIN · Randomization in Clinical Trials: Theory and Practice

ROSS · Introduction to Probability and Statistics for Engineers and Scientists

ROUSSEEUW and LEROY · Robust Regression and Outlier Detection

RUBIN · Multiple Imputation for Nonresponse in Surveys

RUBINSTEIN · Simulation and the Monte Carlo Method

RUBINSTEIN and MELAMED · Modern Simulation and Modeling

RYAN · Modern Regression Methods

RYAN · Statistical Methods for Quality Improvement, *Second Edition*

SALTELLI, CHAN, and SCOTT (editors) · Sensitivity Analysis

*SCHEFFE · The Analysis of Variance

SCHIMEK · Smoothing and Regression: Approaches, Computation, and Application

SCHOTT · Matrix Analysis for Statistics

SCHOUTENS · Levy Processes in Finance: Pricing Financial Derivatives

SCHUSS · Theory and Applications of Stochastic Differential Equations

SCOTT · Multivariate Density Estimation: Theory, Practice, and Visualization

*SEARLE · Linear Models

SEARLE · Linear Models for Unbalanced Data

SEARLE · Matrix Algebra Useful for Statistics

SEARLE, CASELLA, and McCULLOCH · Variance Components

SEARLE and WILLETT · Matrix Algebra for Applied Economics

SEBER and LEE · Linear Regression Analysis, *Second Edition*

SEBER · Multivariate Observations

SEBER and WILD · Nonlinear Regression

SENNOTT · Stochastic Dynamic Programming and the Control of Queueing Systems

*SERFLING · Approximation Theorems of Mathematical Statistics

SHAFER and VOVK · Probability and Finance: It's Only a Game!

SMALL and McLEISH · Hilbert Space Methods in Probability and Statistical Inference

SRIVASTAVA · Methods of Multivariate Statistics

STAPLETON · Linear Statistical Models

STAUDTE and SHEATHER · Robust Estimation and Testing

STOYAN, KENDALL, and MECKE · Stochastic Geometry and Its Applications, *Second Edition*

STOYAN and STOYAN · Fractals, Random Shapes and Point Fields: Methods of Geometrical Statistics

STYAN · The Collected Papers of T. W. Anderson: 1943–1985

SUTTON, ABRAMS, JONES, SHELDON, and SONG · Methods for Meta-Analysis in Medical Research

TANAKA · Time Series Analysis: Nonstationary and Noninvertible Distribution Theory

THOMPSON · Empirical Model Building

THOMPSON · Sampling, *Second Edition*

THOMPSON · Simulation: A Modeler's Approach

THOMPSON and SEBER · Adaptive Sampling

THOMPSON, WILLIAMS, and FINDLAY · Models for Investors in Real World Markets

TIAO, BISGAARD, HILL, PEÑA, and STIGLER (editors) · Box on Quality and Discovery: with Design, Control, and Robustness

TIERNEY · LISP-STAT: An Object-Oriented Environment for Statistical Computing and Dynamic Graphics

TSAY · Analysis of Financial Time Series

UPTON and FINGLETON · Spatial Data Analysis by Example, Volume II: Categorical and Directional Data

VAN BELLE · Statistical Rules of Thumb

VESTRUP · The Theory of Measures and Integration

VIDAKOVIC · Statistical Modeling by Wavelets

*Now available in a lower priced paperback edition in the Wiley Classics Library.

WEISBERG · Applied Linear Regression, *Second Edition*

WELSH · Aspects of Statistical Inference

WESTFALL and YOUNG · Resampling-Based Multiple Testing: Examples and Methods for *p*-Value Adjustment

WHITTAKER · Graphical Models in Applied Multivariate Statistics

WINKER · Optimization Heuristics in Economics: Applications of Threshold Accepting

WONNACOTT and WONNACOTT · Econometrics, *Second Edition*

WOODING · Planning Pharmaceutical Clinical Trials: Basic Statistical Principles

WOOLSON and CLARKE · Statistical Methods for the Analysis of Biomedical Data, *Second Edition*

WU and HAMADA · Experiments: Planning, Analysis, and Parameter Design Optimization

YANG · The Construction Theory of Denumerable Markov Processes

*ZELLNER · An Introduction to Bayesian Inference in Econometrics

ZHOU, OBUCHOWSKI, and McCLISH · Statistical Methods in Diagnostic Medicine

*Now available in a lower priced paperback edition in the Wiley Classics Library.